STUDENT SOLUTIONS MANUAL
FOR SWOKOWSKI AND COLE'S

FUNDAMENTALS OF ALGEBRA AND TRIGONOMETRY

NINTH EDITION

JEFFERY A. COLE
Anoka-Ramsey Community College

BROOKS/COLE PUBLISHING COMPANY
I(T)P® An International Thomson Publishing Company

Pacific Grove • Albany • Belmont • Bonn • Boston • Cincinnati • Detroit • Johannesburg • London
Madrid • Melbourne • Mexico City • New York • Paris • Singapore • Tokyo • Toronto • Washington

Project Development Editor: *Elizabeth Rammel*
Editorial Assistant: *Melissa Duge*
Production: *Dorothy Bell*

Art Coordination: *Lifland et al., Bookmakers*
Cover Design: *Cassandra Chu*
Printing and Binding: *Malloy Lithographing*

For more information, contact:

BROOKS/COLE PUBLISHING COMPANY
511 Forest Lodge Rd.
Pacific Grove, CA 93950
USA

International Thomson Editores
Seneca 53
Col. Polanco
11560 México, D. F., México

International Thomson Publishing Europe
Berkshire House 168-173
High Holborn
London WC1V 7AA
England

International Thomson Publishing GmbH
Königswinterer Strasse 418
53227 Bonn
Germany

Thomas Nelson Australia
102 Dodds Street
South Melbourne, 3205
Victoria, Australia

International Thomson Publishing Asia
221 Henderson Road
#05-10 Henderson Building
Singapore 0315

Nelson Canada
1120 Birchmount Road
Scarborough, Ontario
Canada M1K 5G4

International Thomson Publishing Japan
Hirakawacho Kyowa Building, 3F
2-2-1 Hirakawacho
Chiyoda-ku, Tokyo 102
Japan

Printed in the United States of America

10 9 8 7 6 5 4 3 2

ISBN 0-534-34624-3

PREFACE

This *Student's Solutions Manual* contains selected solutions and strategies for solving typical exercises in the text, *Fundamentals of Algebra and Trigonometry, Ninth Edition*, by Earl W. Swokowski and Jeffery A. Cole.

In each exercise set, nearly all odd-numbered solutions are included, with an emphasis on the solutions of the applied "word" problems. In the review exercise sections at the end of each chapter, odd- and even-numbered solutions are included. For the discussion exercises at the end of each chapter, all odd-numbered solutions are included. I have tried to illustrate enough solutions so that the student will be able to obtain an understanding of all types of problems in each section.

A significant number of today's students are involved in various outside activities, and find it difficult, if not impossible, to attend all class sessions. This manual should help meet the needs of these students. In addition, it is my hope that this manual's solutions will enhance the understanding of all readers of the material and provide insights to solving other exercises.

I would appreciate any feedback concerning errors, solution correctness, solution style, or manual style—comments from students using previous editions have greatly strengthened the text's supplements as well as the text itself.. These and any other comments may be sent directly to me at the address below or in care of the publisher.

I would like to thank: Elizabeth Rammel, of Brooks/Cole Publishing Company, for her patience and support; Gary Rockswold, of Mankato State University, for supplying solutions for many of the new applied problems and calculator exercises; Joan Cole, my wife, for proofing various features of the manual; George Morris, of Scientific Illustrators, for creating the mathematically precise art package; and Sally Lifland and Gail Magin, of Lifland et al. Bookmakers, for assembling the final manuscript. I dedicate this book to my children, Becky and Brad.

Jeffery A. Cole

Anoka-Ramsey Community College

11200 Mississippi Blvd. NW

Coon Rapids, MN 55433

Table of Contents

Table of Contents

Table of Contents

To the Student

This manual is a text supplement and should be read along *with* the text. Read all exercise solutions in this manual since explanations of concepts are given and then appear in subsequent solutions. All concepts necessary to solve a particular problem are not reviewed for every exercise. If you are having difficulty with a previously covered concept, look back to the section where it was covered for more complete help. The writing style I have used in this manual reflects the way I explain concepts to my own students. It is not as mathematically precise as that of the text, including phrases such as "goes down" or "touches and turns around." My students have told me that these terms help them understand difficult concepts with ease.

Lengthier explanations and more steps are given for the more difficult problems. Additional information that my students have found helpful is included—see page 29. The guidelines given in the text are followed for some solutions—see page 119.

In the review sections, the solutions are somewhat abbreviated since more detailed solutions were given in previous sections. However, this is not true for the word problems in these sections since they are unique. In easier groups of exercises, representative solutions are shown. Occasionally, alternate solutions are also given.

All figures have been plotted using computer software, offering a high degree of precision. The calculator graphs are from the TI-82 screen, and any specific instructions are for the TI-81/82/83. When possible, each piece of art was made with the same scale to show a realistic and consistent graph.

This manual was done using EXP: *The Scientific Word Processor.* I have used a variety of display formats for the mathematical equations, including centering, vertical alignment, and flushing text to the right. I hope that these make reading and comprehending the material easier for you.

Notations

The following notations are used in the manual.

Note: { Notes to the student pertaining to hints on solutions, common mistakes, or

conventions to follow. }

{ }   { comments to the reader are in braces }

LS   { Left Side of an equation }

RS   { Right Side of an equation }

$\approx$ { approximately equal to }

$\Rightarrow$ { implies, next equation, logically follows }

$\Leftrightarrow$ { if and only if, is equivalent to }

•   { bullet, used to separate problem statement from solution or explanation }

★   { used to identify the answer to the problem }

§   { *section* references }

$\forall$ { For all, i.e., $\forall x$ means "for all x". }

$\mathbb{R} - \{a\}$ { The set of all real numbers except a. }

$\therefore$ { therefore }

QI–QIV   { quadrants I, II, III, IV }

Chapter 1: Fundamental Concepts of Algebra

$\boxed{1}$ (a) Since x and y have opposite signs, the product xy is negative.

(b) Since $x^2 > 0$ and $y > 0$, $x^2 y > 0$.

(c) Since $x < 0$ { x is negative } and $y > 0$ { y is positive }, $\frac{x}{y}$ is negative.

Thus, $\frac{x}{y} + x$ is the sum of two negatives, which is *negative*.

(d) Since $y > 0$ and $x < 0$, $y - x > 0$.

$\boxed{3}$ (a) Since -7 is to the left of -4 on a coordinate line, $-7\boxed{<}-4$.

(b) Using a calculator, we see that $\frac{\pi}{2} \approx 1.5708$. Hence, $\frac{\pi}{2}\boxed{>}1.57$.

(c) $\sqrt{225}\boxed{=}15$ *Note:* $\sqrt{225} \neq \pm 15$

$\boxed{5}$ (a) Since $\frac{1}{11} = 0.\overline{09}$, $\frac{1}{11}\boxed{>}0.09$.   (b) Since $\frac{2}{3} = 0.\overline{6}$, $\frac{2}{3}\boxed{>}0.6666$.

(c) Since $\frac{22}{7} = 3.\overline{142857}$ and $\pi \approx 3.141593$, $\frac{22}{7}\boxed{>}\pi$.

Note: An informal definition of absolute value that may be helpful is

$$| \, something \, | = \begin{cases} itself & \text{if } itself \text{ is positive or zero} \\ -(itself) & \text{if } itself \text{ is negative} \end{cases}$$

$\boxed{9}$ (a) $|-3-2| = |-5| = -(-5)$ { since $-5 < 0$ } $= 5$

(b) $|-5| - |2| = -(-5) - 2 = 5 - 2 = 3$

(c) $|7| + |-4| = 7 + [-(-4)] = 7 + 4 = 11$

$\boxed{11}$ (a) $(-5)|3-6| = (-5)|-3| = (-5)[-(-3)] = (-5)(3) = -15$

(b) $|-6|/(-2) = -(-6)/(-2) = 6/(-2) = -3$

(c) $|-7| + |4| = -(-7) + 4 = 7 + 4 = 11$

$\boxed{13}$ (a) Since $(4 - \pi)$ is positive, $|4 - \pi| = 4 - \pi$.

(b) Since $(\pi - 4)$ is negative, $|\pi - 4| = -(\pi - 4) = 4 - \pi$.

(c) Since $(\sqrt{2} - 1.5)$ is negative, $|\sqrt{2} - 1.5| = -(\sqrt{2} - 1.5) = 1.5 - \sqrt{2}$.

$\boxed{17}$ (a) $d(A, B) = |1 - (-9)| = |10| = 10$ (b) $d(B, C) = |10 - 1| = |9| = 9$

(c) $d(C, B) = d(B, C) = 9$ (d) $d(A, C) = |10 - (-9)| = |19| = 19$

Note: Exer. 19–24: Since $|a| = |-a|$, the answers could have a different form.

For example, $|-3 - x| \geq 8$ is equivalent to $|x + 3| \geq 8$.

$\boxed{19}$ $d(A, B) = |7 - x| \Rightarrow |7 - x| < 5$

$\boxed{21}$ $d(A, B) = |-3 - x| \Rightarrow |-3 - x| \geq 8$

$\boxed{25}$ Pick an arbitrary value for x that is less than -3, say -5.

Since $3 + (-5) = -2$ is negative, we conclude that if $x < -3$, then $3 + x$ is negative.

Hence, $|3 + x| = -(3 + x) = -x - 3$.

$\boxed{27}$ If $x < 2$, then $2 - x > 0$, and $|2 - x| = 2 - x$.

29 If $a < b$, then $a - b < 0$, and $|a - b| = -(a - b) = b - a$.

31 Since $x^2 + 4 > 0$ for every x, $|x^2 + 4| = x^2 + 4$.

33 LS $= \dfrac{ab + ac}{a} = \dfrac{ab}{a} + \dfrac{ac}{a} = b + c \boxed{\neq}$ RS $(b + ac)$.

35 LS $= \dfrac{b + c}{a} = \dfrac{b}{a} + \dfrac{c}{a} \boxed{=}$ RS.

37 LS $= (a \div b) \div c = \dfrac{a}{b} \cdot \dfrac{1}{c} = \dfrac{a}{bc}$. RS $= a \div (b \div c) = a \div \dfrac{b}{c} = a \cdot \dfrac{c}{b} = \dfrac{ac}{b}$. LS $\boxed{\neq}$ RS

39 LS $= \dfrac{a - b}{b - a} = \dfrac{-(b - a)}{b - a} = -1 \boxed{=}$ RS.

43 (a) $\dfrac{1.2 \times 10^3}{3.1 \times 10^2 + 1.52 \times 10^3} \approx 0.6557 = 6.557 \times 10^{-1}$ *Note:* For the TI-82,

         use 1.2E3/(3.1E2 + 1.52E3), where E is obtained by pressing $\boxed{\text{2nd}}$ $\boxed{\text{EE}}$.

     (b) $(1.23 \times 10^{-4}) + \sqrt{4.5 \times 10^3} \approx 67.08 = 6.708 \times 10^1$

45 Construct a right triangle with sides of lengths $\sqrt{2}$ and 1. The hypotenuse will have length $\sqrt{(\sqrt{2})^2 + 1^2} = \sqrt{3}$. Next construct a right triangle with sides of lengths $\sqrt{3}$ and $\sqrt{2}$. The hypotenuse will have length $\sqrt{(\sqrt{3})^2 + (\sqrt{2})^2} = \sqrt{5}$.

47 The large rectangle has area = width × length = $a(b + c)$. The sum of the areas of the two small rectangles is $ab + ac$. Since the areas are the same, we have $a(b + c) = ab + ac$.

49 (a) Since the decimal point is 5 places to the right of the first nonzero digit,
$$427,000 = 4.27 \times 10^5.$$

     (b) Since the decimal point is 8 places to the left of the first nonzero digit,
$$0.000\,000\,098 = 9.8 \times 10^{-8}.$$

51 (a) Moving the decimal point 5 places to the right, we have $8.3 \times 10^5 = 830,000$.

     (b) Moving the decimal point 12 places to the left, we have
$$2.9 \times 10^{-12} = 0.000\,000\,000\,002\,9.$$

55 It is helpful to write the units of any fraction, and then "cancel" those units to determine the units of the final answer.

$$\dfrac{186,000 \text{ miles}}{\text{second}} \cdot \dfrac{60 \text{ seconds}}{1 \text{ minute}} \cdot \dfrac{60 \text{ minutes}}{1 \text{ hour}} \cdot \dfrac{24 \text{ hours}}{1 \text{ day}} \cdot \dfrac{365 \text{ days}}{1 \text{ year}} \cdot 1 \text{ year} \approx 5.87 \times 10^{12} \text{ mi}$$

57 $\dfrac{\dfrac{1.01 \text{ grams}}{\text{mole}}}{\dfrac{6.02 \times 10^{23} \text{ atoms}}{\text{mole}}} \cdot 1 \text{ atom} = \dfrac{1.01 \text{ grams}}{6.02 \times 10^{23}} \approx 0.1678 \times 10^{-23} \text{ g} = 1.678 \times 10^{-24} \text{ g}$

59 $\dfrac{24 \text{ frames}}{\text{second}} \cdot \dfrac{60 \text{ seconds}}{1 \text{ minute}} \cdot \dfrac{60 \text{ minutes}}{1 \text{ hour}} \cdot 48 \text{ hours} = 4.1472 \times 10^6 \text{ frames}$

                                           1.0368×10^{18} calculations

61 (a) IQ $= \dfrac{\text{mental age (MA)}}{\text{chronological age (CA)}} \times 100 = \dfrac{15}{12} \times 100 = 125.$

61 (b) $CA = 15$ and $IQ = 140 \Rightarrow 140 = MA/15 \times 100 \Rightarrow MA = 21$.

63 (a) $1 \text{ ft}^2 = 144 \text{ in.}^2 \Rightarrow 144 \text{ in.}^2 \times 1.4 \text{ lb/in.}^2 = 201.6 \text{ lb}$.

(b) $40 \times 8 = 320 \text{ ft}^2 = 46{,}080 \text{ in.}^2$; $46{,}080 \times 1.4 = 64{,}512 \text{ lb}$;

$$64{,}512 \text{ lb}/(2000 \text{ lb/ton}) = 32.256 \text{ tons}$$

1.2 Exercises

1 $\left(-\frac{2}{3}\right)^4 = \left(-\frac{2}{3}\right) \cdot \left(-\frac{2}{3}\right) \cdot \left(-\frac{2}{3}\right) \cdot \left(-\frac{2}{3}\right) = \frac{16}{81}$

Note: Do not confuse $(-x)^4$ and $-x^4$ since $(-x)^4 = x^4$ and $-x^4$ is the negative of x^4.

3 $\dfrac{2^{-3}}{3^{-2}} = \dfrac{3^2}{2^3} = \dfrac{9}{8}$ *Note:* Remember that negative exponents don't necessarily equate to

negative results—that is, $2^{-3} = \dfrac{1}{2^3} = \dfrac{1}{8}$, not $-\dfrac{1}{8}$.

5 $-2^4 + 3^{-1} = -16 + \frac{1}{3} = -\frac{48}{3} + \frac{1}{3} = \frac{-47}{3}$

7 $16^{-3/4} = 1/16^{3/4} = 1/(\sqrt[4]{16})^3 = 1/2^3 = \frac{1}{8}$

9 $(-0.008)^{2/3} = (\sqrt[3]{-0.008})^2 = (-0.2)^2 = 0.04 = \frac{4}{100} = \frac{1}{25}$

13 A common mistake is to write $x^3 x^2 = x^6$, and another is to write $(x^2)^3 = x^5$.

The following solution illustrates the proper use of the exponent rules.

$$\frac{(2x^3)(3x^2)}{(x^2)^3} = \frac{(2 \cdot 3)x^{3+2}}{x^{2 \cdot 3}} = \frac{6x^5}{x^6} = 6x^{5-6} = 6x^{-1} = \frac{6}{x}$$

17 $\dfrac{(6x^3)^2}{(2x^2)^3} = \dfrac{36x^6}{8x^6} = \dfrac{9}{2}$

19 $(3u^7 v^3)(4u^4 v^{-5}) = 12u^{7+4}v^{3+(-5)} = 12u^{11}v^{-2} = \dfrac{12u^{11}}{v^2}$

23 $\left(\frac{1}{3}x^4 y^{-3}\right)^{-2} = \left(\frac{1}{3}\right)^{-2}(x^4)^{-2}(y^{-3})^{-2} = \left(\frac{3}{1}\right)^2 x^{-8} y^6 = 3^2 x^{-8} y^6 = \dfrac{9y^6}{x^8}$

25 $(3y^3)^4 (4y^2)^{-3} = 81y^{12} \cdot 4^{-3} y^{-6} = 81y^6 \cdot \dfrac{1}{64} = \dfrac{81}{64}y^6$

27 $(-2r^4 s^{-3})^{-2} = (-2)^{-2} r^{-8} s^6 = \dfrac{s^6}{(-2)^2 r^8} = \dfrac{s^6}{4r^8}$

31 $\left(\dfrac{3x^5 y^4}{x^0 y^{-3}}\right)^2$ {remember that $x^0 = 1$} $= \dfrac{9x^{10} y^8}{y^{-6}} = 9x^{10} y^{14}$

35 $(3x^{5/6})(8x^{2/3}) = 24x^{(5/6)+(4/6)} = 24x^{9/6} = 24x^{3/2}$

37 $(27a^6)^{-2/3} = 27^{-2/3} a^{-12/3} = \dfrac{1}{(\sqrt[3]{27})^2 a^4} = \dfrac{1}{9a^4}$

41 $\left(\dfrac{-8x^3}{y^{-6}}\right)^{2/3} = \dfrac{(-8)^{2/3}(x^3)^{2/3}}{(y^{-6})^{2/3}} = \dfrac{(\sqrt[3]{-8})^2\, x^{(3)(2/3)}}{y^{(-6)(2/3)}} = \dfrac{(-2)^2 x^2}{y^{-4}} = \dfrac{4x^2}{y^{-4}} = 4x^2 y^4$

45 $\dfrac{(x^6 y^3)^{-1/3}}{(x^4 y^2)^{-1/2}} = \dfrac{(x^6)^{-1/3}(y^3)^{-1/3}}{(x^4)^{-1/2}(y^2)^{-1/2}} = \dfrac{x^{-2}y^{-1}}{x^{-2}y^{-1}} = 1$

49 $\sqrt[3]{(a+b)^2} = [(a+b)^2]^{1/3} = (a+b)^{2/3}$

53 (a) $4x^{3/2} = 4x^1 x^{1/2} = 4x\sqrt{x}$

(b) $(4x)^{3/2} = (4x)^1 (4x)^{1/2} = (4x)^1\, 4^{1/2} x^{1/2} = 4x \cdot 2 \cdot x^{1/2} = 8x\sqrt{x}$

59 $\sqrt[5]{-64} = \sqrt[5]{-32}\,\sqrt[5]{2} = \sqrt[5]{(-2)^5}\,\sqrt[5]{2} = -2\sqrt[5]{2}$

61 In the denominator, you would like to have $\sqrt[3]{2^3}$. How do you get it? Multiply by

$\sqrt[3]{2^2}$, or, equivalently, $\sqrt[3]{4}$. $\quad \dfrac{1}{\sqrt[3]{2}} = \dfrac{1}{\sqrt[3]{2}} \cdot \dfrac{\sqrt[3]{4}}{\sqrt[3]{4}} = \dfrac{\sqrt[3]{4}}{\sqrt[3]{8}} = \dfrac{\sqrt[3]{4}}{2} = \tfrac{1}{2}\sqrt[3]{4}$

63 $\sqrt{9x^{-4}y^6} = (9x^{-4}y^6)^{1/2} = 9^{1/2}(x^{-4})^{1/2}(y^6)^{1/2} = 3x^{-2}y^3 = \dfrac{3y^3}{x^2}$

Note: For exercises similar to those in 67–74, pick a multiplier that will make

all of the exponents of the terms in the denominator a multiple of the index.

67 The index is 2. Choose the multiplier to be $\sqrt{2y}$ so that the denominator contains

only terms with even exponents. $\quad \sqrt{\dfrac{3x}{2y^3}} = \sqrt{\dfrac{3x}{2y^3}} \cdot \dfrac{\sqrt{2y}}{\sqrt{2y}} = \sqrt{\dfrac{6xy}{4y^4}} = \dfrac{\sqrt{6xy}}{2y^2}$, or $\dfrac{1}{2y^2}\sqrt{6xy}$

69 The index is 3. Choose the multiplier to be $\sqrt[3]{3x^2}$ so that the denominator contains

only terms with exponents that are multiples of 3.

$\sqrt[3]{\dfrac{2x^4 y^4}{9x}} = \sqrt[3]{\dfrac{2x^4 y^4}{9x}} \cdot \dfrac{\sqrt[3]{3x^2}}{\sqrt[3]{3x^2}} = \dfrac{\sqrt[3]{6x^6 y^4}}{\sqrt[3]{27x^3}} = \dfrac{\sqrt[3]{x^6 y^3}\,\sqrt[3]{6y}}{3x} = \dfrac{x^2 y\,\sqrt[3]{6y}}{3x} = \dfrac{xy}{3}\sqrt[3]{6y}$

71 The index is 4. Choose the multiplier to be $\sqrt[4]{3x^2}$ so that the denominator contains

only terms with exponents that are multiples of 4.

$\sqrt[4]{\dfrac{5x^8 y^3}{27x^2}} = \sqrt[4]{\dfrac{5x^8 y^3}{27x^2}} \cdot \dfrac{\sqrt[4]{3x^2}}{\sqrt[4]{3x^2}} = \dfrac{\sqrt[4]{15x^{10} y^3}}{\sqrt[4]{81x^4}} = \dfrac{\sqrt[4]{x^8}\,\sqrt[4]{15x^2 y^3}}{3x} = \dfrac{x^2\,\sqrt[4]{15x^2 y^3}}{3x}$

$\qquad\qquad\qquad = \tfrac{x}{3}\sqrt[4]{15x^2 y^3}$

73 The index is 5. Choose the multiplier to be $\sqrt[5]{4x^2}$ so that the denominator contains

only terms with exponents that are multiples of 5.

$\sqrt[5]{\dfrac{5x^7 y^2}{8x^3}} = \sqrt[5]{\dfrac{5x^7 y^2}{8x^3}} \cdot \dfrac{\sqrt[5]{4x^2}}{\sqrt[5]{4x^2}} = \dfrac{\sqrt[5]{x^5}\,\sqrt[5]{20x^4 y^2}}{2x} = \dfrac{x\,\sqrt[5]{20x^4 y^2}}{2x} = \tfrac{1}{2}\sqrt[5]{20x^4 y^2}$

$\boxed{75}$ $\sqrt[4]{(3x^5y^{-2})^4} = 3x^5y^{-2} = \dfrac{3x^5}{y^2}$

$\boxed{77}$ $\sqrt[5]{\dfrac{8x^3}{y^4}} \sqrt[5]{\dfrac{4x^4}{y^2}} = \sqrt[5]{\dfrac{8x^3}{y^4}} \sqrt[5]{\dfrac{4x^4}{y^2}} \cdot \dfrac{\sqrt[5]{y^4}}{\sqrt[5]{y^4}} = \dfrac{\sqrt[5]{32x^5}\,\sqrt[5]{x^2y^4}}{y^2} = \dfrac{2x}{y^2}\sqrt[5]{x^2y^4}$

$\boxed{79}$ $\sqrt[3]{3t^4v^2}\,\sqrt[3]{-9t^{-1}v^4} = \sqrt[3]{-27t^3v^6} = -3tv^2$

$\boxed{81}$ $\sqrt{x^6y^4} = \sqrt{(x^3)^2(y^2)^2} = \sqrt{(x^3)^2}\,\sqrt{(y^2)^2} = |\,x^3\,|\,|\,y^2\,| = |\,x^3\,|\,y^2$ since y^2 is always

nonnegative. *Note:* $|\,x^3\,|$ could be written as $x^2\,|\,x\,|$.

$\boxed{83}$ $\sqrt[4]{x^8(y-1)^{12}} = \sqrt[4]{(x^2)^4((y-1)^3)^4} = |\,x^2\,|\,|\,(y-1)^3\,| = x^2\,|\,(y-1)^3\,|$,

or $x^2(y-1)^2\,|\,(y-1)\,|$

$\boxed{85}$ $(a^r)^2 = a^{2r}\,\boxed{\neq}\,a^{(r^2)}$ since $2r \neq r^2$ for every r.

$\boxed{87}$ $(ab)^{xy} = a^{xy}b^{xy}\,\boxed{\neq}\,a^xb^y$ for every x and y.

$\boxed{91}$ (a) $(-3)^{2/5} = [(-3)^2]^{1/5} = 9^{1/5} \approx 1.5518$

     (b) $(-5)^{4/3} = [(-5)^4]^{1/3} = 625^{1/3} \approx 8.5499$

$\boxed{97}$ $W = 230$ kg $\Rightarrow L = 0.46\sqrt[3]{W} = 0.46\sqrt[3]{230} \approx 2.82$ m

$\boxed{99}$ $b = 75$ and $w = 180 \Rightarrow W = \dfrac{w}{\sqrt[3]{b-35}} = \dfrac{180}{\sqrt[3]{75-35}} \approx 52.6.$

     $b = 120$ and $w = 250 \Rightarrow W = \dfrac{w}{\sqrt[3]{b-35}} = \dfrac{250}{\sqrt[3]{120-35}} \approx 56.9.$

It is interesting to note that the 75-kg lifter can lift 2.4 times his/her body weight

and the 120-kg lifter can lift approximately 2.08 times his/her body weight, but

the formula ranks the 120-kg lifter as the superior lifter.

$\boxed{101}$ $W = 0.1166h^{1.7}$

Height	64	65	66	67	68	69	70	71
Weight	137	141	145	148	152	156	160	164
Height	72	73	74	75	76	77	78	79
Weight	168	172	176	180	184	188	192	196

1.3 Exercises

$\boxed{5}$ $(2x+5)(3x-7) = (2x)(3x) + (2x)(-7) + (5)(3x) + (5)(-7)$

     $= 6x^2 - 14x + 15x - 35$

     $= 6x^2 + x - 35$

$\boxed{9}$ $(2u+3)(u-4) + 4u(u-2) = (2u^2 - 5u - 12) + (4u^2 - 8u) = 6u^2 - 13u - 12$

$\boxed{11}$ $(3x+5)(2x^2+9x-5)$ $= 3x(2x^2+9x-5)+5(2x^2+9x-5)$
$$= (6x^3+27x^2-15x)+(10x^2+45x-25)$$
$$= 6x^3+37x^2+30x-25$$

$\boxed{13}$ $(t^2+2t-5)(3t^2-t+2)$ $= t^2(3t^2-t+2)+2t(3t^2-t+2)+(-5)(3t^2-t+2)$
$$= (3t^4-t^3+2t^2)+(6t^3-2t^2+4t)+(-15t^2+5t-10)$$
$$= 3t^4+5t^3-15t^2+9t-10$$

$\boxed{15}$ $(x+1)(2x^2-2)(x^3+5)= 2\,[(x+1)(x^2-1)](x^3+5)$
$$= 2(x^3+x^2-x-1)(x^3+5)$$
$$= 2(x^6+x^5-x^4+4x^3+5x^2-5x-5)$$
$$= 2x^6+2x^5-2x^4+8x^3+10x^2-10x-10$$

$\boxed{19}$ $\dfrac{3u^3v^4-2u^5v^2+(u^2v^2)^2}{u^3v^2}=\dfrac{3u^3v^4}{u^3v^2}-\dfrac{2u^5v^2}{u^3v^2}+\dfrac{u^4v^4}{u^3v^2}=3v^2-2u^2+uv^2$

$\boxed{21}$ We recognize this product as the difference of two squares.
$$(2x+3y)(2x-3y)=(2x)^2-(3y)^2=4x^2-9y^2$$

$\boxed{25}$ $(x^2+9)(x^2-4)=x^4-4x^2+9x^2-36=x^4+5x^2-36$

$\boxed{27}$ $(3x+2y)^2=(3x)^2+2(3x)(2y)+(2y)^2=9x^2+12xy+4y^2$

$\boxed{29}$ $(x^2-3y^2)^2=(x^2)^2-2(x^2)(3y^2)+(3y^2)^2=x^4-6x^2y^2+9y^4$

$\boxed{31}$ We could expand $(x+2)^2$ and $(x-2)^2$ and then multiply the resulting expressions, but the following solution is simpler.
$(x+2)^2(x-2)^2$ $= [(x+2)(x-2)]^2$
$$= (x^2-4)^2$$
$$= (x^2)^2-2(x^2)(4)+(4)^2$$
$$= x^4-8x^2+16$$

$\boxed{35}$ $(x^{1/3}-y^{1/3})(x^{2/3}+x^{1/3}y^{1/3}+y^{2/3})$
$$= x^{1/3}(x^{2/3}+x^{1/3}y^{1/3}+y^{2/3})-y^{1/3}(x^{2/3}+x^{1/3}y^{1/3}+y^{2/3})$$
$$= x+x^{2/3}y^{1/3}+x^{1/3}y^{2/3}-x^{2/3}y^{1/3}-x^{1/3}y^{2/3}-y = x-y$$

This exercise illustrates how the difference of *any* two terms can be factored as the difference of cubes. Another example of this concept is
$$x-5=(\sqrt[3]{x}-\sqrt[3]{5})(\sqrt[3]{x^2}+\sqrt[3]{5x}+\sqrt[3]{25}).$$

$\boxed{37}$ Use Product Formula (3) on text page 32.
$$(x-2y)^3=(x)^3-3(x)^2(2y)+3(x)(2y)^2-(2y)^3=x^3-6x^2y+12xy^2-8y^3$$

$\boxed{39}$ $(2x+3y)^3$ $= (2x)^3+3(2x)^2(3y)+3(2x)(3y)^2+(3y)^3$
$$= 8x^3+3(4x^2)(3y)+3(2x)(9y^2)+27y^3$$
$$= 8x^3+36x^2y+54xy^2+27y^3$$

Note: Exer. 41–44: Treat these as "the sum of the squares plus twice the product of all possible pairs of terms", that is, $(x + y + z)^2 = x^2 + y^2 + z^2 + 2xy + 2xz + 2yz$.

$\boxed{43}$ $(2x + y - 3z)^2 = (2x)^2 + (y)^2 + (-3z)^2 + 2(2x)(y) + 2(2x)(-3z) + 2(y)(-3z)$
$$= 4x^2 + y^2 + 9z^2 + 4xy - 12xz - 6yz$$

$\boxed{47}$ Always factor out the greatest common factor first. $3a^2b^2 - 6a^2b = 3a^2b(b - 2)$

$\boxed{51}$ $15x^3y^5 - 25x^4y^2 + 10x^6y^4 = 5x^3y^2(3y^3 - 5x + 2x^3y^2)$

$\boxed{53}$ We recognize this as a trinomial that may be able to be factored into the product of two binomials. $8x^2 - 53x - 21 = (8x + 3)(x - 7)$

$\boxed{55}$ The factors for $x^2 + 3x + 4$ would have to be of the form $(x + \_)$ and $(x + \_)$.

The factors of 4 are 1 & 4 and 2 & 2, but their sums are 5 and 4, respectively.

Thus, $x^2 + 3x + 4$ is *irreducible*.

$\boxed{61}$ $4x^2 - 20x + 25 = (2x - 5)(2x - 5) = (2x - 5)^2$

$\boxed{65}$ $45x^2 + 38xy + 8y^2 = (5x + 2y)(9x + 4y)$

$\boxed{69}$ $z^4 - 64w^2 = (z^2)^2 - (8w)^2 = (z^2 + 8w)(z^2 - 8w)$

$\boxed{71}$ $x^4 - 4x^2 = x^2(x^2 - 4) = x^2(x^2 - 2^2) = x^2(x + 2)(x - 2)$

$\boxed{73}$ $x^2 + 25$ is irreducible. *Note:* A common mistake is to confuse the *sum* of two squares with the *difference* of two squares.

$\boxed{75}$ $75x^2 - 48y^2 = 3(25x^2 - 16y^2) = 3\left[(5x)^2 - (4y)^2\right] = 3(5x + 4y)(5x - 4y)$

$\boxed{77}$ $64x^3 + 27 = (4x)^3 + (3)^3 = (4x + 3)\left[(4x)^2 - (4x)(3) + (3)^2\right] = (4x + 3)(16x^2 - 12x + 9)$

$\boxed{79}$ We recognize this expression as the difference of two cubes.
$$64x^3 - y^6 = (4x)^3 - (y^2)^3$$
$$= (4x - y^2)\left[(4x)^2 + (4x)(y^2) + (y^2)^2\right]$$
$$= (4x - y^2)(16x^2 + 4xy^2 + y^4)$$

$\boxed{81}$ We recognize this expression as the sum of two cubes.
$$343x^3 + y^9 = (7x)^3 + (y^3)^3$$
$$= (7x + y^3)\left[(7x)^2 - (7x)(y^3) + (y^3)^2\right]$$
$$= (7x + y^3)(49x^2 - 7xy^3 + y^6)$$

$\boxed{83}$ Since there are more than 3 terms, we will try to factor by grouping first.
$$2ax - 6bx + ay - 3by = 2x(a - 3b) + y(a - 3b) = (2x + y)(a - 3b)$$

$\boxed{85}$ $3x^3 + 3x^2 - 27x - 27 = 3(x^3 + x^2 - 9x - 9) =$
$$3[x^2(x + 1) - 9(x + 1)] = 3(x^2 - 9)(x + 1) = 3(x + 3)(x - 3)(x + 1)$$

$\boxed{87}$ Since there are more than 3 terms, we will try to factor by grouping first.
$$x^4 + 2x^3 - x - 2 = x^3(x + 2) - 1(x + 2) = (x^3 - 1)(x + 2) = (x - 1)(x + 2)(x^2 + x + 1)$$

$\boxed{89}$ $a^3 - a^2b + ab^2 - b^3 = a^2(a - b) + b^2(a - b) = (a^2 + b^2)(a - b)$

[91] We could treat this expression as the difference of two squares or the difference of two cubes. Factoring as the difference of two squares and then as the sum and difference of two cubes leads to the following:

$$
\begin{aligned}
a^6 - b^6 &= (a^3)^2 - (b^3)^2 \\
&= (a^3 + b^3)(a^3 - b^3) \\
&= (a + b)(a - b)(a^2 - ab + b^2)(a^2 + ab + b^2)
\end{aligned}
$$

[93] We might first try to factor this expression by grouping since it has more than 3 terms, but this would prove to be unsuccessful. Instead, we will group the terms containing x and the constant term together, and then proceed as in Example 10(c).

$$
x^2 + 4x + 4 - 9y^2 = (x + 2)^2 - (3y)^2 = (x + 2 + 3y)(x + 2 - 3y)
$$

[95] We will group the terms containing y and the constant term together, and then proceed as in Example 10(c).

$$
\begin{aligned}
y^2 - x^2 + 8y + 16 &= (y^2 + 8y + 16) - x^2 \\
&= (y + 4)^2 - (x)^2 \\
&= (y + 4 + x)(y + 4 - x)
\end{aligned}
$$

[97] We should first note that one of the two variable terms, y^6, is the square of the other, y^3. Thus, we may treat this expression as a simple trinomial that can be factored into the product of two binomials.

$$
y^6 + 7y^3 - 8 = (y^3 + 8)(y^3 - 1) = (y + 2)(y^2 - 2y + 4)(y - 1)(y^2 + y + 1)
$$

[99] $$
\begin{aligned}
x^{16} - 1 = (x^8 + 1)(x^8 - 1) &= (x^8 + 1)(x^4 + 1)(x^4 - 1) \\
&= (x^8 + 1)(x^4 + 1)(x^2 + 1)(x^2 - 1) \\
&= (x^8 + 1)(x^4 + 1)(x^2 + 1)(x + 1)(x - 1)
\end{aligned}
$$

[101] The dimensions of I are (x) and $(x - y)$. The area of I is $(x - y)x$, and the area of II is $(x - y)y$. The area $A = \underline{x^2 - y^2} = (x - y)x + (x - y)y = \underline{(x - y)(x + y)}$.

1.4 Exercises

[1] $\dfrac{3}{50} + \dfrac{7}{30} = \dfrac{3}{2 \cdot 5^2} + \dfrac{7}{2 \cdot 3 \cdot 5} = \dfrac{3 \cdot 3 + 7 \cdot 5}{2 \cdot 3 \cdot 5^2} = \dfrac{9 + 35}{2 \cdot 3 \cdot 5^2} = \dfrac{44}{2 \cdot 3 \cdot 5^2} = \dfrac{22}{3 \cdot 5^2} = \dfrac{22}{75}$

[3] $\dfrac{5}{24} - \dfrac{3}{20} = \dfrac{5}{2^3 \cdot 3} - \dfrac{3}{2^2 \cdot 5} = \dfrac{5 \cdot 5 - 3(2 \cdot 3)}{2^3 \cdot 3 \cdot 5} = \dfrac{25 - 18}{2^3 \cdot 3 \cdot 5} = \dfrac{7}{2^3 \cdot 3 \cdot 5} = \dfrac{7}{120}$

[7] $\dfrac{y^2 - 25}{y^3 - 125} = \dfrac{(y + 5)(y - 5)}{(y - 5)(y^2 + 5y + 25)} = \dfrac{y + 5}{y^2 + 5y + 25}$

[9] $\dfrac{12 + r - r^2}{r^3 + 3r^2} = \dfrac{(3 + r)(4 - r)}{r^2(r + 3)} = \dfrac{4 - r}{r^2}$

$\boxed{11}$ $\dfrac{9x^2-4}{3x^2-5x+2} \cdot \dfrac{9x^4-6x^3+4x^2}{27x^4+8x} = \dfrac{(3x+2)(3x-2)}{(3x-2)(x-1)} \cdot \dfrac{x^2(9x^2-6x+4)}{x(3x+2)(9x^2-6x+4)} = \dfrac{x}{x-1}$

$\boxed{13}$ $\dfrac{5a^2+12a+4}{a^4-16} \div \dfrac{25a^2+20a+4}{a^2-2a} = \dfrac{(5a+2)(a+2)}{(a^2+4)(a+2)(a-2)} \cdot \dfrac{a(a-2)}{(5a+2)(5a+2)} =$

$$\dfrac{a}{(a^2+4)(5a+2)}$$

$\boxed{15}$ $\dfrac{6}{x^2-4} - \dfrac{3x}{x^2-4} = \dfrac{6-3x}{x^2-4} = \dfrac{3(2-x)}{(x+2)(x-2)} = \dfrac{-3}{x+2}.$ Since $2-x=-(x-2),$ we

canceled the two factors, $2-x$ and $x-2$, and replaced them with -1. In general,
you may do this whenever you encounter factors of the form $a-b$ and $b-a$ in the
numerator and the denominator, respectively, of a fractional expression.

$\boxed{17}$ $\dfrac{2}{3s+1} - \dfrac{9}{(3s+1)^2} = \dfrac{2(3s+1)-9}{(3s+1)^2} = \dfrac{6s-7}{(3s+1)^2}$

$\boxed{19}$ $\dfrac{2}{x} + \dfrac{3x+1}{x^2} - \dfrac{x-2}{x^3} = \dfrac{2x^2+(3x+1)x-x+2}{x^3} = \dfrac{5x^2+2}{x^3}$

$\boxed{21}$ $\dfrac{3t}{t+2} + \dfrac{5t}{t-2} - \dfrac{40}{t^2-4} = \dfrac{3t}{t+2} + \dfrac{5t}{t-2} - \dfrac{40}{(t+2)(t-2)}$

$$= \dfrac{3t(t-2)}{(t+2)(t-2)} + \dfrac{5t(t+2)}{(t+2)(t-2)} - \dfrac{40}{(t+2)(t-2)}$$

$$= \dfrac{3t^2-6t+5t^2+10t-40}{(t+2)(t-2)}$$

$$= \dfrac{8t^2+4t-40}{(t+2)(t-2)}$$

$$= \dfrac{4(2t+5)(t-2)}{(t+2)(t-2)} = \dfrac{4(2t+5)}{t+2}$$

$\boxed{23}$ $\dfrac{4x}{3x-4} + \dfrac{8}{3x^2-4x} + \dfrac{2}{x} = \dfrac{4x(x)+8+2(3x-4)}{x(3x-4)} = \dfrac{4x^2+6x}{x(3x-4)} = \dfrac{2x(2x+3)}{x(3x-4)} = \dfrac{2(2x+3)}{3x-4}$

$\boxed{25}$ $\dfrac{2x}{x+2} - \dfrac{8}{x^2+2x} + \dfrac{3}{x} = \dfrac{2x(x)-8+3(x+2)}{x(x+2)} = \dfrac{2x^2+3x-2}{x(x+2)} = \dfrac{(2x-1)(x+2)}{x(x+2)} = \dfrac{2x-1}{x}$

$\boxed{27}$ $\dfrac{p^4+3p^3-8p-24}{p^3-2p^2-9p+18} = \dfrac{p^3(p+3)-8(p+3)}{p^2(p-2)-9(p-2)} = \dfrac{(p^3-8)(p+3)}{(p^2-9)(p-2)} =$

$$\dfrac{(p-2)(p^2+2p+4)(p+3)}{(p+3)(p-3)(p-2)} = \dfrac{p^2+2p+4}{p-3}$$

$\boxed{29}$ $3 + \dfrac{5}{u} + \dfrac{2u}{3u+1} = \dfrac{3u(3u+1)+5(3u+1)+2u(u)}{u(3u+1)} = \dfrac{11u^2+18u+5}{u(3u+1)}$

$\boxed{31}$ $\dfrac{2x+1}{x^2+4x+4} - \dfrac{6x}{x^2-4} + \dfrac{3}{x-2} = \dfrac{(2x+1)(x-2) - 6x(x+2) + 3(x^2+4x+4)}{(x+2)^2(x-2)} =$

$$\dfrac{-x^2-3x+10}{(x+2)^2(x-2)} = -\dfrac{x^2+3x-10}{(x+2)^2(x-2)} = -\dfrac{(x+5)(x-2)}{(x+2)^2(x-2)} = -\dfrac{x+5}{(x+2)^2}$$

$\boxed{33}$ The lcd of the entire expression is ab.

Thus, we will multiply both the numerator and denominator by ab.

$$\dfrac{\frac{b}{a} - \frac{a}{b}}{\frac{1}{a} - \frac{1}{b}} = \dfrac{\left(\frac{b}{a} - \frac{a}{b}\right) \cdot ab}{\left(\frac{1}{a} - \frac{1}{b}\right) \cdot ab} = \dfrac{b^2 - a^2}{b - a} = \dfrac{(b+a)(b-a)}{b-a} = a+b$$

$\boxed{35}$ $\dfrac{\frac{x}{y^2} - \frac{y}{x^2}}{\frac{1}{y^2} - \frac{1}{x^2}} = \dfrac{\left(\frac{x}{y^2} - \frac{y}{x^2}\right) \cdot x^2 y^2}{\left(\frac{1}{y^2} - \frac{1}{x^2}\right) \cdot x^2 y^2} = \dfrac{x^3 - y^3}{x^2 - y^2} = \dfrac{(x-y)(x^2+xy+y^2)}{(x+y)(x-y)} = \dfrac{x^2+xy+y^2}{x+y}$

$\boxed{37}$ $\dfrac{y^{-1} + x^{-1}}{(xy)^{-1}} = \dfrac{\frac{1}{y} + \frac{1}{x}}{\frac{1}{xy}} = \dfrac{\left(\frac{1}{y} + \frac{1}{x}\right) \cdot xy}{\left(\frac{1}{xy}\right) \cdot xy} = \dfrac{x+y}{1} = x+y$

$\boxed{39}$ $\dfrac{\frac{5}{x+1} + \frac{2x}{x+3}}{\frac{x}{x+1} + \frac{7}{x+3}} = \dfrac{\dfrac{5(x+3) + 2x(x+1)}{(x+1)(x+3)}}{\dfrac{x(x+3) + 7(x+1)}{(x+1)(x+3)}} = \dfrac{5x+15+2x^2+2x}{x^2+3x+7x+7} = \dfrac{2x^2+7x+15}{x^2+10x+7}$

$\boxed{41}$ $\dfrac{\frac{3}{x-1} - \frac{3}{a-1}}{x-a} = \dfrac{\dfrac{3(a-1) - 3(x-1)}{(x-1)(a-1)}}{x-a} = \dfrac{3a-3x}{(x-1)(a-1)(x-a)} = \dfrac{3(a-x)}{(x-1)(a-1)(x-a)} =$

$$-\dfrac{3}{(x-1)(a-1)}$$

$\boxed{43}$ $\dfrac{(x+h)^2 - 3(x+h) - (x^2 - 3x)}{h} = \dfrac{2xh + h^2 - 3h}{h} = \dfrac{h(2x+h-3)}{h} = 2x+h-3$

$\boxed{45}$ $\dfrac{\frac{1}{(x+h)^3} - \frac{1}{x^3}}{h} = \dfrac{\dfrac{x^3 - (x+h)^3}{(x+h)^3 x^3}}{h} = \dfrac{x^3 - (x+h)^3}{hx^3(x+h)^3} =$

$$\dfrac{[x - (x+h)][x^2 + x(x+h) + (x+h)^2]}{hx^3(x+h)^3} = \dfrac{-h(3x^2 + 3xh + h^2)}{hx^3(x+h)^3} = -\dfrac{3x^2 + 3xh + h^2}{x^3(x+h)^3}$$

$\boxed{47}$ $\dfrac{\frac{4}{3x+3h-1} - \frac{4}{3x-1}}{h} = \dfrac{\dfrac{4(3x-1) - 4(3x+3h-1)}{(3x+3h-1)(3x-1)}}{h} = \dfrac{12x-4-12x-12h+4}{h(3x+3h-1)(3x-1)} =$

$$\dfrac{-12h}{h(3x+3h-1)(3x-1)} = \dfrac{-12}{(3x+3h-1)(3x-1)}$$

51 $\dfrac{81x^2 - 16y^2}{3\sqrt{x} - 2\sqrt{y}} = \dfrac{81x^2 - 16y^2}{3\sqrt{x} - 2\sqrt{y}} \cdot \dfrac{3\sqrt{x} + 2\sqrt{y}}{3\sqrt{x} + 2\sqrt{y}}$

$$= \dfrac{(9x + 4y)(9x - 4y)(3\sqrt{x} + 2\sqrt{y})}{9x - 4y} = (9x + 4y)(3\sqrt{x} + 2\sqrt{y})$$

53 $\dfrac{1}{\sqrt[3]{a} - \sqrt[3]{b}} = \dfrac{1}{\sqrt[3]{a} - \sqrt[3]{b}} \cdot \dfrac{\sqrt[3]{a^2} + \sqrt[3]{ab} + \sqrt[3]{b^2}}{\sqrt[3]{a^2} + \sqrt[3]{ab} + \sqrt[3]{b^2}} = \dfrac{\sqrt[3]{a^2} + \sqrt[3]{ab} + \sqrt[3]{b^2}}{a - b}$

55 $\dfrac{\sqrt{a} - \sqrt{b}}{a^2 - b^2} = \dfrac{\sqrt{a} - \sqrt{b}}{a^2 - b^2} \cdot \dfrac{\sqrt{a} + \sqrt{b}}{\sqrt{a} + \sqrt{b}} = \dfrac{a - b}{(a + b)(a - b)(\sqrt{a} + \sqrt{b})} = \dfrac{1}{(a + b)(\sqrt{a} + \sqrt{b})}$

57 $\dfrac{\sqrt{2(x + h) + 1} - \sqrt{2x + 1}}{h} = \dfrac{\sqrt{2(x + h) + 1} - \sqrt{2x + 1}}{h} \cdot \dfrac{\sqrt{2(x + h) + 1} + \sqrt{2x + 1}}{\sqrt{2(x + h) + 1} + \sqrt{2x + 1}}$

$$= \dfrac{(2x + 2h + 1) - (2x + 1)}{h(\sqrt{2(x + h) + 1} + \sqrt{2x + 1})} = \dfrac{2}{\sqrt{2(x + h) + 1} + \sqrt{2x + 1}}$$

59 $\dfrac{\sqrt{1 - x - h} - \sqrt{1 - x}}{h} = \dfrac{\sqrt{1 - x - h} - \sqrt{1 - x}}{h} \cdot \dfrac{\sqrt{1 - x - h} + \sqrt{1 - x}}{\sqrt{1 - x - h} + \sqrt{1 - x}}$

$$= \dfrac{(1 - x - h) - (1 - x)}{h(\sqrt{1 - x - h} + \sqrt{1 - x})} = \dfrac{-1}{\sqrt{1 - x - h} + \sqrt{1 - x}}$$

61 $\dfrac{4x^2 - x + 5}{x^{2/3}} = \dfrac{4x^2}{x^{2/3}} - \dfrac{x}{x^{2/3}} + \dfrac{5}{x^{2/3}} = 4x^{4/3} - x^{1/3} + 5x^{-2/3}$

63 $\dfrac{(x^2 + 2)^2}{x^5} = \dfrac{x^4 + 4x^2 + 4}{x^5} = \dfrac{x^4}{x^5} + \dfrac{4x^2}{x^5} + \dfrac{4}{x^5} = x^{-1} + 4x^{-3} + 4x^{-5}$

Note: Exercises 65–82 are worked using the factoring concept given as the third method of simplification in Example 9.

65 The *smallest* exponent that appears on the factor x is -3.

$$x^{-3} + x^2 \ \{\text{factor out } x^{-3}\} \ = x^{-3}(1 + x^{2 - (-3)}) = x^{-3}(1 + x^5) = \dfrac{1 + x^5}{x^3}$$

67 $x^{-1/2} - x^{3/2} = x^{-1/2}(1 - x^2) = \dfrac{1 - x^2}{x^{1/2}}$

69 $(2x^2 - 3x + 1)(4)(3x + 2)^3(3) + (3x + 2)^4(4x - 3) =$

$$(3x + 2)^3[12(2x^2 - 3x + 1) + (3x + 2)(4x - 3)] = (3x + 2)^3(36x^2 - 37x + 6)$$

[71] The smallest exponent that appears on the factor $(x^2 - 4)$ is $-\frac{1}{2}$ and the smallest exponent that appears on the factor $(2x + 1)$ is 2. Thus, we will factor out $(x^2 - 4)^{-1/2}(2x + 1)^2$.

$$(x^2 - 4)^{1/2}(3)(2x + 1)^2(2) + (2x + 1)^3(\tfrac{1}{2})(x^2 - 4)^{-1/2}(2x)$$
$$= (x^2 - 4)^{-1/2}(2x + 1)^2[6(x^2 - 4) + x(2x + 1)]$$

If you are unsure of this factoring, it is easy to visually check at this stage by merely multiplying the expression—that is, we mentally add the exponents on the factor $(x^2 - 4)$, $-\frac{1}{2}$ and 1, and we get $\frac{1}{2}$, which is the exponent we started with.

Proceeding, we may simplify as follows:

$$= \frac{(2x + 1)^2(8x^2 + x - 24)}{(x^2 - 4)^{1/2}}$$

[73] $(3x + 1)^6(\tfrac{1}{2})(2x - 5)^{-1/2}(2) + (2x - 5)^{1/2}(6)(3x + 1)^5(3) =$
$$(3x + 1)^5(2x - 5)^{-1/2}[(3x + 1) + 18(2x - 5)] = \frac{(3x + 1)^5(39x - 89)}{(2x - 5)^{1/2}}$$

[75] $\dfrac{(6x + 1)^3(27x^2 + 2) - (9x^3 + 2x)(3)(6x + 1)^2(6)}{(6x + 1)^6} =$
$$\frac{(6x + 1)^2[(6x + 1)(27x^2 + 2) - 18(9x^3 + 2x)]}{(6x + 1)^6} = \frac{27x^2 - 24x + 2}{(6x + 1)^4}$$

[77] $\dfrac{(x^2 + 2)^3(2x) - x^2(3)(x^2 + 2)^2(2x)}{[(x^2 + 2)^3]^2} = \dfrac{(x^2 + 2)^2(2x)[(x^2 + 2)^1 - x^2(3)]}{(x^2 + 2)^6} =$
$$\frac{2x(x^2 + 2 - 3x^2)}{(x^2 + 2)^4} = \frac{2x(2 - 2x^2)}{(x^2 + 2)^4} = \frac{4x(1 - x^2)}{(x^2 + 2)^4}$$

[79] $\dfrac{(x^2 + 4)^{1/3}(3) - (3x)(\tfrac{1}{3})(x^2 + 4)^{-2/3}(2x)}{[(x^2 + 4)^{1/3}]^2} = \dfrac{(x^2 + 4)^{-2/3}[3(x^2 + 4) - 2x^2]}{(x^2 + 4)^{2/3}} = \dfrac{x^2 + 12}{(x^2 + 4)^{4/3}}$

[81] $\dfrac{(4x^2 + 9)^{1/2}(2) - (2x + 3)(\tfrac{1}{2})(4x^2 + 9)^{-1/2}(8x)}{[(4x^2 + 9)^{1/2}]^2} =$
$$\frac{(4x^2 + 9)^{-1/2}[2(4x^2 + 9) - 4x(2x + 3)]}{(4x^2 + 9)} = \frac{18 - 12x}{(4x^2 + 9)^{3/2}} = \frac{6(3 - 2x)}{(4x^2 + 9)^{3/2}}$$

83 Table $Y_1 = \dfrac{113x^3 + 280x^2 - 150x}{22x^3 + 77x^2 - 100x - 350}$ and $Y_2 = \dfrac{3x}{2x+7} + \dfrac{4x^2}{1.1x^2 - 5}$.

x	Y_1	Y_2
1	-0.6923	-0.6923
2	-26.12	-26.12
3	8.0392	8.0392
4	5.8794	5.8794
5	5.3268	5.3268

The values for Y_1 and Y_2 agree. Therefore, the two expressions might be equal.

Chapter 1 Review Exercises

4 (c) $|3^{-1} - 2^{-1}| = \left|\frac{1}{3} - \frac{1}{2}\right| = \left|\frac{2}{6} - \frac{3}{6}\right| = \left|-\frac{1}{6}\right| = -\left(-\frac{1}{6}\right) = \frac{1}{6}$

6 (a) $(x+y)^2 = x^2 + 2xy + y^2 \boxed{\neq} x^2 + y^2$ for every nonzero x and nonzero y.

(b) $\dfrac{1}{\sqrt{x+y}} = \dfrac{1}{\sqrt{x}} + \dfrac{1}{\sqrt{y}}$ is not true if $x = y = 1$ { we need only find one set of values

of the variables for which the expression is false }.

(c) $\dfrac{1}{\sqrt{c} - \sqrt{d}} = \dfrac{1}{\sqrt{c} - \sqrt{d}} \cdot \dfrac{\sqrt{c} + \sqrt{d}}{\sqrt{c} + \sqrt{d}} \boxed{=} \dfrac{\sqrt{c} + \sqrt{d}}{c - d}$

10 If $2 < x < 3$, then $x - 2 > 0$ { $x - 2$ is positive } and $x - 3 < 0$ { $x - 3$ is negative }.

Thus, $(x-2)(x-3) < 0$ { positive times negative is negative }, and since the absolute

value of an expression that is negative is the negative of the expression,

$$|(x-2)(x-3)| = -(x-2)(x-3), \text{ or, equivalently, } (2-x)(x-3).$$

11 $-3^2 + 2^0 + 27^{-2/3} = -9 + 1 + \dfrac{1}{(\sqrt[3]{27})^2} = -8 + \dfrac{1}{3^2} = -\dfrac{72}{9} + \dfrac{1}{9} = \dfrac{-71}{9}$

15 $\dfrac{(3x^2 y^{-3})^{-2}}{x^{-5}y} = \dfrac{3^{-2}x^{-4}y^6}{x^{-5}y} = \dfrac{x^5 y^5}{3^2 x^4} = \dfrac{xy^5}{9}$

18 $c^{-4/3}c^{3/2}c^{1/6} = c^{-8/6}c^{9/6}c^{1/6} = c^{(-8+9+1)/6} = c^{2/6} = c^{1/3}$

21 $\left[(a^{2/3}b^{-2})^3\right]^{-1} = (a^2 b^{-6})^{-1} = a^{-2}b^6 = \dfrac{b^6}{a^2}$

24 Do not expand $(u+v)^3$ since it can be combined with $(u+v)^{-2}$.

$$(u+v)^3(u+v)^{-2} = (u+v)^1 = u + v$$

29 Since $\sqrt[3]{4} = \sqrt[3]{2^2}$, we need to multiply the numerator and the denominator by $\sqrt[3]{2}$ to

obtain a cube in the radicand of the denominator. $\dfrac{1}{\sqrt[3]{4}} = \dfrac{1}{\sqrt[3]{4}} \cdot \dfrac{\sqrt[3]{2}}{\sqrt[3]{2}} = \dfrac{\sqrt[3]{2}}{\sqrt[3]{8}} = \frac{1}{2}\sqrt[3]{2}$

$\boxed{32}$ $\sqrt[4]{(-4a^3b^2c)^2} = \sqrt[4]{16a^6b^4c^2} = \sqrt[4]{2^4a^4b^4}\sqrt[4]{a^2c^2} = 2ab\sqrt[4]{(ac)^2} = 2ab\sqrt{ac}$

$\boxed{33}$ $\dfrac{1}{\sqrt{t}}\left(\dfrac{1}{\sqrt{t}} - 1\right) = \dfrac{1}{\sqrt{t}}\left(\dfrac{1}{\sqrt{t}} - \dfrac{\sqrt{t}}{\sqrt{t}}\right) = \dfrac{1}{\sqrt{t}}\left(\dfrac{1 - \sqrt{t}}{\sqrt{t}}\right) = \dfrac{1 - \sqrt{t}}{t}$

$\boxed{37}$ $\sqrt[3]{\dfrac{1}{2\pi^2}} = \dfrac{1}{\sqrt[3]{2\pi^2}} \cdot \dfrac{\sqrt[3]{4\pi}}{\sqrt[3]{4\pi}} = \dfrac{\sqrt[3]{4\pi}}{\sqrt[3]{8\pi^3}} = \dfrac{\sqrt[3]{4\pi}}{2\pi}$, or $\dfrac{1}{2\pi}\sqrt[3]{4\pi}$

$\boxed{40}$ $\dfrac{1}{\sqrt{a} + \sqrt{a-2}} = \dfrac{1}{\sqrt{a} + \sqrt{a-2}} \cdot \dfrac{\sqrt{a} - \sqrt{a-2}}{\sqrt{a} - \sqrt{a-2}} = \dfrac{\sqrt{a} - \sqrt{a-2}}{a - (a-2)} = \dfrac{\sqrt{a} - \sqrt{a-2}}{2}$

$\boxed{47}$ $(3y^3 - 2y^2 + y + 4)(y^2 - 3)$ $= (3y^3 - 2y^2 + y + 4)y^2 + (3y^3 - 2y^2 + y + 4)(-3)$

$\qquad\qquad\qquad\qquad\qquad\qquad = (3y^5 - 2y^4 + y^3 + 4y^2) + (-9y^3 + 6y^2 - 3y - 12)$

$\qquad\qquad\qquad\qquad\qquad\qquad = 3y^5 - 2y^4 - 8y^3 + 10y^2 - 3y - 12$

$\boxed{50}$ $\dfrac{9p^4q^3 - 6p^2q^4 + 5p^3q^2}{3p^2q^2} = \dfrac{9p^4q^3}{3p^2q^2} - \dfrac{6p^2q^4}{3p^2q^2} + \dfrac{5p^3q^2}{3p^2q^2} = 3p^2q - 2q^2 + \frac{5}{3}p$

$\boxed{54}$ $\qquad\qquad (a^3 - a^2)^2 = (a^3)^2 - 2(a^3)(a^2) + (a^2)^2 = a^6 - 2a^5 + a^4$

Alternatively, we could factor out a^2 first.

$\qquad (a^3 - a^2)^2 = \left[a^2(a - 1)\right]^2 = (a^2)^2(a - 1)^2 = a^4(a^2 - 2a + 1) = a^6 - 2a^5 + a^4$

$\boxed{57}$ $(3x + 2y)^2(3x - 2y)^2 = \left[(3x + 2y)(3x - 2y)\right]^2 = (9x^2 - 4y^2)^2 = 81x^4 - 72x^2y^2 + 16y^4$

$\boxed{64}$ $2c^3 - 12c^2 + 3c - 18 = 2c^2(c - 6) + 3(c - 6) = (2c^2 + 3)(c - 6)$

$\boxed{67}$ $p^8 - q^8$ $= (p^4)^2 - (q^4)^2$

$\qquad\qquad = (p^4 + q^4)(p^4 - q^4)$

$\qquad\qquad = (p^4 + q^4)(p^2 + q^2)(p^2 - q^2)$

$\qquad\qquad = (p^4 + q^4)(p^2 + q^2)(p + q)(p - q)$

$\boxed{69}$ $w^6 + 1 = (w^2)^3 + (1)^3 = (w^2 + 1)(w^4 - w^2 + 1)$, which cannot be factored any further.

$\boxed{72}$ $x^2 - 49y^2 - 14x + 49 = (x^2 - 14x + 49) - 49y^2$

$\qquad\qquad\qquad\qquad\qquad = (x - 7)^2 - (7y)^2$

$\qquad\qquad\qquad\qquad\qquad = (x - 7 + 7y)(x - 7 - 7y)$

$\boxed{76}$ $\dfrac{r^3 - t^3}{r^2 - t^2} = \dfrac{(r - t)(r^2 + rt + t^2)}{(r + t)(r - t)} = \dfrac{r^2 + rt + t^2}{r + t}$

$\boxed{80}$ $\dfrac{x + x^{-2}}{1 + x^{-2}} = \dfrac{x + \dfrac{1}{x^2}}{1 + \dfrac{1}{x^2}} = \dfrac{\left(x + \dfrac{1}{x^2}\right) \cdot x^2}{\left(1 + \dfrac{1}{x^2}\right) \cdot x^2} = \dfrac{x^3 + 1}{x^2 + 1}$. We could factor the numerator,

but since it doesn't lead to a reduction of the fraction, we leave it in this form.

81 $\dfrac{1}{x} - \dfrac{2}{x^2+x} - \dfrac{3}{x+3} = \dfrac{1(x+1)(x+3) - 2(x+3) - 3x(x+1)}{x(x+1)(x+3)}$

$\qquad\qquad = \dfrac{x^2+4x+3 - 2x - 6 - 3x^2 - 3x}{x(x+1)(x+3)} = \dfrac{-2x^2 - x - 3}{x(x+1)(x+3)}$

83 $\dfrac{x+2-\dfrac{3}{x+4}}{\dfrac{x}{x+4}+\dfrac{1}{x+4}} = \dfrac{\dfrac{(x+2)(x+4)-3}{x+4}}{\dfrac{x+1}{x+4}} = \dfrac{x^2+6x+5}{x+1} = \dfrac{(x+1)(x+5)}{x+1} = x+5$

85 $(x^2+1)^{3/2}(4)(x+5)^3 + (x+5)^4(\frac{3}{2})(x^2+1)^{1/2}(2x)$

$\qquad = (x^2+1)^{1/2}(x+5)^3[4(x^2+1) + 3x(x+5)]$

$\qquad = (x^2+1)^{1/2}(x+5)^3(7x^2+15x+4)$

86 $\dfrac{(4-x^2)(\frac{1}{3})(6x+1)^{-2/3}(6) - (6x+1)^{1/3}(-2x)}{(4-x^2)^2} = \dfrac{(6x+1)^{-2/3}[2(4-x^2) + 2x(6x+1)]}{(4-x^2)^2}$

$\qquad\qquad\qquad = \dfrac{10x^2+2x+8}{(6x+1)^{2/3}(4-x^2)^2}$

$\qquad\qquad\qquad = \dfrac{2(5x^2+x+4)}{(6x+1)^{2/3}(4-x^2)^2}$

87 $(5.5 \text{ liters})\left(10^6 \dfrac{\text{mm}^3}{\text{liter}}\right)\left(5\times 10^6 \dfrac{\text{cells}}{\text{mm}^3}\right) = 2.75\times 10^{13}$ red blood cells $\{\, 27.5 \textit{ trillion} \,\}$

88 $\dfrac{70 \text{ (or 90) beats}}{\text{minute}} \cdot \dfrac{60 \text{ minutes}}{1 \text{ hour}} \cdot \dfrac{24 \text{ hours}}{1 \text{ day}} \cdot \dfrac{365 \text{ days}}{1 \text{ year}} \cdot 80 \text{ years} =$

$\qquad\qquad\qquad\qquad 2.94336\times 10^9 \text{ (or } 3.78432\times 10^9) \text{ beats}$

89 $h = 86$ cm and $w = 13$ kg $\Rightarrow$

$\qquad S = (0.007184)w^{0.425}h^{0.725} = (0.007184)(13)^{0.425}(86)^{0.725} \approx 0.54 \text{ m}^2.$

90 $p = 40$ dyne/cm^2 and $v = 60$ cm$^3 \Rightarrow c = pv^{-1.4} = 40(60)^{-1.4} \approx 0.13$ dyne-cm.

Chapter 1 Discussion Exercises

1 1 gallon ≈ 0.13368 ft^3 is a conversion factor that would help. The volume of the tank is 10,000 gallons ≈ 1336.8 ft^3. Use $V = \frac{4}{3}\pi r^3$ to determine the radius. $1336.8 = \frac{4}{3}\pi r^3 \Rightarrow r^3 = \dfrac{1002.6}{\pi} \Rightarrow r \approx 6.83375$ ft. Then use $S = 4\pi r^2$ to find the surface area. $S = 4\pi(6.83375)^2 \approx 586.85$ ft^2.

3 We first need to determine the term that needs to be added and subtracted. Since $25 = \underline{5}^2$, it makes sense to add and subtract $2 \cdot \underline{5}x = 10x$. Then we will obtain the square of a binomial—i.e., $(x^2 + 10x + 25) - 10x = (x+5)^2 - 10x$. We can now factor this expression as the difference of two squares,

$\qquad (x+5)^2 - 10x = (x+5)^2 - (\sqrt{10x})^2 = (x+5+\sqrt{10x})(x+5-\sqrt{10x}).$

[5] Try $\dfrac{3x^2 - 4x + 7}{8x^2 + 9x - 100}$ with $x = 10^3$, 10^4, and 10^5. You get 0.374, 0.3749, and 0.37499. The numbers seem to be getting closer to 0.375, which is the decimal representation for $\frac{3}{8}$, which is the ratio of the coefficients of the x^2 terms. In general, the quotients of this form get close to the ratio of leading coefficients as x gets larger.

Chapter 2: Equations and Inequalities

⑤ $4(2y + 5) = 3(5y - 2)$ given

     $8y + 20 = 15y - 6$ multiply terms

     $26 = 7y$ get constants on one side, variables on the other

     $y = \frac{26}{7}$ solve for y

⑦ $\frac{1}{5}x + 2 = 3 - \frac{2}{7}x$ given

     $(\frac{1}{5}x + 2) \cdot 35 = (3 - \frac{2}{7}x) \cdot 35$ multiply by the lcd, 35

     $7x + 70 = 105 - 10x$ simplify

     $17x = 35$ simplify

     $x = \frac{35}{17}$ solve for x

⑨ Decimal values can be thought of as equivalent rationals, i.e., $0.3 = \frac{3}{10}$. We then multiply all terms by the lcd, 10 in this case, just as we multiplied by 35 in Exercise 7. A common mistake is to multiply 10 times a term such as $0.3(x + 1)$ and get $3(10x + 10)$. However, this is just like multiplying 10 times ab, which is $10ab$. Thus, $10[0.3(x + 1)] = 10(0.3)(x + 1) = 3(x + 1)$. With that in mind, we proceed with the problem: $[0.3(3 + 2x) + 1.2x = 3.2] \cdot 10 \Rightarrow 3(3 + 2x) + 12x = 32 \Rightarrow$

$$9 + 6x + 12x = 32 \Rightarrow 18x = 23 \Rightarrow x = \frac{23}{18}$$

Note: You may have solved equations such as those in Exercise 13 using a process called *cross-multiplication* in the past. This method is sufficient for problems of the form $\frac{P}{Q} = \frac{R}{S}$, but the guidelines for solving an equation containing rational expressions (page 58 in the text) apply to rational equations of a more complex form.

⑬ The lcd is $4(4x + 1)$ and we need to remember that x cannot equal $-\frac{1}{4}$.

$$\left[\frac{13 + 2x}{4x + 1} = \frac{3}{4} \right] \cdot 4(4x + 1) \Rightarrow 4(13 + 2x) = 3(4x + 1) \Rightarrow 52 + 8x = 12x + 3 \Rightarrow 49 = 4x \Rightarrow$$

$$x = \frac{49}{4}. \text{ Since } x \neq -\frac{1}{4}, \ x = \frac{49}{4} \text{ is a solution.}$$

⑮ The lcd is x and we need to remember that x cannot equal 0.

$$\left[8 - \frac{5}{x} = 2 + \frac{3}{x} \right] \cdot x \Rightarrow 8x - 5 = 2x + 3 \Rightarrow 6x = 8 \Rightarrow x = \frac{4}{3}$$

⑰ $(3x - 2)^2 = (x - 5)(9x + 4) \Rightarrow 9x^2 - 12x + 4 = 9x^2 - 41x - 20 \Rightarrow 29x = -24 \Rightarrow$

$$x = -\frac{24}{29}$$

$\boxed{21}$ $\left[\dfrac{3x+1}{6x-2} = \dfrac{2x+5}{4x-13}\right] \cdot (6x-2)(4x-13) \Rightarrow 12x^2 - 35x - 13 = 12x^2 + 26x - 10 \Rightarrow$

$$-3 = 61x \Rightarrow x = -\tfrac{3}{61} \ \{\text{note that } x \neq \tfrac{1}{3}, \tfrac{13}{4}\}$$

$\boxed{25}$ Since $2x - 4 = 2(x-2)$ and $3x - 6 = 3(x-2)$, the lcd is $2 \cdot 3 \cdot 5(x-2) = 30(x-2)$.

$$\left[\dfrac{3}{2x-4} - \dfrac{5}{3x-6} = \dfrac{3}{5}\right] \cdot 30(x-2) \Rightarrow 3(15) - 5(10) = 3(6)(x-2) \Rightarrow 18x = 31 \Rightarrow x = \tfrac{31}{18}$$

$\boxed{27}$ $2 - \dfrac{5}{3x-7} = 2 \Rightarrow \dfrac{5}{3x-7} = 0 \Rightarrow$ *no solution* since the numerator is never 0.

$\boxed{29}$ $\dfrac{1}{2x-1} = \dfrac{4}{8x-4} \Rightarrow \dfrac{1}{2x-1} = \dfrac{4}{4(2x-1)} \Rightarrow \dfrac{1}{2x-1} = \dfrac{1}{2x-1}$. This is an identity,

and the solutions consist of every number in the domains of the given expressions.

Thus, the solutions are all real numbers except $\frac{1}{2}$, which we denote by $\mathbb{R} - \{\frac{1}{2}\}$.

$\boxed{35}$ $\left[\dfrac{9x}{3x-1} = 2 + \dfrac{3}{3x-1}\right] \cdot (3x-1) \Rightarrow 9x = 2(3x-1) + 3 \Rightarrow 9x = 6x + 1 \Rightarrow 3x = 1 \Rightarrow$

$x = \frac{1}{3}$, which is not in the domain of the given expressions. No solution

$\boxed{37}$ $\left[\dfrac{1}{x+4} + \dfrac{3}{x-4} = \dfrac{3x+8}{x^2-16}\right] \cdot (x+4)(x-4) \Rightarrow x - 4 + 3(x+4) = 3x + 8 \Rightarrow$

$$4x + 8 = 3x + 8 \Rightarrow x = 0$$

$\boxed{39}$ $\left[\dfrac{4}{x+2} + \dfrac{1}{x-2} = \dfrac{5x-6}{x^2-4}\right] \cdot (x+2)(x-2) \Rightarrow 4(x-2) + x + 2 = 5x - 6 \Rightarrow 0 = 0,$

indicating an identity. $\mathbb{R} - \{\pm 2\}$

$\boxed{41}$ $\left[\dfrac{2}{2x+1} - \dfrac{3}{2x-1} = \dfrac{-2x+7}{4x^2-1}\right] \cdot (2x+1)(2x-1) \Rightarrow 2(2x-1) - 3(2x+1) = -2x + 7 \Rightarrow$

$$-2x - 5 = -2x + 7 \Rightarrow -5 = 7, \text{ a contradiction. No solution}$$

$\boxed{43}$ $\left[\dfrac{5}{2x+3} + \dfrac{4}{2x-3} = \dfrac{14x+3}{4x^2-9}\right] \cdot (2x+3)(2x-3) \Rightarrow 5(2x-3) + 4(2x+3) = 14x + 3 \Rightarrow$

$$18x - 3 = 14x + 3 \Rightarrow 4x = 6 \Rightarrow x = \tfrac{3}{2},$$

which is not in the domain of the given expressions. No solution

Note: For Exercises 45–50, we must show that LS = RS.

$\boxed{47}$ $\text{LS} = \dfrac{x^2-9}{x+3} = \dfrac{(x+3)(x-3)}{x+3} = x - 3 = \text{RS}, \ \forall x \text{ except } x = -3.$

$\boxed{51}$ Substituting -2 for x in $4x + 1 + 2c = 5c - 3x + 6$ yields $-7 + 2c = 5c + 12 \Rightarrow$

$$3c = -19 \Rightarrow c = -\tfrac{19}{3}.$$

$\boxed{53}$ (a) $\dfrac{7x}{x-5} = \dfrac{42}{x-5} \Rightarrow 7x = 42 \Rightarrow x = 6,$

and the two equations are equivalent since they have the same solution.

(b) No, 5 is not a solution of the first equation since it is undefined if $x = 5$.

$\boxed{55}$ Substituting $\frac{5}{3}$ for x yields $\frac{5}{3}a + b = 0$, or, equivalently, $b = -\frac{5}{3}a$.

Choose any a and b such that $b = -\frac{5}{3}a$. For example, let $a = 3$ and $b = -5$.

$\boxed{57}$ Going from the second line to the third line, we must remember that division by the

variable expression $x - 2$ is not allowed. $\bigstar$ $x + 1 = x + 2$

63 $F = g\dfrac{mM}{d^2} \Rightarrow Fd^2 = gmM \Rightarrow m = \dfrac{Fd^2}{gM}$

65 $P = 2l + 2w$ { get all terms containing w on one side } $\Rightarrow P - 2l = 2w \Rightarrow w = \dfrac{P - 2l}{2}$

67 $A = \frac{1}{2}(b_1 + b_2)h \Rightarrow \dfrac{2A}{h} = b_1 + b_2 \Rightarrow b_1 = \dfrac{2A}{h} - b_2$, or $b_1 = \dfrac{2A - hb_2}{h}$

69

$$S = \dfrac{p}{q + p(1 - q)} \qquad\qquad \text{given}$$

$$Sq + Sp(1 - q) = p \qquad\qquad \text{eliminate the fraction}$$

$$Sq + Sp - Spq = p \qquad\qquad \text{multiply terms}$$

$$Sq - Spq = p - Sp \qquad\qquad \text{isolate terms containing } q$$

$$Sq(1 - p) = p(1 - S) \qquad\qquad \text{factor}$$

$$q = \dfrac{p(1 - S)}{S(1 - p)} \qquad\qquad \text{solve for } q$$

71 $\dfrac{1}{f} = \dfrac{1}{p} + \dfrac{1}{q}$ { multiply by the lcd fpq } $\Rightarrow$

$$pq = fq + fp \Rightarrow pq - fq = fp \Rightarrow q(p - f) = fp \Rightarrow q = \dfrac{fp}{p - f}$$

73 The y-value decreases 1.2 units for each 1 unit increase in the x-value. The data is best described by equation (1), $y = -1.2x + 2$. Don't forget to go through Appendix I if you are using a TI-82/83 graphing calculator in your course.

2.2 Exercises

---

3 Let x denote the gross pay. Gross pay $-$ deductions $=$ Net (take-home) pay $\Rightarrow$

$$x - 0.40x = 492 \Rightarrow 0.60x = 492 \Rightarrow x = 820.$$

5 Let x denote the number of months needed to recover the cost of the insulation. The savings in one month is 10% of $60 = $6. $6x = 1080 \Rightarrow x = 180$ months (or 15 yr).

7 Let x denote the amount invested in the 8% account and $100{,}000 - x$ the amount invested in the 6.4% account. Use the formula $I = Prt$

(interest $=$ principal $\times$ rate $\times$ time) with $t = 1$.

$\text{Interest}_{\text{first account}} + \text{Interest}_{\text{second account}} = \text{Interest}_{\text{total}} \Rightarrow$

$x(0.08) + (100{,}000 - x)(0.064) = 7500 \Rightarrow 0.08x + 6400 - 0.064x = 7500 \Rightarrow$

$0.016x = 1100 \Rightarrow x = 68{,}750.$ Since only $50,000 can be insured in the 8% account,

we *cannot* fully insure the money and earn annual interest of $7,500.

9 Let x denote the number of children and $600 - x$ the number of adults.

$\text{Receipts}_{\text{children}} + \text{Receipts}_{\text{adults}} = \text{Receipts}_{\text{total}} \Rightarrow x(2) + (600 - x)(5) = 2400 \Rightarrow$

$$-3x = -600 \Rightarrow x = 200 \text{ children.}$$

$\boxed{11}$ Let x denote the number of ounces of glucose solution and $7 - x$ the number of ounces of water. $\text{Glucose}_{30\%} + \text{Water}_{0\%} = \text{Glucose}_{20\%} \Rightarrow$

$x(0.30) + (7 - x)(0) = 7(0.20) \Rightarrow 0.3x = 1.4 \Rightarrow x = \frac{14}{3}$.

Use $\frac{14}{3}$ oz of the 30% glucose solution and $7 - \frac{14}{3} = \frac{7}{3}$ oz of water.

$\boxed{13}$ Let x denote the number of grams of British sterling silver and $200 - x$ the number of grams of pure copper. We will compare the amounts of pure copper.

{ The percentages are 7.5%, 100%, and 10% or, equivalently, 0.075, 1, and 0.10. }

$\text{Copper}_{\text{British sterling silver}} + \text{Copper}_{\text{pure}} = \text{Copper}_{\text{alloy}} \Rightarrow$

$(0.075)x + 1(200 - x) = (0.10)(200) \Rightarrow 180 = 0.925x \Rightarrow$

$x = 194.6$. Use 194.6 g of British sterling silver and 5.4 g of copper.

$\boxed{15}$ (a) They will meet when the sum of their distances is 224. Let t denote the desired number of seconds. Using distance $=$ rate $\times$ time, we have

$$1.5t + 2t = 224 \Rightarrow 3.5t = 224 \Rightarrow t = 64 \text{ sec}$$

(b) $64(1.5) = 96$ m and $64(2) = 128$ m, respectively

$\boxed{17}$ Let r denote the rate of the snowplow. At 8:30 A.M., the car has traveled 15 miles and the snowplow has been traveling for $2\frac{1}{2} = \frac{5}{2}$ hours.

Using the relationship time $\times$ rate $=$ distance, $\frac{5}{2}r = 15 \Rightarrow r = 6$ mi/hr.

$\boxed{19}$ (a) Let r denote the rate of the river's current.

The rates of the boat upstream and downstream are $5 - r$ and $5 + r$, respectively.

$\text{Distance}_{\text{upstream}} = \text{Distance}_{\text{downstream}} \Rightarrow (5 - r)\frac{15}{60} = (5 + r)\frac{12}{60} \Rightarrow$

$5(5 - r) = 4(5 + r) \Rightarrow 25 - 5r = 20 + 4r \Rightarrow 5 = 9r \Rightarrow r = \frac{5}{9}$ mi/hr.

(b) The distance upstream is $(5 - \frac{5}{9})\frac{1}{4} = \frac{10}{9}$. The total distance is $2 \cdot \frac{10}{9} = \frac{20}{9}$, or $2\frac{2}{9}$ mi.

$\boxed{21}$ Let x denote the distance to the target. We know the total time involved and need a formula for time. Solving $d = rt$ for t gives us $t = d/r$.

$\text{Time}_{\text{to target}} + \text{Time}_{\text{from target}} = \text{Time}_{\text{total}} \Rightarrow$

$\frac{x}{3300} + \frac{x}{1100} = 1.5$ { multiply by the lcd, 3300 } $\Rightarrow$

$x + 3x = 1.5(3300) \Rightarrow 4x = 4950 \Rightarrow x = 1237.5$ ft.

$\boxed{23}$ Let l denote the length of the side parallel to the river bank. $P = 2w + l$

(a) $P = 2w + 2w = 4w;\ 4w = 180 \Rightarrow w = 45$ ft and $A = (45)(90) = 4050$ ft^2.

(b) $P = 2w + \frac{1}{2}w = \frac{5}{2}w;\ \frac{5}{2}w = 180 \Rightarrow w = 72$ ft and $A = (72)(36) = 2592$ ft^2.

(c) $P = 2w + w = 3w;\ 3w = 180 \Rightarrow w = 60$ ft and $A = (60)(60) = 3600$ ft^2.

$\boxed{25}$ $\text{Area}_{\text{semicircle}} + \text{Area}_{\text{rectangle}} = \text{Area}_{\text{total}} \Rightarrow \frac{1}{2}\pi r^2 + lw = 24 \Rightarrow$

$\frac{1}{2}\pi(\frac{3}{2})^2 + (h - \frac{3}{2})3 = 24 \Rightarrow (h - \frac{3}{2})3 = 24 - \frac{9\pi}{8} \Rightarrow h - \frac{3}{2} = 8 - \frac{3\pi}{8} \Rightarrow h = \frac{19}{2} - \frac{3\pi}{8} \approx 8.32$ ft.

27 Let h_1 denote the height of the cylinder. $V = \frac{2}{3}\pi r^3 + \pi r^2 h_1 = 11{,}250\pi$ and $r = 15 \Rightarrow$

$2250\pi + 225\pi h_1 = 11{,}250\pi \Rightarrow h_1 = 40$. The total height is 40 ft + 15 ft = 55 ft.

29 Let x denote the desired time. Using the rates (in minutes), $\frac{1}{90} + \frac{1}{60} = \frac{1}{x} \Rightarrow$

{ multiply by the lcd, $180x$ } $2x + 3x = 180 \Rightarrow 5x = 180 \Rightarrow x = 36$ min.

31 Let x denote the desired time.

Using the rates (in minutes), $\frac{1}{45} + \frac{1}{x} = \frac{1}{20} \Rightarrow 4x + 180 = 9x \Rightarrow x = 36$ min.

33 Let x denote the number of additional games. The team has won $0.650(100) = 65$

games and will win $0.5x$ more. $\dfrac{\text{games won}}{\text{games played}} = \text{record} \Rightarrow \dfrac{65 + 0.5x}{100 + x} = 0.600 \Rightarrow$

$65 + 0.5x = 60 + 0.6x \Rightarrow 5 = 0.1x \Rightarrow x = 50$.

35 (a) $h = 5280$ and $T_0 = 70 \Rightarrow T = 70 - \left(\dfrac{5.5}{1000}\right)5280 = 40.96°\text{F}$.

(b) $T = 32 \Rightarrow 32 = 70 - \left(\dfrac{5.5}{1000}\right)h \Rightarrow h = (70 - 32)\left(\dfrac{1000}{5.5}\right) \approx 6909$ ft.

2.3 Exercises

3 $15x^2 - 12 = -8x$ { get all terms on one side of the equals sign, zero on the other side }

$\Rightarrow 15x^2 + 8x - 12 = 0$ { factor } $\Rightarrow (5x + 6)(3x - 2) = 0 \Rightarrow x = -\frac{6}{5}, \frac{2}{3}$

5 A common mistake for this exercise is to write $2x = 27$ or $4x + 15 = 27$.

However, remember that you want to get 0 on one side of the equals sign.

$2x(4x + 15) = 27 \Rightarrow 8x^2 + 30x - 27 = 0 \Rightarrow (2x + 9)(4x - 3) = 0 \Rightarrow x = -\frac{9}{2}, \frac{3}{4}$

7 Divide both sides by a nonzero constant whenever possible. In this case, 5 divides

into both sides. $75x^2 + 35x - 10 = 0$ { divide by 5 } $\Rightarrow 15x^2 + 7x - 2 = 0$ { factor } $\Rightarrow$

$(3x + 2)(5x - 1) = 0$ { zero factor theorem } $\Rightarrow x = -\frac{2}{3}, \frac{1}{5}$

9 $12x^2 + 60x + 75 = 0$ { factor out the gcf, 3 } $\Rightarrow$

$3(4x^2 + 20x + 25) = 0$ { divide by 3 } $\Rightarrow$

$4x^2 + 20x + 25 = 0 \Rightarrow (2x + 5)^2 = 0 \Rightarrow x = -\frac{5}{2}$

11 We will use the same process for solving rational equations as outlined in §2.1.

Here, the lcd is $x(x + 3)$ and we need to remember that $x \neq 0, -3$.

$\left[\dfrac{2x}{x + 3} + \dfrac{5}{x} - 4 = \dfrac{18}{x^2 + 3x}\right] \cdot x(x + 3) \Rightarrow 2x(x) + 5(x + 3) - 4(x^2 + 3x) = 18 \Rightarrow$

$0 = 2x^2 + 7x + 3 \Rightarrow (2x + 1)(x + 3) = 0 \Rightarrow$

$x = -\frac{1}{2}$ { -3 is not in the domain of the given expressions }

$\boxed{13}$ $\left[\dfrac{5x}{x-3} + \dfrac{4}{x+3} = \dfrac{90}{x^2-9}\right] \cdot (x+3)(x-3) \Rightarrow 5x(x+3) + 4(x-3) = 90 \Rightarrow$

$$5x^2 + 19x - 102 = 0 \Rightarrow (5x+34)(x-3) = 0 \Rightarrow$$

$$x = -\tfrac{34}{5} \ \{\, 3 \text{ is not in the domain of the given expressions} \,\}$$

$\boxed{15}$ (a) The first equation, $x^2 = 16$, has solutions $x = \pm 4$. The equations are not

equivalent since -4 is not a solution of the second equation, $x = 4$.

(b) First note that $x = \sqrt{9} = 3$.

Thus, the equations are equivalent since they have exactly the same solutions.

$\boxed{17}$ Using the special quadratic equation in this section, $x^2 = 169 \Rightarrow$

$$x = \pm\sqrt{169} = \pm 13. \text{ Note that this is } not \text{ the same as saying } \sqrt{169} = \pm 13.$$

$\boxed{19}$ $25x^2 = 9 \Rightarrow x^2 = \tfrac{9}{25} \Rightarrow x = \pm\sqrt{\tfrac{9}{25}} = \pm\tfrac{3}{5}$

$\boxed{21}$ $(x-3)^2 = 17 \Rightarrow x - 3 = \pm\sqrt{17} \Rightarrow x = 3 \pm \sqrt{17}$

$\boxed{23}$ $4(x+2)^2 = 11 \Rightarrow (x+2)^2 = \tfrac{11}{4} \Rightarrow x + 2 = \pm\sqrt{\tfrac{11}{4}} \Rightarrow x = -2 \pm \tfrac{1}{2}\sqrt{11}$

$\boxed{25}$ For this exercise, consider the general expression $x^2 + bx + c$.

(a) In general, $d = (\tfrac{1}{2}b)^2$. In this case, $d = \left[\tfrac{1}{2}(9)\right]^2 = \tfrac{81}{4}$.

(b) As in part (a), $d = (\tfrac{1}{2}b)^2 = \left[\tfrac{1}{2}(-8)\right]^2 = 16$. *Note:* It is appropriate to use 8 or -8.

(c) In general, $d = 2(\pm\sqrt{c})$ for $c > 0$.

In this case, $c = 36 \Rightarrow \sqrt{c} = 6$, and $d = 2(\pm 6) = \pm 12$.

(d) $c = \tfrac{49}{4} \Rightarrow \sqrt{c} = \tfrac{7}{2}$, and $d = 2(\pm\tfrac{7}{2}) = \pm 7$.

$\boxed{27}$ $x^2 + 6x + 7 = 0$ { add 9 to both sides to complete the square with $x^2 + 6x$ } $\Rightarrow$

$$x^2 + 6x + \underline{\,9\,} = -7 + \underline{\,9\,} \Rightarrow (x+3)^2 = 2 \Rightarrow x + 3 = \pm\sqrt{2} \Rightarrow x = -3 \pm \sqrt{2}$$

$\boxed{29}$

$4x^2 - 12x - 11 = 0$	given
$x^2 - 3x - \tfrac{11}{4} = 0$	divide by 4
$x^2 - 3x = \tfrac{11}{4}$	isolate x^2 and x terms
$x^2 - 3x + \tfrac{9}{4} = \tfrac{11}{4} + \tfrac{9}{4}$	add $(\tfrac{1}{2} \cdot 3)^2 = \tfrac{9}{4}$
$(x - \tfrac{3}{2})^2 = 5$	factor and simplify
$x - \tfrac{3}{2} = \pm\sqrt{5}$	take the square root
$x = \tfrac{3}{2} \pm \sqrt{5}$	solve for x

$\boxed{31}$ $6x^2 - x = 2 \Rightarrow 6x^2 - x - 2 = 0.$

Use the quadratic formula, $x = \dfrac{-b \pm \sqrt{b^2 - 4ac}}{2a}$, with $a = 6$, $b = -1$, and $c = -2$.

$$x = \dfrac{-(-1) \pm \sqrt{(-1)^2 - 4(6)(-2)}}{2(6)} = \dfrac{1 \pm \sqrt{1 + 48}}{12} = \dfrac{1 \pm 7}{12} = -\tfrac{1}{2}, \tfrac{2}{3}$$

$\boxed{35}$ $2x^2 - 3x - 4 = 0 \Rightarrow x = \dfrac{-(-3) \pm \sqrt{(-3)^2 - 4(2)(-4)}}{2(2)} = \dfrac{3 \pm \sqrt{9 + 32}}{4} = \dfrac{3}{4} \pm \dfrac{1}{4}\sqrt{41}$

Note: A common mistake is to not divide 4 into both terms of the numerator.

$\boxed{37}$ $\frac{3}{2}z^2 - 4z - 1 = 0$ { multiply by 2 } $\Rightarrow 3z^2 - 8z - 2 = 0 \Rightarrow$

$$z = \frac{8 \pm \sqrt{64 + 24}}{6} = \frac{8 \pm \sqrt{88}}{6} = \frac{2(4 \pm \sqrt{22})}{2 \cdot 3} = \frac{4}{3} \pm \frac{1}{3}\sqrt{22}$$

$\boxed{39}$ Multiply by the lcd, w^2. $\left[\dfrac{5}{w^2} - \dfrac{10}{w} + 2 = 0\right] \cdot w^2 \Rightarrow 5 - 10w + 2w^2 = 0 \Rightarrow$

$$w = \frac{10 \pm \sqrt{100 - 40}}{4} = \frac{10 \pm 2\sqrt{15}}{4} = \frac{5}{2} \pm \frac{1}{2}\sqrt{15}$$

$\boxed{41}$ $4x^2 + 81 = 36x \Rightarrow 4x^2 - 36x + 81 = 0 \Rightarrow x = \dfrac{36 \pm \sqrt{1296 - 1296}}{8} = \dfrac{36}{8} = \dfrac{9}{2}$

$\boxed{43}$ $\dfrac{5x}{x^2 + 9} = -1 \Rightarrow 5x = -x^2 - 9 \Rightarrow x^2 + 5x + 9 = 0 \Rightarrow x = \dfrac{-5 \pm \sqrt{25 - 36}}{2}$.

Since the discriminant is negative, there are no real solutions.

$\boxed{45}$ (a) For this exercise, we must recognize the equation as a quadratic in x, that is,

$$(A)x^2 + (B)x + (C) = 0,$$

where A is the coefficient of x^2, B is the coefficient of x, and C is the collection of all terms that do not contain x^2 or x.

$4x^2 - 4xy + 1 - y^2 = 0 \Rightarrow (4)x^2 + (-4y)x + (1 - y^2) = 0 \Rightarrow$

$$x = \frac{4y \pm \sqrt{16y^2 - 16(1 - y^2)}}{2(4)} = \frac{4y \pm \sqrt{16\left[y^2 - (1 - y^2)\right]}}{2(4)} =$$

$$\frac{4y \pm 4\sqrt{2y^2 - 1}}{2(4)} = \frac{y \pm \sqrt{2y^2 - 1}}{2}$$

(b) Similar to part (a), we must now recognize the equation as a quadratic equation in y. $4x^2 - 4xy + 1 - y^2 = 0 \Rightarrow (-1)y^2 + (-4x)y + (4x^2 + 1) = 0 \Rightarrow$

$$y = \frac{4x \pm \sqrt{16x^2 + 4(4x^2 + 1)}}{-2} = \frac{4x \pm 2\sqrt{8x^2 + 1}}{-2} = -2x \pm \sqrt{8x^2 + 1}$$

$\boxed{47}$ $K = \frac{1}{2}mv^2 \Rightarrow v^2 = \dfrac{2K}{m} \Rightarrow v = \pm\sqrt{\dfrac{2K}{m}} \Rightarrow v = \sqrt{\dfrac{2K}{m}}$ since $v > 0$.

$\boxed{49}$ $A = 2\pi r(r + h) \Rightarrow A = 2\pi r^2 + 2\pi rh \Rightarrow (2\pi)r^2 + (2\pi h)r - A = 0$ { a quadratic in r } $\Rightarrow$

$$r = \frac{-(2\pi h) \pm \sqrt{(2\pi h)^2 - 4(2\pi)(-A)}}{2(2\pi)} = \frac{-2\pi h \pm \sqrt{4\pi^2 h^2 + 8\pi A}}{2(2\pi)} =$$

$$\frac{-2\pi h \pm 2\sqrt{\pi^2 h^2 + 2\pi A}}{2(2\pi)} = \frac{-\pi h \pm \sqrt{\pi^2 h^2 + 2\pi A}}{2\pi}.$$

Since $r > 0$, we must use the plus sign, and $r = \dfrac{-\pi h + \sqrt{\pi^2 h^2 + 2\pi A}}{2\pi}$.

51 $V = V_{\max}\left[1 - \left(\frac{r}{r_0}\right)^2\right] \Rightarrow \frac{V}{V_{\max}} = 1 - \left(\frac{r}{r_0}\right)^2 \Rightarrow$

$\left(\frac{r}{r_0}\right)^2 = 1 - (V/V_{\max}) \Rightarrow r^2 = r_0^2[1 - (V/V_{\max})] \{r > 0\} \Rightarrow r = r_0\sqrt{1 - (V/V_{\max})}$

53 Using $V = \pi r^2 h$ with $V = 3000$ and $h = 20$ gives us:

$$3000 = \pi r^2(20) \Rightarrow r^2 = 150/\pi \Rightarrow r = \sqrt{150/\pi} \approx 6.9 \text{ cm}$$

55 (a) $s = 48 \Rightarrow -16t^2 + 64t = 48 \Rightarrow t^2 - 4t + 3 = 0 \Rightarrow (t-1)(t-3) = 0 \Rightarrow t = 1, 3.$

After 1 sec and after 3 sec

(b) It will hit the ground when $s = 0$.

$$s = 0 \Rightarrow -16t^2 + 64t = 0 \Rightarrow t(t-4) = 0 \Rightarrow t = 0, 4. \text{ After 4 seconds.}$$

57 (a) $T = 98 \Rightarrow h = 1000(100 - T) + 580(100 - T)^2 = 1000(2) + 580(2)^2 = 4320 \text{ m.}$

(b) If $x = 100 - T$ and $h = 8840$, then $8840 = 1000x + 580x^2 \Rightarrow$

$29x^2 + 50x - 442 = 0 \Rightarrow x = \dfrac{-25 \pm \sqrt{13{,}443}}{29} \approx -4.86, 3.14.$

$x = -4.86 \Rightarrow T = 100 - x = 104.86\,°\text{C}$, which is outside the allowable range of T.

$x = 3.14 \Rightarrow T = 100 - x = 96.86\,°\text{C}$ for $95 \leq T \leq 100$.

59 Let x denote the width of the walk. The area (including the walk) has dimensions

$(26 + x + x)$ by $(30 + x + x)$ or, equivalently, $(26 + 2x)$ by $(30 + 2x)$.

$\text{Area}_{\text{plot}} + \text{Area}_{\text{walk}} = \text{Area}_{\text{total}} \Rightarrow 26 \cdot 30 + 240 = (26 + 2x)(30 + 2x) \Rightarrow$

$26 \cdot 30 + 240 = 26 \cdot 30 + 52x + 60x + 4x^2 \Rightarrow 240 = 4x^2 + 112x \Rightarrow$

$$x^2 + 28x - 60 = 0 \Rightarrow (x + 30)(x - 2) = 0 \Rightarrow x = 2 \text{ ft, since } x \text{ is positive.}$$

61 Let x denote the length of one side. $\text{Cost}_{\text{preparation}} + \text{Cost}_{\text{fence}} = \text{Cost}_{\text{total}} \Rightarrow$

$x^2(\$0.50) + 4x(\$1) = \$120 \Rightarrow x^2 + 8x - 240 = 0 \Rightarrow (x + 20)(x - 12) = 0 \Rightarrow x = 12.$

The size of the garden should be 12 ft by 12 ft.

63 Let $d(A, P) = x$ and $d(P, B) = 6 - x$.

$x^2 + (6 - x)^2 = 5^2 \Rightarrow 2x^2 - 12x + 11 = 0 \Rightarrow x = 3 \pm \frac{1}{2}\sqrt{14} \approx 4.9, 1.1 \text{ mi.}$

There are 4 possible roads since P could be on either side of segment AB.

65 (a) The distances of the northbound and eastbound planes are $100 + 200t$ and $400t$,

respectively. Using the Pythagorean theorem,

$$d = \sqrt{(100 + 200t)^2 + (400t)^2} = \sqrt{100^2(1 + 2t)^2 + 100^2(4t)^2} = 100\sqrt{20t^2 + 4t + 1}.$$

(b) $d = 500 \Rightarrow 500 = 100\sqrt{20t^2 + 4t + 1} \Rightarrow 5 = \sqrt{20t^2 + 4t + 1} \Rightarrow$

$5^2 = 20t^2 + 4t + 1 \Rightarrow 5t^2 + t - 6 = 0 \Rightarrow$

$$(5t + 6)(t - 1) = 0 \Rightarrow t = 1 \text{ hour after 2:30 P.M., or 3:30 P.M.}$$

$\boxed{67}$ Let x denote the outer width of the box, $x - 2$ the inner width.

Since the base is square, $(x-2)^2 = 144 \Rightarrow x = 14$.

The length is $3(1) + 2(x-2) = 2x - 1$. Thus, the size is 14 in. by 27 in.

$\boxed{69}$ Let x denote the rate of the canoeist in still water. $x - 5$ is the rate upstream and

$x + 5$ is the rate downstream. $\text{Time}_{\text{up}} = \text{Time}_{\text{down}} + \frac{1}{2} \Rightarrow \left\{ t = \frac{d}{r} \right\} \frac{1.2}{x-5} = \frac{1.2}{x+5} + \frac{1}{2}$

$\Rightarrow 2.4(x+5) = 2.4(x-5) + x^2 - 25 \Rightarrow x^2 = 49 \Rightarrow x = 7$ mi/hr.

$\boxed{71}$ Let x denote the number of pairs ordered. The price per pair is the discount
subtracted from \$40. Since the discount is \$0.04 times the number ordered, x, the
cost per pair is $40 - 0.04x$. Cost = (# of pairs)(cost per pair) $\Rightarrow$

$8400 = x(40 - 0.04x) \Rightarrow \frac{1}{25}x^2 - 40x + 8400 = 0$ { multiply by 25 } $\Rightarrow$

$x^2 - 1000x + 210,000 = 0 \Rightarrow (x - 300)(x - 700) = 0 \Rightarrow x = 300$ for $0 \le x \le 600$.

$\boxed{73}$ The total surface area is the sum of the surface area of the cylinder and that of the
top and bottom. $S = 2\pi rh + 2\pi r^2 \Rightarrow 10\pi = 8\pi r + 2\pi r^2$ { divide by 2π } $\Rightarrow$

$r^2 + 4r - 5 = 0 \Rightarrow (r+5)(r-1) = 0 \Rightarrow r = 1$, and the diameter is 2 ft.

$\boxed{75}$ V is 95% of $V_0 \Rightarrow V = 0.95V_0 \Rightarrow \frac{V}{V_0} = 0.95$.

$0.95 = 0.8197 + 0.007752t + 0.0000281t^2 \Rightarrow 0.281t^2 + 77.52t - 1303 = 0 \Rightarrow$

$t = \dfrac{-77.52 \pm \sqrt{(77.52)^2 - 4(0.281)(-1303)}}{2(0.281)} \approx -291.76,\ 15.89.$

Thus, the volume of the fireball will be 95% of the maximum volume

approximately 15.89 seconds after the explosion.

$\boxed{77}$ (a) $x = \dfrac{-4,500,000 \pm \sqrt{4,500,000^2 - 4(1)(-0.96)}}{2} \approx 0$ and $-4,500,000$

(b) $x = \dfrac{-b \pm \sqrt{b^2 - 4ac}}{2a} \cdot \dfrac{-b \mp \sqrt{b^2 - 4ac}}{-b \mp \sqrt{b^2 - 4ac}} = \dfrac{b^2 - (b^2 - 4ac)}{2a(-b \mp \sqrt{b^2 - 4ac})} =$

$\dfrac{4ac}{2a(-b \mp \sqrt{b^2 - 4ac})} = \dfrac{2c}{-b \mp \sqrt{b^2 - 4ac}}.$ The root near zero was obtained in part

(a) using the plus sign, In the second formula, it corresponds to the minus sign.

$$x = \dfrac{2(-0.96)}{-4,500,000 - \sqrt{4,500,000^2 - 4(1)(-0.96)}} \approx 2.13 \times 10^{-7}$$

$\boxed{79}$ (a) Let $Y_1 = T_1 = -1.09L + 96.01$ and $Y_2 = T_2 = -0.011L^2 - 0.126L + 81.45$.

Table each equation and compare them to the actual temperatures.

x (L)	Y_1	Y_2	S. Hem.
85	3.36	−8.74	−5
75	14.26	10.13	10
65	25.16	26.79	27
55	36.06	41.25	42
45	46.96	53.51	53
35	57.86	63.57	65
25	68.76	71.43	75
15	79.66	77.09	78
5	90.56	80.55	79

Comparing Y_1 (T_1) with Y_2 (T_2), we can see that the linear equation T_1 is not as accurate as the quadratic equation T_2.

(b) $L = 50 \Rightarrow T_2 = -0.011(50)^2 - 0.126(50) + 81.45 = 47.65°\,\text{F}.$

2.4 Exercises

$\boxed{5}$ $(3 + 5i)(2 - 7i) = (3 + 5i)2 + (3 + 5i)(-7i) = 6 + 10i - 21i - 35i^2$ {group the real parts and the imaginary parts} $= (6 - 35i^2) + (10 - 21)i = (6 + 35) - 11i = 41 - 11i$

$\boxed{9}$ $(5 - 2i)^2 = 5^2 - 2(5)(2i) + (2i)^2 = (25 - 4) - 20i = 21 - 20i$

$\boxed{11}$ $i(3 + 4i)^2 = i\big[(9 - 16) + 2(3)(4i)\big] = i(-7 + 24i) = -24 - 7i$

$\boxed{13}$ $(3 + 4i)(3 - 4i)$ { note that this difference of squares ... } $=$
$$3^2 - (4i)^2 = 9 - (-16) = \{\dots \text{becomes a "sum of squares"}\}\ 9 + 16 = 25$$

$\boxed{17}$ Since $i^k = 1$ if k is a multiple of 4, we will write i^{73} as $i^{72}i^1$,
knowing that i^{72} will reduce to 1. $i^{73} = i^{72}i = (i^4)^{18}i = 1^{18}i = i$

$\boxed{21}$ Multiply by the conjugate of the denominator to eliminate all i's in the denominator. The new denominator is the sum of the squares of the coefficients—in this case, 6^2 and 2^2. $\quad \dfrac{1 - 7i}{6 - 2i} \cdot \dfrac{6 + 2i}{6 + 2i} = \dfrac{(6 + 14) + (2 - 42)i}{36 - (-4)} = \dfrac{20 - 40i}{40} = \dfrac{1}{2} - i$

$\boxed{25}$ Multiplying the denominator by i will eliminate the i's in the denominator.
$$\dfrac{4 - 2i}{-5i} = \dfrac{4 - 2i}{-5i} \cdot \dfrac{i}{i} = \dfrac{4i - 2i^2}{-5i^2} = \dfrac{2 + 4i}{5} = \dfrac{2}{5} + \dfrac{4}{5}i$$

$\boxed{27}$ $(2 + 5i)^3 = (2)^3 + 3(2)^2(5i) + 3(2)(5i)^2 + (5i)^3 = (8 + 150i^2) + (60i + 125i^3) =$
$$(8 - 150) + (60 - 125)i = -142 - 65i$$

[29] A common mistake is to multiply $\sqrt{-4}\,\sqrt{-16}$ and obtain $\sqrt{64}$, or 8.

The correct procedure is $\sqrt{-4}\,\sqrt{-16} = \sqrt{4}\,i \cdot \sqrt{16}\,i = (2i)(4i) = 8i^2 = -8$.

$$(2 - \sqrt{-4})(3 - \sqrt{-16}) = (2 - 2i)(3 - 4i) = (6 - 8) + (-6i - 8i) = -2 - 14i$$

[31] $\dfrac{4 + \sqrt{-81}}{7 - \sqrt{-64}} = \dfrac{4 + 9i}{7 - 8i} \cdot \dfrac{7 + 8i}{7 + 8i} = \dfrac{(28 - 72) + (32 + 63)i}{49 - (-64)} = \dfrac{-44 + 95i}{113} = -\dfrac{44}{113} + \dfrac{95}{113}i$

[33] $\dfrac{\sqrt{-36}\,\sqrt{-49}}{\sqrt{-16}} = \dfrac{(6i)(7i)}{4i} \cdot \dfrac{-i}{-i} = \dfrac{(-42)(-i)}{-4i^2} = \dfrac{42i}{4} = \dfrac{21}{2}i$

[35] We need to equate the real parts and the imaginary parts on each side of " $=$ ".

$8 + (3x + y)i = 2x - 4i \Rightarrow 2x = 8$ and $3x + y = -4 \Rightarrow x = 4,\ y = -16$

[37] $(3x + 2y) - y^3 i = 9 - 27i \Rightarrow y^3 = 27\ \{y = 3\}$ and $3x + 2y = 9 \Rightarrow x = 1,\ y = 3$

[39] $x^2 - 6x + 13 = 0 \Rightarrow$

$$x = \frac{-(-6) \pm \sqrt{(-6)^2 - 4(1)(13)}}{2(1)} = \frac{6 \pm \sqrt{36 - 52}}{2} = \frac{6 \pm \sqrt{-16}}{2} = \frac{6 \pm 4i}{2} = 3 \pm 2i$$

[41] $x^2 + 4x + 13 = 0 \Rightarrow x = \dfrac{-4 \pm \sqrt{16 - 52}}{2} = \dfrac{-4 \pm 6i}{2} = -2 \pm 3i$

[47] Solving $x^3 = -125$ would only give us the solution $x = -5$. So first we need to factor

$x^3 + 125$ as the sum of cubes. $x^3 + 125 = 0 \Rightarrow (x + 5)(x^2 - 5x + 25) = 0 \Rightarrow$

$x = -5$ or $x = \dfrac{5 \pm \sqrt{25 - 100}}{2} = \dfrac{5 \pm 5\sqrt{3}\,i}{2}$. The three solutions are $-5,\ \dfrac{5}{2} \pm \dfrac{5}{2}\sqrt{3}\,i$.

[49] $x^4 = 256 \Rightarrow x^4 - 256 = 0 \Rightarrow (x^2 - 16)(x^2 + 16) = 0 \Rightarrow x = \pm 4,\ \pm 4i$

[51] $4x^4 + 25x^2 + 36 = 0 \Rightarrow (x^2 + 4)(4x^2 + 9) = 0 \Rightarrow x = \pm 2i,\ \pm\frac{3}{2}i$

[53] $x^3 + 3x^2 + 4x = 0 \Rightarrow x(x^2 + 3x + 4) = 0 \Rightarrow x = 0,\ -\frac{3}{2} \pm \frac{1}{2}\sqrt{7}\,i$

[55] If $z = a + bi$ and $w = c + di$, then

$$\overline{z + w} = \overline{(a + bi) + (c + di)} \qquad \text{definition of } z \text{ and } w$$
$$= \overline{(a + c) + (b + d)i} \qquad \text{write in complex number form}$$
$$= (a + c) - (b + d)i \qquad \text{definition of conjugate}$$
$$= (a - bi) + (c - di) \qquad \text{rearrange terms}$$
$$= \bar{z} + \bar{w}. \qquad \text{definition of conjugates of } z \text{ and } w$$

In the step described by "rearrange terms",

we are really looking ahead to the terms we want to obtain, $\bar{z}$ and $\bar{w}$.

[59] If $\bar{z} = z$, then $a - bi = a + bi$ and hence $-bi = bi$, or $2bi = 0$.

Thus, $b = 0$ and $z = a$ is real. Conversely, if z is real, then $b = 0$ and hence

$$\bar{z} = \overline{a + 0i} = a - 0i = a + 0i = z.$$

$\boxed{3}$ We must first isolate the absolute value term before proceeding.

$|3x - 2| + 3 = 7 \Rightarrow |3x - 2| = 4 \Rightarrow 3x - 2 = 4$ or $3x - 2 = -4 \Rightarrow$

$$3x = 6 \text{ or } 3x = -2 \Rightarrow x = 2 \text{ or } x = -\tfrac{2}{3}$$

$\boxed{5}$ $3|x + 1| - 2 = -11 \Rightarrow 3|x + 1| = -9 \Rightarrow |x + 1| = -3.$

Since the absolute value of an expression is nonnegative, $|x + 1| = -3$ has no solution.

$\boxed{7}$ Since there are four terms, we first try factoring by grouping.

$9x^3 - 18x^2 - 4x + 8 = 0 \Rightarrow 9x^2(x - 2) - 4(x - 2) = 0 \Rightarrow (9x^2 - 4)(x - 2) = 0 \Rightarrow$

$$x = \pm\tfrac{2}{3},\ 2$$

$\boxed{9}$ Notice that we can factor an x out of each term.

$4x^4 + 10x^3 = 6x^2 + 15x \Rightarrow x(4x^3 + 10x^2 - 6x - 15) = 0 \Rightarrow$

$$x\left[2x^2(2x + 5) - 3(2x + 5)\right] = 0 \Rightarrow x(2x^2 - 3)(2x + 5) = 0 \Rightarrow x = 0,\ \pm\tfrac{1}{2}\sqrt{6},\ -\tfrac{5}{2}$$

Note: The comment on page 95 about raising both sides to a reciprocal power addresses what is often a difficult concept for many students. The following illustration may help in understanding this concept. Note that if m is even, we have to use the $\pm$ symbol.

Problem	**Solution**
$x^{1/2} = 4$	$(x^{1/2})^{2/1} = 4^{2/1} \Rightarrow x = 16$
$x^{-1/2} = 5$	$(x^{-1/2})^{-2/1} = 5^{-2/1} \Rightarrow x = \frac{1}{25}$
$x^{3/4} = 8$	$(x^{3/4})^{4/3} = 8^{4/3} \Rightarrow x = 16$
$x^{4/3} = 16$	$(x^{4/3})^{3/4} = 16^{3/4} \Rightarrow x = \pm 8$

This principle is used in many exercises, especially Exercises 51 and 52.

$\boxed{11}$ $y^{3/2} = 5y \Rightarrow y^{3/2} - 5y = 0 \Rightarrow y(y^{1/2} - 5) = 0 \Rightarrow y = 0$ or $y^{1/2} = 5.$

$$y^{1/2} = 5 \Rightarrow (y^{1/2})^2 = 5^2 \Rightarrow y = 25. \quad y = 0,\ 25$$

Note: The following guidelines may be helpful when solving radical equations.

> ### Guidelines for Solving a Radical Equation

(1) Isolate the radical. If we cannot get the radical isolated on one side of the equals sign because there is more than one radical, then we will split up the radical terms as evenly as possible on each side of the equals sign. For example, if there are 2 radicals, we put one on each side; if there are 3 radicals, we put 2 on one side and 1 on the the other.

(2) Raise both sides to the same power as the root index. *Note:* Remember here that

$$\boxed{(a + b\sqrt{n})^2 = a^2 + 2ab\sqrt{n} + b^2n}$$ and that it is *not* $a^2 + b^2n$.

(3) If your equation contains no radicals, proceed to part (4). If there are still radicals in the equation, go back to part (1).

(4) Solve the resulting equation.

(5) Check the answers found in part (4) in the original equation to determine the valid solutions. *Note:* You may check the solutions in any equivalent equation of the original equation, i.e., an equation which occurs prior to raising both sides to a power. Also, extraneous solutions are introduced when raising both sides to an even power. Hence, all solutions *must* be checked in this case. Checking solutions when raising each side to an odd power is up to the individual instructor.

$\boxed{15}$ $2 + \sqrt[3]{1 - 5t} = 0 \Rightarrow \sqrt[3]{1 - 5t} = -2 \Rightarrow (\sqrt[3]{1 - 5t})^3 = (-2)^3 \Rightarrow 1 - 5t = -8 \Rightarrow t = \frac{9}{5}$

$\boxed{19}$ $\sqrt{7 - x} = x - 5$ { square both sides } $\Rightarrow 7 - x = x^2 - 10x + 25$ { set equal to zero } $\Rightarrow$
$$x^2 - 9x + 18 = 0 \text{ \{ factor \}} \Rightarrow (x - 3)(x - 6) = 0 \Rightarrow x = 3, 6.$$
Check $x = 6$: LS $= \sqrt{7 - 6} = 1$; RS $= 6 - 5 = 1$.

Since both sides have the same value, $x = 6$ is a valid solution.

Check $x = 3$: LS $= \sqrt{7 - 3} = 2$; RS $= 3 - 5 = -2$.

Since both sides do not have the same value, $x = 3$ is an extraneous solution.

$\boxed{21}$ $3\sqrt{2x-3} + 2\sqrt{7-x} = 11$ given

$\qquad\qquad 3\sqrt{2x-3} = 11 - 2\sqrt{7-x}$ split radicals evenly

$\qquad\qquad 9(2x-3) = 121 - 44\sqrt{7-x} + 4(7-x)$ square both sides

$\qquad\qquad 44\sqrt{7-x} = -22x + 176$ isolate radical again and simplify

$\qquad\qquad 2\sqrt{7-x} = 8 - x$ divide by gcf, 22

$\qquad\qquad 4(7-x) = 64 - 16x + x^2$ square both sides

$\qquad\qquad x^2 - 12x + 36 = 0$ collect terms on one side

$\qquad\qquad (x-6)^2 = 0$ factor

$\qquad\qquad x - 6 = 0$ take the square root

$\qquad\qquad x = 6$ solve for x

Check $x = 6$: LS $= 3(3) + 2(1) = 11 = $ RS $\Rightarrow x = 6$ is the solution.

$\boxed{23}$ Remember to isolate the radical term first.

$\qquad x = 4 + \sqrt{4x-19} \Rightarrow x - 4 = \sqrt{4x-19} \Rightarrow x^2 - 8x + 16 = 4x - 19 \Rightarrow$

$\qquad\qquad\qquad x^2 - 12x + 35 = 0 \Rightarrow (x-5)(x-7) = 0 \Rightarrow x = 5, 7$

$\boxed{25}$ $x + \sqrt{5x+19} = -1 \Rightarrow \sqrt{5x+19} = -x - 1 \Rightarrow 5x + 19 = x^2 + 2x + 1 \Rightarrow$

$\qquad\qquad\qquad x^2 - 3x - 18 = 0 \Rightarrow (x-6)(x+3) = 0 \Rightarrow x = -3, 6.$

Check $x = -3$: LS $= -3 + 2 = -1 = $ RS $\Rightarrow x = -3$ is a solution.

Check $x = 6$: LS $= 6 + 7 = 13 \neq $ RS $\Rightarrow x = 6$ is an extraneous solution.

$\boxed{27}$ $\sqrt{7-2x} - \sqrt{5+x} = \sqrt{4+3x}$ $\{$square both sides$\}$ $\Rightarrow$

$\qquad (7-2x) - 2\sqrt{(7-2x)(5+x)} + (5+x) = 4 + 3x$ $\{$isolate the radical$\}$ $\Rightarrow$

$\qquad -4x + 8 = 2\sqrt{-2x^2 - 3x + 35}$ $\{$divide by 2$\}$ $\Rightarrow$

$\qquad -2x + 4 = \sqrt{-2x^2 - 3x + 35}$ $\{$square both sides$\}$ $\Rightarrow$

$\qquad 4x^2 - 16x + 16 = -2x^2 - 3x + 35$ $\{$simplify$\}$ $\Rightarrow$

$\qquad\qquad\qquad 6x^2 - 13x - 19 = 0 \Rightarrow (x+1)(6x-19) = 0 \Rightarrow x = -1, \frac{19}{6}.$

Check $x = -1$: LS $= 3 - 2 = 1 = $ RS $\Rightarrow x = -1$ is a solution.

Check $x = \frac{19}{6}$: LS $= \sqrt{\frac{2}{3}} - \sqrt{\frac{49}{6}}$ $\{$note that this is negative$\} \neq \sqrt{\frac{27}{2}} = $ RS $\Rightarrow$

$\qquad\qquad\qquad\qquad\qquad\qquad x = \frac{19}{6}$ is an extraneous solution.

$\boxed{29}$ $\sqrt{11+8x} + 1 = \sqrt{9+4x} \Rightarrow (11+8x) + 2\sqrt{11+8x} + 1 = 9 + 4x \Rightarrow$

$\qquad 2\sqrt{8x+11} = -4x - 3 \Rightarrow 4(8x+11) = 16x^2 + 24x + 9 \Rightarrow 16x^2 - 8x - 35 = 0 \Rightarrow$

$\qquad\qquad (4x-7)(4x+5) = 0 \Rightarrow x = -\frac{5}{4}$ and $\frac{7}{4}$ is an extraneous solution.

$\boxed{31}$ $\sqrt{2\sqrt{x+1}} = \sqrt{3x-5} \Rightarrow 2\sqrt{x+1} = 3x-5 \Rightarrow 4(x+1) = 9x^2 - 30x + 25 \Rightarrow$
$$9x^2 - 34x + 21 = 0 \Rightarrow (x-3)(9x-7) = 0 \Rightarrow x = 3, \tfrac{7}{9}.$$

Check $x = 3$: LS $= 2 =$ RS $\Rightarrow x = 3$ is a solution.

Check $x = \tfrac{7}{9}$: LS $= \sqrt{2 \cdot \tfrac{4}{3}} = \sqrt{\tfrac{8}{3}} \neq \sqrt{-\tfrac{8}{3}} =$ RS $\Rightarrow x = \tfrac{7}{9}$ is an extraneous solution.

$\boxed{33}$ $\sqrt{1 + 4\sqrt{x}} = \sqrt{x} + 1 \Rightarrow 1 + 4\sqrt{x} = x + 2\sqrt{x} + 1 \Rightarrow 2\sqrt{x} = x \Rightarrow 4x = x^2 \Rightarrow$
$$x(4-x) = 0 \Rightarrow x = 0, 4$$

$\boxed{35}$ $x^4 - 25x^2 + 144 = 0 \Rightarrow (x^2 - 9)(x^2 - 16) = 0 \Rightarrow x = \pm 3, \pm 4$

$\boxed{37}$ $5y^4 - 7y^2 + 1 = 0 \Rightarrow y^2 = \dfrac{7 \pm \sqrt{29}}{10} \cdot \dfrac{10}{10} = \dfrac{70 \pm 10\sqrt{29}}{100} \Rightarrow y = \pm\tfrac{1}{10}\sqrt{70 \pm 10\sqrt{29}}$

Alternatively, let $u = y^2$ and solve $5u^2 - 7u + 1 = 0$.

$\boxed{39}$ $36x^{-4} - 13x^{-2} + 1 = 0 \Rightarrow (4x^{-2} - 1)(9x^{-2} - 1) = 0 \Rightarrow x^{-2} = \tfrac{1}{4}, \tfrac{1}{9} \Rightarrow x^2 = 4, 9 \Rightarrow$
$$x = \pm 2, \pm 3$$

Alternatively, let $u = x^{-2}$ and solve $36u^2 - 13u + 1 = 0$.

$\boxed{41}$ $3x^{2/3} + 4x^{1/3} - 4 = 0 \Rightarrow (3x^{1/3} - 2)(x^{1/3} + 2) = 0 \Rightarrow \sqrt[3]{x} = \tfrac{2}{3}, -2 \Rightarrow x = \tfrac{8}{27}, -8$

Alternatively, let $u = x^{1/3}$ and solve $3u^2 + 4u - 4 = 0$.

$\boxed{43}$ $6w - 23w^{1/2} + 20 = 0 \Rightarrow (2w^{1/2} - 5)(3w^{1/2} - 4) = 0 \Rightarrow \sqrt{w} = \tfrac{5}{2}, \tfrac{4}{3} \Rightarrow w = \tfrac{25}{4}, \tfrac{16}{9}$

Alternatively, let $u = w^{1/2}$ and solve $6u^2 - 23u + 20 = 0$.

$\boxed{45}$ $\left(\dfrac{t}{t+1}\right)^2 - \dfrac{2t}{t+1} - 8 = 0 \Rightarrow \left(\dfrac{t}{t+1} - 4\right)\left(\dfrac{t}{t+1} + 2\right) = 0 \Rightarrow \dfrac{t}{t+1} = 4, -2 \Rightarrow$
$$t = 4t + 4, -2t - 2 \Rightarrow t = -\tfrac{4}{3}, -\tfrac{2}{3}$$

Alternatively, let $u = \dfrac{t}{t+1}$ and solve $u^2 - 2u - 8 = 0$.

$\boxed{47}$ For this exercise, we must note that the variable terms are both cubed, and then combine them into one term.
$$27x^3 = (x+5)^3 \Rightarrow \left(\dfrac{x+5}{x}\right)^3 = 27 \Rightarrow \dfrac{x+5}{x} = 3 \Rightarrow x + 5 = 3x \Rightarrow x = \tfrac{5}{2}$$

$\boxed{49}$ The least common multiple of 3 and 4 is 12—so by raising both sides to the 12th power we will eliminate the radicals. $\sqrt[3]{x} = 2\sqrt[4]{x} \Rightarrow (\sqrt[3]{x})^{12} = (2\sqrt[4]{x})^{12} \Rightarrow$
$$x^4 = 2^{12}x^3 \Rightarrow x^4 - 4096x^3 = 0 \Rightarrow x^3(x - 4096) = 0 \Rightarrow x = 0, 4096.$$

Check $x = 0$: LS $= 0 =$ RS $\Rightarrow x = 0$ is a solution.

Check $x = 4096 = 2^{12}$: LS $= \sqrt[3]{2^{12}} = 2^4$; RS $= 2\sqrt[4]{2^{12}} = 2 \cdot 2^3 = 2^4 \Rightarrow$
$$x = 4096 \text{ is a solution.}$$

51 See the note before the solution for Exercise 11.

     (a) $x^{5/3} = 32 \Rightarrow (x^{5/3})^{3/5} = (32)^{3/5} \Rightarrow x = (\sqrt[5]{32})^3 = 2^3 = 8$

     (b) $x^{4/3} = 16 \Rightarrow (x^{4/3})^{3/4} = \pm(16)^{3/4} \Rightarrow x = \pm(\sqrt[4]{16})^3 = \pm 2^3 = \pm 8$

     (c) $x^{2/3} = -36 \Rightarrow (x^{2/3})^{3/2} = \pm(-36)^{3/2} \Rightarrow x = \pm(\sqrt{-36})^3,$

                                which are not real numbers. No real solutions

     (d) $x^{3/4} = 125 \Rightarrow (x^{3/4})^{4/3} = (125)^{4/3} \Rightarrow x = (\sqrt[3]{125})^4 = 5^4 = 625$

     (e) $x^{3/2} = -27 \Rightarrow (x^{3/2})^{2/3} = (-27)^{2/3} \Rightarrow x = (\sqrt[3]{-27})^2 = (-3)^2 = 9,$

                                which is an extraneous solution. No real solutions

53 $T = 2\pi\sqrt{\dfrac{l}{g}} \Rightarrow \dfrac{T}{2\pi} = \sqrt{\dfrac{l}{g}} \Rightarrow \dfrac{T^2}{4\pi^2} = \dfrac{l}{g} \Rightarrow l = \dfrac{gT^2}{4\pi^2}$

55 $S = \pi r\sqrt{r^2 + h^2} \Rightarrow \dfrac{S}{\pi r} = \sqrt{r^2 + h^2} \Rightarrow \dfrac{S^2}{\pi^2 r^2} = h^2 + r^2 \Rightarrow \dfrac{S^2}{\pi^2 r^2} - r^2 = h^2 \Rightarrow$

       $h^2 = \dfrac{1}{\pi^2 r^2}(S^2 - \pi^2 r^4) \Rightarrow h = \pm\dfrac{1}{\pi r}\sqrt{S^2 - \pi^2 r^4} \Rightarrow h = \dfrac{1}{\pi r}\sqrt{S^2 - \pi^2 r^4}$ since $h > 0$

57 $P = 0.31 E D^2 V^3 \Rightarrow V = \left(\dfrac{P}{0.31 E D^2}\right)^{1/3} = \left(\dfrac{10{,}000}{(0.31)(0.42)10^2}\right)^{1/3} \approx 9.16$ ft/sec.

                               This is about $9.16 \cdot \dfrac{15}{22} \approx 6.24$ mi/hr.

59 $k = 10^5$ and $c = \frac{1}{2} \Rightarrow Q = kP^{-c} = 10^5 P^{-1/2} \Rightarrow Q = 10^5/\sqrt{P} \Rightarrow$

       $\sqrt{P} = \dfrac{10^5}{Q} \Rightarrow P = \left(\dfrac{10^5}{Q}\right)^2 = \left(\dfrac{100{,}000}{5000}\right)^2 = (20)^2 = 400$ cents, or, \$4.00.

61 $V = 144$ and $V = \frac{1}{3}\pi r^2 h \Rightarrow 144 = \frac{1}{3}\pi r^3$ {since $r = h$} $\Rightarrow r^3 = 432/\pi \Rightarrow$

                       $r = \sqrt[3]{432/\pi}$, and the diameter is $2\sqrt[3]{432/\pi} \approx 10.3$ cm.

63 $y = 60\% \Rightarrow \dfrac{x^3}{x^3 + (1-x)^3} = \dfrac{3}{5} \Rightarrow 5x^3 = 3x^3 + 3(1-x)^3 \Rightarrow 2x^3 = 3(1-x)^3 \Rightarrow$

     $\left(\dfrac{x}{1-x}\right)^3 = \dfrac{3}{2} \Rightarrow \dfrac{x}{1-x} = \sqrt[3]{1.5} \Rightarrow x = \sqrt[3]{1.5} - \sqrt[3]{1.5}\,x \Rightarrow x + \sqrt[3]{1.5}\,x = \sqrt[3]{1.5} \Rightarrow$

                       $(1 + \sqrt[3]{1.5})x = \sqrt[3]{1.5} \Rightarrow x = \dfrac{\sqrt[3]{1.5}}{1 + \sqrt[3]{1.5}} \approx 0.534$, or 53.4%.

65 $\text{Cost}_{\text{underwater}} + \text{Cost}_{\text{overland}} = \text{Cost}_{\text{total}} \Rightarrow$

     $7500 \cdot (\text{underwater miles}) + 6000 \cdot (\text{overland miles}) = 35{,}000 \Rightarrow$

     $7500\sqrt{x^2 + 1} + 6000(5 - x) = 35{,}000 \Rightarrow 15\sqrt{x^2 + 1} = 12x + 10 \Rightarrow$

     $225(x^2 + 1) = 144x^2 + 240x + 100 \Rightarrow 81x^2 - 240x + 125 = 0 \Rightarrow$

     $x = \dfrac{240 \pm \sqrt{17{,}100}}{162} = \dfrac{40 \pm 5\sqrt{19}}{27} \approx 2.2887,\ 0.6743$ mi. There are two possible routes.

[67] (a) Let $Y_1 = D_1 = 6.096L + 685.7$ and

$Y_2 = D_2 = 0.00178L^3 - 0.072L^2 + 4.37L + 719$.

Table each equation and compare them to the actual values.

x (L)	Y_1	Y_2	Summer
0	686	719	720
10	747	757	755
20	808	792	792
30	869	833	836
40	930	893 ·	892
50	991	980	978
60	1051	1106	1107

Comparing Y_1 (D_1) with Y_2 (D_2) we can see that the linear equation D_1 is not as accurate as the cubic equation D_2.

(b) $L = 35 \Rightarrow D_2 = 0.00178(35)^3 - 0.072(35)^2 + 4.37(35) + 719 \approx 860$ min.

[69] The volume of the box is $V = hw^2 = 25$, where h is the height and w is the length of a side of the square base. The amount of cardboard will be minimized when the surface area of the box is a minimum. The surface area is given by $S = w^2 + 4wh$. Since $h = 25/w^2$, we have $S = w^2 + 100/w$. Form a table for w and S.

w	S	w	S
3.4	40.972	3.7	40.717
3.5	40.821	3.8	40.756
3.6	40.738	3.9	40.851

The minimum surface area is $S \approx 40.717$ when $w \approx 3.7$ and $h = 25/w^2 \approx 1.8$.

2.6 Exercises

Note: The bracket symbols "[" and "]", are used with $\leq$ or $\geq$ to denote that the end point of the interval is part of the solution. Parentheses, "(" and ")" are used with $<$ or $>$ and denote that the end point is *not* part of the solution.

[3] $x < -2 \Leftrightarrow (-\infty, -2)$

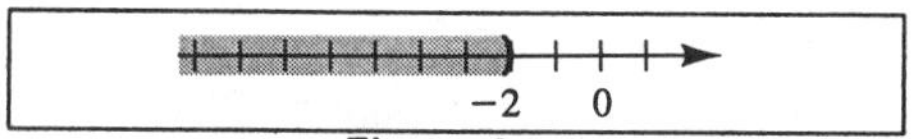

Figure 3

[7] $-2 < x \leq 4 \Leftrightarrow (-2, 4]$

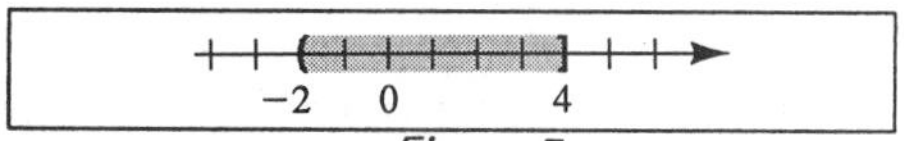

Figure 7

[9] $3 \leq x \leq 7 \Leftrightarrow [3, 7]$

Figure 9

[11] $5 > x \geq -2 \Rightarrow -2 \leq x < 5 \Leftrightarrow [-2, 5)$

Figure 11

13 $(-5, 8] \Leftrightarrow -5 < x \le 8$ 15 $[-4, -1] \Leftrightarrow -4 \le x \le -1$

17 $[4, \infty) \Leftrightarrow x \ge 4$ 19 $(-\infty, -5) \Leftrightarrow x < -5$

23 $-2 - 3x \ge 2 \Rightarrow -3x \ge 4$ { Remember to change the direction of the inequality when
   multiplying or dividing by a negative value. } $\Rightarrow x \le -\frac{4}{3} \Leftrightarrow (-\infty, -\frac{4}{3}]$

25 $2x + 5 < 3x - 7 \Rightarrow -x < -12 \Rightarrow x > 12 \Leftrightarrow (12, \infty)$

27 $\left[9 + \frac{1}{3}x \ge 4 - \frac{1}{2}x\right] \cdot 6$ { multiply by the lcd, 6 } $\Rightarrow 54 + 2x \ge 24 - 3x \Rightarrow 5x \ge -30 \Rightarrow$
$$x \ge -6 \Leftrightarrow [-6, \infty)$$

29 $-3 < 2x - 5 < 7$ { add 5 to all three expressions } $\Rightarrow$
$$2 < 2x < 12 \text{ { divide all three expressions by 2 } } \Rightarrow 1 < x < 6 \Leftrightarrow (1, 6)$$

31 $\left[3 \le \frac{2x - 3}{5} < 7\right] \cdot 5$ { given inequality } $\Rightarrow$

$15 \le 2x - 3 < 35$ { multiply by the lcd, 5 } $\Rightarrow$

$18 \le 2x < 38$ { add 3 to all three parts } $\Rightarrow$

$9 \le x < 19$ { divide all three parts by 2 } $\Leftrightarrow [9, 19)$ { equivalent interval notation }

33 $4 > \frac{2 - 3x}{7} \ge -2$ { multiply all three expressions by 7 } $\Rightarrow 28 > 2 - 3x \ge -14 \Rightarrow$

$26 > -3x \ge -16$ { divide by -3 and change directions of *both* inequality signs } $\Rightarrow$
$$-\frac{26}{3} < x \le \frac{16}{3} \Leftrightarrow (-\frac{26}{3}, \frac{16}{3}]$$

37 $(2x - 3)(4x + 5) \le (8x + 1)(x - 7) \Rightarrow 8x^2 - 2x - 15 \le 8x^2 - 55x - 7 \Rightarrow 53x \le 8 \Rightarrow$
$$x \le \frac{8}{53} \Leftrightarrow (-\infty, \frac{8}{53}]$$

41 By the law of signs, a quotient is positive if the sign of the numerator and the sign of
   the denominator are the same. Since the numerator is positive, $\frac{4}{3x + 2} > 0 \Rightarrow$

   $3x + 2 > 0 \Rightarrow x > -\frac{2}{3} \Leftrightarrow (-\frac{2}{3}, \infty)$. The expression is never equal to 0 since the
   numerator is never 0. Thus, the solution of $\frac{4}{3x + 2} \ge 0$ is $(-\frac{2}{3}, \infty)$.

43 $\frac{-2}{4 - 3x} > 0 \Rightarrow 4 - 3x < 0$ { denominator must also be negative } $\Rightarrow x > \frac{4}{3} \Leftrightarrow (\frac{4}{3}, \infty)$

45 $(1 - x)^2 > 0 \ \forall x$ except 1. Thus, $\frac{2}{(1 - x)^2} > 0$ has solution $\mathbb{R} - \{1\}$.

47 When reading this inequality, think of the concept "I want all the numbers that lie
   less than 3 units from the origin," and your answer should make some common sense.
$$|x| < 3 \Rightarrow -3 < x < 3 \Leftrightarrow (-3, 3)$$

49 When reading this inequality, think of the concept "I want all the numbers that lie
   at least 5 units from the origin," and your answer should make some common sense.
$$|x| \ge 5 \Rightarrow x \ge 5 \text{ or } x \le -5 \Leftrightarrow (-\infty, -5] \cup [5, \infty)$$

51 $|x + 3| < 0.01 \Rightarrow -0.01 < x + 3 < 0.01 \Rightarrow -3.01 < x < -2.99 \Leftrightarrow (-3.01, -2.99)$

$\boxed{53}$ $|x+2|+0.1 \geq 0.2$ {isolate the absolute value expression} $\Rightarrow$ $|x+2| \geq 0.1 \Rightarrow$
$$x+2 \geq 0.1 \text{ or } x+2 \leq -0.1 \Rightarrow x \geq -1.9 \text{ or } x \leq -2.1 \Leftrightarrow (-\infty, -2.1] \cup [-1.9, \infty)$$

$\boxed{57}$ $-\frac{1}{3}|6-5x|+2 \geq 1 \Rightarrow -\frac{1}{3}|6-5x| \geq -1 \Rightarrow |6-5x| \leq 3 \Rightarrow -3 \leq 6-5x \leq 3 \Rightarrow$
$$-9 \leq -5x \leq -3 \Rightarrow \frac{9}{5} \geq x \geq \frac{3}{5} \Leftrightarrow [\frac{3}{5}, \frac{9}{5}]$$

$\boxed{59}$ Since $|7x+2| \geq 0 \;\forall x$, $|7x+2| > -2$ has solution $(-\infty, \infty)$.

$\boxed{61}$ $|3x-9| > 0 \;\forall x$ except when $3x-9 = 0$, or $x = 3$. The solution is $(-\infty, 3) \cup (3, \infty)$.

$\boxed{63}$ $\left|\dfrac{2-3x}{5}\right| \geq 2 \Rightarrow \dfrac{|2-3x|}{|5|} \geq 2 \Rightarrow |2-3x| \geq 10 \Rightarrow 2-3x \geq 10 \text{ or } 2-3x \leq -10 \Rightarrow$
$$-3x \geq 8 \text{ or } -3x \leq -12 \Rightarrow x \leq -\frac{8}{3} \text{ or } x \geq 4 \Leftrightarrow (-\infty, -\frac{8}{3}] \cup [4, \infty)$$

$\boxed{65}$ Since $|5-2x| \geq 0 \;\forall x$, we can multiply the inequality by $|5-2x|$ without changing the direction of the inequality sign. We must exclude $x = \frac{5}{2}$ from the solution since it makes the original inequality undefined.

$\dfrac{3}{|5-2x|} < 2 \Rightarrow |5-2x| > \frac{3}{2} \Rightarrow 5-2x > \frac{3}{2} \text{ or } 5-2x < -\frac{3}{2} \Rightarrow$

$-2x > -\frac{7}{2} \text{ or } -2x < -\frac{13}{2} \Rightarrow x < \frac{7}{4} \text{ or } x > \frac{13}{4}$ {$\frac{5}{2}$ doesn't fall in this region} $\Leftrightarrow$
$$(-\infty, \tfrac{7}{4}) \cup (\tfrac{13}{4}, \infty)$$

$\boxed{67}$ $-2 < |x| < 4 \Rightarrow -2 < |x| \; and \; |x| < 4$.

Since -2 is always less than $|x|$ {because $|x| \geq 0$}, we only need to consider
$$|x| < 4. \quad |x| < 4 \Rightarrow -4 < x < 4 \Leftrightarrow (-4, 4)$$

$\boxed{69}$ From the definition of absolute value, $|x-2| = $ either $x-2$ or $-(x-2)$.

Thus, $1 < |x-2| < 4 \Rightarrow 1 < x-2 < 4 \text{ or } 1 < -(x-2) < 4 \Rightarrow$
$$1 < x-2 < 4 \text{ or } -1 > x-2 > -4 \Rightarrow 3 < x < 6 \text{ or } 1 > x > -2 \Leftrightarrow (-2, 1) \cup (3, 6).$$

An alternative method is to rewrite the inequality as $|x-2| > 1 \; and \; |x-2| < 4$.

Solving independently gives us

$x-2 > 1 \text{ or } x-2 < -1 \Rightarrow x > 3 \text{ or } x < 1 \; and \; -4 < x-2 < 4 \Rightarrow -2 < x < 6.$

Taking the *intersection* of these intervals gives $(-2, 1) \cup (3, 6)$.

$\boxed{71}$ (a) $|x+5| = 3 \Rightarrow x+5 = 3 \text{ or } x+5 = -3 \Rightarrow x = -2 \text{ or } x = -8$.

(b) $|x+5| < 3$ has solutions between the values found in part (a), that is, $(-8, -2)$.

(c) The solutions of $|x+5| > 3$ are the portions of the real line that are not in parts (a) and (b), that is, $(-\infty, -8) \cup (-2, \infty)$.

$\boxed{73}$ We could think of this statement as "the difference between w and 148 is at most 2." In symbols, we have $|w-148| \leq 2$. Intuitively, we know that this inequality must describe the weights from 146 to 150.

[75] The difference of two temperatures T_1 and T_2 can be represented by $T_1 - T_2$.

Since there is no indication as to whether T_1 is larger than T_2, or vice versa,

we will use $|T_1 - T_2|$. $5 < |T_1 - T_2| < 10$

[77] $30 \le C \le 40 \Rightarrow 30 \le \frac{5}{9}(F - 32) \le 40 \Rightarrow 54 \le F - 32 \le 72 \Rightarrow 86 \le F \le 104$

[79] Since $V = 110$, $R = \frac{110}{I}$, or equivalently, $I = \frac{110}{R}$. If the current is not to exceed 10,

we want to solve the inequality $I \le 10$. $I \le 10 \Rightarrow \frac{110}{R} \le 10 \Rightarrow 110 \le 10R$ { $R > 0$, so

we may multiply by R without changing the direction of the inequality } $\Rightarrow R \ge 11$

[81] We want to know what condition will assure us that an object's image is at least 3

times as large as the object, or, equivalently, when $M \ge 3$.

$M \ge 3$ { $f = 6$ } $\Rightarrow \frac{6}{6 - p} \ge 3 \Rightarrow 6 \ge 18 - 3p$ { since $6 - p > 0$, we can multiply by

$6 - p$ and not change the direction of the inequality } $\Rightarrow 3p \ge 12 \Rightarrow$

$p \ge 4$, but $p < 6$ since $p < f$. Thus, $4 \le p < 6$.

[83] Let x denote the number of years before A becomes more economical than B.

The costs are the initial costs plus the yearly costs times the number of years.

$\text{Cost}_A < \text{Cost}_B \Rightarrow 50,000 + 4000x < 40,000 + 5500x \Rightarrow 10,000 < 1500x \Rightarrow$

$x > \frac{20}{3}$, or $6\frac{2}{3}$ yr.

[85] (a) 5 ft 9 in = 69 in. In a 40 year period, a person's height will decrease by
$40 \times 0.024 = 0.96$ in ≈ 1 in. The person will be approximately one inch shorter,
or 5 ft 8 in. at age 70.

(b) 5 ft 6 in = 66 in. In 20 years, a person's height $(h = 66)$ will change by
$0.024 \times 20 = 0.48$ in. Thus, $66 - 0.48 \le h \le 66 + 0.48 \Rightarrow 65.52 \le h \le 66.48$.

2.7 Exercises

Note: Many solutions for exercises involving inequalities contain a sign diagram. You
may want to read Example 3 again if you have trouble interpreting the sign
diagrams.

[1] $(3x + 1)(5 - 10x) > 0$ has solutions in the interval $(-\frac{1}{3}, \frac{1}{2})$. See *Diagram 1* for details
concerning the signs of the individual factors and the resulting sign.

Resulting sign:	$\ominus$	$\oplus$	$\ominus$
Sign of $5 - 10x$:	$+$	$+$	$-$
Sign of $3x + 1$:	$-$	$+$	$+$
x values:	$-1/3$	$1/2$	

Diagram 1

3 $(x+2)(x-1)(4-x) \leq 0$; $[-2, 1] \cup [4, \infty)$

Resulting sign:	$\oplus$	$\ominus$	$\oplus$	$\ominus$
Sign of $4-x$:	$+$	$+$	$+$	$-$
Sign of $x-1$:	$-$	$-$	$+$	$+$
Sign of $x+2$:	$-$	$+$	$+$	$+$
x values:	-2		1	4

Diagram 3

5 $x^2 - x - 6 < 0 \Rightarrow (x-3)(x+2) < 0$; $(-2, 3)$

Resulting sign:	$\oplus$	$\ominus$	$\oplus$
Sign of $x-3$:	$-$	$-$	$+$
Sign of $x+2$:	$-$	$+$	$+$
x values:	-2		3

Diagram 5

7 $x^2 - 2x - 5 > 3 \Rightarrow x^2 - 2x - 8 > 0 \Rightarrow (x-4)(x+2) > 0$; $(-\infty, -2) \cup (4, \infty)$

Resulting sign:	$\oplus$	$\ominus$	$\oplus$
Sign of $x-4$:	$-$	$-$	$+$
Sign of $x+2$:	$-$	$+$	$+$
x values:	-2		4

Diagram 7

9 $x(2x+3) \geq 5 \Rightarrow 2x^2 + 3x - 5 \geq 0 \Rightarrow (2x+5)(x-1) \geq 0$; $(-\infty, -\frac{5}{2}] \cup [1, \infty)$

Resulting sign:	$\oplus$	$\ominus$	$\oplus$
Sign of $x-1$:	$-$	$-$	$+$
Sign of $2x+5$:	$-$	$+$	$+$
x values:	$-5/2$		1

Diagram 9

11 $6x - 8 > x^2 \Rightarrow x^2 - 6x + 8 < 0 \Rightarrow (x-2)(x-4) < 0$; $(2, 4)$

Resulting sign:	$\oplus$	$\ominus$	$\oplus$
Sign of $x-4$:	$-$	$-$	$+$
Sign of $x-2$:	$-$	$+$	$+$
x values:	2		4

Diagram 11

Note: Solving $x^2 < $ (or $>$) a^2 for $a > 0$ may be solved using factoring, that is,

$x^2 - a^2 < 0 \Rightarrow (x+a)(x-a) < 0 \Rightarrow -a < x < a$; or by taking the square

root of each side, that is, $\sqrt{x^2} < \sqrt{a^2} \Rightarrow |x| < a \Rightarrow -a < x < a$.

13 *Note:* The most common mistake is not remembering that $\sqrt{x^2} = |x|$.

$$x^2 < 16 \Rightarrow |x| < 4 \Rightarrow -4 < x < 4 \Leftrightarrow (-4, 4).$$

17 $16x^2 \geq 9x \Rightarrow x(16x - 9) \geq 0$; $(-\infty, 0] \cup [\frac{9}{16}, \infty)$

Resulting sign:	$\oplus$	$\ominus$	$\oplus$
Sign of $16x-9$:	$-$	$-$	$+$
Sign of x:	$-$	$+$	$+$
x values:	0		$9/16$

Diagram 17

$\boxed{19}$ $x^4 + 5x^2 \geq 36 \Rightarrow x^4 + 5x^2 - 36 \geq 0 \Rightarrow (x^2 + 9)(x^2 - 4) \geq 0 \Rightarrow x^2 - 4 \geq 0 \; \{ x^2 + 9 > 0 \}$

$\Rightarrow x^2 \geq 4 \Rightarrow |x| \geq 2 \Rightarrow x \geq 2 \text{ or } x \leq -2 \Leftrightarrow (-\infty, -2] \cup [2, \infty)$

$\boxed{21}$ $x^3 + 2x^2 - 4x - 8 \geq 0 \Rightarrow x^2(x + 2) - 4(x + 2) \geq 0 \Rightarrow (x^2 - 4)(x + 2) \geq 0 \Rightarrow$

$(x - 2)(x + 2)^2 \geq 0$. The expression $(x + 2)^2$ is positive except when $x = -2$.

The sign is determined by the sign of $x - 2$, which is positive if $x > 2$. Since $x = \pm 2$

make the expression zero, the solution is $x \geq 2$ or $x = -2 \Leftrightarrow \{-2\} \cup [2, \infty)$.

$\boxed{23}$ $\dfrac{x^2(x + 2)}{(x + 2)(x + 1)} \leq 0 \Rightarrow \dfrac{x^2}{x + 1} \leq 0 \;$ {we will exclude $x = -2$ since it makes the original

expression undefined} $\Rightarrow \dfrac{1}{x + 1} \leq 0 \;$ {we can divide by x^2 since $x^2 \geq 0$ and we will

include $x = 0$ since it makes x^2 equal to zero and we want all solutions less than *or*

equal to zero} $\Rightarrow x + 1 < 0 \;$ {the fraction cannot equal zero and $x + 1$ must be

negative so that the fraction is negative} $\Rightarrow x < -1; \; (-\infty, -2) \cup (-2, -1) \cup \{0\}$

$\boxed{25}$ $\dfrac{x^2 - x}{x^2 + 2x} \leq 0 \Rightarrow \dfrac{x(x - 1)}{x(x + 2)} \leq 0 \Rightarrow \dfrac{x - 1}{x + 2} \leq 0 \;$ {we will exclude $x = 0$ from the solution };

$(-2, 0) \cup (0, 1]$

Resulting sign:	$\oplus$	$\ominus$	$\oplus$
Sign of $x - 1$:	−	−	+
Sign of $x + 2$:	−	+	+
x values:	−2		1

Diagram 25

$\boxed{27}$ $\dfrac{x - 2}{x^2 - 3x - 10} \geq 0 \Rightarrow \dfrac{x - 2}{(x - 5)(x + 2)} \geq 0 \;$ {$x = 2$ is a solution since it makes the fraction

equal to zero, $x = 5$ and $x = -2$ are excluded since these values make the fraction

undefined }; $(-2, 2] \cup (5, \infty)$

Resulting sign:	$\ominus$	$\oplus$	$\ominus$	$\oplus$
Sign of $x - 5$:	−	−	−	+
Sign of $x - 2$:	−	−	+	+
Sign of $x + 2$:	−	+	+	+
x values:	−2		2	5

Diagram 27

$\boxed{29}$ $\dfrac{-3x}{x^2 - 9} > 0 \Rightarrow \dfrac{x}{(x + 3)(x - 3)} < 0 \;$ {divide by -3 }; $(-\infty, -3) \cup (0, 3)$

Resulting sign:	$\ominus$	$\oplus$	$\ominus$	$\oplus$
Sign of $x - 3$:	−	−	−	+
Sign of x:	−	−	+	+
Sign of $x + 3$:	−	+	+	+
x values:	−3		0	3

Diagram 29

31 $\dfrac{x+1}{2x-3} > 2 \Rightarrow \dfrac{x+1-2(2x-3)}{2x-3} > 0 \Rightarrow \dfrac{-3x+7}{2x-3} > 0.$

From *Diagram 31*, the solution is $(\frac{3}{2}, \frac{7}{3})$. Note that you should *not* multiply by the factor $2x-3$ as we did with rational *equations* because $2x-3$ may be positive or negative, and multiplying by it would require solving two inequalities. This method of solution tends to be more difficult than the sign diagram method.

Resulting sign:	$\ominus$	$\oplus$	$\ominus$
Sign of $-3x+7$:	+	+	−
Sign of $2x-3$:	−	+	+
x values:	3/2	7/3	

Diagram 31

33 $\dfrac{1}{x-2} \geq \dfrac{3}{x+1} \Rightarrow \dfrac{1(x+1)-3(x-2)}{(x-2)(x+1)} \geq 0 \Rightarrow \dfrac{-2x+7}{(x-2)(x+1)} \geq 0;\ (-\infty, -1) \cup (2, \frac{7}{2}]$

Resulting sign:	$\oplus$	$\ominus$	$\oplus$	$\ominus$
Sign of $-2x+7$:	+	+	+	−
Sign of $x-2$:	−	−	+	+
Sign of $x+1$:	−	+	+	+
x values:	−1	2	7/2	

Diagram 33

35 $\dfrac{4}{3x-2} \leq \dfrac{2}{x+1} \Rightarrow \dfrac{4(x+1)-2(3x-2)}{(3x-2)(x+1)} \leq 0 \Rightarrow \dfrac{-2x+8}{(3x-2)(x+1)} \leq 0;\ (-1, \frac{2}{3}) \cup [4, \infty)$

Resulting sign:	$\oplus$	$\ominus$	$\oplus$	$\ominus$
Sign of $-2x+8$:	+	+	+	−
Sign of $3x-2$:	−	−	+	+
Sign of $x+1$:	−	+	+	+
x values:	−1	2/3	4	

Diagram 35

37 $\dfrac{x}{3x-5} \leq \dfrac{2}{x-1} \Rightarrow \dfrac{x(x-1)-2(3x-5)}{(3x-5)(x-1)} \leq 0 \Rightarrow \dfrac{(x-2)(x-5)}{(3x-5)(x-1)} \leq 0;\ (1, \frac{5}{3}) \cup [2, 5]$

Res. sign:	$\oplus$	$\ominus$	$\oplus$	$\ominus$	$\oplus$
$x-5$:	−	−	−	−	+
$x-2$:	−	−	−	+	+
$3x-5$:	−	−	+	+	+
$x-1$:	−	+	+	+	+
x values:	1	5/3	2	5	

Diagram 37

39 $x^3 > x \Rightarrow x^3 - x > 0 \Rightarrow x(x^2-1) > 0 \Rightarrow x(x+1)(x-1) > 0;\ (-1, 0) \cup (1, \infty)$

Resulting sign:	$\ominus$	$\oplus$	$\ominus$	$\oplus$
Sign of $x-1$:	−	−	−	+
Sign of x:	−	−	+	+
Sign of $x+1$:	−	+	+	+
x values:	−1	0	1	

Diagram 39

$\boxed{41}$ $v \geq k \Rightarrow t^3 - 3t^2 - 4t + 20 \geq 8 \Rightarrow t^3 - 3t^2 - 4t + 12 \geq 0 \Rightarrow t^2(t-3) - 4(t-3) \geq 0 \Rightarrow$

$(t^2 - 4)(t-3) \geq 0 \Rightarrow (t+2)(t-2)(t-3) \geq 0.$ For $[0, 5]$, we have $[0, 2] \cup [3, 5]$.

Resulting sign:	$\ominus$	$\oplus$	$\ominus$	$\oplus$
Sign of $t - 3$:	$-$	$-$	$-$	$+$
Sign of $t - 2$:	$-$	$-$	$+$	$+$
Sign of $t + 2$:	$-$	$+$	$+$	$+$
t values:		-2	2	3

Diagram 41

$\boxed{43}$ $s > 9 \Rightarrow -16t^2 + 24t + 1 > 9 \Rightarrow -16t^2 + 24t - 8 > 0 \Rightarrow$

$2t^2 - 3t + 1 < 0$ { divide by -8 } $\Rightarrow (2t-1)(t-1) < 0 \Rightarrow \frac{1}{2} < t < 1.$

The dog is more than 9 ft off the ground for $1 - \frac{1}{2} = \frac{1}{2}$ sec.

$\boxed{45}$ $d < 75 \Rightarrow v + \frac{1}{20}v^2 < 75$ { multiply by 20 } $\Rightarrow 20v + v^2 < 1500 \Rightarrow$

$v^2 + 20v - 1500 < 0 \Rightarrow (v+50)(v-30) < 0$ { use a sign diagram } $\Rightarrow$

$-50 < v < 30 \Rightarrow 0 \leq v < 30$ { since $v \geq 0$ }

$\boxed{47}$ $R > S \Rightarrow \frac{4500\,S}{S+500} > S \Rightarrow \frac{S(S-4000)}{S+500} < 0$ { use a sign diagram } $\Rightarrow$

$S < -500$ or $0 < S < 4000 \Rightarrow 0 < S < 4000$ { since $S > 0$ }

$\boxed{49}$ $W < 5 \Rightarrow 125\left(\frac{6400}{6400+x}\right)^2 < 5 \Rightarrow \left(\frac{6400}{6400+x}\right)^2 < \left(\frac{1}{5}\right)^2$ { take the square root } $\Rightarrow$

$\frac{6400}{6400+x} < \frac{1}{5}$ { since $\frac{6400}{6400+x} > 0$ } $\Rightarrow 32{,}000 < x + 6400 \Rightarrow x > 25{,}600$ km.

$\boxed{51}$ $7500 \leq 0.00334V^2S \leq 10{,}000 \Rightarrow 7500 \leq 0.00334(210)V^2 \leq 10{,}000 \Rightarrow$

$\frac{7500}{0.7014} \leq V^2 \leq \frac{10{,}000}{0.7014} \Rightarrow \sqrt{\frac{7500}{0.7014}} \leq V \leq \sqrt{\frac{10{,}000}{0.7014}} \Rightarrow 103.4 \leq V \leq 119.4$ ft/sec $\Rightarrow$

$70.5 \leq V \leq 81.4$ mi/hr. To convert ft/sec to mi/hr, multiply by $\frac{60}{88}$, which is the

reduced form of $\frac{1 \text{ foot}}{1 \text{ second}} \times \frac{3600 \text{ seconds}}{1 \text{ hour}} \times \frac{1 \text{ mile}}{5280 \text{ feet}}.$

$\boxed{53}$ By using a table it can be shown that the expression is equal to zero when $x = -3$, -2, 2, 4. The expression $Y_1 = x^4 - x^3 - 16x^2 + 4x + 48$ is negative when $x \in (-3, -2) \cup (2, 4)$.

x	Y_1	x	Y_1
-3.5	30.938	1.0	36
-3.0	0	1.5	19.688
-2.5	-7.313	2.0	0
-2.0	0	2.5	-18.56
-1.5	14.438	3.0	-30
-1.0	30	3.5	-26.81
-0.5	42.188	4.0	0
0.0	48	4.5	60.938
0.5	45.938	5.0	168

Chapter 2 Review Exercises

$\boxed{3}$ $\left[\dfrac{2}{x+5} - \dfrac{3}{2x+1} = \dfrac{5}{6x+3}\right] \cdot 3(x+5)(2x+1) \Rightarrow 6(2x+1) - 9(x+5) = 5(x+5) \Rightarrow$

$$3x - 39 = 5x + 25 \Rightarrow -2x = 64 \Rightarrow x = -32$$

$\boxed{4}$ $\left[\dfrac{7}{x-2} - \dfrac{6}{x^2-4} = \dfrac{3}{2x+4}\right] \cdot 2(x+2)(x-2) \Rightarrow 14(x+2) - 12 = 3(x-2) \Rightarrow 11x = -22$

$\Rightarrow x = -2$, which is not in the domain of the given expressions. No solution

$\boxed{5}$ $LS = \dfrac{1}{\sqrt{x}} - 2 = \dfrac{1 - 2\sqrt{x}}{\sqrt{x}} = RS$, an identity.

The given equation is true for every $x > 0$.

$\boxed{9}$ $(x-2)(x+1) = 3 \Rightarrow x^2 - x - 2 = 3 \Rightarrow x^2 - x - 5 = 0 \Rightarrow x = \dfrac{1 \pm \sqrt{1+20}}{2} = \dfrac{1}{2} \pm \dfrac{1}{2}\sqrt{21}$

$\boxed{11}$ $x^{2/3} - 2x^{1/3} - 15 = 0 \Rightarrow (x^{1/3} + 3)(x^{1/3} - 5) = 0 \Rightarrow \sqrt[3]{x} = -3, 5 \Rightarrow x = -27, 125$

$\boxed{15}$ $6x^4 + 29x^2 + 28 = 0 \Rightarrow (2x^2 + 7)(3x^2 + 4) = 0 \Rightarrow x^2 = -\dfrac{7}{2}, -\dfrac{4}{3} \Rightarrow$

$$x = \pm\tfrac{1}{2}\sqrt{14}i, \ \pm\tfrac{2}{3}\sqrt{3}i$$

$\boxed{16}$ $x^4 - 3x^2 + 1 = 0 \Rightarrow x^2 = \dfrac{3 \pm \sqrt{5}}{2} \cdot \dfrac{2}{2} = \dfrac{6 \pm 2\sqrt{5}}{4} \Rightarrow$

$$x = \pm\tfrac{1}{2}\sqrt{6 \pm 2\sqrt{5}} \approx \pm 1.62, \ \pm 0.62$$

$\boxed{18}$ $2|2x+1| + 1 = 19 \Rightarrow 2|2x+1| = 18 \Rightarrow |2x+1| = 9 \Rightarrow$

$$2x + 1 = 9 \text{ or } 2x + 1 = -9 \Rightarrow 2x = 8 \text{ or } 2x = -10 \Rightarrow x = 4 \text{ or } x = -5$$

$\boxed{19}$ $\left[\dfrac{1}{x} + 6 = \dfrac{5}{\sqrt{x}} \right] \cdot x \Rightarrow 1 + 6x = 5\sqrt{x} \Rightarrow$

$6x - 5\sqrt{x} + 1 = 0$ { factoring or substituting would be appropriate } $\Rightarrow$

$$(2\sqrt{x} - 1)(3\sqrt{x} - 1) = 0 \Rightarrow \sqrt{x} = \tfrac{1}{2}, \tfrac{1}{3} \Rightarrow x = \tfrac{1}{4}, \tfrac{1}{9}$$

Check $x = \tfrac{1}{4}$: LS $= 4 + 6 = 10$; RS $= 5/\tfrac{1}{2} = 10 \Rightarrow x = \tfrac{1}{4}$ is a solution.

Check $x = \tfrac{1}{9}$: LS $= 9 + 6 = 15$; RS $= 5/\tfrac{1}{3} = 15 \Rightarrow x = \tfrac{1}{9}$ is a solution.

$\boxed{23}$ $\sqrt{3x+1} - \sqrt{x+4} = 1 \Rightarrow \sqrt{3x+1} = 1 + \sqrt{x+4} \Rightarrow 3x + 1 = 1 + 2\sqrt{x+4} + x + 4 \Rightarrow$

$2\sqrt{x+4} = 2x - 4 \Rightarrow \sqrt{x+4} = x - 2 \Rightarrow (\sqrt{x+4})^2 = (x-2)^2 \Rightarrow$

$$x + 4 = x^2 - 4x + 4 \Rightarrow x^2 - 5x = 0 \Rightarrow x(x-5) = 0 \Rightarrow x = 0, 5.$$

Check $x = 0$: LS $= 1 - 2 = -1 \neq$ RS $\Rightarrow x = 0$ is an extraneous solution.

Check $x = 5$: LS $= 4 - 3 = 1 =$ RS $\Rightarrow x = 5$ is a solution.

$\boxed{24}$ $x^{4/3} = 16 \Rightarrow (x^{4/3})^{3/4} = \pm 16^{3/4} \Rightarrow x = \pm(\sqrt[4]{16})^3 = \pm 2^3 = \pm 8$

$\boxed{25}$ $3x^2 - 12x + 3 = 0 \Rightarrow x^2 - 4x + 1 = 0 \Rightarrow x^2 - 4x + 4 = -1 + 4 \Rightarrow (x-2)^2 = 3 \Rightarrow$

$$x - 2 = \pm\sqrt{3} \Rightarrow x = 2 \pm \sqrt{3}$$

$\boxed{27}$ The expression $(x-3)^2$ is never less than 0, but it is equal to 0 when $x = 3$.

$$\text{Thus, } (x-3)^2 \le 0 \text{ has solution } x = 3.$$

$\boxed{31}$ $\dfrac{6}{10x+3} < 0 \Rightarrow 10x + 3 < 0$ { since $6 > 0$ } $\Rightarrow x < -\tfrac{3}{10} \Leftrightarrow (-\infty, -\tfrac{3}{10})$

$\boxed{33}$ $2\,|\,3 - x\,| + 1 > 5 \Rightarrow 2\,|\,3 - x\,| > 4 \Rightarrow |\,3 - x\,| > 2 \Rightarrow$

$$3 - x > 2 \text{ or } 3 - x < -2 \Rightarrow 1 > x \text{ or } 5 < x \Rightarrow x < 1 \text{ or } x > 5 \Leftrightarrow (-\infty, 1) \cup (5, \infty)$$

$\boxed{35}$ $|\,16 - 3x\,| \ge 5 \Rightarrow 16 - 3x \ge 5 \text{ or } 16 - 3x \le -5 \Rightarrow -3x \ge -11 \text{ or } -3x \le -21 \Rightarrow$

$$x \le \tfrac{11}{3} \text{ or } x \ge 7 \Leftrightarrow (-\infty, \tfrac{11}{3}] \cup [7, \infty)$$

$\boxed{38}$ $x(x-3) \le 10 \Rightarrow x^2 - 3x - 10 \le 0 \Rightarrow (x-5)(x+2) \le 0$; $[-2, 5]$

Resulting sign:	$\oplus$	$\ominus$	$\oplus$
Sign of $x - 5$:	$-$	$-$	$+$
Sign of $x + 2$:	$-$	$+$	$+$
x values:	-2		5

Diagram 38

$\boxed{39}$ $\dfrac{x^2(3-x)}{x+2} \le 0 \Rightarrow \dfrac{3-x}{x+2} \le 0$ { include 0 }; $(-\infty, -2) \cup \{\,0\,\} \cup [3, \infty)$

Resulting sign:	$\ominus$	$\oplus$	$\ominus$
Sign of $3 - x$:	$+$	$+$	$-$
Sign of $x + 2$:	$-$	$+$	$+$
x values:	-2		3

Diagram 39

[41] $\dfrac{3}{2x+3} < \dfrac{1}{x-2} \Rightarrow \dfrac{3(x-2)-1(2x+3)}{(2x+3)(x-2)} < 0 \Rightarrow \dfrac{x-9}{(2x+3)(x-2)} < 0;\ (-\infty,\ -\tfrac{3}{2}) \cup (2,\ 9)$

Resulting sign:	$\ominus$	$\oplus$	$\ominus$	$\oplus$
Sign of $x-9$:	−	−	−	+
Sign of $x-2$:	−	−	+	+
Sign of $2x+3$:	−	+	+	+
x values:	−3/2		2	9

Diagram 41

[43] $x^3 > x^2 \Rightarrow x^2(x-1) > 0\ \{x^2 \ge 0\} \Rightarrow x-1 > 0 \Rightarrow x > 1 \Leftrightarrow (1,\ \infty)$

[44] $(x^2 - x)(x^2 - 5x + 6) < 0 \Rightarrow$

$x(x-1)(x-2)(x-3) < 0;\ (0,\ 1) \cup (2,\ 3)$

Res. sign:	$\oplus$	$\ominus$	$\oplus$	$\ominus$	$\oplus$
$x-3$:	−	−	−	−	+
$x-2$:	−	−	−	+	+
$x-1$:	−	−	+	+	+
x:	−	+	+	+	+
x values:	0		1	2	3

Diagram 44

[47] $c = \sqrt{4h(2R-h)} \Rightarrow c^2 = 8Rh - 4h^2 \Rightarrow 4h^2 - 8Rh + c^2 = 0 \Rightarrow$

$$h = \dfrac{8R \pm \sqrt{64R^2 - 16c^2}}{8} = \dfrac{8R \pm 4\sqrt{4R^2 - c^2}}{8} = R \pm \tfrac{1}{2}\sqrt{4R^2 - c^2}$$

[48] $V = \tfrac{1}{3}\pi h(r^2 + R^2 + rR) \Rightarrow r^2 + Rr + R^2 - \dfrac{3V}{\pi h} = 0 \Rightarrow$

$(\pi h)r^2 + (\pi h R)r + (\pi h R^2 - 3V) = 0 \Rightarrow$

$$r = \dfrac{-\pi h R \pm \sqrt{\pi^2 h^2 R^2 - 4\pi h(\pi h R^2 - 3V)}}{2\pi h} = \dfrac{-\pi h R \pm \sqrt{12\pi h V - 3\pi^2 h^2 R^2}}{2\pi h}.$$

Since $r > 0$, we must use the plus sign, and $r = \dfrac{-\pi h R + \sqrt{12\pi h V - 3\pi^2 h^2 R^2}}{2\pi h}.$

[52] $\dfrac{1}{9 - \sqrt{-4}} = \dfrac{1}{9 - 2i} = \dfrac{1}{9 - 2i} \cdot \dfrac{9 + 2i}{9 + 2i} = \dfrac{9 + 2i}{81 + 4} = \dfrac{9}{85} + \dfrac{2}{85}i$

[56] Let P denote the principal that will be invested, and r the yield rate of the stock fund. $\text{Income}_{\text{stocks}} - 28\%$ federal tax $- 7\%$ state tax $= \text{Income}_{\text{bonds}} \Rightarrow$
$(Pr) - 0.28(Pr) - 0.07(Pr) = 0.07186P$ { divide by P } $\Rightarrow$
$1r - 0.28r - 0.07r = 0.07186 \Rightarrow 0.65r = 0.07186 \Rightarrow r \approx 0.11055$, or, 11.055%.

[57] Let x denote the number of cm^3 of gold. $\text{Grams}_{\text{gold}} + \text{Grams}_{\text{silver}} = \text{Grams}_{\text{total}} \Rightarrow$
$x(19.3) + (5 - x)(10.5) = 80 \Rightarrow 8.8x = 27.5 \Rightarrow x = 3.125.$

The number of grams of gold is $19.3x = 60.3125 \approx 60.3.$

[59] Let x denote the number of grams of 95% ethyl alcohol solution used, $400 - x$ the number of grams of water. $95(x) + 0(400 - x) = 75(400)$ { all in % } $\Rightarrow$
$95x = 75(400) \Rightarrow x = \dfrac{6000}{19} \approx 315.8.$ Use 315.8 g of ethyl alcohol and 84.2 g of water.

$\boxed{60}$ Let x denote the number of gallons of 20% solution, $120 - x$ the number of gallons of 50% solution. $20(x) + 50(120 - x) = 30(120)$ { all in % } $\Rightarrow 20 \cdot 120 = 30x \Rightarrow x = 80$.

Use 80 gal of the 20% solution and 40 gal of the 50% solution.

$\boxed{61}$ Let x denote the distance upstream. 10 gallons of gas @ 16 mi/gal $= 160$ miles. At 20 mi/hr, there is enough fuel for 8 hours of travel. The rate of the boat upstream is 15 mi/hr and the rate downstream is 25 mi/hr. $\text{Time}_{\text{up}} + \text{Time}_{\text{down}} = \text{Time}_{\text{total}} \Rightarrow$

$$\left[\frac{x}{15} + \frac{x}{25} = 8 \right] \cdot 75 \Rightarrow 5x + 3x = 600 \Rightarrow 8x = 600 \Rightarrow x = 75 \text{ mi.}$$

$\boxed{64}$ Let $50 + r$ denote the rate the automobile, that is, r is the rate over 50 mi/hr. The automobile must travel $40 + 20 = 60$ ft more than the truck (traveling at 50 mi/hr) in 5 seconds. Since 1 mi/hr $= \frac{5280}{3600} = \frac{22}{15}$ ft/sec, the automobile's rate in *excess* of 50 mi/hr is $\frac{22}{15}r$. Thus, $d = rt \Rightarrow 60 = (\frac{22}{15}r)(5) \Rightarrow r = \frac{90}{11}$.

The rate is $50 + \frac{90}{11} = \frac{640}{11} \approx 58.2$ mi/hr.

$\boxed{65}$ Let x denote the number of hours needed to fill an empty bin.

Using the hourly rates, $\left[\frac{1}{2} - \frac{1}{5} = \frac{1}{x} \right] \cdot 10x \Rightarrow 5x - 2x = 10 \Rightarrow 3x = 10 \Rightarrow x = \frac{10}{3}$ hr.

Since the bin was half-full at the start, $\frac{1}{2}x = \frac{1}{2} \cdot \frac{10}{3} = \frac{5}{3}$ hr, or, 1 hr 40 min.

$\boxed{67}$ Let d denote the distance from the center of the city to a corner and $2x$ denote the length of one side of the city. $x^2 + x^2 = d^2 \Rightarrow d = \sqrt{2}x$.

$A = $ area of the city $= (2x)^2 = 4x^2$, or $2d^2$. Currently: $d = 10 \Rightarrow A = 200$. One decade ago: $A = 150 \Rightarrow d = \sqrt{75} = 5\sqrt{3}$. The change in d is $10 - 5\sqrt{3} \approx 1.34$ mi.

$\boxed{69}$ (a) The eastbound car has distance $20t$ and the southbound car has distance

$$(-2 + 50t). \quad d^2 = (20t)^2 + (-2 + 50t)^2 \Rightarrow d = \sqrt{2900t^2 - 200t + 4}$$

(b) $104 = \sqrt{2900t^2 - 200t + 4} \Rightarrow 2900t^2 - 200t - 10{,}812 = 0 \Rightarrow$

$$725t^2 - 50t - 2703 = 0 \Rightarrow t = \frac{50 \pm \sqrt{7{,}841{,}200}}{1450} \, \{\, t > 0 \,\} = \frac{5 + 2\sqrt{19{,}603}}{145} \approx 1.97,$$

or approximately 11:58 A.M.

$\boxed{71}$ Let x denote the length of one side of an end.

(a) $V = lwh \Rightarrow 48 = 6 \cdot x \cdot x \Rightarrow x^2 = 8 \Rightarrow x = 2\sqrt{2}$ ft

(b) $S = lw + 2wh + 2lh \Rightarrow 44 = 6x + 2(x^2) + 2(6x) \Rightarrow 44 = 2x^2 + 18x \Rightarrow$

$$x^2 + 9x - 22 = 0 \Rightarrow (x + 11)(x - 2) = 0 \Rightarrow x = 2 \text{ ft}$$

$\boxed{73}$ Let x denote the width of the tiled area, $2x$ the length.

The bathing area has measurements $x - 2$ and $2x - 2$. For the bathing area, width $\cdot$ length $=$ area $\Rightarrow (x - 2)(2x - 2) = 40 \Rightarrow 2x^2 - 6x + 4 = 40 \Rightarrow$

$x^2 - 3x + 2 = 20 \Rightarrow x^2 - 3x - 18 = 0 \Rightarrow (x - 6)(x + 3) = 0 \Rightarrow x = 6 \, \{ x > 0 \}$.

The tiled area is 12 ft by 6 ft and the bathing area is 10 ft by 4 ft.

[75] $pv = 200 \Rightarrow v = \frac{200}{p}$. $25 \le v \le 50 \Rightarrow 25 \le \frac{200}{p} \le 50 \Rightarrow \frac{1}{25} \ge \frac{p}{200} \ge \frac{1}{50} \Rightarrow$

$$8 \ge p \ge 4 \Rightarrow 4 \le p \le 8$$

[76] Let x denote the amount of yearly business (in dollars).

$\text{Pay}_B > \text{Pay}_A \Rightarrow \$20{,}000 + 0.10x > \$25{,}000 + 0.05x \Rightarrow 0.05x > \$5000 \Rightarrow x > \$100{,}000$

[78] $T = 2\pi\sqrt{\frac{l}{980}} \Rightarrow l = \frac{980\,T^2}{4\pi^2}$. $98 \le l \le 100 \Rightarrow 98 \le \frac{980\,T^2}{4\pi^2} \le 100 \Rightarrow$

$$\frac{2\pi^2}{5} \le T^2 \le \frac{20\pi^2}{49} \Rightarrow \frac{10\pi^2}{25} \le T^2 \le \frac{20\pi^2}{49} \Rightarrow \frac{\pi}{5}\sqrt{10} \le T \le \frac{2\pi}{7}\sqrt{5} \ \{T \ge 0\},$$

or, approximately, $1.987 \le T \le 2.007$ sec.

[80] $P = 2l + 2w \Rightarrow 100 = 2l + 2w \Rightarrow l = 50 - w$. $A \ge 600 \Rightarrow lw \ge 600 \Rightarrow$

$(50 - w)w \ge 600 \Rightarrow -w^2 + 50w - 600 \ge 0 \Rightarrow w^2 - 50w + 600 \le 0 \Rightarrow$

$(w - 20)(w - 30) \le 0 \Rightarrow 20 \le w \le 30$. If w is greater than 25,

it would be the length. Hence, the desired values of w are between 20 and 25.

[81] Let x denote the number of trees over 24. Then $24 + x$ represents the total number

of trees planted per acre, and $600 - 12x$ represents the number of apples per tree.

Total apples $= (\#$ of trees$)(\#$ of apples per tree$) = (24 + x)(600 - 12x) =$

$-12x^2 + 312x + 14{,}400$. Apples $\ge 16{,}416 \Rightarrow -12x^2 + 312x + 14{,}400 \ge 16{,}416 \Rightarrow$

$-12x^2 + 312x - 2016 \ge 0 \Rightarrow x^2 - 26x + 168 \le 0 \Rightarrow (x - 12)(x - 14) \le 0 \Rightarrow$

$12 \le x \le 14$. Hence, 36 to 38 trees per acre should be planted.

[82] Let x denote the number of \$10 increases in rent. Then the

number of occupied apartments is $180 - 5x$ and the rent per apartment is $300 + 10x$.

Total income $= (\#$ of occupied apartments$)($rent per apartment$) =$

$(180 - 5x)(300 + 10x) = -50x^2 + 300x + 54{,}000$. Income $\ge 54{,}400 \Rightarrow$

$-50x^2 + 300x + 54{,}000 \ge 54{,}400 \Rightarrow -50x^2 + 300x - 400 \ge 0 \Rightarrow x^2 - 6x + 8 \le 0 \Rightarrow$

$(x - 2)(x - 4) \le 0 \Rightarrow 2 \le x \le 4$. Hence, the rent charged should be \$320 to \$340.

Chapter 2 Discussion Exercises

[1] We need to solve the equation $x^2 - xy + y^2 = 0$ for x.

Use the quadratic formula with $a = 1$, $b = -y$, and $c = y^2$.

$$x = \frac{-(-y) \pm \sqrt{(-y)^2 - 4(1)(y^2)}}{2(1)} = \frac{y \pm \sqrt{y^2 - 4y^2}}{2} = \frac{y \pm \sqrt{-3y^2}}{2} = \frac{y \pm |y|\sqrt{3}\,i}{2}.$$

Since this equation has imaginary solutions, $x^2 - xy + y^2$ is not factorable over the

reals. A similar argument holds for $x^2 + xy + y^2$.

$\boxed{3}$ (a) $\dfrac{1}{\dfrac{a+bi}{c+di}} = \dfrac{c+di}{a+bi} \cdot \dfrac{a-bi}{a-bi} = \dfrac{ac+bd+(ad-bc)i}{a^2+b^2} = \dfrac{ac+bd}{a^2+b^2} + \dfrac{ad-bc}{a^2+b^2}i = p+qi$

(b) Yes, try an example such as $3/4$. Let $a=3$, $b=0$, $c=4$, and $d=0$. Then, from part (a), $p+qi = \frac{12}{9} + \frac{0}{9}i = \frac{12}{9} = \frac{4}{3}$, which is the multiplicative inverse of $3/4$.

(c) a and b cannot both be 0 because then the denominator would be 0.

$\boxed{5}$ *Hint*: Try these examples to help you get to the general solution.

(1) $x^2 + 1 \geq 0$ {In this case, $a > 0$, $D = -4 < 0$, and by examining a sign chart with
$x^2 + 1$ as the only factor, we see that the solution is $x \in \mathbb{R}$. }

(2) $x^2 - 2x - 3 \geq 0$

(3) $-x^2 - 4 \geq 0$

(4) $-x^2 - 2x - 1 \geq 0$

(5) $-x^2 + 2x + 3 \geq 0$

General solutions categorized by a and D:

(1) $a > 0$, $D \leq 0$: solution is $x \in \mathbb{R}$

(2) $a > 0$, $D > 0$: let $x_1 = (-b - \sqrt{D})/(2a)$ and $x_2 = (-b + \sqrt{D})/(2a) \Rightarrow$
solution is $(-\infty, x_1] \cup [x_2, \infty)$

(3) $a < 0$, $D < 0$: solution is $\{\ \}$

(4) $a < 0$, $D = 0$: solution is $x = -b/(2a)$

(5) $a < 0$, $D > 0$: solution is $[x_1, x_2]$

Chapter 3: Functions and Graphs

$\boxed{7}$ (a) $x = -2$ is the line parallel to the y-axis that intersects the x-axis at $(-2, 0)$.

(b) $y = 3$ is the line parallel to the x-axis that intersects the y-axis at $(0, 3)$.

(c) $x \geq 0$ { x is zero or positive }

is the set of all points to the right of and on the y-axis.

(d) $xy > 0$ { x and y have the same sign, that is, either both are positive or both are

negative } is the set of all points in quadrants I and III.

(e) $y < 0$ { y is negative } is the set of all points below the x-axis.

(f) $x = 0$ is the set of all points on the y-axis.

$\boxed{9}$ (a) $A(4, -3)$, $B(6, 2) \Rightarrow d(A, B) = \sqrt{(6-4)^2 + [2-(-3)]^2} = \sqrt{4+25} = \sqrt{29}$

(b) $M_{AB} = \left(\dfrac{4+6}{2}, \dfrac{-3+2}{2}\right) = (5, -\tfrac{1}{2})$

$\boxed{11}$ (a) $A(-5, 0)$, $B(-2, -2) \Rightarrow d(A, B) = \sqrt{[-2-(-5)]^2 + (-2-0)^2} = \sqrt{9+4} = \sqrt{13}$

(b) $M_{AB} = \left(\dfrac{-5+(-2)}{2}, \dfrac{0+(-2)}{2}\right) = (-\tfrac{7}{2}, -1)$

$\boxed{15}$ We need to show that the sides satisfy the Pythagorean theorem. Finding the distances, we have $d(A, B) = \sqrt{98}$, $d(B, C) = \sqrt{32}$, and $d(A, C) = \sqrt{130}$. Since $d(A, C)$ is the largest of the three values, it must be the hypotenuse, hence, we need to check if $d(A, C)^2 = d(A, B)^2 + d(B, C)^2$. Since $(\sqrt{130})^2 = (\sqrt{98})^2 + (\sqrt{32})^2$, we know that $\triangle ABC$ is a right triangle. The area of a triangle is given by $A = \tfrac{1}{2}(\text{base})(\text{height})$. We can use $d(B, C)$ for the base and $d(A, B)$ for the height. Hence, area $= \tfrac{1}{2}bh = \tfrac{1}{2}(\sqrt{32})(\sqrt{98}) = \tfrac{1}{2}(4\sqrt{2})(7\sqrt{2}) = \tfrac{1}{2}(28)(2) = 28$.

$\boxed{17}$ We need to show that all 4 sides are the same length. Checking, we find that $d(A, B) = d(B, C) = d(C, D) = d(D, A) = \sqrt{29}$. This guarantees that we have a rhombus { a parallelogram with 4 equal sides }. Thus, we also need to show that adjacent sides meet at right angles. This can be done by showing that two adjacent sides and a diagonal form a right triangle. Using $\triangle ABC$, we see that $d(A, C) = \sqrt{58}$ and hence $d(A, C)^2 = d(A, B)^2 + d(B, C)^2$. We conclude that $ABCD$ is a square.

$\boxed{19}$ Let $B = (x, y)$. $A(-3, 8) \Rightarrow M_{AB} = \left(\dfrac{-3+x}{2}, \dfrac{8+y}{2}\right)$. $M_{AB} = C(5, -10) \Rightarrow$

$-3 + x = 2(5)$ and $8 + y = 2(-10) \Rightarrow x = 13$ and $y = -28$. $B = (13, -28)$.

[21] The perpendicular bisector of AB is the line that passes through the midpoint of segment AB and intersects segment AB at a right angle. The points on the perpendicular bisector are all equidistant from A and B. Thus, we need to show that $d(A,\ C) = d(B,\ C)$. Since each of these is $\sqrt{145}$, we conclude that C is on the perpendicular bisector of AB.

[23] We must have $d(A,\ P) = d(B,\ P)$. $\sqrt{(x+4)^2 + (y+3)^2} = \sqrt{(x-6)^2 + (y-1)^2} \Rightarrow$
$x^2 + 8x + 16 + y^2 + 6y + 9 = x^2 - 12x + 36 + y^2 - 2y + 1 \Rightarrow$
$$8x + 6y + 25 = -12x - 2y + 37 \Rightarrow 20x + 8y = 12 \Rightarrow 5x + 2y = 3$$

[25] Let $O(0,\ 0)$ represent the origin. Applying the distance formula with O and $P(x,\ y)$, we have $d(O,\ P) = 5 \Rightarrow \sqrt{(x-0)^2 + (y-0)^2} = 5 \Rightarrow \sqrt{x^2 + y^2} = 5$.

This is a circle of radius 5 with center at the origin.

[27] Let $Q(0,\ y)$ be an arbitrary point on the y-axis.

Applying the distance formula with Q and $P(5,\ 3)$, we have $6 = d(P,\ Q) \Rightarrow$
$6 = \sqrt{(0-5)^2 + (y-3)^2} \Rightarrow 36 = 25 + y^2 - 6y + 9 \Rightarrow$
$y^2 - 6y - 2 = 0 \Rightarrow y = 3 \pm \sqrt{11}$. The points are $(0,\ 3 + \sqrt{11})$ and $(0,\ 3 - \sqrt{11})$.

[29] $5 = \sqrt{(2a-1)^2 + (a-3)^2} \Rightarrow 25 = 4a^2 - 4a + 1 + a^2 - 6a + 9 \Rightarrow$
$5a^2 - 10a - 15 = 0 \Rightarrow a^2 - 2a - 3 = 0 \Rightarrow (a-3)(a+1) = 0 \Rightarrow a = 3,\ -1$. Since the y-coordinate is negative in the third quadrant, $a = -1$, and $(2a,\ a) = (-2,\ -1)$.

[31] With $P(a,\ 3)$ and $Q(5,\ 2a)$, $d(P,\ Q) > \sqrt{26} \Rightarrow \sqrt{(5-a)^2 + (2a-3)^2} > \sqrt{26} \Rightarrow$
$25 - 10a + a^2 + 4a^2 - 12a + 9 > 26 \Rightarrow 5a^2 - 22a + 8 > 0 \Rightarrow (5a - 2)(a - 4) > 0$.

Using *Diagram 31*, we see that $a < \frac{2}{5}$ or $a > 4$ will assure us that $d(P,\ Q) > \sqrt{26}$.

Resulting sign:	$\oplus$	$\ominus$	$\oplus$
Sign of $5a - 2$:	$-$	$+$	$+$
Sign of $a - 4$:	$-$	$-$	$+$
a values:	2/5	4	

Diagram 31

[33] Let M be the midpoint of the hypotenuse. Then $M = (\frac{1}{2}a,\ \frac{1}{2}b)$.

Show that $d(A,\ M) = d(B,\ M) = d(O,\ M) = \frac{1}{2}\sqrt{a^2 + b^2}$.

[35–38]   For point plotting instructions on the TI-82/83, see Example 2 in Appendix I.

[35] Plot $A(-5,\ -3.5)$, $B(-2,\ 2)$, $C(1,\ 0.5)$, $D(4,\ 1)$, and $E(7,\ 2.5)$. See *Figure 35*.

[−10, 10] by [−10, 10]

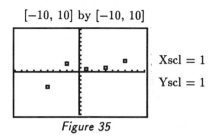

Xscl = 1

Yscl = 1

$[1988, 1996]$ by $[54 \times 10^6, 61 \times 10^6]$

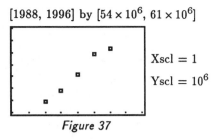

Xscl = 1

Yscl = 10^6

Figure 35   Figure 37

[37] (a) Plot (1990, 54,871,330), (1991, 55,786,390), (1992, 57,211,600),

(1993, 58,834,440), and (1994, 59,332,200).

(b) The number of cable subscribers is increasing each year.

3.2 Exercises

[1] As in Example 1, we expect the graph to be a line.

Creating a table of values similar to those in the text, we have:

x	−2	−1	0	1	2
y	−7	−5	−3	−1	1

By plotting these points and connecting them, we obtain *Figure 1*.

To find the x-intercept, let $y = 0$ in $y = 2x - 3$, and solve for x. (1.5, 0)

To find the y-intercept, let $x = 0$ in $y = 2x - 3$, and solve for y. (0, −3)

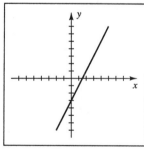

Figure 1

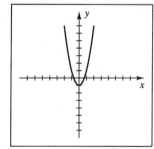

Figure 7

[7] $y = 2x^2 - 1$ • x-intercepts: $(\pm \frac{1}{2}\sqrt{2},\ 0)$ y-intercept: (0, −1)

Since we can substitute $-x$ for x in the equation and obtain an equivalent equation,

we know the graph is symmetric with respect to the y-axis. We will make use of this

fact when constructing our table. As in Example 2, we obtain a parabola.

x	± 2	$\pm \frac{3}{2}$	± 1	$\pm \frac{1}{2}$	0
y	7	$\frac{7}{2}$	1	$-\frac{1}{2}$	−1

$\boxed{11}$ $x = -y^2 + 3$ • x-intercept: $(3, 0)$ y-intercepts: $(0, \pm\sqrt{3})$

Since we can substitute $-y$ for y in the equation and obtain an equivalent equation, we know the graph is symmetric with respect to the x-axis. We will make use of this fact when constructing our table. As in Example 4, we obtain a parabola.

x	-13	-6	-1	2	3
y	± 4	± 3	± 2	± 1	0

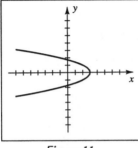

Figure 11

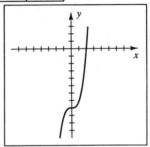

Figure 15

$\boxed{15}$ $y = x^3 - 8$ • x-intercept: $(2, 0)$ y-intercept: $(0, -8)$

x	-2	-1	0	1	2
y	-16	-9	-8	-7	0

$\boxed{19}$ $y = \sqrt{x} - 4$ • x-intercept: $(16, 0)$ y-intercept: $(0, -4)$

x	0	1	4	9	16
y	-4	-3	-2	-1	0

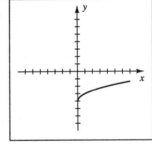

Figure 19

$\boxed{21}$ You may be able to do this exercise mentally. For example (using Exercise 1 with $y = 2x - 3$), we see that substituting $-x$ for x gives us $y = -2x - 3$; substituting $-y$ for y gives us $-y = 2x - 3$ or, equivalently, $y = -2x + 3$; and substituting $-x$ for x and $-y$ for y gives us $-y = -2x - 3$ or, equivalently, $y = 2x + 3$.   None of the resulting equations are equivalent to the original equation, so there is no symmetry with respect to the y-axis, x-axis, or the origin.

(a) The graphs of the equations in Exercises 5 and 7 are symmetric with respect to the y-axis. (continued)

(b) The graphs of the equations in Exercises 9 and 11 are symmetric with respect to the x-axis.

(c) The graph of the equation in Exercise 13 is symmetric with respect to the origin.

25 $(x+3)^2 + (y-2)^2 = 9$ is a circle of radius $r = \sqrt{9} = 3$ with center $C(-3, 2)$.

See *Figure 25.*

29 $4x^2 + 4y^2 = 25 \Rightarrow x^2 + y^2 = \frac{25}{4}$ is a circle of radius $r = \sqrt{\frac{25}{4}} = \frac{5}{2}$ with center $C(0, 0)$.

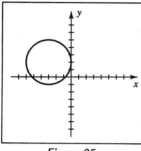

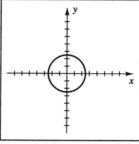

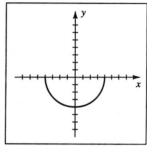

| *Figure 25* | *Figure 29* | *Figure 31* |

31 As in Example 8, $y = -\sqrt{16 - x^2}$ is the lower half of the circle $x^2 + y^2 = 16$.

35 Center $C(2, -3)$, radius 5 $\quad\bullet\quad (x-2)^2 + (y+3)^2 = 5^2 = 25$

37 Center $C(\frac{1}{4}, 0)$, radius $\sqrt{5}$ $\quad\bullet\quad (x-\frac{1}{4})^2 + y^2 = (\sqrt{5})^2 = 5$

39 An equation of a circle with center $C(-4, 6)$ is $(x+4)^2 + (y-6)^2 = r^2$.

Since the circle passes through $P(1, 2)$, we know that $x = 1$ and $y = 2$ is one solution of the general equation. Letting $x = 1$ and $y = 2$ yields $5^2 + (-4)^2 = r^2 \Rightarrow r^2 = 41$.

An equation is $(x+4)^2 + (y-6)^2 = 41$.

41 "Tangent to the y-axis" means that the circle will intersect the y-axis at exactly one point. The distance from the center $C(-3, 6)$ to this point of tangency is 3 units—this is the length of the radius of the circle. An equation is $(x+3)^2 + (y-6)^2 = 9$.

43 Since the radius is 4 and $C(h, k)$ is in QII, $h = -4$ and $k = 4$.

An equation is $(x+4)^2 + (y-4)^2 = 16$.

45 The center of the circle is the midpoint M of $A(4, -3)$ and $B(-2, 7)$. $M = (1, 2)$.

The radius of the circle is $\frac{1}{2} \cdot d(A, B) = \frac{1}{2}\sqrt{136} = \sqrt{34}$.

An equation is $(x-1)^2 + (y-2)^2 = 34$.

47 $x^2 + y^2 - 4x + 6y - 36 = 0$ { complete the square on x and y } $\Rightarrow$

$x^2 - 4x + \underline{4} + y^2 + 6y + \underline{9} = 36 + \underline{4} + \underline{9} \Rightarrow$

$(x-2)^2 + (y+3)^2 = 49.$ $\qquad$ This is a circle with center $C(2, -3)$ and radius $r = 7$.

[51] $2x^2 + 2y^2 - 12x + 4y - 15 = 0$ { add 15 to both sides and divide by 2 } $\Rightarrow$

$x^2 + y^2 - 6x + 2y = \frac{15}{2}$ { complete the square on x and y } $\Rightarrow$

$x^2 - 6x + \underline{\;9\;} + y^2 + 2y + \underline{\;1\;} = \frac{15}{2} + \underline{\;9\;} + \underline{\;1\;} \Rightarrow$

$(x-3)^2 + (y+1)^2 = \frac{35}{2}$.   This is a circle with center $C(3,\,-1)$ and radius $r = \frac{1}{2}\sqrt{70}$.

[53] $x^2 + y^2 + 4x - 2y + 5 = 0 \Rightarrow x^2 + 4x + \underline{\;4\;} + y^2 - 2y + \underline{\;1\;} = -5 + \underline{\;4\;} + \underline{\;1\;} \Rightarrow$

$\qquad\qquad\qquad (x+2)^2 + (y-1)^2 = 0.\;\; C(-2,\,1);\; r = 0$ (a point)

[55] $x^2 + y^2 - 2x - 8y + 19 = 0 \Rightarrow x^2 - 2x + \underline{\;1\;} + y^2 - 8y + \underline{\;16\;} = -19 + \underline{\;1\;} + \underline{\;16\;} \Rightarrow$

$\qquad\qquad (x-1)^2 + (y-4)^2 = -2.$ This is not a circle since r^2 cannot equal -2.

[57] See Example 8 in the text.

[59] To obtain equations for the upper and lower halves, we solve the given equation for y

in terms of x. $(x-2)^2 + (y+1)^2 = 49 \Rightarrow (y+1)^2 = 49 - (x-2)^2 \Rightarrow$

$y + 1 = \pm\sqrt{49 - (x-2)^2} \Rightarrow y = -1 \pm \sqrt{49 - (x-2)^2}.$

The upper half is $y = -1 + \sqrt{49 - (x-2)^2}$ and

the lower half is $y = -1 - \sqrt{49 - (x-2)^2}.$

To obtain equations for the right and left halves, we solve for x in terms of y.

$(x-2)^2 + (y+1)^2 = 49 \Rightarrow (x-2)^2 = 49 - (y+1)^2 \Rightarrow$

$x - 2 = \pm\sqrt{49 - (y+1)^2} \Rightarrow x = 2 \pm \sqrt{49 - (y+1)^2}.$ The right half is

$x = 2 + \sqrt{49 - (y+1)^2}$ and the left half is $x = 2 - \sqrt{49 - (y+1)^2}.$

[61] We need to determine if the distance from P to C is *less than* r, *greater than* r, or

equal to r and hence, P will be *inside* the circle, *outside* the circle, or *on* the circle,

respectively.

(a) $P(2,\,3)$, $C(4,\,6) \Rightarrow d(P,\,C) = \sqrt{4+9} = \sqrt{13} < r$ { $r = 4$ } $\Rightarrow P$ is *inside* C.

(b) $P(4,\,2)$, $C(1,\,-2) \Rightarrow d(P,\,C) = \sqrt{9+16} = 5 = r$ { $r = 5$ } $\Rightarrow P$ is *on* C.

(c) $P(-3,\,5)$, $C(2,\,1) \Rightarrow d(P,\,C) = \sqrt{25+16} = \sqrt{41} > r$ { $r = 6$ } $\Rightarrow P$ is *outside* C.

[63] (a) To find the x-intercepts, let $y = 0$ and solve the resulting equation for x.

$\qquad\qquad\qquad x^2 - 4x + 4 = 0 \Rightarrow (x-2)^2 = 0 \Rightarrow x = 2.$

(b) To find the y-intercepts, let $x = 0$ and solve the resulting equation for y.

$$y^2 - 6y + 4 = 0 \Rightarrow y = \frac{6 \pm \sqrt{36 - 16}}{2} = 3 \pm \sqrt{5}.$$

[65] $x^2 + y^2 + 4x - 6y + 4 = 0 \Leftrightarrow (x+2)^2 + (y-3)^2 = 9.$ This is a circle with

center $C(-2,\,3)$ and radius 3. The circle we want has the same center, $C(-2,\,3)$,

and radius that is equal to the distance from C to $P(2,\,6)$.

$\qquad\quad d(P,\,C) = \sqrt{16+9} = 5$ and an equation is $(x+2)^2 + (y-3)^2 = 25.$

67 The equation of circle C_2 is $(x-h)^2 + (y-2)^2 = 2^2$. If we draw a line from the origin to the center of C_2, we form a right triangle with hypotenuse $5-2$ $\{C_2$ radius $-C_1$ radius$\} = 3$ and sides of length 2 and h. Thus, $h^2 + 2^2 = 3^2 \Rightarrow h = \sqrt{5}$.

69 Assuming that the x- and y-values of each point of intersection are integers, and that the tics each represent one unit, we see that the intersection points are $(-3, -5)$ and $(2, 0)$. The viewing rectangle (VR) is $[-15, 15]$ by $[-10, 10]$.

$$Y_1 < Y_2 \text{ on } [-15, -3) \cup (2, 15].$$

71 VR: $[-5, 5]$ by $[-1.3, 5.3]$. The intersection points are $(-1, 1)$, $(0, 0)$, and $(1, 1)$.

$$Y_1 > Y_2 \text{ on } [-5, -1) \cup (1, 5]. \quad Y_1 < Y_2 \text{ on } (-1, 0) \cup (0, 1).$$

73 The viewing rectangles significantly affect the shape of the circle. Viewing rectangle (2) results in a graph that most looks like a circle.

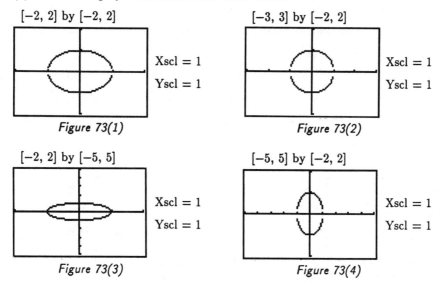

$[-2, 2]$ by $[-2, 2]$ Xscl $= 1$ Yscl $= 1$

Figure 73(1)

$[-3, 3]$ by $[-2, 2]$ Xscl $= 1$ Yscl $= 1$

Figure 73(2)

$[-2, 2]$ by $[-5, 5]$ Xscl $= 1$ Yscl $= 1$

Figure 73(3)

$[-5, 5]$ by $[-2, 2]$ Xscl $= 1$ Yscl $= 1$

Figure 73(4)

75 Assign $x^3 - \frac{9}{10}x^2 - \frac{43}{25}x + \frac{24}{25}$ to Y_1. After trying a standard viewing rectangle, we see that the x-intercepts are near the origin and we choose the viewing rectangle $[-6, 6]$ by $[-4, 4]$. This is simply one choice, not necessarily the best choice. For most $\boxed{c}$ exercises, we have selected viewing rectangles that are in a $3:2$ proportion (horizontal : vertical) to maintain a true proportion. From the graph, there are three x-intercepts. Use a zoom or root feature to determine that they are approximately -1.2, 0.5, and 1.6. See *Figure 75* on the next page.

[−6, 6] by [−4, 4]

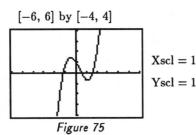

Xscl = 1
Yscl = 1

Figure 75

[−3, 3] by [−2, 2]

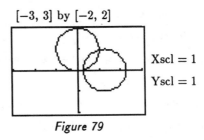

Xscl = 1
Yscl = 1

Figure 79

79 Depending on the type of graphing utility used, you may need to solve for y first.

$$x^2 + (y-1)^2 = 1 \Rightarrow y = 1 \pm \sqrt{1-x^2}; \qquad (x-\tfrac{5}{4})^2 + y^2 = 1 \Rightarrow y = \pm \sqrt{1-(x-\tfrac{5}{4})^2}.$$

Make the assignments $Y_1 = \sqrt{1-x^2}$, $Y_2 = 1 + Y_1$, $Y_3 = 1 - Y_1$, $Y_4 = \sqrt{1-(x-\tfrac{5}{4})^2}$,
and $Y_5 = -Y_4$. If a Y_5 is not available, you will need to use other function
assignments or alternate methods. For example, on the TI-81, you can graph Y_5 by
using DrawF $-Y_4$. Be sure to "turn off" Y_1 before graphing. From the graph, there
are two points of intersection.

They are approximately $(0.999, 0.968)$ and $(0.251, 0.032)$.

81 The cars are initially 4 miles apart. Their distance decreases to 0 when they meet on
the highway after 2 minutes. Then, their distance starts to increase until it is 4 miles
after a total of 4 minutes.

[0, 4] by [0, 4]

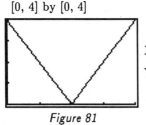

Xscl = 1
Yscl = 1

Figure 81

[−50, 50] by [900, 1200]

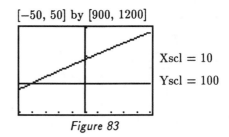

Xscl = 10
Yscl = 100

Figure 83

83 (a) $v = 1087\sqrt{(20+273)/273} \approx 1126$ ft/sec.

(b) Algebraically: $v = 1087\sqrt{\dfrac{T+273}{273}} \Rightarrow 1000 = 1087\sqrt{\dfrac{T+273}{273}} \Rightarrow$

$$T = \frac{1000^2 \times 273}{1087^2} - 273 \approx -42°\,\mathrm{C}.$$

Graphically: Graph $Y_1 = 1087\sqrt{(T+273)/273}$ and $Y_2 = 1000$.

At the point of their intersection, $T \approx -42°\,\mathrm{C}.$

$\boxed{1}$ $A(-3, 2)$, $B(5, -4) \Rightarrow m_{AB} = \dfrac{(-4) - 2}{5 - (-3)} = \dfrac{-6}{8} = -\dfrac{3}{4}$

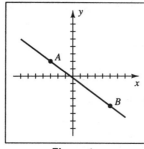

Figure 1

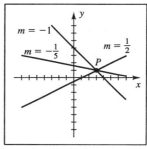

Figure 15

$\boxed{7}$ Show that the slopes of opposite sides are equal.

$A(-3, 1)$, $B(5, 3)$, $C(3, 0)$, $D(-5, -2) \Rightarrow m_{AB} = \frac{1}{4} = m_{DC}$ and $m_{DA} = \frac{3}{2} = m_{CB}$.

$\boxed{9}$ Show that the slopes of opposite sides are equal (parallel lines) and the slopes of two

adjacent sides are negative reciprocals (perpendicular lines). $A(6, 15)$,

$B(11, 12)$, $C(-1, -8)$, $D(-6, -5) \Rightarrow m_{DA} = \frac{5}{3} = m_{CB}$ and $m_{AB} = -\frac{3}{5} = m_{DC}$.

$\boxed{11}$ $A(-1, -3)$ is 5 units to the left and 5 units down from $B(4, 2)$. D will have the

same relative position from $C(-7, 5)$, that is, $(-7 - 5, 5 - 5) = (-12, 0)$.

$\boxed{15}$ $P(3, 1)$; $m = \frac{1}{2}, -1, -\frac{1}{5}$ •   See *Figure 15* above.

$\boxed{19}$ (a) Parallel to the y-axis implies the equation is of the form $x = k$.

The x-value of $A(5, -2)$ is 5, hence $x = 5$ is the equation.

(b) Perpendicular to the y-axis implies the equation is of the form $y = k$.

The y-value of $A(5, -2)$ is -2, hence $y = -2$ is the equation.

$\boxed{21}$ Using the point-slope form, the equation of the line through $A(5, -3)$ with slope -4

is $y + 3 = -4(x - 5) \Rightarrow y + 3 = -4x + 20 \Rightarrow 4x + y = 17$.

$\boxed{23}$ $A(4, 0)$; slope -3 { use the point-slope form of a line } $\Rightarrow$

$y - 0 = -3(x - 4) \Rightarrow y = -3x + 12 \Rightarrow 3x + y = 12$.

$\boxed{25}$ $A(4, -5)$, $B(-3, 6) \Rightarrow m_{AB} = -\frac{11}{7}$.

$y + 5 = -\frac{11}{7}(x - 4) \Rightarrow 7(y + 5) = -11(x - 4) \Rightarrow 7y + 35 = -11x + 44 \Rightarrow 11x + 7y = 9$.

$\boxed{27}$ $5x - 2y = 4 \Leftrightarrow y = \frac{5}{2}x - 2$. Using the same slope, $\frac{5}{2}$, with $A(2, -4)$, gives us

$y + 4 = \frac{5}{2}(x - 2) \Rightarrow 2(y + 4) = 5(x - 2) \Rightarrow 2y + 8 = 5x - 10 \Rightarrow 5x - 2y = 18$.

$\boxed{29}$ $2x - 5y = 8 \Leftrightarrow y = \frac{2}{5}x - \frac{8}{5}$. Using the negative reciprocal of $\frac{2}{5}$ for the slope,

$y + 3 = -\frac{5}{2}(x - 7) \Rightarrow 2(y + 3) = -5(x - 7) \Rightarrow 2y + 6 = -5x + 35 \Rightarrow 5x + 2y = 29$.

$\boxed{31}$ $A(4,\,0)$, $B(0,\,-3) \Rightarrow m = \frac{3}{4}$.

Using the slope-intercept form with $b = -3$ gives us $y = \frac{3}{4}x - 3$.

$\boxed{33}$ $A(5,\,2)$, $B(-1,\,4) \Rightarrow m = -\frac{1}{3}$.

$$y - 2 = -\tfrac{1}{3}(x - 5) \Rightarrow y = -\tfrac{1}{3}x + \tfrac{5}{3} + 2 \Rightarrow y = -\tfrac{1}{3}x + \tfrac{11}{3}.$$

$\boxed{35}$ We need the line through the midpoint of segment AB that is perpendicular to

segment AB. $A(3,\,-1)$, $B(-2,\,6) \Rightarrow M_{AB} = (\frac{1}{2}, \frac{5}{2})$ and $m_{AB} = -\frac{7}{5}$.

$$y - \tfrac{5}{2} = \tfrac{5}{7}(x - \tfrac{1}{2}) \Rightarrow 7(y - \tfrac{5}{2}) = 5(x - \tfrac{1}{2}) \Rightarrow 7y - \tfrac{35}{2} = 5x - \tfrac{5}{2} \Rightarrow 5x - 7y = -15.$$

$\boxed{37}$ An equation of the line with slope -1 through the origin is

$$y - 0 = -1(x - 0),\ \text{or}\ y = -x.$$

$\boxed{39}$ We can solve the given equation for y to obtain the slope-intercept form, $y = mx + b$.

$2x = 15 - 3y \Rightarrow 3y = -2x + 15 \Rightarrow y = -\frac{2}{3}x + 5$; $m = -\frac{2}{3}$, $b = 5$

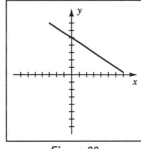

Figure 39

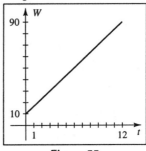

Figure 55

$\boxed{43}$ (a) An equation of the horizontal line with y-intercept 3 is $y = 3$.

(b) An equation of the line through the origin with slope $-\frac{1}{2}$ is $y = -\frac{1}{2}x$.

(c) An equation of the line with slope $-\frac{3}{2}$ and y-intercept 1 is $y = -\frac{3}{2}x + 1$.

(d) An equation of the line through $(3,\,-2)$ with slope -1 is $y + 2 = -(x - 3)$.

Alternatively, we have a slope of -1 and a y-intercept of 1, i.e., $y = -x + 1$.

$\boxed{45}$ Since we want to obtain a "1" on the right side of the equation, we will divide by 6.

$$\left[4x - 2y = 6\right] \cdot \tfrac{1}{6} \Rightarrow \frac{4x}{6} - \frac{2y}{6} = \frac{6}{6} \Rightarrow \frac{2x}{3} - \frac{y}{3} = 1 \Rightarrow \frac{x}{\frac{3}{2}} + \frac{y}{-3} = 1$$

The x-intercept is $(\frac{3}{2},\,0)$ and the y-intercept is $(0,\,-3)$.

$\boxed{47}$ The radius of the circle is the vertical distance from the center of the circle to the line

$y = 5$, that is, $r = 5 - (-2) = 7$. An equation is $(x - 3)^2 + (y + 2)^2 = 49$.

$\boxed{49}$ $L = 28 \Rightarrow 1.53t - 6.7 = 28 \Rightarrow t = \dfrac{28 + 6.7}{1.53} \approx 22.68$, or approximately 23 weeks.

$\boxed{51}$ (a) $L = 40 \Rightarrow W = 1.70(40) - 42.8 = 25.2$ tons

(b) Error in $L = \pm 2 \Rightarrow$ Error in $W = 1.70(\pm 2) = \pm 3.4$ tons

53 (a) $y = mx = \dfrac{\text{change in } y \text{ from the beginning of the season}}{\text{change in } x \text{ from the beginning of the season}}(x) = \dfrac{5 - 0}{14 - 0}x = \dfrac{5}{14}x.$

   (b) $x = 162 \Rightarrow y = \frac{5}{14}(162) \approx 58.$

55 (a) Using the slope-intercept form, $W = mt + b = mt + 10.$

   $W = 30$ when $t = 3 \Rightarrow 30 = 3m + 10 \Rightarrow m = \frac{20}{3}$ and $W = \frac{20}{3}t + 10.$

   (b) $t = 6 \Rightarrow W = \frac{20}{3}(6) + 10 \Rightarrow W = 50$ lb

   (c) $W = 70 \Rightarrow 70 = \frac{20}{3}t + 10 \Rightarrow 60 = \frac{20}{3}t \Rightarrow t = 9$ years old

   (d) The graph has end points at $(0, 10)$ and $(12, 90)$. See *Figure 55* (preceding page).

57 Using $(10, 2480)$ and $(25, 2440)$,

   we have $H - 2440 = \dfrac{2440 - 2480}{25 - 10}(T - 25),$ or $H = -\frac{8}{3}T + \frac{7520}{3}.$

59 (a) Using the slope-intercept form with $m = 0.032$ and $b = 13.5,$

   we have $T = 0.032t + 13.5.$

   (b) $t = 2000 - 1915 = 85 \Rightarrow T = 0.032(85) + 13.5 = 16.22\,°C.$

61 (a) Expenses $(E) = (\$1000) + (5\% \text{ of } R) + (\$2600) + (50\% \text{ of } R) \Rightarrow$

   $E = 1000 + 0.05R + 2600 + 0.50R = 0.55R + 3600.$

   (b) Profit $(P) =$ Revenue $(R) -$ Expenses $(E) \Rightarrow$

   $P = R - (0.55R + 3600) = R - 0.55R - 3600 = 0.45R - 3600.$

   (c) *Break even* means P would be 0. $P = 0 \Rightarrow 0 = 0.45R - 3600 \Rightarrow 0.45R = 3600 \Rightarrow$

   $R = 3600(\frac{100}{45}) = \$8000/\text{month}$

63 The targets are on the x-axis { which is the line $y = 0$ }.

   To determine if a target is hit, set $y = 0$ and solve for $x.$

   (a) $y - 2 = -1(x - 1) \Rightarrow x + y = 3.$ $y = 0 \Rightarrow x = 3$ and a creature is hit.

   (b) $y - \frac{5}{3} = -\frac{4}{9}(x - \frac{3}{2}) \Rightarrow 4x + 9y = 21.$ $y = 0 \Rightarrow x = 5.25$ and no creature is hit.

65 $s = \dfrac{v_2 - v_1}{h_2 - h_1} \Rightarrow 0.07 = \dfrac{v_2 - 22}{185 - 0} \Rightarrow v_2 = 22 + 0.07(185) = 34.95$ mi/hr.

67 The slope of AB is $\dfrac{-1.11905 - (-1.3598)}{-0.55 - (-1.3)} = 0.321.$ Similarly, the slopes of BC and

   CD are also 0.321. Therefore, the points all lie on the same line. Since the common

   slope is 0.321, let $a = 0.321.$ $y = 0.321x + b \Rightarrow -1.3598 = 0.321(-1.3) + b \Rightarrow$

   $b = -0.9425.$ Thus, the points are linearly related by the equation

   $$y = 0.321x - 0.9425.$$

$\boxed{69}$ $x - 3y = -58 \Leftrightarrow y = (x + 58)/3$ and $3x - y = -70 \Leftrightarrow y = 3x + 70$. Assign $(x + 58)/3$ to Y_1 and $3x + 70$ to Y_2. Using a standard viewing rectangle, we don't see the lines. Zooming out gives us an indication where the lines intersect and by tracing and zooming in or using an intersect feature, we find that the lines intersect at $(-19, 13)$.

[−30, 3] by [−2, 20]

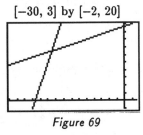

Xscl = 2
Yscl = 2

[−15, 15] by [−10, 10]

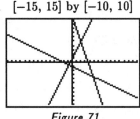

Xscl = 1
Yscl = 1

Figure 69   *Figure 71*

$\boxed{71}$ From the graph, we can see that the points of intersection are $A(-0.8, -0.6)$, $B(4.8, -3.4)$, and $C(2, 5)$. The lines intersecting at A are perpendicular since they have slopes of 2 and $-\frac{1}{2}$. Since $d(A, B) = \sqrt{39.2}$ and $d(A, C) = \sqrt{39.2}$, the triangle is isosceles. Thus, the polygon is a right isosceles triangle.

$\boxed{73}$ The data appear to be linear. Using the two arbitrary points $(-7, -25)$ and $(4.6, 12.2)$, the slope of the line is $\dfrac{12.2 - (-25)}{4.6 - (-7)} \approx 3.2$. An equation of the line is $y + 25 = 3.2(x + 7) \Rightarrow y = 3.2x - 2.6$.

[−8, 5] by [−27, 15]

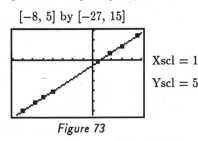

Xscl = 1
Yscl = 5

[1980, 1988] by [300, 625]

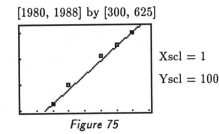

Xscl = 1
Yscl = 100

Figure 73   *Figure 75*

$\boxed{75}$ (a) See *Figure 75*.

(b) To find a first approximation for the line use the arbitrary points $(1982, 325)$ and $(1987, 600)$. The resulting line is $y = 55x - 108{,}685$. Adjustments may be made to this equation.

(c) Let $y = 55x - 108{,}685$. When $x = 1984$, $y = 435$ and when $x = 1995$, $y = 1040$.

3.4 Exercises

$\boxed{3}$ $f(x) = \sqrt{x - 4} - 3x \Rightarrow f(4) = \sqrt{4 - 4} - 3(4) = \sqrt{0} - 12 = 0 - 12 = -12$. Similarly, $f(8) = -22$ and $f(13) = -36$. Note that $f(a)$, with $a < 4$, would be undefined.

⑤ (a) $f(x) = 5x - 2 \Rightarrow f(a) = 5(a) - 2 = 5a - 2$ (b) $f(-a) = 5(-a) - 2 = -5a - 2$

(c) $-f(a) = -1 \cdot (5a - 2) = -5a + 2$ (d) $f(a + h) = 5(a + h) - 2 = 5a + 5h - 2$

(e) $f(a) + f(h) = (5a - 2) + (5h - 2) = 5a + 5h - 4$

(f) $\dfrac{f(a + h) - f(a)}{h} = \dfrac{(5a + 5h - 2) - (5a - 2)}{h} = \dfrac{5h}{h} = 5$

⑦ (a) $f(x) = x^2 - x + 3 \Rightarrow f(a) = (a)^2 - (a) + 3 = a^2 - a + 3$

(b) $f(-a) = (-a)^2 - (-a) + 3 = a^2 + a + 3$

(c) $-f(a) = -1 \cdot (a^2 - a + 3) = -a^2 + a - 3$

(d) $f(a + h) = (a + h)^2 - (a + h) + 3 = a^2 + 2ah + h^2 - a - h + 3$

(e) $f(a) + f(h) = (a^2 - a + 3) + (h^2 - h + 3) = a^2 + h^2 - a - h + 6$

(f) $\dfrac{f(a + h) - f(a)}{h} = \dfrac{(a^2 + 2ah + h^2 - a - h + 3) - (a^2 - a + 3)}{h} = \dfrac{2ah + h^2 - h}{h} =$

$\dfrac{h(2a + h - 1)}{h} = 2a + h - 1$

⑨ $\dfrac{f(x + h) - f(x)}{h} = \dfrac{[(x + h)^2 + 5] - [x^2 + 5]}{h} = \dfrac{(x^2 + 2xh + h^2 + 5) - (x^2 + 5)}{h} =$

$\dfrac{2xh + h^2}{h} = \dfrac{h(2x + h)}{h} = 2x + h$

⑪ $\dfrac{f(x) - f(a)}{x - a} = \dfrac{\sqrt{x - 3} - \sqrt{a - 3}}{x - a} = \dfrac{\sqrt{x - 3} - \sqrt{a - 3}}{x - a} \cdot \dfrac{\sqrt{x - 3} + \sqrt{a - 3}}{\sqrt{x - 3} + \sqrt{a - 3}} =$

$\dfrac{(x - 3) - (a - 3)}{(x - a)(\sqrt{x - 3} + \sqrt{a - 3})} = \dfrac{x - a}{(x - a)(\sqrt{x - 3} + \sqrt{a - 3})} = \dfrac{1}{\sqrt{x - 3} + \sqrt{a - 3}}$

⑬ (a) $g\left(\dfrac{1}{a}\right) = 4\left(\dfrac{1}{a}\right)^2 = \dfrac{4}{a^2}$ (b) $\dfrac{1}{g(a)} = \dfrac{1}{4(a)^2} = \dfrac{1}{4a^2}$

(c) $g(\sqrt{a}) = 4(\sqrt{a})^2 = 4a$ (d) $\sqrt{g(a)} = \sqrt{4a^2} = 2\,|\,a\,| = 2a$ since $a > 0$

⑮ (a) $g(x) = \dfrac{2x}{x^2 + 1} \Rightarrow g\left(\dfrac{1}{a}\right) = \dfrac{2(1/a)}{(1/a)^2 + 1} = \dfrac{2/a}{1/a^2 + 1} \cdot \dfrac{a^2}{a^2} = \dfrac{2a}{1 + a^2} = \dfrac{2a}{a^2 + 1}$

(b) $\dfrac{1}{g(a)} = \dfrac{1}{\dfrac{2a}{a^2 + 1}} = \dfrac{a^2 + 1}{2a}$ (c) $g(\sqrt{a}) = \dfrac{2\sqrt{a}}{(\sqrt{a})^2 + 1} = \dfrac{2\sqrt{a}}{a + 1}$

(d) $\sqrt{g(a)} = \sqrt{\dfrac{2a}{a^2 + 1}} \cdot \dfrac{\sqrt{a^2 + 1}}{\sqrt{a^2 + 1}} = \dfrac{\sqrt{2a(a^2 + 1)}}{a^2 + 1}$, or, equivalently, $\dfrac{\sqrt{2a^3 + 2a}}{a^2 + 1}$

⑰ (a) The domain of a function f is the set of all x-values for which the function is defined. In this case, the graph extends from $x = -3$ to $x = 4$. Hence, the domain is $[-3, 4]$.

(b) The range of a function f is the set of all y-values that the function takes on. In this case, the graph includes all values from $y = -2$ to $y = 2$. Hence, the range is $[-2, 2]$.

(c) $f(1)$ is the y-value of f corresponding to $x = 1$. In this case, $f(1) = 0$.

(d) If we were to draw the horizontal line $y = 1$ on the same coordinate plane, it would intersect the graph at $x = -1$, $\frac{1}{2}$, and 2. Hence, $f(x) = 1 \Rightarrow x = -1, \frac{1}{2}, 2$.

(e) The function is above 1 between $x = -1$ and $x = \frac{1}{2}$, and also to the right of $x = 2$. Hence, $f(x) > 1 \Rightarrow x \in (-1, \frac{1}{2}) \cup (2, 4]$.

$\boxed{19\text{--}30}$ We need to make sure that the radicand { the expression under the radical sign } is greater than or equal to zero and that the denominator is not equal to zero.

$\boxed{21}$ $f(x) = \sqrt{9 - x^2}$ • $9 - x^2 \geq 0 \Rightarrow 3 \geq |x| \Rightarrow -3 \leq x \leq 3$ ★ $[-3, 3]$

$\boxed{23}$ $f(x) = \dfrac{x + 1}{x^3 - 4x}$ • $x^3 - 4x = 0 \Rightarrow x(x + 2)(x - 2) = 0$ ★ $\mathbb{R} - \{\pm 2, 0\}$

$\boxed{25}$ $f(x) = \dfrac{\sqrt{2x - 3}}{x^2 - 5x + 4}$ •   For this function we must have the radicand greater than or equal to 0 *and* the denominator not equal to 0. The radicand is greater than or equal to 0 if $2x - 3 \geq 0$, or, equivalently, $x \geq \frac{3}{2}$. The denominator is $(x - 1)(x - 4)$, so $x \neq 1, 4$. The solution is then all real numbers greater than or equal to $\frac{3}{2}$, excluding 4. In interval notation, we have $[\frac{3}{2}, 4) \cup (4, \infty)$.

$\boxed{27}$ $f(x) = \dfrac{x - 4}{\sqrt{x - 2}}$ • $x - 2 > 0 \Rightarrow x > 2$

{ $>$ must be used since the denominator cannot equal 0 }   ★ $(2, \infty)$

$\boxed{29}$ $f(x) = \sqrt{x + 2} + \sqrt{2 - x}$ • $x + 2 \geq 0 \Rightarrow x \geq -2$; $2 - x \geq 0 \Rightarrow x \leq 2$.

The domain is the intersection of $x \geq -2$ and $x \leq 2$, that is, $[-2, 2]$.

$\boxed{33}$ (a) To sketch the graph of $f(x) = 4 - x^2$, we can make use of the symmetry with respect to the y-axis. See *Figure 33* on the next page.

x	± 4	± 3	± 2	± 1	0
y	-12	-5	0	3	4

(b) Since we can substitute any number for x, the domain is all real numbers, that is, $D = \mathbb{R}$. By examining *Figure 33*, we see that the values of y are at most 4. Hence, the range of f is all reals less than or equal to 4, that is, $R = (-\infty, 4]$.

(c) A common mistake is to confuse the function values, the y's, with the input values, the x's. We are not interested in the specific y-values for determining if the function is increasing, decreasing, or constant. We are only interested if the y-values are going up, going down, or staying the same. For the function $f(x) = 4 - x^2$, we say f *is increasing on* $(-\infty, 0]$ since the y-values are getting larger as we move from left to right over the x-values from $-\infty$ to 0. Also, f *is decreasing on* $[0, \infty)$ since the y-values are getting smaller as we move from left to right over the x-values from 0 to ∞. Note that this answer would have been the same if the function was $f(x) = 500 - x^2$, $f(x) = -300 - x^2$, or any function of the form $f(x) = a - x^2$, where a is any real number.

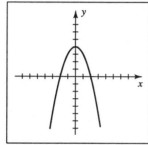

Figure 33

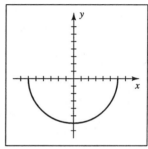

Figure 39

[39] (a) We recognize $y = f(x) = -\sqrt{36 - x^2}$ as the lower half of the circle $x^2 + y^2 = 36$.

(b) To find the domain, we solve $36 - x^2 \geq 0$. $36 - x^2 \geq 0 \Rightarrow x^2 \leq 36 \Rightarrow$

$|x| \leq 6 \Rightarrow D = [-6, 6]$. From *Figure 39*, we see that the y-values vary from

$y = -6$ to $y = 0$. Hence, the range R is $[-6, 0]$.

(c) As we move from left to right, $x = -6$ to $x = 0$, the y-values are decreasing.

From $x = 0$ to $x = 6$, the y-values increase.

Hence, f is decreasing on $[-6, 0]$ and increasing on $[0, 6]$.

[41] As in Example 7, $a = \dfrac{2 - 1}{3 - (-3)} = \dfrac{1}{6}$ and f has the form $f(x) = \frac{1}{6}x + b$.

$f(3) = \frac{1}{6}(3) + b = \frac{1}{2} + b$. But $f(3) = 2$, so $\frac{1}{2} + b = 2 \Rightarrow b = \frac{3}{2}$, and $f(x) = \frac{1}{6}x + \frac{3}{2}$.

[43–52] *Note:* A good question to consider is "Given a particular value of x, can a unique value of y be found?" If the answer is yes, the value of y (general formula) is given. If no, two ordered pairs satisfying the relation having x in the first position are given.

[43] $2y = x^2 + 5 \Rightarrow y = \dfrac{x^2 + 5}{2}$, a function

[45] $x^2 + y^2 = 4 \Rightarrow y^2 = 4 - x^2 \Rightarrow y = \pm\sqrt{4 - x^2}$, not a function, $(0, \pm 2)$

[47] $y = 3$ is a function since for any x,

$(x,\ 3)$ is the only ordered pair in W having x in the first position.

[49] Any ordered pair with x-coordinate 0 satisfies $xy = 0$.

Two such ordered pairs are $(0,\ 0)$ and $(0,\ 1)$. Not a function

[53] $V = lwh = (30 - 2x)(20 - 2x)(x) = 4x(15 - x)(10 - x)$

[55] (a) The formula for the area of a rectangle is $A = lw$ { Area = length $\times$ width }.

$$A = 500 \Rightarrow xy = 500 \Rightarrow y = \frac{500}{x}$$

(b) We need to determine the number of linear feet (P) first. There are two walls of length y, two walls of length $(x - 3)$, and one wall of length x.

Thus, $P = $ Linear feet of wall $= x + 2(y) + 2(x - 3) = 3x + 2\left(\dfrac{500}{x}\right) - 6$.

The cost C is 100 times P, so $C = 100P = 300x + \dfrac{100{,}000}{x} - 600$.

[57] The expression $(h - 25)$ represents the number of feet *above* 25 feet.

$$S(h) = 6(h - 25) + 100 = 6h - 150 + 100 = 6h - 50.$$

[59] (a) Using $(6,\ 48)$ and $(7,\ 50.5)$, we have

$$y - 48 = \frac{50.5 - 48}{7 - 6}(t - 6),\ \text{or}\ y = 2.5t + 33.$$

(b) The slope represents the yearly increase in height, 2.5 in./yr.

(c) $t = 10 \Rightarrow y = 2.5(10) + 33 = 58$ in.

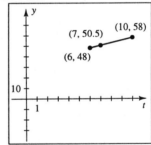

Figure 59

[61] The height of the balloon is $2t$. Using the Pythagorean theorem,

$$d^2 = 100^2 + (2t)^2 \Rightarrow d^2 = 2^2(50)^2 + 2^2 t^2 \Rightarrow d = 2\sqrt{t^2 + 2500}.$$

[63] (a) CTP forms a right angle, so the Pythagorean theorem may be applied.

$$(CT)^2 + (PT)^2 = (PC)^2 \Rightarrow r^2 + y^2 = (h + r)^2 \Rightarrow r^2 + y^2 = h^2 + 2hr + r^2 \Rightarrow$$
$$y^2 = h^2 + 2hr \ \{y > 0\} \Rightarrow y = \sqrt{h^2 + 2hr}$$

(b) $y = \sqrt{(200)^2 + 2(4000)(200)} = \sqrt{(200)^2(1 + 40)} = 200\sqrt{41} \approx 1280.6$ mi

[65] Form a right triangle with the control booth and the beginning of the runway. Let y denote the distance from the control booth to the beginning of the runway and apply the Pythagorean theorem. $y^2 = 300^2 + 20^2 \Rightarrow y^2 = 90{,}400$. Now form a right triangle, in a different plane, with sides y and x and hypotenuse d.

Then $d^2 = y^2 + x^2 \Rightarrow d^2 = 90{,}400 + x^2 \Rightarrow d = \sqrt{90{,}400 + x^2}$.

[67] (b) The maximum y-value of 0.75 occurs when $x \approx 0.55$ and

the minimum y-value of -0.75 occurs when $x \approx -0.55$.

Therefore, the range of f is approximately $[-0.75, 0.75]$.

(c) f is decreasing on $[-2, -0.55]$ and on $[0.55, 2]$. f is increasing on $[-0.55, 0.55]$.

$[-2, 2]$ by $[-2, 2]$ $[-0.7, 1.4]$ by $[-1.1, 1]$

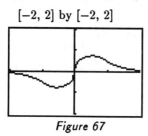   Xscl $= 1$ 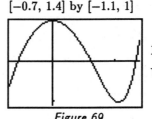   Xscl $= 1$

Yscl $= 1$ Yscl $= 1$

Figure 67   *Figure 69*

[69] (b) The maximum y-value of 1 occurs when $x = 0$ and

the minimum y-value of -1.03 occurs when $x \approx 1.06$.

Therefore, the range of f is approximately $[-1.03, 1]$.

(c) f is decreasing on $[0, 1.06]$. f is increasing on $[-0.7, 0]$ and on $[1.06, 1.4]$.

[71] For each of (a)–(e), an assignment to Y_1, an appropriate viewing rectangle, and the
solution(s) are listed.

(a) $Y_1 = (x\verb|^|5)\verb|^|(1/3)$, VR: $[-40, 40]$ by $[-40, 40]$, $x = 8$

(b) $Y_1 = (x\verb|^|4)\verb|^|(1/3)$, VR: $[-20, 20]$ by $[-20, 20]$, $x = \pm 8$

(c) $Y_1 = (x\verb|^|2)\verb|^|(1/3)$, VR: $[-40, 40]$ by $[-40, 40]$,   no real solutions

(d) $Y_1 = (x\verb|^|3)\verb|^|(1/4)$, VR: $[0, 650]$ by $[0, 650]$, $x = 625$

(e) $Y_1 = (x\verb|^|3)\verb|^|(1/2)$, VR: $[-30, 30]$ by $[-30, 30]$,   no real solutions

[73] (a) $95 \times 63 = 5985$

(b) If a function is graphed in dot mode, only one pixel in each column of pixels on
the screen can be darkened. Therefore, there are at most 95 pixels darkened.
Note: In connected mode this may not be true.

[75] (a) First, we must determine the equation of the line that passes through the points
(1985, 11,450) and (1994, 20,021).

$$y - 11{,}450 = \frac{20{,}021 - 11{,}450}{1994 - 1985}(x - 1985) = \frac{2857}{3}(x - 1985) \Rightarrow$$

$$y = \frac{2857}{3}x - \frac{5{,}636{,}795}{3}. \text{ Thus, let } f(x) = \frac{2857}{3}x - \frac{5{,}636{,}795}{3} \text{ and graph } f.$$

See *Figure 75(a)* on the next page.

(b) The average annual increase in the price paid for a new car is equal to the slope:

$$\frac{2857}{3} \approx \$952.33.$$

(c) Graph $y = \dfrac{2857}{3}x - \dfrac{5,636,795}{3}$ and $y = 25,000$ on the same coordinate axes. Their point of intersection is approximately $(1999.2, 25,000)$. Thus, according to this model, in the year 1999 the average price paid for a new car will be \$25,000.

[1984, 2005] by [10,000, 30,000]

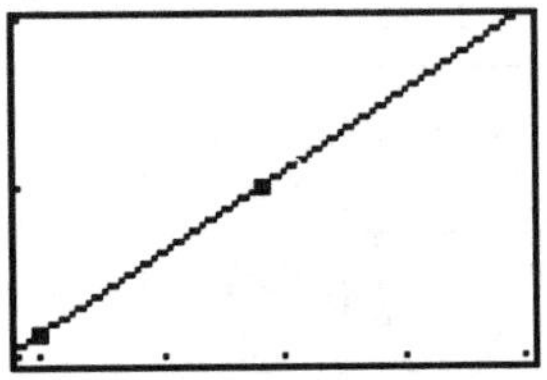

[1984, 2005] by [10,000, 30,000]

Xscl = 5

Yscl = 10,000

Figure 75(a)

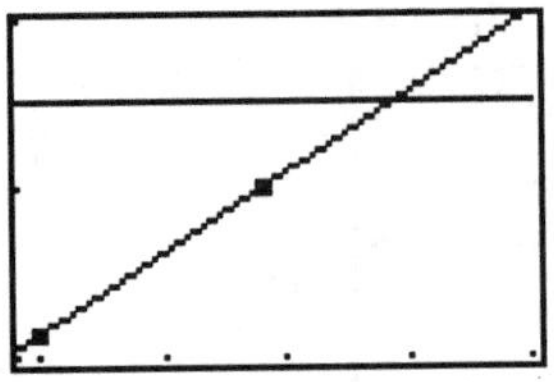

Xscl = 5

Yscl = 10,000

Figure 75(c)

3.5 Exercises

3 $f(x) = 3x^4 + 2x^2 - 5 \Rightarrow f(-x) = 3(-x)^4 + 2(-x)^2 - 5 = 3x^4 + 2x^2 - 5 = f(x)$

Since $f(-x) = f(x)$, f is even and its graph is symmetric with respect to the y-axis.

Note that this means if (a, b) is a point on the graph of f,

then the point $(-a, b)$ is also on the graph.

5 $f(x) = 8x^3 - 3x^2 \Rightarrow f(-x) = 8(-x)^3 - 3(-x)^2 = -8x^3 - 3x^2$

$-f(x) = -1 \cdot f(x) = -1(8x^3 - 3x^2) = -8x^3 + 3x^2$

Since $f(-x) \neq f(x)$ and $f(-x) \neq -f(x)$, f is neither even nor odd.

9 $f(x) = \sqrt[3]{x^3 - x} \Rightarrow f(-x) = \sqrt[3]{(-x)^3 - (-x)} = \sqrt[3]{-x^3 + x} = \sqrt[3]{-1(x^3 - x)} =$

$\sqrt[3]{-1}\sqrt[3]{x^3 - x} = -\sqrt[3]{x^3 - x}; \ -f(x) = -1 \cdot f(x) = -1 \cdot \sqrt[3]{x^3 - x} = -\sqrt[3]{x^3 - x}$

Since $f(-x) = -f(x)$, f is odd and its graph is symmetric with respect to the origin.

Note that this means if (a, b) is a point on the graph of f,

then the point $(-a, -b)$ is also on the graph.

15 $f(x) = 2\sqrt{x} + c$, $c = -3, 0, 2$ • The graph of $y^2 = x$ is shown in Figure 11 in Section 3.2 in the text. The top half of this graph is the graph of the **square root function**, $h(x) = \sqrt{x}$. The second value of c, 0, gives us the graph of $g(x) = 2\sqrt{x}$, which is a vertical stretching of h by a factor of 2. The effect of *adding* -3 and 2 is to vertically shift g down 3 units and up 2 units, respectively. See *Figure 15* on the next page.

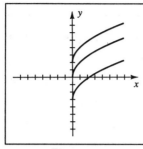

Figure 15

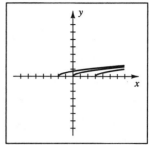

Figure 17

17 $f(x) = \frac{1}{2}\sqrt{x-c}$, $c = -2$, 0, 3 • The graph of $g(x) = \frac{1}{2}\sqrt{x}$ is a vertical compression of the square root function by a factor of $1/(1/2) = 2$. The effect of *subtracting* -2 and 3 from x will be to horizontally shift g left 2 units and right 3 units, respectively. If you forget which way to shift the graph, it is helpful to find the domain of the function. For example, if $h(x) = \sqrt{x-2}$, then $x-2$ must be nonnegative. $x - 2 \geq 0 \Rightarrow x \geq 2$, which also indicates a shift of 2 units to the right.

19 $f(x) = c\sqrt{4 - x^2}$, $c = -2$, 1, 3 • For $c = 1$, the graph of $g(x) = \sqrt{4 - x^2}$ is the upper half of the circle $x^2 + y^2 = 4$. For $c = -2$, reflect g through the x-axis and vertically stretch it by a factor of 2. For $c = 3$, vertically stretch g by a factor of 3.

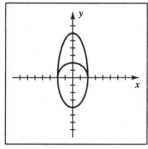

Figure 19

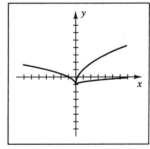

Figure 23

23 $f(x) = \sqrt{cx} - 1$, $c = -1$, $\frac{1}{9}$, 4 • If $c = 1$, then the graph of $g(x) = \sqrt{x} - 1$ is the graph of the square root function vertically shifted down one unit. For $c = -1$, reflect g through the y-axis. For $c = \frac{1}{9}$, horizontally stretch g by a factor $1/(1/9) = 9$ { x-intercept changes from 1 to 9 }. For $c = 4$, horizontally compress g by a factor 4 { x-intercept changes from 1 to $\frac{1}{4}$ }.

[25] To determine what happens to a point P under this transformation, think of how you would evaluate $y = 2f(x-4)+1$ for a particular value of x. You would first subtract 4 from x and then put that value into the function, obtaining a corresponding y-value. Next, you would multiply that y-value by 2 and finally, add 1. Summarizing these steps using the given point P, we have the following:

$$P(3, -2) \quad \{x - 4 \text{ [add 4 to the } x\text{-coordinate]}\} \qquad \to (7, -2)$$
$$\{\times 2 \text{ [multiply the } y\text{-coordinate by 2]}\} \quad \to (7, -4)$$
$$\{+1 \text{ [add 1 to the } y\text{-coordinate]}\} \qquad \to (7, -3)$$

[27] (a) $y = f(x+3)$ •   shift f left 3 units

    (b) $y = f(x-3)$ •   shift f right 3 units

    (c) $y = f(x)+3$ •   shift f up 3 units

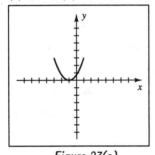
Figure 27(a)

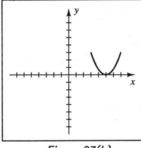
Figure 27(b)

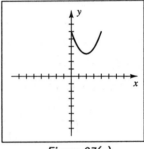
Figure 27(c)

    (d) $y = f(x)-3$ •   shift f down 3 units

    (e) $y = -3f(x)$ •

                  reflect f through the x-axis and vertically stretch it by a factor of 3

    (f) $y = -\frac{1}{3}f(x)$ •   reflect f through the x-axis { the effect of the negative sign in front of $\frac{1}{3}$} and vertically compress it by a factor of $1/(1/3) = 3$

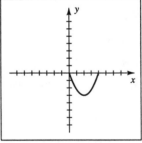
Figure 27(d)

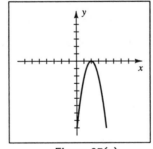
Figure 27(e)

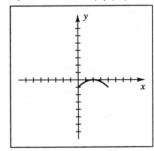
Figure 27(f)

(g) $y = f(-\frac{1}{2}x)$ • reflect f through the y-axis {the effect of the negative sign

inside the parentheses} and horizontally stretch it by a factor of $1/(1/2) = 2$

(h) $y = f(2x)$ • horizontally compress f by a factor of 2

(i) $y = -f(x+2) - 3$ • shift f left 2 units, reflect it about the x-axis,

and then shift it down 3 units

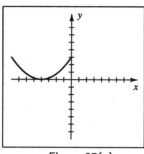

Figure 27(g)   *Figure 27(h)*   *Figure 27(i)*

(j) $y = f(x-2) + 3$ • shift f right 2 units and up 3

(k) $y = |f(x)|$ • since no portion of the graph lies below the x-axis,

the graph is unchanged

(l) $y = f(|x|)$ • include the reflection of the given graph through the y-axis

since all points have positive x-coordinates

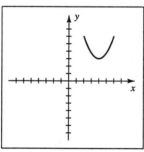
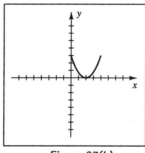
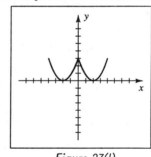

Figure 27(j)   *Figure 27(k)*   *Figure 27(l)*

29 (a) The minimum point on $y = f(x)$ is $(2, -1)$.

On the graph labeled (a), the minimum point is $(-7, 0)$.

It has been shifted left 9 units and up 1. Hence, $y = f(x+9) + 1$.

(b) f is reflected about the x-axis $\Rightarrow y = -f(x)$

(c) f is reflected about the x-axis and shifted left 7 units and down 1 $\Rightarrow$

$$y = -f(x+7) - 1$$

$\boxed{31}$ (a) f is shifted left 4 units $\Rightarrow y = f(x+4)$

(b) f is shifted up 1 unit $\Rightarrow y = f(x)+1$

(c) f is reflected about the y-axis $\Rightarrow y = f(-x)$

$\boxed{33}$ $f(x) = \begin{cases} 3 & \text{if } x \le -1 \\ -2 & \text{if } x > -1 \end{cases}$

We can think of f as 2 functions: If $x \le -1$, then $y = 3$ { include the point $(-1,\,3)$ },

and if $x > -1$, then $y = -2$ { exclude the point $(-1,\,-2)$ }.

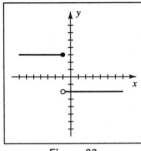

Figure 33

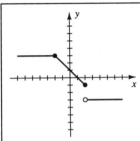

Figure 35

$\boxed{35}$ $f(x) = \begin{cases} 3 & \text{if } x < -2 \\ -x+1 & \text{if } |x| \le 2 \\ -3 & \text{if } x > 2 \end{cases}$

For the second part of the function, we have $|x| \le 2$, or, equivalently, $-2 \le x \le 2$.

On this part of the domain, we want to graph $f(x) = -x+1$, a line with slope -1

and y-intercept 1. Include both end points, $(-2,\,3)$ and $(2,\,-1)$.

$\boxed{37}$ $f(x) = \begin{cases} x+2 & \text{if } x \le -1 \\ x^3 & \text{if } |x| < 1 \\ -x+3 & \text{if } x \ge 1 \end{cases}$

If $x \le -1$, we want the graph of $y = x+2$. To determine the end point of this part

of the graph, merely substitute $x = -1$ in $y = x+2$, obtaining $y = 1$. If $|x| < 1$, or,

equivalently, $-1 < x < 1$, we want the graph of $y = x^3$. We do not include the end

points $(-1,\,-1)$ and $(1,\,1)$. If $x \ge 1$, we want the graph of $y = -x+3$ and include its

end point $(1,\,2)$.

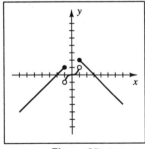

Figure 37

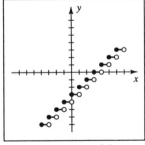

Figure 39(a)

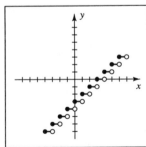

Figure 39(b)

39 (a) $f(x) = [\![x - 3]\!]$ • shift $g(x) = [\![x]\!]$ right 3 units { see figure on page 68 }

(b) $f(x) = [\![x]\!] - 3$ • shift g down 3 units, which is the same graph as in part (a).

(c) $f(x) = 2[\![x]\!]$ • vertically stretch g by a factor of 2

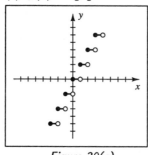
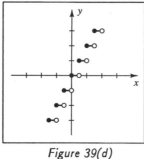
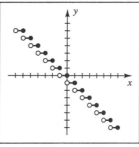

| Figure 39(c) | Figure 39(d) | Figure 39(e) |

(d) $f(x) = [\![2x]\!]$ • horizontally compress g by a factor of 2

Alternatively, we could determine the pattern of "steps" for this function by finding the values of x that make $f(x)$ change from 0 to 1, then from 1 to 2, etc. If $2x = 0$, then $x = 0$, and if $2x = 1$, then $x = \frac{1}{2}$.

Thus, the function will equal 0 from $x = 0$ to $x = \frac{1}{2}$ and then jump to 1 at $x = \frac{1}{2}$. If $2x = 2$, then $x = 1$. The pattern is established: each step will be $\frac{1}{2}$ unit long.

(e) $f(x) = [\![-x]\!]$ • reflect g through the y-axis

41 A question you can ask to help determine if a relationship is a function is "If x is a particular value, can I find a unique y-value?" In this case, if x was 16, then $16 = y^2 \Rightarrow y = \pm 4$. Since we cannot find a unique y-value, this is not a function. Graphically, { see Figure 11 in §3.2 in the text } given any x-value greater than 0, there are two points on the graph and a vertical line intersects the graph in more than one point.

43 Reflect each portion of the graph that is below the x-axis through the x-axis.

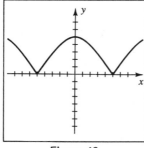

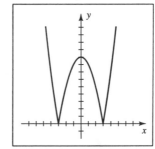

| Figure 43 | Figure 45 |

45 $y = |\,9 - x^2\,|$ • First sketch $y = 9 - x^2$,

then reflect the portions of the graph below the x-axis through the x-axis.

49 (a) For $y = -2f(x)$, multiply the y-coordinates by -2.

The domain { x-coordinates } remains the same. $D = [-2, 6]$, $R = [-16, 8]$

(b) For $y = f(\frac{1}{2}x)$, multiply the x-coordinates by 2.

The range { y-coordinates } remains the same. $D = [-4, 12]$, $R = [-4, 8]$

(c) For $y = f(x - 3) + 1$, add 3 to the x-coordinates and add 1 to the y-coordinates.

$D = [1, 9]$, $R = [-3, 9]$

(d) For $y = f(x + 2) - 3$, subtract 2 from the x-coordinates and subtract 3 from the y-coordinates. $D = [-4, 4]$, $R = [-7, 5]$

(e) For $y = f(-x)$, negate all x-coordinates. $D = [-6, 2]$, $R = [-4, 8]$

(f) For $y = -f(x)$, negate all y-coordinates. $D = [-2, 6]$, $R = [-8, 4]$

(g) $y = f(|x|)$ • Graphically, we can reflect all points with positive x-coordinates through the y-axis, so the domain $[-2, 6]$ becomes $[-6, 6]$. Algebraically, we are replacing x with $|x|$, so $-2 \le x \le 6$ becomes $-2 \le |x| \le 6$, which is equivalent $|x| \le 6$, or, equivalently, $-6 \le x \le 6$. The range stays the same because of the given assumptions: $f(2) = 8$ and $f(6) = -4$; that is, the full range is taken on for $x \ge 0$. Note that the range could not be determined if $f(-2)$ was equal to 8. $D = [-6, 6]$, $R = [-8, 4]$

(h) $y = |f(x)|$ • The points with y-coordinates having values from -4 to 0 will have values from 0 to 4, so the range will be $[0, 8]$. $D = [-2, 6]$, $R = [0, 8]$

51 If $x \le 20,000$, then $T(x) = 0.15x$. If $x > 20,000$, then the tax is 15% of the first 20,000, which is 3000, plus 20% of the amount over 20,000, which is $(x - 20,000)$. We may summarize and simplify as follows:

$$T(x) = \begin{cases} 0.15x & \text{if } x \le 20,000 \\ 3000 + 0.20(x - 20,000) & \text{if } x > 20,000 \end{cases} = \begin{cases} 0.15x & \text{if } x \le 20,000 \\ 0.20x - 1000 & \text{if } x > 20,000 \end{cases}$$

53 The author receives $1.20 on the first 10,000 copies, $1.50 on the next 5000, and $1.80 on each additional copy.

$$R(x) = \begin{cases} 1.20x & \text{if } 0 \le x \le 10,000 \\ 12,000 + 1.50(x - 10,000) & \text{if } 10,000 < x \le 15,000 \\ 19,500 + 1.80(x - 15,000) & \text{if } x > 15,000 \end{cases}$$

$$= \begin{cases} 1.20x & \text{if } 0 \le x \le 10,000 \\ 1.50x - 3000 & \text{if } 10,000 < x \le 15,000 \\ 1.80x - 7500 & \text{if } x > 15,000 \end{cases}$$

$\boxed{57}$ Assign ABS$(1.2x^2 - 10.8)$ to Y_1 and $1.36x + 4.08$ to Y_2. The standard viewing rectangle $[-15, 15]$ by $[-10, 10]$ shows intersection points at approximately 1.87 and 4.13. The solution is $(-\infty, -3) \cup (-3, 1.87) \cup (4.13, \infty)$.

$\boxed{61}$ Since $g(x) = f(\frac{1}{2}x)$, the graph of g can be obtained by stretching the graph of f horizontally by a factor of 2.

[−12, 12] by [−8, 8]

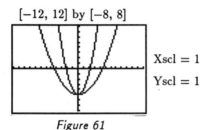

Xscl = 1

Yscl = 1

[−12, 12] by [−8, 8]

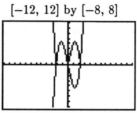

Xscl = 1

Yscl = 1

Figure 61   Figure 63

$\boxed{63}$ Since $g(x) = |f(x)|$, the graph of g is the same as the graph of f if f is non-negative. If $f(x) < 0$, then the graph of f will be reflected about the x–axis.

$\boxed{65}$ (a) Option I gives $C_1 = 4(\$29.95) + \$0.25(500 - 200) = 119.80 + 75.00 = \$194.80$.

        Option II gives $C_2 = 4(\$39.95) + \$0.15(500) = 159.80 + 75.00 = \$234.80$.

(b) Let x represent the mileage. The cost function for Option I is the piecewise linear function

$$C_1(x) = \begin{cases} 119.80 & \text{if } 0 \le x \le 200 \\ 119.80 + 0.25(x - 200) & \text{if } x > 200 \end{cases}$$

        Option II is the linear function $C_2(x) = 159.8 + 0.15x$ for $x \ge 0$.

(c) Let $C_1 = Y_1$ and $C_2 = Y_2$.

        Table $Y_1 = 119.80 + 0.25(x - 200)*(x > 200)$ and $Y_2 = 159.80 + 0.15x$

x	Y_1	Y_2	x	Y_1	Y_2
100	119.8	174.8	700	244.8	264.8
200	119.8	189.8	800	269.8	279.8
300	144.8	204.8	900	294.8	294.8
400	169.8	219.8	1000	319.8	309.8
500	194.8	234.8	1100	344.8	324.8
600	219.8	249.8	1200	369.8	339.8

(d) From the table, we see that the options are equal in cost for $x = 900$ miles. Option I is preferable if $x \in [0, 900)$ and Option II is preferable if $x > 900$.

3.6 Exercises

$\boxed{1}$ We can use the standard equation of a parabola with a vertical axis.

$$V(-3, 1) \Rightarrow y = a[x - (-3)]^2 + 1 \Rightarrow y = a(x + 3)^2 + 1.$$

$\boxed{5}$ $f(x) = -x^2 - 4x - 8$ { given }

$ = -(x^2 + 4x + \underline{\hphantom{0}}) - 8 + \underline{\hphantom{0}}$ { factor out -1 from $-x^2 - 4x$ }

$ = -(x^2 + 4x + \underline{4}) - 8 + \underline{4}$ { complete the square for $x^2 + 4x$ }

$ = -(x + 2)^2 - 4$ { equivalent equation }

$\boxed{7}$ Examples 3 and 4 in the text and Exercise 5 above illustrate a method for expressing $f(x)$ in the standard form. The approach shown here is slightly different in that we *divide both sides by the leading coefficient*—it makes completing the square easier, but you do have to remember to multiply both sides by the same coefficient in the end. Either method is fine.

$f(x) = 2x^2 - 12x + 22$ { given }

$\tfrac{1}{2}f(x) = (x^2 - 6x + \underline{\hphantom{0}}) + 11 - \underline{\hphantom{0}}$ { divide the equation by 2 }

$\phantom{\tfrac{1}{2}f(x)} = (x^2 - 6x + \underline{9}) + 11 - \underline{9}$ { complete the square for $x^2 - 6x$ }

$\phantom{\tfrac{1}{2}f(x)} = (x - 3)^2 + 2$ { equivalent equation }

$f(x) = 2(x - 3)^2 + 4$ { the desired standard form }

$\boxed{11}$ $f(x) = -\tfrac{3}{4}x^2 + 9x - 34$ { divide by $-\tfrac{3}{4}$ and proceed as in Exercise 7 } $\Rightarrow$

$-\tfrac{4}{3}f(x) = x^2 - 12x + \tfrac{136}{3} = x^2 - 12x + \underline{36} + \tfrac{136}{3} - \underline{36} = (x - 6)^2 + \tfrac{28}{3} \Rightarrow$

$$f(x) = (-\tfrac{3}{4})(x - 6)^2 + (-\tfrac{3}{4}) \cdot (\tfrac{28}{3}) \Rightarrow \underline{f(x) = -\tfrac{3}{4}(x - 6)^2 - 7}$$

$\boxed{13}$ (a) $x^2 - 4x = 0$ { $a = 1,\ b = -4,\ c = 0$ } $\Rightarrow x = \dfrac{4 \pm \sqrt{16 - 0}}{2} = 0,\ 4$

(b) $f(x) = x^2 - 4x \Rightarrow -\dfrac{b}{2a} = -\dfrac{-4}{2(1)} = 2.$ $f(2) = -4$ is a minimum since $a > 0$.

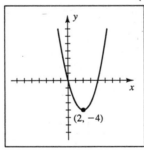

Figure 13

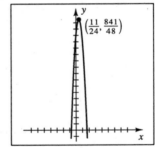

Figure 15

$\boxed{15}$ (a) $f(x) = -12x^2 + 11x + 15 = 0 \Rightarrow x = \dfrac{-11 \pm \sqrt{121 + 720}}{-24} = \dfrac{-11 \pm 29}{-24} = -\dfrac{3}{4},\ \dfrac{5}{3}$

(b) The x-coordinate of the vertex is given by $x = -\dfrac{b}{2a} = -\dfrac{11}{2(-12)} = \dfrac{11}{24}$.

Note that this value is easily obtained from part (a). The y-coordinate of the vertex is then $f(\tfrac{11}{24}) = \tfrac{841}{48} \approx 17.52$. This is a maximum since $a = -12 < 0$.

$\boxed{19}$ (a) $f(x) = x^2 + 4x + 9 = 0 \Rightarrow x = \dfrac{-4 \pm \sqrt{16-36}}{2} = -2 \pm \sqrt{5}\, i.$

The imaginary part indicates that there are *no* x-intercepts.

(b) $-\dfrac{b}{2a} = -\dfrac{4}{2(1)} = -2.$ $f(-2) = 5$ is a minimum since $a = 1 > 0.$

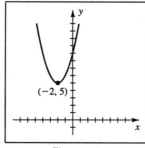

Figure 19

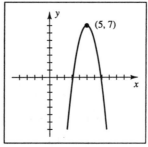

Figure 21

$\boxed{21}$ (a) $-2x^2 + 20x - 43 = 0 \Rightarrow x = \dfrac{-20 \pm \sqrt{400 - 344}}{-4} = 5 \pm \tfrac{1}{2}\sqrt{14} \approx 6.87,\ 3.13$

(b) $f(x) = -2x^2 + 20x - 43 \Rightarrow -\dfrac{b}{2a} = -\dfrac{20}{2(-2)} = 5.$

$f(5) = 7$ is a maximum since $a < 0.$

$\boxed{23}$ $V(4, -1) \Rightarrow y = a(x-4)^2 - 1.$ $x = 0,\ y = 1 \Rightarrow 1 = a(0-4)^2 - 1 \Rightarrow 2 = 16a \Rightarrow a = \tfrac{1}{8}.$

Hence, $y = \tfrac{1}{8}(x-4)^2 - 1.$

$\boxed{25}$ $V(-2, 4) \Rightarrow y = a(x+2)^2 + 4.$ $x = 1,\ y = 0 \Rightarrow 0 = a(1+2)^2 + 4 \Rightarrow -4 = 9a \Rightarrow$

$a = -\tfrac{4}{9}.$ Hence, $y = -\tfrac{4}{9}(x+2)^2 + 4.$

$\boxed{27}$ $V(0, -2) \Rightarrow (h, k) = (0, -2).$ $x = 3,\ y = 25 \Rightarrow 25 = a(3-0)^2 - 2 \Rightarrow$

$27 = 9a \Rightarrow a = 3.$ Hence, $y = 3(x-0)^2 - 2,$ or $y = 3x^2 - 2.$

$\boxed{29}$ $V(3, 5) \Rightarrow y = a(x-3)^2 + 5$ (*). If the x-intercept is 0, then the point $(0, 0)$ is on

the parabola. Substituting $x = 0$ and $y = 0$ into (*) gives us $0 = a(0-3)^2 + 5 \Rightarrow$

$-5 = 9a \Rightarrow a = -\tfrac{5}{9}.$ Hence, $y = -\tfrac{5}{9}(x-3)^2 + 5.$

$\boxed{31}$ Since the x-intercepts are -3 and 5, the x-coordinate of the vertex of the parabola is

1 { the average of the x-intercept values }. Since the highest point has y-coordinate 4,

the vertex is $(1, 4).$ $V(1, 4) \Rightarrow y = a(x-1)^2 + 4.$ We can use the point $(-3, 0)$ since

there is an x-intercept of $-3.$ $x = -3,\ y = 0 \Rightarrow 0 = a(-3-1)^2 + 4 \Rightarrow -4 = 16a \Rightarrow$

$a = -\tfrac{1}{4}.$ Hence, $y = -\tfrac{1}{4}(x-1)^2 + 4.$

$\boxed{33}$ Let d denote the distance between the parabola and the line.

$$d = (\text{parabola}) - (\text{line}) = (-2x^2 + 4x + 3) - (x - 2) = -2x^2 + 3x + 5.$$

This relation is quadratic and the x-value of its maximum value is

$$-\frac{b}{2a} = -\frac{3}{2(-2)} = \frac{3}{4}. \text{ Thus, maximum } d = -2\left(\frac{3}{4}\right)^2 + 3\left(\frac{3}{4}\right) + 5 = \frac{49}{8} = 6.125.$$

Note that the maximum value *of the parabola* is $f(1) = 5$,

which is not the same as the maximum value of the distance d.

Note: We may find the vertex of a parabola in the following problems using either:

(1) the complete the square method,

(2) the formula method, or

(3) the fact that the vertex lies halfway between the x-intercepts.

$\boxed{35}$ The vertex is located at $h = \frac{-b}{2a} = \frac{-2.867}{2(-0.058)} \approx 24.72$ km. Since $a < 0$,

this will produce a maximum value.

$\boxed{37}$ Since the x-intercepts of $y = cx(21 - x)$ are 0 and 21, the maximum will occur

halfway between them, that is, when the infant weighs 10.5 lb.

$\boxed{39}$ (a) $s(t) = -16t^2 + 144t + 100$ will be a maximum when $t = \frac{-b}{2a} = \frac{-144}{2(-16)} = \frac{9}{2}$.

$$s\left(\frac{9}{2}\right) = -16\left(\frac{9}{2}\right)^2 + 144\left(\frac{9}{2}\right) + 100 = -324 + 648 + 100 = 424 \text{ ft.}$$

(b) When $t = 0$, $s(t) = 100$ ft, which is the height of the building.

$\boxed{41}$ Let x and $40 - x$ denote the numbers, and their product P is $x(40 - x)$.

P has zeros at 0 and 40 and is a maximum (since $a < 0$) when $x = \frac{0 + 40}{2} = 20$.

The product will be a maximum when both numbers are 20.

$\boxed{43}$ (a) The 1000 ft of fence is made up of 3 sides of length x and 4 sides of length y.

To express y as a function of x, we need to solve $3x + 4y = 1000$ for y.

$$3x + 4y = 1000 \Rightarrow 4y = 1000 - 3x \Rightarrow y = 250 - \tfrac{3}{4}x.$$

(b) Using the value of y from part (a), $A = xy = x(250 - \tfrac{3}{4}x) = -\tfrac{3}{4}x^2 + 250x.$

(c) A will be a maximum when $x = \frac{-b}{2a} = \frac{-250}{2(-3/4)} = \frac{500}{3} = 166\tfrac{2}{3}$ ft. Using part (a)

to find the corresponding value of y, $y = 250 - \tfrac{3}{4}(\tfrac{500}{3}) = 250 - 125 = 125$ ft.

$\boxed{45}$ The parabola has vertex $V(\tfrac{9}{2}, 3)$. Hence, the equation has the form $y = a(x - \tfrac{9}{2})^2 + 3$.

Using the point $(9, 0)$ { or $(0, 0)$ }, we have $0 = a(9 - \tfrac{9}{2})^2 + 3 \Rightarrow a = -\tfrac{4}{27}$.

Thus, the path may be described by $y = -\tfrac{4}{27}(x - \tfrac{9}{2})^2 + 3.$

47 (a) Since the vertex is at $(0, 10)$, an equation for the parabola is $y = ax^2 + 10$.

The points $(200, 90)$ and $(-200, 90)$ are on the parabola.

Substituting $(200, 90)$ for (x, y) yields $90 = a(200)^2 + 10 \Rightarrow a = \frac{80}{40,000} = \frac{1}{500}$.

Hence, $y = \frac{1}{500}x^2 + 10$.

(b) The cables are spaced 40 ft apart. Using $y = \frac{1}{500}x^2 + 10$ with

$x = 40, 80, 120$, and 160, we get $y = \frac{66}{5}, \frac{114}{5}, \frac{194}{5}$, and $\frac{306}{5}$, respectively.

There is one cable of length 10 ft and 2 cables of each of the other lengths.

Thus, the total length is $10 + 2(\frac{66}{5} + \frac{114}{5} + \frac{194}{5} + \frac{306}{5}) = 282$ ft.

49 An equation describing the doorway is $y = ax^2 + 9$. Since the doorway is 6 feet wide

at the base, $x = 3$ when $y = 0 \Rightarrow 0 = 9a + 9 \Rightarrow a = -1$. Thus, the equation is

$y = -x^2 + 9$. To fit an 8 foot high box through the doorway, we must find x when

$y = 8$. $y = 8 \Rightarrow 8 = -x^2 + 9 \Rightarrow x = \pm 1$. Hence, the box can only be 2 feet wide.

51 Let x denote the number of pairs of shoes that are ordered.

$$A(x) = \begin{cases} 40x & \text{if } x < 50 \\ (40 - 0.04x)x & \text{if } 50 \le x \le 600 \end{cases}$$

The maximum value of the first part of A is $(\$40)(49) = \$1960$. For the second part

of A, $A = -0.04x^2 + 40x$ has a maximum when $x = \frac{-b}{2a} = \frac{-40}{2(-0.04)} = 500$ pairs.

$A(500) = 10,000 > 1960$, so $x = 500$ produces a maximum for both parts of A.

53 (a) Let y denote the number of \$1 decreases in the monthly charge.

$R(y)$ = (# of customers)(monthly charge per customer)

= $(5000 + 500y)(20 - y)$

= $500(10 + y)(20 - y)$

Now let x denote the monthly charge, which is $20 - y$.

R becomes $500[10 + (20 - x)](x) = 500x(30 - x)$.

(b) R has x-intercepts at 0 and 30, and must have its

vertex halfway between them at $x = 15$.

Note that this gives us $y = 5$, and we have

7500 customers for a revenue of \$112,500.

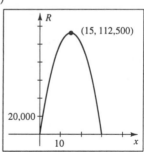

Figure 53

55 From the graph, there are three points of intersection.

   Their coordinates are approximately $(-0.57, 0.64)$, $(0.02, -0.27)$, and $(0.81, -0.41)$.

[−3, 3] by [−2, 2]   [−8, 4] by [−1, 7]

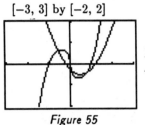   Xscl = 1   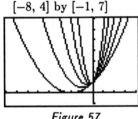   Xscl = 1

                              Yscl = 1   Yscl = 1

      Figure 55   *Figure 57*

57 Since $a > 0$, all parabolas open upward. From the graph, we can see that smaller values of a result in the parabola opening wider while larger values of a result in the parabola becoming narrower.

59 (a) Let January correspond to 1, February to 2, ... , and December to 12.

   (b) Let $f(x) = a(x - h)^2 + k$. The vertex appears to occur near $(7, 0.8)$ Thus, $h = 7$ and $k = 0.8$. Using trial and error, a reasonable value for a is 0.17. Thus, let $f(x) = 0.17(x - 7)^2 + 0.8$.

   (c) $f(4) = 2.33$, compared to the actual value of 2.4 in.

[0.5, 12] by [0, 8]   [−800, 800] by [−100, 200]

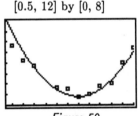   Xscl = 1   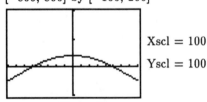   Xscl = 100

                              Yscl = 1   Yscl = 100

      Figure 59   *Figure 61*

61 (a) The equation of the line passing through $A(-800, -48)$ and $B(-500, 0)$ is $y = \frac{4}{25}x + 80$. The equation of the line passing through $D(500, 0)$ and $E(800, -48)$ is $y = -\frac{4}{25}x + 80$. Let $y = a(x - h)^2 + k$ be the equation of the parabola passing through the points $B(-500, 0)$, $C(0, 40)$, and $D(500, 0)$. The vertex is located at $(0, 40)$ so $y = a(x - 0)^2 + 40$. Since $D(500, 0)$ is on the graph, $0 = a(500 - 0)^2 + 40 \Rightarrow a = -\frac{1}{6250}$ and $y = -\frac{1}{6250}x^2 + 40$. Thus, let

$$f(x) = \begin{cases} \frac{4}{25}x + 80 & \text{if } -800 \leq x < -500 \\ -\frac{1}{6250}x^2 + 40 & \text{if } -500 \leq x \leq 500 \\ -\frac{4}{25}x + 80 & \text{if } 500 < x \leq 800 \end{cases}$$

(b) Graph the equations:

$$Y_1 = (4/25*x + 80)/(x < -500),$$

$$Y_2 = (-1/6250*x^2 + 40)/(x \geq -500 \text{ and } x \leq 500),$$

$$Y_3 = (-4/25*x + 80)/(x > 500)$$

63 (a) f must have zeros of 0 and 150. Thus, $f(x) = a(x - 0)(x - 150)$. Also, f will have a maximum of 100 occurring at $x = 75$. (The vertex will be midway between the zeros of f.) $a(75 - 0)(75 - 150) = 100 \Rightarrow a = \frac{100}{(75)(-75)} = -\frac{4}{225}$.
$f(x) = -\frac{4}{225}(x)(x - 150) = -\frac{4}{225}x^2 + \frac{8}{3}x$.

(c) The value of k affects both the distance and the height traveled by the object. The distance and height decrease by a factor of $\frac{1}{k}$ when $k > 1$ and increase by a factor of $\frac{1}{k}$ when $0 < k < 1$.

[0, 180] by [0, 120]

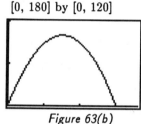

Xscl = 50

Yscl = 50

Figure 63(b)

[0, 600] by [0, 400]

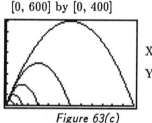

Xscl = 50

Yscl = 50

Figure 63(c)

3.7 Exercises

1   (a) $(f + g)(3) = f(3) + g(3) = 6 + 9 = 15$ (b) $(f - g)(3) = f(3) - g(3) = 6 - 9 = -3$

(c) $(fg)(3) = f(3) \cdot g(3) = 6 \cdot 9 = 54$ (d) $(f/g)(3) = f(3)/g(3) = 6/9 = \frac{2}{3}$

3   (a) $(f + g)(x) = f(x) + g(x) = (x^2 + 2) + (2x^2 - 1) = 3x^2 + 1;$

$(f - g)(x) = f(x) - g(x) = (x^2 + 2) - (2x^2 - 1) = 3 - x^2;$

$(fg)(x) = f(x) \cdot g(x) = (x^2 + 2) \cdot (2x^2 - 1) = 2x^4 + 3x^2 - 2;$

$\left(\frac{f}{g}\right)(x) = \frac{f(x)}{g(x)} = \frac{x^2 + 2}{2x^2 - 1}$

(b) The domain of $f + g$, $f - g$, and fg is the set of all real numbers, $\mathbb{R}$.

(c) The domain of f/g is the same as in (b), except we must exclude the zeros of g.

Hence, the domain of f/g is all real numbers except $\pm\frac{1}{2}\sqrt{2}$.

5   (a) $(f + g)(x) = f(x) + g(x) = \sqrt{x + 5} + \sqrt{x + 5} = 2\sqrt{x + 5};$

$(f - g)(x) = f(x) - g(x) = \sqrt{x + 5} - \sqrt{x + 5} = 0;$

$(fg)(x) = f(x) \cdot g(x) = \sqrt{x + 5} \cdot \sqrt{x + 5} = x + 5; \left(\frac{f}{g}\right)(x) = \frac{f(x)}{g(x)} = \frac{\sqrt{x + 5}}{\sqrt{x + 5}} = 1$

(b) $[-5, \infty)$ (c) $(-5, \infty)$

7̲ (a) $(f+g)(x) = f(x) + g(x) = \dfrac{2x}{x-4} + \dfrac{x}{x+5} = \dfrac{2x(x+5) + x(x-4)}{(x-4)(x+5)} = \dfrac{3x^2 + 6x}{(x-4)(x+5)}$;

$(f-g)(x) = f(x) - g(x) = \dfrac{2x}{x-4} - \dfrac{x}{x+5} = \dfrac{2x(x+5) - x(x-4)}{(x-4)(x+5)} = \dfrac{x^2 + 14x}{(x-4)(x+5)}$;

$(fg)(x) = f(x) \cdot g(x) = \dfrac{2x}{x-4} \cdot \dfrac{x}{x+5} = \dfrac{2x^2}{(x-4)(x+5)}$;

$\left(\dfrac{f}{g}\right)(x) = \dfrac{f(x)}{g(x)} = \dfrac{2x/(x-4)}{x/(x+5)} = \dfrac{2(x+5)}{x-4}$

(b) The domain of f is $\mathbb{R} - \{4\}$ and the domain of g is $\mathbb{R} - \{-5\}$. The intersection of these two domains, $\mathbb{R} - \{-5, 4\}$, is the domain of the three functions.

(c) To determine the domain of the quotient f/g, we also exclude any values that make the denominator g equal to zero. Hence, we exclude $x = 0$ and the domain of the quotient is all real numbers except -5, 0, and 4, that is, $\mathbb{R} - \{-5, 0, 4\}$.

9̲ (a) $(f \circ g)(x) = f(g(x)) = f(-x^2) = 2(-x^2) - 1 = -2x^2 - 1$

(b) $(g \circ f)(x) = g(f(x)) = g(2x - 1) = -(2x - 1)^2 = -(4x^2 - 4x + 1) = -4x^2 + 4x - 1$

(c) $(f \circ f)(x) = f(f(x)) = f(2x - 1) = 2(2x - 1) - 1 = (4x - 2) - 1 = 4x - 3$

(d) $(g \circ g)(x) = g(g(x)) = g(-x^2) = -(-x^2)^2 = -(x^4) = -x^4$

Note: Let $h(x) = (f \circ g)(x) = f(g(x))$ and $k(x) = (g \circ f)(x) = g(f(x))$.

$h(-2)$ and $k(3)$ could be worked two ways, as in Example 3(c) in the text.

11̲ (a) $h(x) = f(3x + 7) = 2(3x + 7) - 5 = 6x + 9$

(b) $k(x) = g(2x - 5) = 3(2x - 5) + 7 = 6x - 8$

(c) Using the result from part (a), $h(-2) = 6(-2) + 9 = -12 + 9 = -3$.

(d) Using the result from part (b), $k(3) = 6(3) - 8 = 18 - 8 = 10$.

15̲ (a) $h(x) = f(2x - 1) = 2(2x - 1)^2 + 3(2x - 1) - 4 = 8x^2 - 2x - 5$

(b) $k(x) = g(2x^2 + 3x - 4) = 2(2x^2 + 3x - 4) - 1 = 4x^2 + 6x - 9$

(c) $h(-2) = 8(-2)^2 - 2(-2) - 5 = 32 + 4 - 5 = 31$

(d) $k(3) = 4(3)^2 + 6(3) - 9 = 36 + 18 - 9 = 45$

19̲ (a) $h(x) = f(-7) = |-7| = 7$ (b) $k(x) = g(|x|) = -7$

(c) $h(-2) = 7$ since h(any value) $= 7$ (d) $k(3) = -7$ since k(any value) $= -7$

21̲ (a) $h(x) = f(\sqrt{x+2}) = (\sqrt{x+2})^2 - 3(\sqrt{x+2}) = x + 2 - 3\sqrt{x+2}$. The domain of $f \circ g$ is the set of all x in the domain of g, $x \geq -2$, such that $g(x)$ is in the domain of f. Since the domain of f is $\mathbb{R}$, any value of $g(x)$ is in its domain.

Thus, the domain is all x such that $x \geq -2$.

(b) $k(x) = g(x^2 - 3x) = \sqrt{(x^2 - 3x) + 2} = \sqrt{x^2 - 3x + 2}$. The domain of $g \circ f$ is the

set of all x in the domain of f, $\mathbf{R}$, such that $f(x)$ is in the domain of g.

Since the domain of g is $x \geq -2$, we must solve $f(x) \geq -2$. $x^2 - 3x \geq -2 \Rightarrow$

$x^2 - 3x + 2 \geq 0 \Rightarrow (x - 1)(x - 2) \geq 0 \Rightarrow x \in (-\infty, 1] \cup [2, \infty)$ { use a sign

diagram }. Thus, the domain is all x such that $x \in (-\infty, 1] \cup [2, \infty)$.

$\boxed{23}$ (a) $h(x) = f(\sqrt{3x}) = (\sqrt{3x})^2 - 4 = 3x - 4$.

Domain of $g = [0, \infty)$. Domain of $f = \mathbf{R}$. Since $g(x)$ is always in the domain of

f, the domain of $f \circ g$ is the same as the domain of g, $[0, \infty)$.

(b) $k(x) = g(x^2 - 4) = \sqrt{3(x^2 - 4)} = \sqrt{3x^2 - 12}$.

Domain of $f = \mathbf{R}$. Domain of $g = [0, \infty)$.

$f(x) \geq 0 \Rightarrow x^2 - 4 \geq 0 \Rightarrow x^2 \geq 4 \Rightarrow |x| \geq 2 \Rightarrow x \in (-\infty, -2] \cup [2, \infty)$.

$\boxed{25}$ (a) $h(x) = f(\sqrt{x + 5}) = \sqrt{\sqrt{x + 5} - 2}$. Domain of $g = [-5, \infty)$. Domain of

$f = [2, \infty)$. $g(x) \geq 2 \Rightarrow \sqrt{x + 5} \geq 2 \Rightarrow x + 5 \geq 4 \Rightarrow x \geq -1$ or $x \in [-1, \infty)$.

(b) $k(x) = g(\sqrt{x - 2}) = \sqrt{\sqrt{x - 2} + 5}$. Domain of $f = [2, \infty)$.

Domain of $g = [-5, \infty)$. $f(x) \geq -5 \Rightarrow \sqrt{x - 2} \geq -5$. This is always true since

the result of a square root is nonnegative. The domain is $[2, \infty)$.

$\boxed{27}$ (a) $h(x) = f(\sqrt{x^2 - 16}) = \sqrt{3 - \sqrt{x^2 - 16}}$. Domain of $g = (-\infty, -4] \cup [4, \infty)$.

Domain of $f = (-\infty, 3]$. $g(x) \leq 3 \Rightarrow \sqrt{x^2 - 16} \leq 3 \Rightarrow x^2 - 16 \leq 9 \Rightarrow x^2 \leq 25 \Rightarrow$

$x \in [-5, 5]$. But $|x| \geq 4$ from the domain of g.

Hence, the domain of $f \circ g$ is $[-5, -4] \cup [4, 5]$.

(b) $k(x) = g(\sqrt{3 - x}) = \sqrt{(\sqrt{3 - x})^2 - 16} = \sqrt{3 - x - 16} = \sqrt{-x - 13}$.

Domain of $f = (-\infty, 3]$. Domain of $g = (-\infty, -4] \cup [4, \infty)$.

$f(x) \geq 4$ { $f(x)$ cannot be less than 0 } $\Rightarrow \sqrt{3 - x} \geq 4 \Rightarrow 3 - x \geq 16 \Rightarrow x \leq -13$.

$\boxed{29}$ (a) $h(x) = f\left(\dfrac{2x - 5}{3}\right) = \dfrac{3\left(\dfrac{2x - 5}{3}\right) + 5}{2} = \dfrac{2x - 5 + 5}{2} = \dfrac{2x}{2} = x$.

Domain of $g = \mathbf{R}$. Domain of $f = \mathbf{R}$. All values of $g(x)$ are in the domain of f.

Hence, the domain of $f \circ g$ is $\mathbf{R}$.

(b) $k(x) = g\left(\dfrac{3x + 5}{2}\right) = \dfrac{2\left(\dfrac{3x + 5}{2}\right) - 5}{3} = \dfrac{3x + 5 - 5}{3} = \dfrac{3x}{3} = x$.

Domain of $f = \mathbf{R}$. Domain of $g = \mathbf{R}$. All values of $f(x)$ are in the domain of g.

Hence, the domain of $g \circ f$ is $\mathbf{R}$.

$\boxed{31}$ (a) $h(x) = f\left(\dfrac{1}{x^3}\right) = \left(\dfrac{1}{x^3}\right)^2 = \dfrac{1}{x^6}$. Domain of $g = \mathbb{R} - \{0\}$. Domain of $f = \mathbb{R}$.

All values of $g(x)$ are in the domain of f. Hence, the domain of $f \circ g$ is $\mathbb{R} - \{0\}$.

(b) $k(x) = g(x^2) = \dfrac{1}{(x^2)^3} = \dfrac{1}{x^6}$. Domain of $f = \mathbb{R}$. Domain of $g = \mathbb{R} - \{0\}$.

All values of $f(x)$ are in the domain of g except for 0.

Since f is 0 when x is 0, the domain of $f \circ g$ is $\mathbb{R} - \{0\}$.

$\boxed{33}$ (a) $h(x) = f\left(\dfrac{x-3}{x-4}\right) = \dfrac{\dfrac{x-3}{x-4} - 1}{\dfrac{x-3}{x-4} - 2} \cdot \dfrac{x-4}{x-4} = \dfrac{x-3-1(x-4)}{x-3-2(x-4)} = \dfrac{1}{5-x}$.

Domain of $g = \mathbb{R} - \{4\}$. Domain of $f = \mathbb{R} - \{2\}$.

$g(x) \neq 2 \Rightarrow \dfrac{x-3}{x-4} \neq 2 \Rightarrow x-3 \neq 2x-8 \Rightarrow x \neq 5$. The domain is $\mathbb{R} - \{4, 5\}$.

(b) $k(x) = g\left(\dfrac{x-1}{x-2}\right) = \dfrac{\dfrac{x-1}{x-2} - 3}{\dfrac{x-1}{x-2} - 4} \cdot \dfrac{x-2}{x-2} = \dfrac{x-1-3(x-2)}{x-1-4(x-2)} = \dfrac{-2x+5}{-3x+7}$.

Domain of $f = \mathbb{R} - \{2\}$. Domain of $g = \mathbb{R} - \{4\}$.

$f(x) \neq 4 \Rightarrow \dfrac{x-1}{x-2} \neq 4 \Rightarrow x-1 \neq 4x-8 \Rightarrow x \neq \frac{7}{3}$. The domain is $\mathbb{R} - \{2, \frac{7}{3}\}$.

$\boxed{35}$ $(f \circ g)(x) = f(g(x)) = f(x+3) = (x+3)^2 - 2$.

$(f \circ g)(x) = 0 \Rightarrow (x+3)^2 - 2 = 0 \Rightarrow (x+3)^2 = 2 \Rightarrow x+3 = \pm\sqrt{2} \Rightarrow x = -3 \pm \sqrt{2}$

$\boxed{37}$ (a) $(f \circ g)(6) = f(g(6)) = f(8) = 5$ (b) $(g \circ f)(6) = g(f(6)) = g(7) = 6$

    (c) $(f \circ f)(6) = f(f(6)) = f(7) = 6$ (d) $(g \circ g)(6) = g(g(6)) = g(8) = 5$

$\boxed{39}$ $(D \circ R)(x) = D(R(x)) = D(20x) =$

$$\sqrt{400 + (20x)^2} = \sqrt{400 + 400x^2} = \sqrt{400(1+x^2)} = 20\sqrt{x^2+1}$$

$\boxed{41}$ We need to examine $(fg)(-x)$—if we obtain $(fg)(x)$, then fg is an even function, whereas if we obtain $-(fg)(x)$, then fg is an odd function.

$(fg)(-x) = f(-x)g(-x)$ {definition of the product of two functions}

         $= -f(x)g(x)$ {f is odd, so $f(-x) = -f(x)$; g is even, so $g(-x) = g(x)$}

         $= -(fg)(x)$ {definition of the product of two functions}

Since $(fg)(-x) = -(fg)(x)$, fg is an odd function.

$\boxed{43}$ (ROUND2 $\circ$ SSTAX)(437.21) $=$ ROUND2(SSTAX(437.21))

                      $=$ ROUND2$(0.0715 \cdot 437.21)$

                      $=$ ROUND2(31.260515) $= 31.26$

$\boxed{45}$ $r = 6t$ and $A = \pi r^2 \Rightarrow A = \pi(6t)^2 = 36\pi t^2$ ft^2.

$\boxed{47}$ $V = \frac{1}{3}\pi r^2 h = \frac{1}{3}\pi r^3 \Rightarrow r = \sqrt[3]{\frac{3V}{\pi}}$. $V = 243\pi t \Rightarrow r = \sqrt[3]{729t} = 9\sqrt[3]{t}$ ft.

$\boxed{49}$ Let l denote the length of the rope. At $t = 0$, $l = 20$.

At time t, $l = 20 + 5t$, *not* just $5t$. We have a right triangle with sides 20, h, and l.

$$h^2 + 20^2 = l^2 \Rightarrow h = \sqrt{(20 + 5t)^2 - 20^2} = \sqrt{25t^2 + 200t} = \sqrt{25(t^2 + 8t)} = 5\sqrt{t^2 + 8t}.$$

$\boxed{51}$ From Exercise 65 of Section 3.4, $d = \sqrt{90{,}400 + x^2}$. The distance x of the plane from the control tower is 500 feet plus 150 feet per second, that is, $x = 500 + 150t$.

$$\text{Thus, } d = \sqrt{90{,}400 + (500 + 150t)^2} = 10\sqrt{225t^2 + 1500t + 3404}.$$

$\boxed{53}$ $y = (x^2 + 3x)^{1/3}$ • Suppose you were to find the value of y if x was equal to 3. Using a calculator, you might compute the value of $x^2 + 3x$ first, and then raise that result to the $\frac{1}{3}$ power. Thus, we would choose $y = u^{1/3}$ and $u = x^2 + 3x$.

$\boxed{59}$ For $y = \dfrac{\sqrt{x + 4} - 2}{\sqrt{x + 4} + 2}$, there is not a "simple" choice for y as in previous exercises.

One choice for u is $u = x + 4$. Then y would be $\dfrac{\sqrt{u} - 2}{\sqrt{u} + 2}$.

Another choice for u is $u = \sqrt{x + 4}$. Then y would be $\dfrac{u - 2}{u + 2}$.

$\boxed{61}$ $(f \circ g)(x) = f(g(x)) = f(x^3 + 1) = \sqrt{x^3 + 1} - 1$. We will multiply this expression by

$\dfrac{\sqrt{x^3 + 1} + 1}{\sqrt{x^3 + 1} + 1}$, treating it as though it was one factor of a difference of two squares.

$$\left(\sqrt{x^3 + 1} - 1\right) \times \frac{\sqrt{x^3 + 1} + 1}{\sqrt{x^3 + 1} + 1} = \frac{\left(\sqrt{x^3 + 1}\right)^2 - 1^2}{\sqrt{x^3 + 1} + 1} = \frac{x^3}{\sqrt{x^3 + 1} + 1}$$

$$\text{Thus, } (f \circ g)(0.0001) = f(g(10^{-4})) \approx \frac{(10^{-4})^3}{2} = 5 \times 10^{-13}.$$

$\boxed{63}$ *Note:* If we relate this problem to the composite function $g(f(u))$, then Y_2 can be considered to be f, Y_1 {which can be considered to be u} is the function of x we are substituting into f, and Y_3 {which can be considered to be g} is accepting Y_2's output values as its input values. Hence, Y_3 is a function of a function of a function.

(a) $y = -2f(x)$; $Y_1 = x$, $Y_2 = 3\sqrt{(Y_1 + 2)(6 - Y_1)} - 4$, graph $Y_3 = -2Y_2$

{turn off Y_1 and Y_2, leaving only Y_3 on}; $D = [-2, 6]$, $R = [-16, 8]$

Note: The graphs often do not show the correct end points—you need to change the viewing rectangle or zoom in to actually view them on the screen.

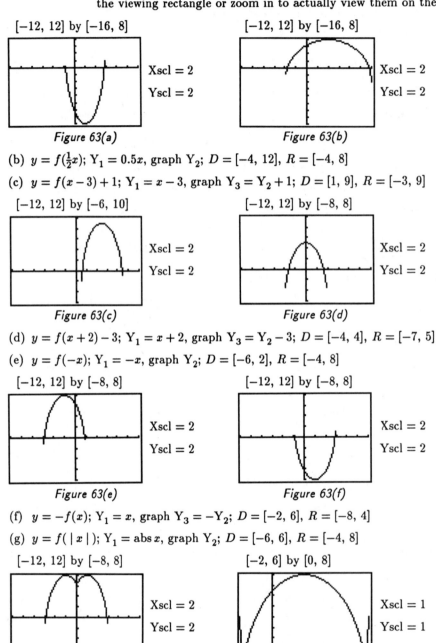

$[-12, 12]$ by $[-16, 8]$

Figure 63(a)

Xscl = 2
Yscl = 2

$[-12, 12]$ by $[-16, 8]$

Figure 63(b)

Xscl = 2
Yscl = 2

(b) $y = f(\frac{1}{2}x)$; $Y_1 = 0.5x$, graph Y_2; $D = [-4, 12]$, $R = [-4, 8]$

(c) $y = f(x-3)+1$; $Y_1 = x-3$, graph $Y_3 = Y_2 + 1$; $D = [1, 9]$, $R = [-3, 9]$

$[-12, 12]$ by $[-6, 10]$

Figure 63(c)

Xscl = 2
Yscl = 2

$[-12, 12]$ by $[-8, 8]$

Figure 63(d)

Xscl = 2
Yscl = 2

(d) $y = f(x+2)-3$; $Y_1 = x+2$, graph $Y_3 = Y_2 - 3$; $D = [-4, 4]$, $R = [-7, 5]$

(e) $y = f(-x)$; $Y_1 = -x$, graph Y_2; $D = [-6, 2]$, $R = [-4, 8]$

$[-12, 12]$ by $[-8, 8]$

Figure 63(e)

Xscl = 2
Yscl = 2

$[-12, 12]$ by $[-8, 8]$

Figure 63(f)

Xscl = 2
Yscl = 2

(f) $y = -f(x)$; $Y_1 = x$, graph $Y_3 = -Y_2$; $D = [-2, 6]$, $R = [-8, 4]$

(g) $y = f(\,|\,x\,|\,)$; $Y_1 = \text{abs}\,x$, graph Y_2; $D = [-6, 6]$, $R = [-4, 8]$

$[-12, 12]$ by $[-8, 8]$

Figure 63(g)

Xscl = 2
Yscl = 2

$[-2, 6]$ by $[0, 8]$

Figure 63(h)

Xscl = 1
Yscl = 1

(h) $y = |\,f(x)\,|$; $Y_1 = x$, graph $Y_3 = \text{abs}\,Y_2$; $D = [-2, 6]$, $R = [0, 8]$

Note: To help determine if you should try to prove the function is one-to-one or look for a counterexample to show that it is not one-to-one, consider the question: "If y was a particular value, could I find a unique x?" If the answer is yes, try to prove the function is one-to-one. Also, consider the Horizontal Line Test listed on page 214 in the text.

1   If y was a particular value, say 5, we would have $5 = 3x - 7$. Trying to solve for x would yield $12 = 3x \Rightarrow x = 4$. Since we could find a unique x, we will try to prove that the function is one-to-one. **Proof** Suppose that $f(a) = f(b)$ for some numbers a and b in the domain. This gives us $3a - 7 = 3b - 7 \Rightarrow 3a = 3b \Rightarrow a = b$. Since $f(a) = f(b)$ implies that $a = b$, we conclude that f is one-to-one.

3   If y was a particular value, say 7, we would have $7 = x^2 - 9$. Trying to solve for x would yield $16 = x^2 \Rightarrow x = \pm 4$. Since we could not find a *unique* x, we will show that two different numbers have the same function value. Using the information already obtained, $f(4) = 7$ and $f(-4) = 7$, but $4 \neq -4$ and hence, f is *not* one-to-one.

5   Suppose $f(a) = f(b)$ with $a, b \geq 0$. $\sqrt{a} = \sqrt{b} \Rightarrow (\sqrt{a})^2 = (\sqrt{b})^2 \Rightarrow a = b$.

f is one-to-one.

9   For $f(x) = \sqrt{4 - x^2}$, $f(-1) = \sqrt{3} = f(1)$. f is *not* one-to-one.

Note: For Exercises 13–16, we need to show that $f(g(x)) = x = g(f(x))$.

15    $f(g(x)) = -(\sqrt{3 - x})^2 + 3 = -(3 - x) + 3 = x$.

$$g(f(x)) = \sqrt{3 - (-x^2 + 3)} = \sqrt{x^2} = |x| = x \ \{\text{since } x \geq 0\}.$$

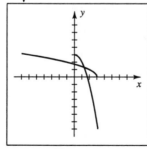

Figure 15

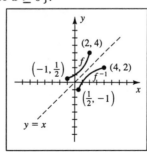

Figure 35

19    $f(x) = \dfrac{1}{3x - 2} \Rightarrow 3xy - 2y = 1 \Rightarrow 3xy = 2y + 1 \Rightarrow x = \dfrac{2y + 1}{3y} \Rightarrow f^{-1}(x) = \dfrac{2x + 1}{3x}$

21    $f(x) = \dfrac{3x + 2}{2x - 5} \Rightarrow 2xy - 5y = 3x + 2 \Rightarrow 2xy - 3x = 5y + 2 \Rightarrow$

$$x(2y - 3) = 5y + 2 \Rightarrow x = \dfrac{5y + 2}{2y - 3} \Rightarrow f^{-1}(x) = \dfrac{5x + 2}{2x - 3}$$

$\boxed{23}$ $f(x) = 2 - 3x^2$, $x \le 0 \Rightarrow y + 3x^2 = 2 \Rightarrow$

$$x^2 = \frac{2-y}{3} \Rightarrow x = \pm\sqrt{\frac{2-y}{3}} \text{ \{ choose minus since } x \le 0 \} \Rightarrow f^{-1}(x) = -\sqrt{\frac{2-x}{3}}$$

$\boxed{27}$ $f(x) = \sqrt{3-x} \Rightarrow y^2 = 3 - x \Rightarrow$

$$x = 3 - y^2 \text{ \{ Since } y \ge 0 \text{ for } f, \ x \ge 0 \text{ for } f^{-1}. \} \Rightarrow f^{-1}(x) = 3 - x^2, \ x \ge 0$$

$\boxed{29}$ $f(x) = \sqrt[3]{x} + 1 \Rightarrow y - 1 = \sqrt[3]{x} \Rightarrow x = (y-1)^3 \Rightarrow f^{-1}(x) = (x-1)^3$

$\boxed{33}$ $f(x) = x^2 - 4 \Rightarrow y = x^2 - 4 \Rightarrow y + 4 = x^2 \Rightarrow x = \pm\sqrt{y+4} \Rightarrow$

$$f^{-1}(x) = \sqrt{x+4} \quad \text{or} \quad f^{-1}(x) = -\sqrt{x+4}.$$

Hence, $f^{-1}(5) = 3$ or $f^{-1}(5) = -3$. Since we are given $f^{-1}(5) = -3$ { and *not* 3 },

$$\text{we choose } f^{-1}(x) = -\sqrt{x+4}.$$

$\boxed{35}$ Remember that the domain of f is the range of f^{-1} and

that the range of f is the domain of f^{-1}. See *Figure 35* on the preceding page.

(b) $D = [-1, 2]$; $R = [\frac{1}{2}, 4]$ (c) $D_1 = R = [\frac{1}{2}, 4]$; $R_1 = D = [-1, 2]$

$\boxed{41}$ (a) $f(x) = -x + b \Rightarrow y = -x + b \Rightarrow x = -y + b$, or $f^{-1}(x) = -x + b$.

(b) $f(x) = \frac{ax+b}{cx-a}$ for $c \ne 0 \Rightarrow y = \frac{ax+b}{cx-a} \Rightarrow cyx - ya = ax + b \Rightarrow$

$$cyx - ax = ay + b \Rightarrow x(cy - a) = ay + b \Rightarrow x = \frac{ay+b}{cy-a}, \text{ or } f^{-1}(x) = \frac{ax+b}{cx-a}.$$

(c) The graph of f is symmetric about the line $y = x$. Thus, $f(x) = f^{-1}(x)$.

$\boxed{43}$ From *Figure 43*, we see that f is always increasing. Thus, f is one-to-one.

[−6, 6] by [−4, 4]   [−12, 12] by [−8, 8]

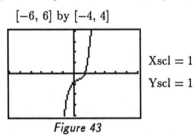
Xscl = 1
Yscl = 1

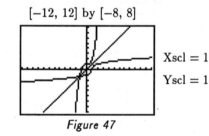
Xscl = 1
Yscl = 1

Figure 43   Figure 47

$\boxed{47}$ The graph of f will be reflected about the line $y = x$.

$$y = \sqrt[3]{x-1} \Rightarrow y^3 = x - 1 \Rightarrow x = y^3 + 1 \Rightarrow f^{-1}(x) = x^3 + 1.$$

$$\text{Graph } Y_1 = \sqrt[3]{x-1}, Y_2 = x^3 + 1, \text{ and } Y_3 = x.$$

$\boxed{49}$ (a) $V(23) = 35(23) = 805 \text{ ft}^3/\text{min}$

(b) $V^{-1}(x) = \frac{1}{35}x$. Given an air circulation of x cubic feet per minute,

$V^{-1}(x)$ computes the maximum number of people that should be in the

restaurant at one time.

(c) $V^{-1}(2350) = \frac{1}{35}(2350) \approx 67.1 \Rightarrow$ the maximum number of people is 67.

$\boxed{1}$ u is directly proportional to $v \Rightarrow u = kv$. If $v = 30$, then $u = 12 \Rightarrow 12 = k(30)$.

Solving for the constant of proportionality, k, we have $12 = 30k \Rightarrow k = \frac{2}{5}$.

$\boxed{3}$ r varies directly as s and inversely as $t \Rightarrow r = k\frac{s}{t}$.

If $s = -2$ and $t = 4$, then $r = 7 \Rightarrow 7 = \dfrac{k(-2)}{4}$.

Solving for the constant of proportionality, k, we have $7 = \dfrac{k(-2)}{4} \Rightarrow k = -14$.

$\boxed{5}$ y is directly proportional to the square of x and inversely proportional to the cube of

$z \Rightarrow y = k\dfrac{x^2}{z^3}$. Substituting $x = 5$, $z = 3$, and $y = 25$ gives us $25 = \dfrac{k(25)}{27}$.

Solving for the constant of proportionality, k, we have $k = 27$.

$\boxed{7}$ z is directly proportional to the product of the square of x and the cube of

$y \Rightarrow z = kx^2y^3$. Substituting $x = 7$, $y = -2$, and $z = 16$ gives us $16 = k(49)(-8)$.

Solving for the constant of proportionality, k, we have $k = -\frac{2}{49}$.

$\boxed{11}$ y is directly proportional to the square root of x and inversely proportional to the

cube of $z \Rightarrow y = k\dfrac{\sqrt{x}}{z^3}$. Substituting $x = 9$, $z = 2$, and $y = 5$ gives us $5 = \dfrac{k(3)}{8}$.

Solving for the constant of proportionality, k, we have $k = \frac{40}{3}$.

$\boxed{13}$ (a) $P = kd$ (b) $118 = k(2) \Rightarrow k = 59$

      (c) $P = 59(5) = 295 \text{ lb/ft}^2$

$\boxed{15}$ (a) $R = k\dfrac{l}{d^2} = \dfrac{kl}{d^2}$ (b) $25 = \dfrac{k(100)}{(0.01)^2} \Rightarrow k = \dfrac{1}{40{,}000}$

      (c) $R = \dfrac{50}{(40{,}000)(0.015)^2} = \dfrac{50}{9}$ ohms

$\boxed{19}$ (a) $T = kd^{3/2}$ (b) $365 = k(93)^{3/2} \Rightarrow k = \dfrac{365}{(93)^{3/2}}$

      (c) $T = \dfrac{365}{(93)^{3/2}} \cdot (67)^{3/2} \approx 223.2$ days

$\boxed{21}$ (a) $V = k\sqrt{L}$ (b) $35 = k\sqrt{50} \Rightarrow k = \frac{7}{2}\sqrt{2}$

      (c) $V = \frac{7}{2}\sqrt{2}(\sqrt{150}) = 35\sqrt{3} \approx 60.6$ mi/hr

$\boxed{23}$ (a) $W = kh^3$ (b) $200 = k(6)^3 \Rightarrow k = \frac{25}{27}$

      (c) 5 feet 6 inches is $\frac{11}{2}$ feet. Hence, $W = \frac{25}{27}(\frac{11}{2})^3 \approx 154.1$ lb, or 154 lb.

$\boxed{25}$ (a) $F = kPr^4 \Rightarrow P = \dfrac{F}{kr^4}$ under normal conditions.

(b) "Normal flow rates triple" means that we will use $3F$ for the flow rate F.

"Radius increases by 10%" means that we will use $(r + 10\%r) = 1.1r$ for the

radius. Hence, $3F = kP(1.1r)^4 \Rightarrow P = \dfrac{3F}{(1.1)^4 kr^4} \approx 2.05\left(\dfrac{F}{kr^4}\right)$,

or about 2.05 times as hard as normal.

$\boxed{27}$ $C = \dfrac{kDE}{Vt} \Rightarrow D = \left(\dfrac{Ct}{k}\right)\dfrac{V}{E}$, where $\dfrac{Ct}{k}$ is constant. If V is twice its original value and

E is 0.8 of its original value (reduced by 20%), then D becomes $D_1 = \left(\dfrac{Ct}{k}\right)\dfrac{2V}{0.8E}$.

Comparing D_1 to D, we have $\dfrac{D_1}{D} = \dfrac{\left(\dfrac{Ct}{k}\right)\dfrac{2V}{0.8E}}{\left(\dfrac{Ct}{k}\right)\dfrac{V}{E}} = \dfrac{2}{0.8} = 2.5 = 250\%$ of its original

value. Thus, D increases by 250%.

$\boxed{29}$ The square of the distance from the origin to the point (x, y) is $x^2 + y^2$. $d = \dfrac{k}{x^2 + y^2}$.

If (x_1, y_1) is the new point that has density d_1, then

$$d_1 = \dfrac{k}{x_1^2 + y_1^2} = \dfrac{k}{(\frac{1}{3}x)^2 + (\frac{1}{3}y)^2} = \dfrac{k}{\frac{1}{9}x^2 + \frac{1}{9}y^2} = \dfrac{k}{\frac{1}{9}(x^2 + y^2)} = 9 \cdot \dfrac{k}{x^2 + y^2} = 9d.$$

The density d is multiplied by 9.

$\boxed{31}$ We need to examine y/x for the given set of data points.

$$\dfrac{y}{x} = \dfrac{0.72}{0.6} = \dfrac{1.44}{1.2} = \dfrac{5.04}{4.2} = \dfrac{8.52}{7.1} = \dfrac{11.16}{9.3} = 1.2 \Rightarrow y = 1.2x.$$

Thus, y varies directly as x with constant of variation $k = 1.2$.

$\boxed{33}$ $x^2y = -10.1$ for each data point.

Thus, y varies inversely as x^2 with constant of variation $k = -10.1$, and $y = -\dfrac{10.1}{x^2}$.

$\boxed{35}$ (a) $D = kS^{2.3} \Rightarrow k = \dfrac{D}{S^{2.3}}$. Using the 6 data points: $\dfrac{33}{20^{2.3}} \approx 0.0336$; $\dfrac{86}{30^{2.3}} \approx 0.0344$;

$\dfrac{167}{40^{2.3}} \approx 0.0345$; $\dfrac{278}{50^{2.3}} \approx 0.0343$; $\dfrac{414}{60^{2.3}} \approx 0.0337$; $\dfrac{593}{70^{2.3}} \approx 0.0338$. Let $k = 0.034$.

Thus, $D = 0.034S^{2.3}$.

(b) Graph the data
together with
$Y_1 = 0.034x^{2.3}$.

[0, 75] by [0, 600]

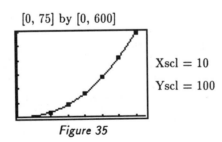

Xscl = 10
Yscl = 100

Figure 35

Chapter 3 Review Exercises

$\boxed{1}$ If $y/x < 0$, then y and x must have opposite signs, and hence,

the set consists of all points in quadrants II and IV.

$\boxed{3}$ (a) $P(-5, 9)$, $Q(-8, -7) \Rightarrow$

$$d(P, Q) = \sqrt{[-8 - (-5)]^2 + (-7 - 9)^2} = \sqrt{9 + 256} = \sqrt{265}.$$

(b) $P(-5, 9)$, $Q(-8, -7) \Rightarrow M_{PQ} = \left(\dfrac{-5 + (-8)}{2}, \dfrac{9 + (-7)}{2}\right) = (-\tfrac{13}{2}, 1).$

(c) Let $R = (x, y)$. $Q = M_{PR} \Rightarrow$

$$(-8, -7) = \left(\frac{-5 + x}{2}, \frac{9 + y}{2}\right) \Rightarrow -8 = \frac{-5 + x}{2} \text{ and } -7 = \frac{9 + y}{2} \Rightarrow$$

$$-5 + x = -16 \text{ and } 9 + y = -14 \Rightarrow x = -11 \text{ and } y = -23 \Rightarrow R = (-11, -23).$$

$\boxed{5}$ With $P(a, 1)$ and $Q(-2, a)$, $d(P, Q) < 3 \Rightarrow \sqrt{(-2 - a)^2 + (a - 1)^2} < 3 \Rightarrow$

$4 + 4a + a^2 + a^2 - 2a + 1 < 9 \Rightarrow 2a^2 + 2a - 4 < 0 \Rightarrow a^2 + a - 2 < 0 \Rightarrow$

$(a + 2)(a - 1) < 0$. Using *Diagram 5*,

we see that $-2 < a < 1$ will assure us that $d(P, Q) < 3$.

Resulting sign:	$\oplus$	$\ominus$	$\oplus$
Sign of $a + 2$:	$-$	$+$	$+$
Sign of $a - 1$:	$-$	$-$	$+$
a values:		-2 $\quad$ 1	

Diagram 5

$\boxed{7}$ The center of the circle is the midpoint of $A(8, 10)$ and $B(-2, -14)$.

$$M_{AB} = \left(\frac{8 + (-2)}{2}, \frac{10 + (-14)}{2}\right) = (3, -2). \text{ The radius of the circle is}$$

$$\tfrac{1}{2} \cdot d(A, B) = \tfrac{1}{2}\sqrt{(-2 - 8)^2 + (-14 - 10)^2} = \tfrac{1}{2}\sqrt{100 + 576} = \tfrac{1}{2} \cdot 26 = 13.$$

An equation is $(x - 3)^2 + (y + 2)^2 = 13^2 = 169.$

$\boxed{11}$ (a) $6x + 2y + 5 = 0 \Leftrightarrow y = -3x - \tfrac{5}{2}$. Using the same slope, -3, with $A(\tfrac{1}{2}, -\tfrac{1}{3})$,

we have $y + \tfrac{1}{3} = -3(x - \tfrac{1}{2}) \Rightarrow 6y + 2 = -18x + 9 \Rightarrow 18x + 6y = 7.$

(b) Using the negative reciprocal of -3 for the slope,

$$y + \tfrac{1}{3} = \tfrac{1}{3}(x - \tfrac{1}{2}) \Rightarrow 6y + 2 = 2x - 1 \Rightarrow 2x - 6y = 3.$$

$\boxed{13}$ The radius of the circle is the distance from the line $x = 4$ to the x-value of the

center $C(-5, -1)$; $r = 4 - (-5) = 9$. An equation is $(x + 5)^2 + (y + 1)^2 = 81.$

$\boxed{16}$ $A(-1, 2)$ and $B(3, -4) \Rightarrow M_{AB} = (1, -1)$ and $m_{AB} = -\tfrac{3}{2}$. We want the

equation of the line through $(1, -1)$ with slope $\tfrac{2}{3}$ { the negative reciprocal of $-\tfrac{3}{2}$ }.

$$y + 1 = \tfrac{2}{3}(x - 1) \Rightarrow 3y + 3 = 2x - 2 \Rightarrow 2x - 3y = 5.$$

$\boxed{17}$ $x^2 + y^2 - 12y + 31 = 0 \Rightarrow x^2 + y^2 - 12y + \underline{36} = -31 + \underline{36} \Rightarrow x^2 + (y-6)^2 = 5.$

$$C(0, 6); \ r = \sqrt{5}$$

$\boxed{19}$ (a) $f(x) = \dfrac{x}{\sqrt{x+3}} \Rightarrow f(1) = \dfrac{1}{\sqrt{4}} = \dfrac{1}{2}$ (b) $f(-1) = -\dfrac{1}{\sqrt{2}}$ (c) $f(0) = \dfrac{0}{\sqrt{3}} = 0$

(d) $f(-x) = \dfrac{-x}{\sqrt{-x+3}} = -\dfrac{x}{\sqrt{3-x}}$ (e) $-f(x) = -1 \cdot f(x) = -\dfrac{x}{\sqrt{x+3}}$

(f) $f(x^2) = \dfrac{x^2}{\sqrt{x^2+3}}$ (g) $[f(x)]^2 = \left(\dfrac{x}{\sqrt{x+3}}\right)^2 = \dfrac{x^2}{x+3}$

$\boxed{21}$ $f(x) = \dfrac{-2(x^2 - 20)(5-x)}{(6-x^2)^{4/3}} \Rightarrow f(4) = \dfrac{(-)(-)(+)}{(+)} = +.$ $f(4)$ is positive.

Note that the factor -2 always contributes one negative sign to the quotient and that the denominator is always positive if $x \neq \pm\sqrt{6}$.

$\boxed{22}$ (a) $3x - 4 \geq 0 \Rightarrow x \geq \frac{4}{3}; \ D = [\frac{4}{3}, \infty).$

Since y is the result of a square root, $y \geq 0$; $R = [0, \infty)$.

(b) $D = $ All real numbers except -3.

Since y is the square of the nonzero term $\dfrac{1}{x+3}$, $y > 0$; $R = (0, \infty)$.

$\boxed{24}$ $\dfrac{f(a+h) - f(a)}{h} = \dfrac{\dfrac{1}{a+h+2} - \dfrac{1}{a+2}}{h} = \dfrac{\dfrac{(a+2) - (a+h+2)}{(a+h+2)(a+2)}}{h} = \dfrac{-h}{(a+h+2)(a+2)h} =$

$$-\dfrac{1}{(a+h+2)(a+2)}$$

$\boxed{25}$ $f(x) = ax + b$ is the desired form. $a = $ slope $= \dfrac{7-2}{3-1} = \dfrac{5}{2}.$ $f(x) = \frac{5}{2}x + b \Rightarrow$

$f(1) = \frac{5}{2} + b$, but $f(1) = 2$, so $\frac{5}{2} + b = 2$, and $b = -\frac{1}{2}$. Thus, $f(x) = \frac{5}{2}x - \frac{1}{2}.$

$\boxed{26}$ (a) $f(x) = \sqrt[3]{x^3 + 4x} \Rightarrow f(-x) = \sqrt[3]{(-x)^3 + 4(-x)} = \sqrt[3]{-1(x^3 + 4x)} = -\sqrt[3]{x^3 + 4x} =$

$$-f(x), \ f \text{ is odd}$$

(b) $f(x) = \sqrt[3]{3x^2 - x^3} \Rightarrow f(-x) = \sqrt[3]{3(-x)^2 - (-x)^3} = \sqrt[3]{3x^2 + x^3} \neq \pm f(x),$

$$f \text{ is neither even nor odd}$$

(c) $f(x) = \sqrt[3]{x^4 + 3x^2 + 5} \Rightarrow f(-x) = \sqrt[3]{(-x)^4 + 3(-x)^2 + 5} = \sqrt[3]{x^4 + 3x^2 + 5} = f(x),$

$$f \text{ is even}$$

$\boxed{29}$ $2y + 5x - 8 = 0 \Leftrightarrow y = -\frac{5}{2}x + 4$, a line with slope $-\frac{5}{2}$ and y-intercept 4;

x-intercept: (1.6, 0) See *Figure 29* on the next page.

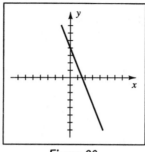

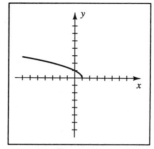

Figure 29   Figure 33

33 $y = \sqrt{1-x}$ • The radicand must be nonnegative for the radical to be defined.

$1 - x \geq 0 \Rightarrow 1 \geq x$ or $x \leq 1$. The domain is $(-\infty, 1]$ and the range is $[0, \infty)$.

x-intercept: $(1, 0)$, y-intercept: $(0, 1)$

37 $x^2 + y^2 - 8x = 0 \Leftrightarrow x^2 - 8x + \underline{16} + y^2 = \underline{16} \Leftrightarrow (x-4)^2 + y^2 = 16$;

$C(4, 0)$, $r = \sqrt{16} = 4$; x-intercepts: $(0, 0)$ and $(8, 0)$, y-intercept: $(0, 0)$

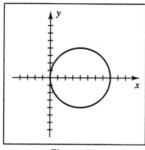

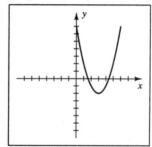

Figure 37   Figure 39

39 $y = (x-3)^2 - 2$ has vertex $(3, -2)$; x-intercepts: $(3 \pm \sqrt{2}, 0)$, y-intercept: $(0, 7)$

43 (a) The graph of $f(x) = |x+3|$ can be thought of as

the graph of $g(x) = |x|$ shifted left 3 units.

(b) The function is defined for all x, so the domain D is the set of all real numbers.

The range is the set of all nonnegative numbers, that is, $R = [0, \infty)$.

(c) The function f is decreasing on $(-\infty, -3]$ and is increasing on $[-3, \infty)$.

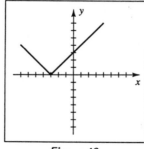

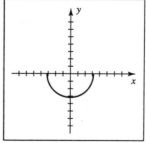

Figure 43   Figure 44

44 (b) $D = [-\sqrt{10}, \sqrt{10}]$; $R = [-\sqrt{10}, 0]$

(c) Decreasing on $[-\sqrt{10}, 0]$, increasing on $[0, \sqrt{10}]$

46 (b) $D = (-\infty, 2]$; $R = [0, \infty)$ (c) Decreasing on $(-\infty, 2]$

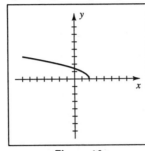

Figure 46

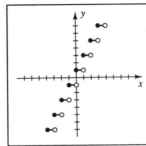

Figure 50

50 (a) The "2" in front of $[\![x]\!]$ has the effect of doubling all the y-values of $g(x) = [\![x]\!]$. The "+1" has the effect of vertically shifting the graph of $h(x) = 2[\![x]\!]$ up 1 unit.

(b) The function is defined for all x, so the domain D is the set of all real numbers. The range of $g(x) = [\![x]\!]$ is the set of integers, that is, $\{\ldots, -2, -1, 0, 1, 2, \ldots\}$. The range of $h(x) = 2[\![x]\!]$ is the set of even integers since we are doubling the values of g—that is, $\{\ldots, -4, -2, 0, 2, 4, \ldots\}$. Since $f(x) = 1 + h(x)$, the range R of f is $\{\ldots, -3, -1, 1, 3, \ldots\}$.

(c) The function f is constant on intervals such as $[0, 1)$, $[1, 2)$, and $[2, 3)$. In general, f is constant on $[n, n+1)$, where n is any integer.

52 (a) $y = f(x - 2)$ •   shift f right 2 units

(b) $y = f(x) - 2$ •   shift f down 2 units

(c) $y = f(-x)$ •   reflect f through the y-axis

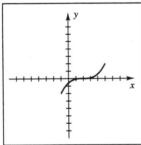

Figure 52(a)

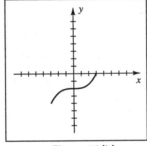

Figure 52(b)

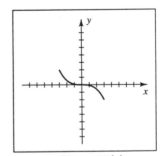

Figure 52(c)

(d) $y = f(2x)$ • horizontally compress f by a factor of 2

(e) $y = f(\frac{1}{2}x)$ • horizontally stretch f by a factor of $1/(1/2) = 2$

(f) $y = f^{-1}(x)$ • reflect f through the line $y = x$

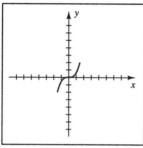

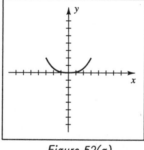

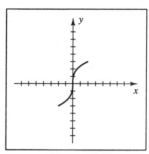

| Figure 52(d) | Figure 52(e) | Figure 52(f) |

(g) $y = |f(x)|$ •

reflect the portion of the graph below the x-axis through the x-axis.

(h) $y = f(|x|)$ • include the reflection of all points with positive x-coordinates

through the y-axis—results in the same graph as in part (g).

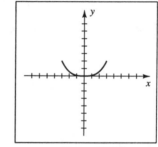

| Figure 52(g) | Figure 52(h) |

53 Using the intercept form of a line with x-intercept 5 and y-intercept -2, an equation

is $\frac{x}{5} + \frac{y}{-2} = 1$. Multiplying by 10 gives us $\left[\frac{x}{5} + \frac{y}{-2} = 1\right] \cdot 10$, or, equivalently,

$$2x - 5y = 10.$$

56 The graph could be made by taking the graph of $y = |x|$, reflecting it through the x-axis $\{y = -|x|\}$, then shifting that graph to the right 2 units $\{y = -|x - 2|\}$, and then shifting that graph down 1 unit, resulting in the graph of the equation $y = -|x - 2| - 1$.

57 $f(x) = 5x^2 + 30x + 49 \Rightarrow -\frac{b}{2a} = -\frac{30}{2(5)} = -3$. $f(-3) = 4$ is a minimum since $a > 0$.

60 $f(x) = 3(x + 2)(x - 10)$ has x-intercepts at -2 and 10. The vertex is halfway

between them at $x = 4$. $f(4) = 3 \cdot 6 \cdot (-6) = -108$ is a minimum.

[62] $V(3, -2) \Rightarrow (h, k) = (3, -2)$ in $y = a(x - h)^2 + k$.

$x = 5, y = 4 \Rightarrow 4 = a(5 - 3)^2 - 2 \Rightarrow 6 = 4a \Rightarrow a = \frac{3}{2}$. Hence, $y = \frac{3}{2}(x - 3)^2 - 2$.

[63] The domain of $f(x) = \sqrt{4 - x^2}$ is $[-2, 2]$. The domain of $g(x) = \sqrt{x}$ is $[0, \infty)$.

(a) The domain of fg is the intersection of those two domains, $[0, 2]$.

(b) The domain of f/g is the same as that of fg,

excluding any values that make g equal to 0. Thus, the domain of f/g is $(0, 2]$.

[66] (a) $(f \circ g)(x) = f(g(x)) = \sqrt{3\left(\dfrac{1}{x^2}\right) + 2} = \sqrt{\dfrac{3 + 2x^2}{x^2}}$

(b) $(g \circ f)(x) = g(f(x)) = \dfrac{1}{(\sqrt{3x + 2})^2} = \dfrac{1}{3x + 2}$

[67] (a) $h(x) = f(\sqrt{x - 3}) = \sqrt{25 - (\sqrt{x - 3})^2} = \sqrt{25 - (x - 3)} = \sqrt{28 - x}$.

Domain of $g = [3, \infty)$. Domain of $f = [-5, 5]$.

$g(x) \le 5$ { $g(x)$ cannot be less than 0 } $\Rightarrow \sqrt{x - 3} \le 5 \Rightarrow$

$x - 3 \le 25 \Rightarrow x \le 28$. $[3, \infty) \cap (-\infty, 28] = [3, 28]$

(b) $k(x) = g(\sqrt{25 - x^2}) = \sqrt{\sqrt{25 - x^2} - 3}$.

Domain of $f = [-5, 5]$. Domain of $g = [3, \infty)$.

$f(x) \ge 3 \Rightarrow \sqrt{25 - x^2} \ge 3 \Rightarrow 25 - x^2 \ge 9 \Rightarrow x^2 \le 16 \Rightarrow x \in [-4, 4]$.

[70] Suppose $f(a) = f(b)$. $2a^3 - 5 = 2b^3 - 5 \Rightarrow 2a^3 = 2b^3 \Rightarrow a^3 = b^3 \Rightarrow a = b$.

Thus, f is a one-to-one function.

[72] $f(x) = 9 - 2x^2, \ x \le 0 \Rightarrow$

$y + 2x^2 = 9 \Rightarrow x^2 = \dfrac{9 - y}{2} \Rightarrow$

$x = \pm\sqrt{\dfrac{9 - y}{2}}$ { choose minus since $x \le 0$ } $\Rightarrow$

$f^{-1}(x) = -\sqrt{\dfrac{9 - x}{2}}$

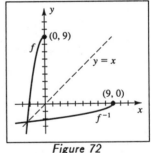

Figure 72

[73] (a) $f(1) = 2$ (b) $(f \circ f)(1) = f(f(1)) = f(2) = 4$

(c) $f(2) = 4$ and f is one-to-one $\Rightarrow f^{-1}(4) = 2$

(d) $f(x) = 4 \Rightarrow x = 2$ (e) $f(x) > 4 \Rightarrow x > 2$

[74] Since f and g are one-to-one functions, we know that $f(2) = 7$, $f(4) = 2$, and

$g(2) = 5$ imply that $f^{-1}(7) = 2$, $f^{-1}(2) = 4$, and $g^{-1}(5) = 2$, respectively.

(a) $(g \circ f^{-1})(7) = g(f^{-1}(7)) = g(2) = 5$ (b) $(f \circ g^{-1})(5) = f(g^{-1}(5)) = f(2) = 7$

(c) $(f^{-1} \circ g^{-1})(5) = f^{-1}(g^{-1}(5)) = f^{-1}(2) = 4$

(d) $(g^{-1} \circ f^{-1})(2) = g^{-1}(f^{-1}(2)) = g^{-1}(4)$, which is not known.

76 The slope of the ramp should be between $\frac{1}{12}$ and $\frac{1}{20}$. If the rise of the ramp is 3 feet, then the run should be between $3 \times 12 = 36$ ft and $3 \times 20 = 60$ ft. The range of the ramp lengths should be from $L = \sqrt{3^2 + 36^2} \approx 36.1$ ft to $L = \sqrt{3^2 + 60^2} \approx 60.1$ ft.

78 (a) $V = at + b$ is the desired form. $V = 89{,}000$ when $t = 0 \Rightarrow V = at + 89{,}000$.

$V = 125{,}000$ when $t = 6 \Rightarrow 125{,}000 = 6a + 89{,}000 \Rightarrow a = \frac{36{,}000}{6} = 6000$ and hence,

$$V = 6000t + 89{,}000.$$

(b) $V = 103{,}000 \Rightarrow 103{,}000 = 6000t + 89{,}000 \Rightarrow t = \frac{7}{3}$, or $2\frac{1}{3}$.

80 (a) $C_1(x) = \left(1.25 \, \frac{\text{dollars}}{\text{gallon}}\right) \div \left(20 \, \frac{\text{miles}}{\text{gallon}}\right) \cdot x \text{ miles} = \frac{1.25}{20}x = 0.0625x$, or $\frac{1}{16}x$.

(b) After the tune-up, the gasoline mileage will be 10% more than 20 mi/gal, that is,

$$22 \text{ mi/gal.} \quad C_2(x) = \frac{1.25}{22}x + 50 = \frac{5}{88}x + 50 \approx 0.0568x + 50.$$

(c) $C_2 < C_1 \Rightarrow \left[\frac{5}{88}x + 50 < \frac{1}{16}x\right] \cdot 16(11) \Rightarrow 10x + 8800 < 11x \Rightarrow x > 8800$ miles.

82 $V = \pi r^2 h$ and $V = 24\pi \Rightarrow h = \frac{24}{r^2}$. $S = \pi r^2 + 2\pi r h = \pi r^2 + 2\pi r \cdot \frac{24}{r^2} = \pi r^2 + \frac{48\pi}{r}$.

$$C = (0.30)(\pi r^2) + (0.10)(\frac{48\pi}{r}) = \frac{3\pi r^2}{10} + \frac{48\pi}{10r} = \frac{3\pi(r^3 + 16)}{10r}.$$

83 (a) $V = (10 \text{ ft}^3 \text{ per minute})(t \text{ minutes}) = 10t$

(b) The height and length of the bottom triangular region are in the proportion 6–60, or 1–10, and the length is 10 times the height. When $0 \le h \le 6$, the volume is

$V = (\text{cross sectional area})(\text{pool width}) = \frac{1}{2}bh(40) = \frac{1}{2}(10h)(h)(40) = 200h^2 \text{ ft}^3$.

When $6 < h \le 9$, the triangular region is full and

$$V = 200(6)^2 + (h - 6)(80)(40) = 7200 + 3200(h - 6).$$

(c) $10t = 200h^2 \Rightarrow h = \sqrt{t/20}$;

$$0 \le h \le 6 \Rightarrow 0 \le \sqrt{t/20} \le 6 \Rightarrow 0 \le t/20 \le 36 \Rightarrow 0 \le t \le 720.$$

$10t = 7200 + 3200(h - 6) \Rightarrow h - 6 = \frac{t - 720}{320} \Rightarrow h = 6 + \frac{t - 720}{320}; \ 6 < h \le 9 \Rightarrow$

$$6 < 6 + \frac{t - 720}{320} \le 9 \Rightarrow 0 < \frac{t - 720}{320} \le 3 \Rightarrow 0 < t - 720 \le 960 \Rightarrow 720 < t \le 1680.$$

84 (a) Using similar triangles, $\frac{r}{x} = \frac{2}{4} \Rightarrow r = \frac{1}{2}x$.

(b) $\text{Volume}_{\text{cone}} + \text{Volume}_{\text{cup}} = \text{Volume}_{\text{total}} \Rightarrow \frac{1}{3}\pi r^2 h + \pi r^2 h = 5 \Rightarrow$

$$\tfrac{1}{3}\pi(\tfrac{1}{2}x)^2(x) + \pi(2)^2(y) = 5 \Rightarrow 5 - \tfrac{\pi}{12}x^3 = 4\pi y \Rightarrow y = \tfrac{5}{4\pi} - \tfrac{1}{48}x^3$$

$\boxed{85}$ (a) $\frac{y}{b} = \frac{y+h}{a} \Rightarrow ay = by + bh \Rightarrow y(a-b) = bh \Rightarrow y = \frac{bh}{a-b}$

(b) $V = \frac{1}{3}\pi a^2(y+h) - \frac{1}{3}\pi b^2 y = \frac{\pi}{3}\left[(a^2-b^2)y + a^2 h\right] =$

$$\frac{\pi}{3}\left[(a^2-b^2)\frac{bh}{a-b} + a^2 h\right] = \frac{\pi}{3}h\left[(a+b)b + a^2\right] = \frac{\pi}{3}h(a^2 + ab + b^2)$$

(c) $a = 6,\ b = 3,\ V = 600 \Rightarrow \frac{\pi}{3}h(6^2 + 6\cdot 3 + 3^2) = 600 \Rightarrow h = \frac{1800}{63\pi} = \frac{200}{7\pi} \approx 9.1$ ft

$\boxed{86}$ Let t denote the time (in hr) after 1:00 P.M. If the starting point for ship B is the origin, then the locations of A and B are $-30 + 15t$ and $-10t$, respectively. Using the Pythagorean theorem, $d^2 = (-30 + 15t)^2 + (-10t)^2 = 325t^2 - 900t + 900$. The time at which the distance between the ships is minimal is the same as the time at which the square of the distance between the ships is minimal.

Thus, $t = -\frac{b}{2a} = -\frac{-900}{2(325)} = \frac{18}{13}$, or about 2:23 P.M.

$\boxed{87}$ Let r denote the radius of the semicircles and x the length of the rectangle.

Perimeter = half-mile $\Rightarrow 2x + 2\pi r = \frac{1}{2} \Rightarrow x = -\pi r + \frac{1}{4}$.

$A = 2rx = 2r(-\pi r + \frac{1}{4}) = -2\pi r^2 + \frac{1}{2}r$. The maximum value of A occurs when

$$r = -\frac{b}{2a} = -\frac{1/2}{2(-2\pi)} = \frac{1}{8\pi} \text{ mi.} \quad x = -\pi\left(\frac{1}{8\pi}\right) + \frac{1}{4} = \frac{1}{8} \text{ mi.}$$

$\boxed{88}$ (a) $g = 32 \Rightarrow f(t) = -16t^2 + 16t$. Solving $f(t) = 0$ gives us $-16t(t-1) \Rightarrow t = 0,\ 1$.

The player is in the air for 1 second.

(b) $t = -\frac{b}{2a} = -\frac{16}{2(-16)} = \frac{1}{2}$. $f(\frac{1}{2}) = 4 \Rightarrow$ the player jumps 4 feet high.

(c) $g = \frac{32}{6} \Rightarrow f(t) = -\frac{8}{3}t^2 + 16t$. Solving $f(t) = 0$ yields $t = 0$ or 6.

The player would be in the air for 6 seconds on the moon.

$$t = -\frac{b}{2a} = -\frac{16}{2(-8/3)} = 3. \quad f(3) = 24 \Rightarrow \text{the player jumps 24 feet high.}$$

$\boxed{89}$ (a) Solving $-0.016x^2 + 1.6x = \frac{1}{5}x$ for x represents the intersection between the parabola and the line. $-0.08x^2 + 8x = x \Rightarrow 7x - \frac{8}{100}x^2 = 0 \Rightarrow x(7 - \frac{8}{100}x) = 0 \Rightarrow$

$x = 0,\ \frac{175}{2}$. The rocket lands at $(\frac{175}{2}, \frac{35}{2}) = (87.5,\ 17.5)$.

(b) The *difference* d between the parabola and the line is to be maximized here.

$d = (-0.016x^2 + 1.6x) - (\frac{1}{5}x) = -0.016x^2 + 1.4x$. d obtains a maximum when

$x = -\frac{b}{2a} = -\frac{1.4}{2(-0.016)} = 43.75$. The maximum height of the rocket

above the ground is $d = -0.016(43.75)^2 + 1.4(43.75) = 30.625$ units.

$\boxed{91}$ $P = kA^2v^3;\ 3000 = k\left[\pi(5)^2\right]^2(20)^3 \Rightarrow k = \frac{3}{5000\pi^2};$

$$P = \frac{3}{5000\pi^2}\left[\pi(5)^2\right]^2(30)^3 = 10{,}125 \text{ watts}$$

1 Graphs of equations of the form $y = x^{p/q}$, where $x \geq 0$, and p and q are positive integers all pass through $(0, 0)$ and $(1, 1)$. If $p/q < 1$, the graph is above $y = x$ for $0 \leq x \leq 1$ and below $y = x$ for $x \geq 1$. The closer p/q is to 1, the closer $y = x^{p/q}$ is to $y = x$. If $p/q > 1$, the graph is below $y = x$ for $0 \leq x \leq 1$ and above $y = x$ for $x \geq 1$.

3 For the graph of $g(x) = \sqrt{f(x)}$, where $f(x) = ax^2 + bx + c$, consider 2 cases:

  (1) $(a > 0)$ If f has 0 or 1 x-intercept(s), the domain of g is $\mathbb{R}$ and its range is $[\sqrt{k}, \infty)$, where k is the y-value of the vertex of f. If f has 2 x-intercepts (say x_1 and x_2 with $x_1 < x_2$), then the domain of g is $(-\infty, x_1] \cup [x_2, \infty)$ and its range is $[0, \infty)$. The general shape is similar to the v-shape of the graph of $y = \sqrt{a} \, |x|$.

  (2) $(a < 0)$ If f has no x-intercepts, there is no graph of g. If f has 1 x-intercept, the graph of g consists of that point. If f has 2 x-intercepts, the domain of g is $[x_1, x_2]$ and the range is $[0, \sqrt{k}]$. The shape of g is that of the top half of an oval.

  The main advantage of graphing g as a composition (say $Y_1 = f$ and $Y_2 = \sqrt{Y_1}$ on a graphing calculator) is to observe the relationship between the range of f and the domain of g.

5 To determine the x-coordinate of R,

we want to start at x_1 and go $\frac{m}{n}$ of the way to x_2. We could write this as

$$x_3 = x_1 + \tfrac{m}{n}\Delta x = x_1 + \tfrac{m}{n}(x_2 - x_1) = x_1 + \tfrac{m}{n}x_2 - \tfrac{m}{n}x_1 = \left(1 - \tfrac{m}{n}\right)x_1 + \tfrac{m}{n}x_2.$$

Similarly, $y_3 = \left(1 - \tfrac{m}{n}\right)y_1 + \tfrac{m}{n}y_2$.

7 The graph from Exercise 40(e) of Section 3.5 ($y = -[\![-x]\!]$) illustrates the concept of one of the most common billing methods with the open and closed end points reversed from those of the greatest integer function. Starting with $y = -[\![-x]\!]$ and adjusting for jumps every 15 minutes gives us $y = -[\![-x/15]\!]$. Since each quarter-hour charge is \$20, we multiply by 20 to obtain $y = -20[\![-x/15]\!]$. Because of the initial \$40 charge, we must add 40 to obtain the function $f(x) = 40 - 20[\![-x/15]\!]$.

9 (a) Let January correspond to 1, February to 2, ... , and December to 12.

[0.5, 12.5] by [0, 5]

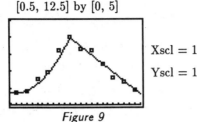

Xscl = 1
Yscl = 1

Figure 9

(b) The data points are (approximately) parabolic on the interval $[1, 6]$ and linear on $[6, 12]$. Let $f_1(x) = a(x - h)^2 + k$ on $[1, 6]$ and $f_2(x) = mx + b$ on $[6, 12]$. On $[1, 6]$, let the vertex $(h, k) = (1, 0.7)$. Since $(6, 4)$ is on the graph of f_1, $f_1(6) = a(6 - 1)^2 + 0.7 = 4 \Rightarrow a = 0.132$. Thus, $f_1(x) = 0.132(x - 1)^2 + 0.7$ on $[1, 6]$. Now, let $f_2(x) = mx + b$ pass through the points $(6, 4)$ and $(12, 0.9)$. An equation of this line is approximately $(y - 4) = -0.517(x - 6)$. Thus, let $f_2(x) = -0.517x + 7.102$ on $[6, 12]$.

$$f(x) = \begin{cases} 0.132(x - 1)^2 + 0.7 & \text{if } 1 \le x \le 6 \\ -0.517x + 7.102 & \text{if } 6 < x \le 12 \end{cases}$$

(c) To plot the piecewise function, let $Y_1 = (0.132(x - 1)\hat{\,}2 + 0.7)/(x \le 6)$ and $Y_2 = (-0.517x + 7.102)/(x > 6)$. These assignments use the concept of Boolean division. For example, when $(x \le 6)$ is false, the expression Y_1 will be undefined (division by 0) and the calculator will not plot any values.

Chapter 4: Polynomial Functions, Rational Functions, and Conic Sections

3 $f(x) = ax^3 + 2$ • The "+2" will shift each graph up 2 units.

(a) The effect of the "2" for a is to vertically stretch $g(x) = x^3 + 2$ by a factor of 2 and make it appear "steeper."

(b) The effect of the "$\frac{1}{3}$" for a is to vertically compress $g(x) = x^3$ by a factor of $1/(1/3) = 3$, making it appear "flatter." The "−" reflects the graph of $h(x) = \frac{1}{3}x^3$ about the x-axis.

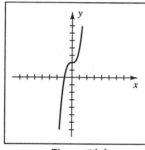

Figure 3(a)

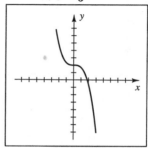

Figure 3(b)

5 We need to show that there is a sign change between $f(3)$ and $f(4)$ for $f(x) = x^3 - 4x^2 + 3x - 2$.

$$f(3) = 27 - 36 + 9 - 2 = -2 \quad \text{and} \quad f(4) = 64 - 64 + 12 - 2 = 10.$$

Since $f(3) = -2 < 0$ and $f(4) = 10 > 0$, the intermediate value theorem for polynomial functions assures us that f takes on every value between -2 and 10 in the interval $[3, 4]$, namely, 0.

7 As in Exercise 5, with $f(x) = -x^4 + 3x^3 - 2x + 1$, $a = 2$, $b = 3$,

we have $f(2) = 5 > 0$ and $f(3) = -5 < 0$.

9 As in Exercise 5, with $f(x) = x^5 + x^3 + x^2 + x + 1$, $a = -\frac{1}{2}$, $b = -1$,

we have $f(-\frac{1}{2}) = \frac{19}{32} > 0$ and $f(-1) = -1 < 0$.

$\boxed{11}$ $f(x) = \frac{1}{4}x^3 - 2 = \frac{1}{4}(x^3 - 8) = \frac{1}{4}(x - 2)(x^2 + 2x + 4)$. The general shape of the graph is that of $f(x) = x^3$. To find the x-intercepts, set $y\ \{f(x)\}$ equal to 0. This means either $x - 2 = 0$ or $x^2 + 2x + 4 = 0$. $x - 2 = 0 \Rightarrow x = 2$ and $x^2 + 2x + 4 = 0 \Rightarrow$ $x = -1 \pm \sqrt{3}\,i$. The imaginary solutions mean that we have no x-intercepts from the factor $x^2 + 2x + 4$ and the only x-intercept is 2. To find the y-intercept, set x equal to 0. $x = 0 \Rightarrow y = -2$. $f(x) > 0$ if $x > 2$, $f(x) < 0$ if $x < 2$

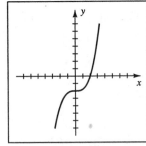

Figure 11

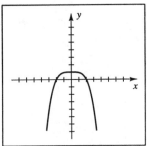

Figure 13

$\boxed{13}$ $f(x) = -\frac{1}{16}x^4 + 1 = -\frac{1}{16}(x^4 - 16) = -\frac{1}{16}(x^2 + 4)(x + 2)(x - 2)$;

$f(x) > 0$ if $|x| < 2$, $f(x) < 0$ if $|x| > 2$

$\boxed{15}$ $f(x) = x^4 - 4x^2 = x^2(x + 2)(x - 2)$; $f(x) > 0$ if $|x| > 2$, $f(x) < 0$ if $0 < |x| < 2$

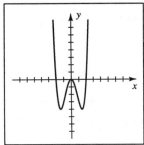

Figure 15

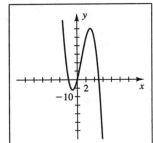

Figure 17

$\boxed{17}$ $f(x) = -x^3 + 3x^2 + 10x = -x(x^2 - 3x - 10) = -x(x + 2)(x - 5)$;

$f(x) > 0$ if $x < -2$ or $0 < x < 5$, $f(x) < 0$ if $-2 < x < 0$ or $x > 5$

19 $f(x) = \frac{1}{6}(x+2)(x-3)(x-4)$ •

The y-intercept is $f(0) = \frac{1}{6}(0+2)(0-3)(0-4) = \frac{1}{6}(2)(-3)(-4) = 4$.   To determine the intervals on which $f(x) > 0$ and $f(x) < 0$, you may want to refer back to the concept of a sign diagram. Below is a sign diagram for this function. Note how the sign of each region correlates to the sign of $f(x)$.

Resulting sign:	$\ominus$	$\oplus$	$\ominus$	$\oplus$
Sign of $x-4$:	−	−	−	+
Sign of $x-3$:	−	−	+	+
Sign of $x+2$:	−	+	+	+
x values:		−2	3	4

Diagram 19

$f(x) > 0$ if $-2 < x < 3$ or $x > 4$, $f(x) < 0$ if $x < -2$ or $3 < x < 4$

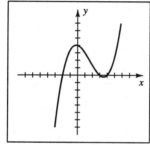

Figure 19

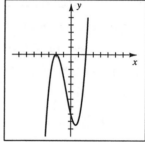

Figure 21

21 $f(x) = x^3 + 2x^2 - 4x - 8 = x^2(x+2) - 4(x+2) = (x+2)^2(x-2)$;

$$f(x) > 0 \text{ if } x > 2, \ f(x) < 0 \text{ if } x < -2 \text{ or } |x| < 2$$

23 $f(x) = x^4 - 6x^2 + 8 = (x^2 - 2)(x+2)(x-2)$ •

The general shape of the graph is that of $g(x) = x^4$.

$$f(x) > 0 \text{ if } |x| > 2 \text{ or } |x| < \sqrt{2}, \ f(x) < 0 \text{ if } \sqrt{2} < |x| < 2$$

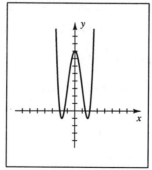

Figure 23

$\boxed{25}$ $f(x) = x^2(x+2)(x-1)^2(x-2)$ •

If an intercept value has a corresponding *linear* factor, the graph will go through that intercept—that is, there will be a sign change from positive to negative or negative to positive in the function. Thus, at $x = -2$ and $x = 2$ the graph of the function "goes through" the point. If an intercept value has a corresponding quadratic factor (as in the case of 0 and x^2; and 1 and $(x-1)^2$), then the function will not change sign. Hence, at $x = 0$ and $x = 1$ the graph of the function "touches the point and turns around."

The above discussion can be generalized as follows:

(1) If an intercept value has a corresponding factor raised to an <u>odd</u> power,

then the function <u>will</u> change sign at that point.

(2) If an intercept value has a corresponding factor raised to an <u>even</u> power,

then the function <u>will not</u> change sign at that point.

$f(x) > 0$ if $|x| > 2$, $f(x) < 0$ if $|x| < 2$, $x \neq 0$, $x \neq 1$

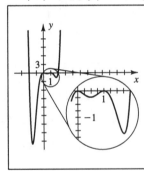

Figure 25

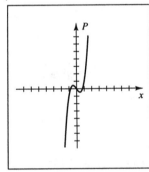

Figure 31

$\boxed{29}$ If a graph contains the point $(-1, 4)$, then $f(-1) = 4$.

For this function, $f(x) = 3x^3 - kx^2 + x - 5k$,

$$f(-1) = 3(-1)^3 - k(-1)^2 + (-1) - 5k = -3 - k - 1 - 5k = -4 - 6k.$$

Thus, $-4 - 6k$ must equal 4. Solving for k yields $-4 - 6k = 4 \Rightarrow k = -\frac{4}{3}$.

$\boxed{31}$ $P(x) = \frac{1}{2}(5x^3 - 3x) = \frac{1}{2}x(5x^2 - 3)$. See *Figure 31* above.

$P(x) = 0 \Rightarrow x = 0, \ \pm\sqrt{\frac{3}{5}}$, or $\pm\frac{1}{5}\sqrt{15}$.

$P(x) > 0$ on $(-\frac{1}{5}\sqrt{15}, 0)$ and $(\frac{1}{5}\sqrt{15}, \infty)$.

$P(x) < 0$ on $(-\infty, -\frac{1}{5}\sqrt{15})$ and $(0, \frac{1}{5}\sqrt{15})$.

33 (a) $V(x) = lwh = (30 - x - x)(20 - x - x)x = x(20 - 2x)(30 - 2x) =$

$$4x(10 - x)(15 - x) = 4x(x - 10)(x - 15).$$

(b) $V(x) > 0$ on $(0, 10)$ and $(15, \infty)$. Allowable values for x are in $(0, 10)$.

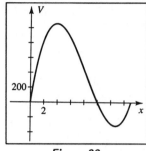

Figure 33

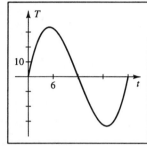

Figure 35

35 We will work part (b) first.

(b) T has the general shape of a cubic with zeros at 0, 12, and 24.

Its sign pattern is negative, positive, negative, positive.

(a) $T = \frac{1}{20}t(t - 12)(t - 24) = 0 \Rightarrow t = 0, 12, 24.$ $T > 0$ for

$0 < t < 12$ {6 A.M. to 6 P.M.}; $T < 0$ for $12 < t < 24$ {6 P.M. to 6 A.M.}.

(c) 12 noon corresponds to $t = 6$, $T(6) = 32.4 > 32°$F and $T(7) = 29.75 < 32°$F

37 (a) $N(t) = -t^4 + 21t^2 + 100$

$$= -(t^4 - 21t^2 - 100)$$

$$= -(t^2 - 25)(t^2 + 4)$$

$$= -(t + 5)(t - 5)(t^2 + 4)$$

If $t > 0$, then $N(t) > 0$ for $0 < t < 5$.

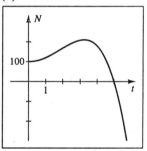

(b) The population becomes extinct when $N = 0$.

This occurs after 5 years.

Figure 37

39 (a) $f(x) = 2x^4$, $g(x) = 2x^4 - 5x^2 + 1$, $h(x) = 2x^4 + 5x^2 - 1$, $k(x) = 2x^4 - x^3 + 2x$

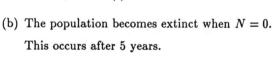

x	$f(x)$	$g(x)$	$h(x)$	$k(x)$
-60	25,920,000	25,902,001	25,937,999	26,135,880
-40	5,120,000	5,112,001	5,127,999	5,183,920
-20	320,000	318,001	321,999	327,960
20	320,000	318,001	321,999	312,040
40	5,120,000	5,112,001	5,127,999	5,056,080
60	25,920,000	25,902,001	25,937,999	25,704,120

(b) As $|x|$ becomes large, the function values become similar.

(c) The term with the highest power of x: $2x^4$.

45 From the graph, f has three zeros. They are approximately -1.88, 0.35, and 1.53.

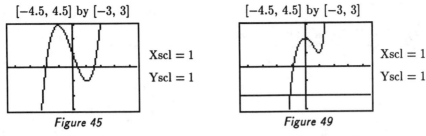

Figure 45   Figure 49

49 If $f(x) = x^5 - 2x^2 + 2$ and $k = -2$, then $f(x) > k$ on $(-1.10, \infty)$.

53 (a) To enter f in your graphing calculator, let $Y_1 = X - 1975$ and then assign

$$0.0014Y_1{}^3 - 0.0388Y_1{}^2 + 0.8783Y_1 + 23.82 \quad \text{to} \quad Y_2.$$

From the graph, we determine that there were approximately 23.82 million recipients in 1975 and 37.066 million recipients in 1995. The number of recipients is increasing quite rapidly.

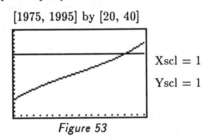

Figure 53

(b) Graph $Y_3 = 34.4$ (shown in *Figure 53*). The graphs of Y_2 and Y_3 intersect when $x \approx 1991.9792 \approx 1992$. There were 34.4 million recipients in 1992.

4.2 Exercises

Note: Refer to the illustration on page 246. We will walk through the steps to obtain the quotient and remainder for this example.

(1) Determine the quotient of the first terms of the dividend $(x^4 - 16)$ and the divisor $(x^2 + 3x + 1)$. $\frac{x^4}{x^2} = x^2$—this is the first term of the resulting quotient.

(2) Multiply the result in step (1), x^2, by the divisor, $x^2 + 3x + 1$. This product is $x^4 + 3x^3 + x^2$.

(3) Subtract the result in step (2) from the dividend. This result is $-3x^3 - x^2$ (there is no need to worry about the lower degree terms here, namely, -16). A common mistake is to forget to subtract <u>all</u> of the terms and obtain (in this case) $3x^3 + x^2$.

(4) Repeat steps (1) through (3) for the remaining dividend until the degree of the remainder is less than the degree of the divisor.

$\boxed{1}$ $f(x) = 2x^4 - x^3 - 3x^2 + 7x - 12;$ $p(x) = x^2 - 3$ •

$$
\begin{array}{r}
2x^2 - x + 3 \\
x^2 - 3 \overline{\smash{\big)}\ 2x^4 - x^3 - 3x^2 + 7x - 12} \\
\underline{2x^4 - 6x^2} \\
-x^3 + 3x^2 \\
\underline{-x^3 + 3x} \\
3x^2 + 4x \\
\underline{3x^2 - 9} \\
4x - 3
\end{array}
$$

★ $2x^2 - x + 3;\ 4x - 3$

$\boxed{5}$ Note that we can't divide $f(x) = 7x + 2$ by $p(x) = 2x^2 - x - 4$ since the degree of the dividend is already less than the degree of the divisor. So the quotient is 0 and the remainder is $7x + 2$.

$\boxed{9}$ We wish to find $f(2)$ by using the remainder theorem. Thus, we need to divide $f(x) = 3x^3 - x^2 + 5x - 4$ by $x - 2$ using either long division or synthetic division. Using synthetic division, we have

$$
\begin{array}{r|rrrr}
2 & 3 & -1 & 5 & -4 \\
& & 6 & 10 & 30 \\
\hline
& 3 & 5 & 15 & 26
\end{array}
$$

The last number in the third row, 26, is our remainder. Hence, $f(2) = 26$.

$\boxed{11}$ We divide by $x + 3$ or $x - (-3)$. Note that we include a 0 for the missing x^3 term in $f(x) = x^4 - 6x^2 + 4x - 8$.

$$
\begin{array}{r|rrrrr}
-3 & 1 & 0 & -6 & 4 & -8 \\
& & -3 & 9 & -9 & 15 \\
\hline
& 1 & -3 & 3 & -5 & 7
\end{array}
$$

Hence, $f(-3) = 7$.

$\boxed{13}$ To show that $x + 3$ is a factor of $f(x) = x^3 + x^2 - 2x + 12$, we must show that $f(-3) = 0$. $f(-3) = -27 + 9 + 6 + 12 = 0$, and, hence, $x + 3$ is a factor of $f(x)$.

$\boxed{17}$ f has degree 3 with zeros -2, 0, 5 $\Rightarrow$

$f(x) = a[x - (-2)](x - 0)(x - 5)$ { let $a = 1$ }

$= x(x + 2)(x - 5) = x(x^2 - 3x - 10) = x^3 - 3x^2 - 10x$

19 f has degree 4 with zeros -2, ± 1, $4 \Rightarrow$

$\quad f(x) = a(x+2)(x+1)(x-1)(x-4)$ { let $a = 1$ }

$\quad\quad = (x^2 - 1)(x^2 - 2x - 8) = x^4 - 2x^3 - 9x^2 + 2x + 8$

21 $\underline{2}$ | $\quad$ 2 $\quad$ -3 $\quad$ 4 $\quad$ -5 $\qquad$ The synthetic division indicates that the

$\qquad\qquad\quad$ 4 $\quad$ 2 $\quad$ 12 $\qquad$ quotient is $2x^2 + x + 6$ and the remainder is 7.

$\qquad\qquad$ 2 $\quad$ 1 $\quad$ 6 $\quad$ 7

23 $\underline{-3}$ | 1 $\quad$ 0 $\quad$ -8 $\quad$ -5 $\qquad$ The synthetic division indicates that the

$\qquad\qquad$ -3 $\quad$ 9 $\quad$ -3 $\qquad$ quotient is $x^2 - 3x + 1$ and the remainder is -8.

$\qquad\qquad$ 1 $\quad$ -3 $\quad$ 1 $\quad$ -8

29 $\underline{3}$ | 2 $\quad$ 3 $\quad$ -4 $\quad$ 4 $\qquad$ The synthetic division indicates that $f(3) = 73$.

$\qquad\qquad$ 6 $\quad$ 27 $\quad$ 69

$\qquad\qquad$ 2 $\quad$ 9 $\quad$ 23 $\quad$ 73

31 $\underline{-0.2}$ | 0.3 $\quad$ 0 $\quad$ 0.04 $\quad$ -0.034

$\qquad\qquad$ -0.06 $\quad$ 0.012 $\quad$ -0.0104

$\qquad\qquad$ 0.3 $\quad$ -0.06 $\quad$ 0.052 $\quad$ -0.0444

The remainder is -0.0444 and the remainder theorem indicates that this is $f(-0.2)$.

33 $\underline{2 + \sqrt{3}}$ | 1 $\quad$ 3 $\quad$ -5

$\qquad\qquad$ $2 + \sqrt{3}$ $\quad$ $13 + 7\sqrt{3}$

$\qquad\qquad$ 1 $\quad$ $5 + \sqrt{3}$ $\quad$ $8 + 7\sqrt{3}$

The remainder is $8 + 7\sqrt{3}$ and the remainder theorem indicates that this is the value

$\qquad\qquad\qquad\qquad\qquad\qquad\qquad\qquad\qquad$ of $f(2 + \sqrt{3})$.

35 -2 is a zero of $f(x)$ if we can show that $f(-2) = 0$.

$\qquad\qquad$ $\underline{-2}$ | 3 $\quad$ 8 $\quad$ -2 $\quad$ -10 $\quad$ 4

$\qquad\qquad\qquad\quad$ -6 $\quad$ -4 $\quad$ 12 $\quad$ -4

$\qquad\qquad\qquad$ 3 $\quad$ 2 $\quad$ -6 $\quad$ 2 $\quad$ 0

$\qquad\qquad\qquad$ Hence, $f(-2) = 0$ and -2 is a zero of $f(x)$.

37 $\underline{\frac{1}{2}}$ | 4 $\quad$ -6 $\quad$ 8 $\quad$ -3

$\qquad\qquad$ 2 $\quad$ -2 $\quad$ 3

$\qquad\qquad$ 4 $\quad$ -4 $\quad$ 6 $\quad$ 0

$\qquad\qquad\qquad$ Hence, $f(\frac{1}{2}) = 0$ and $\frac{1}{2}$ is a zero of $f(x)$.

$\boxed{39}$ $f(-2) = k(-2)^3 + (-2)^2 + k^2(-2) + 3k^2 + 11$

$\quad = -8k + 4 - 2k^2 + 3k^2 + 11 = k^2 - 8k + 15.$

The remainder, $k^2 - 8k + 15$, must be zero if $f(x)$ is to be divisible by $x + 2$.

$$k^2 - 8k + 15 = 0 \Rightarrow (k - 3)(k - 5) = 0 \Rightarrow k = 3, 5.$$

$\boxed{41}$ $x - c$ will not be a factor of $f(x) = 3x^4 + x^2 + 5$ for any real number c if the remainder of dividing $f(x)$ by $x - c$ is not zero for any real number c. The remainder is $f(c)$, or $3c^4 + c^2 + 5$. Since $3c^4 + c^2 + 5$ is greater than or equal to 5 for any real number c, it is never zero and hence, $x - c$ is never a factor of $f(x)$.

$\boxed{45}$ If $f(x) = x^n - y^n$ and n is even, then $f(-y) = (-y)^n - (y)^n = y^n - y^n = 0.$

Hence, $x + y$ is a factor of $x^n - y^n$.

$\boxed{47}$ (a) The radius of the resulting cylinder is x and the height is y, that is, $6 - x$.

$$V = \pi r^2 h = \pi x^2 (6 - x)$$

(b) The volume of the cylinder of radius 1 and altitude 5 is $\pi(1)^2 5 = 5\pi$. To determine another value of x which would result in the same volume, we need to solve the equation $5\pi = \pi x^2(6 - x)$. $5\pi = \pi x^2(6 - x) \Rightarrow 5 = 6x^2 - x^3 \Rightarrow x^3 - 6x^2 + 5 = 0$. We know that $x = 1$ is a solution to this equation. Synthetically dividing, we obtain

$$
\begin{array}{r|rrrr}
1 & 1 & -6 & 0 & 5 \\
  &   & 1 & -5 & -5 \\
\hline
  & 1 & -5 & -5 & 0
\end{array}
$$

The quotient is $x^2 - 5x - 5$. Using the quadratic formula, we see that the solutions are $x = \dfrac{5 \pm \sqrt{45}}{2}$. Since x must be positive in the first quadrant, $\dfrac{5 + \sqrt{45}}{2} \approx 5.85$ would be an allowable value of x. If $x = \frac{1}{2}(5 + \sqrt{45})$, then

$y = 6 - x = 6 - \frac{1}{2}(5 + \sqrt{45}) = \frac{12}{2} - \frac{5}{2} - \frac{1}{2}\sqrt{45} = \frac{1}{2}(7 - \sqrt{45}).$

The point P is $P(x, y) = \left(\frac{1}{2}(5 + \sqrt{45}), \frac{1}{2}(7 - \sqrt{45}) \right) \approx (5.85, 0.15).$

$\boxed{49}$ (a) $A = lw = (2x)(y) = 2x(4 - x^2) = 8x - 2x^3$

(b) This solution is similar to the solution in Exercise 47(b).

$A = 6 \Rightarrow 6 = 8x - 2x^3 \Rightarrow x^3 - 4x + 3 = 0 \Rightarrow (x - 1)(x^2 + x - 3) = 0 \Rightarrow$

$x = 1, \dfrac{-1 \pm \sqrt{13}}{2}. \quad \dfrac{\sqrt{13} - 1}{2}$ would be an allowable value of x.

The base would then be $\sqrt{13} - 1 \approx 2.61.$

51. If $f(x) = x^8 - 7.9x^5 - 0.8x^4 + x^3 + 1.2x - 9.81$,

then the remainder is $f(0.21) \approx -9.55$.

[−1, 1] by [−20, 0]

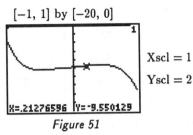

$\text{Xscl} = 1$
$\text{Yscl} = 2$

Figure 51

[−9, 9] by [−3, 9]

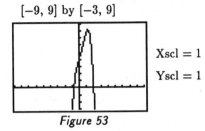

$\text{Xscl} = 1$
$\text{Yscl} = 1$

Figure 53

53. $f(1.6) = -2k^4 + 2.56k^3 + 3.2k + 4.096$. Graph $y = -2k^4 + 2.56k^3 + 3.2k + 4.096$ (that is, $y = -2x^4 + 2.56x^3 + 3.2x + 4.096$).   From the graph, we see that $y = 0$ when $k \approx -0.75$, 1.96. Thus, if k assumes either of these values, $f(1.6) = 0$ and f will be divisible by $x - 1.6$ by the factor theorem.

4.3 Exercises

1. The polynomial is of the form $f(x) = a(x - (-1))(x - 2)(x - 3)$.

$f(-2) = a(-1)(-4)(-5) = 80 \Rightarrow -20a = 80 \Rightarrow a = -4$.

Hence, the polynomial is $-4(x + 1)(x - 2)(x - 3)$, or $-4x^3 + 16x^2 - 4x - 24$.

3. The polynomial is of the form $f(x) = a(x - (-4))(x - 3)(x - 0)$.

$f(2) = a(6)(-1)(2) = -36 \Rightarrow -12a = -36 \Rightarrow a = 3$.

Hence, the polynomial is $3(x + 4)(x - 3)(x)$, or $3x^3 + 3x^2 - 36x$.

5. The polynomial is of the form $f(x) = a(x - (-2i))(x - 2i)(x - 3)$.

$f(1) = a(1 + 2i)(1 - 2i)(-2) = 20 \Rightarrow -10a = 20 \Rightarrow a = -2$.

Hence, the polynomial is $-2(x + 2i)(x - 2i)(x - 3) = -2(x^2 + 4)(x - 3)$,

or $-2x^3 + 6x^2 - 8x + 24$.

7. $f(x) = a(x + 4)^2(x - 3)^2 = (x^2 + x - 12)^2 \{a = 1\} = x^4 + 2x^3 - 23x^2 - 24x + 144$

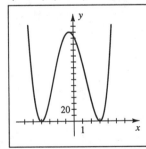

Figure 7

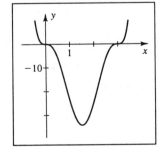

Figure 9

9. $f(x) = a(x)^3(x - 3)^3$ so $f(2) = a(8)(-1) = -8a$. But $f(2) = -24$, so $-8a = -24$, or, $a = 3$. $f(x) = 3(x)^3(x^3 - 9x^2 + 27x - 27) = 3x^6 - 27x^5 + 81x^4 - 81x^3$.

$\boxed{11}$ The graph has x-intercepts at -1, $\frac{3}{2}$, 3, and $f(0) = \frac{7}{2}$.

$f(x) = a(x+1)(x-\frac{3}{2})(x-3)$; $f(0) = a(1)(-\frac{3}{2})(-3) = \frac{7}{2} \Rightarrow \frac{9}{2}a = \frac{7}{2} \Rightarrow a = \frac{7}{9}$.

$$f(x) = \frac{7}{9}(x+1)(x-\frac{3}{2})(x-3)$$

$\boxed{13}$ 3 is a zero of multiplicity one, 1 is a zero of multiplicity two, and $f(0) = 3$.

$f(x) = a(x-1)^2(x-3)$; $f(0) = a(1)(-3) = 3 \Rightarrow a = -1$. $f(x) = -1(x-1)^2(x-3)$.

$\boxed{17}$ $f(x) = 4x^5 + 12x^4 + 9x^3 = x^3(4x^2 + 12x + 9) = x^3(2x+3)^2$

★ $-\frac{3}{2}$ (multiplicity 2); 0 (multiplicity 3)

$\boxed{19}$ $f(x) = (x^2 + x - 12)^3(x^2 - 9)^2 = [(x+4)(x-3)]^3[(x+3)(x-3)]^2$

$= (x+4)^3(x-3)^3(x+3)^2(x-3)^2 = (x+4)^3(x+3)^2(x-3)^5$

★ -4 (multiplicity 3); -3 (multiplicity 2); 3 (multiplicity 5)

$\boxed{23}$ Using synthetic division,

$$
\begin{array}{r|rrrrr}
-3 & 1 & 7 & 13 & -3 & -18 \\
   &   & -3 & -12 & -3 & 18 \\
\hline
   & 1 & 4 & 1 & -6 & 0
\end{array}
$$

$$
\begin{array}{r|rrrr}
-3 & 1 & 4 & 1 & -6 \\
   &   & -3 & -3 & 6 \\
\hline
   & 1 & 1 & -2 & 0
\end{array}
$$

The remaining polynomial, $x^2 + x - 2$, can be factored as $(x+2)(x-1)$.

Hence, $f(x) = (x+3)^2(x+2)(x-1)$.

$\boxed{25}$ *Note:* If the sum of the coefficients of the polynomial is 0, then 1 is a zero of the polynomial. Synthetically dividing 1 five times, we obtain the remaining polynomial, $x+1$. Hence, $f(x) = x^6 - 4x^5 + 5x^4 - 5x^2 + 4x - 1 = (x-1)^5(x+1)$.

Note: For the following exercises, let $f(x)$ denote the polynomial, P, the number of sign changes in $f(x)$, and N, the number of sign changes in $f(-x)$. The types of possible solutions are listed in the order positive, negative, nonreal complex.

$\boxed{27}$ $f(x) = 4x^3 - 6x^2 + x - 3$. The sign pattern for coefficients is $+, -, +, -$. Since there are 3 sign changes in $f(x)$, P = 3. $f(-x) = -4x^3 - 6x^2 - x - 3$. *Note:* Simply negate the coefficients of the odd-powered terms of $f(x)$ to find $f(-x)$. The sign pattern for coefficients is $-, -, -, -$. Since there are no sign changes in $f(-x)$, N = 0. If there are 3 positive solutions, then there are no negative or nonreal complex. If there is 1 positive solution, there are no negative and 2 nonreal complex solutions.

$\boxed{31}$ $f(x) = 3x^4 + 2x^3 - 4x + 2$ and P = 2.

$f(-x) = 3x^4 - 2x^3 + 4x + 2$ and N = 2. ★ 2, 2, 0; 2, 0, 2; 0, 2, 2; 0, 0, 4

35 Synthetically dividing 5 into the polynomial yields a bottom row consisting of all nonnegative numbers (see below). Synthetically dividing −2 into the polynomial yields a bottom row that alternates in sign (see below).

$$
\begin{array}{r|rrrr}
5 & 1 & -4 & -5 & 7 \\
  &   & 5 & 5 & 0 \\
\hline
  & 1 & 1 & 0 & 7
\end{array}
\qquad
\begin{array}{r|rrrr}
-2 & 1 & -4 & -5 & 7 \\
   &   & -2 & 12 & -14 \\
\hline
   & 1 & -6 & 7 & -7
\end{array}
$$

These results indicate that the upper bound is 5 and the lower bound is −2. From the graph of $f(x) = x^3 - 4x^2 - 5x + 7$, we see that the bounds given by the theorem (5 and −2) are indeed the smallest and largest integers that are upper and lower bounds.

37 Synthetically dividing 2 into the polynomial yields a bottom row consisting of all nonnegative numbers (see below). Synthetically dividing −2 into the polynomial yields a bottom row that alternates in sign (see below).

$$
\begin{array}{r|rrrrr}
2 & 1 & -1 & -2 & 3 & 6 \\
  &   & 2 & 2 & 0 & 6 \\
\hline
  & 1 & 1 & 0 & 3 & 12
\end{array}
\qquad
\begin{array}{r|rrrrr}
-2 & 1 & -1 & -2 & 3 & 6 \\
   &   & -2 & 6 & -8 & 10 \\
\hline
   & 1 & -3 & 4 & -5 & 16
\end{array}
$$

These results indicate that the upper bound is 2 and the lower bound is −2. From the graph of $f(x) = x^4 - x^3 - 2x^2 + 3x + 6$, we see that there are *no* real zeros. Remember, the theorem only gives us upper positive and lower negative bounds for the zeros of a polynomial—it doesn't guarantee that there are any zeros.

41 The zero at $x = -1$ must be of even multiplicity since the graph does not cross the x-axis. Since we want f to have minimal degree, we will let the associated factor be $(x+1)^2$. The graph goes through the zero at $x = 1$ without "flattening out," so the associated factor is $(x-1)^1$. Because the graph flattens out at the zero at $x = 2$, we will let its associated factor be $(x-2)^3$. Thus, f has the form

$$ f(x) = a(x+1)^2(x-1)(x-2)^3. $$

Since the y-intercept is $(0, -2)$, we have $f(0) = a(1)(-1)(-8) = 8a$ and $f(0) = -2 \Rightarrow 8a = -2 \Rightarrow a = -\frac{1}{4}$. Hence, $f(x) = -\frac{1}{4}(x+1)^2(x-1)(x-2)^3$.

43 (a) Similar to the discussion in the solution of Exercise 41, we have

$$ f(x) = a(x+3)^3(x+1)(x-2)^2. $$

(b) $a = 1$ and $x = 0 \Rightarrow f(0) = 1(0+3)^3(0+1)(0-2)^2 = 1(3)^3(1)(-2)^2 = 108.$

45 From the graph, we see that $f(x) = x^5 - 16.75x^3 + 12.75x^2 + 49.5x - 54$ has zeros of -4, -2, 1.5, and 3. There is a double root at 1.5. Since the leading coefficient of f is 1, we have $f(x) = 1(x+4)(x+2)(x-1.5)^2(x-3)$.

$[-5, 5]$ by $[-150, 150]$

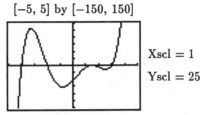

Xscl $= 1$
Yscl $= 25$

Figure 45

47 Since the zeros are -2, 1, 2, and 3, the polynomial must have the form
$$f(x) = a(x+2)(x-1)(x-2)(x-3).$$
Now, $f(0) = a(2)(-1)(-2)(-3) = -12a$ and $f(0) = -24 \Rightarrow -12a = -24 \Rightarrow a = 2$.
Let $f(x) = 2(x+2)(x-1)(x-2)(x-3)$. Since f has been completely determined, we must check the remaining data point(s). $f(-1) = 2(1)(-2)(-3)(-4) = -48 \neq -52$.

Thus, a fourth-degree polynomial *does not* fit the data points.

49 Since the zeros are 2, 5.2, and 10.1, the polynomial must have the form
$$f(x) = a(x-2)(x-5.2)(x-10.1).$$ Now, $f(1.1) = a(-33.21) = -49.815 \Rightarrow a = 1.5$.
Let $f(x) = 1.5(x-2)(x-5.2)(x-10.1)$. Since f has been completely determined, we must check the remaining data points. $f(3.5) = 25.245$ and $f(6.4) = -29.304$.

Thus, a third-degree polynomial does fit the data points.

51 The zeros are 0, 5, 19, 24, and $f(12) = 10$. $f(t) = a(t)(t-5)(t-19)(t-24)$;
$f(12) = a(12)(7)(-7)(-12) = 10 \Rightarrow 7056a = 10 \Rightarrow a = \frac{5}{3528}$.
$$f(t) = \frac{5}{3528}t(t-5)(t-19)(t-24)$$

53 The graph of f does not cross the x-axis at a zero of even multiplicity, but does cross the x-axis at a zero of odd multiplicity. The higher the multiplicity of a zero, the more horizontal the graph of f is near that zero.

$[-3, 3]$ by $[-2, 2]$ $[-3, 3]$ by $[-3, 1]$

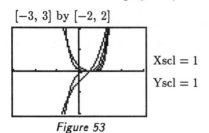   Xscl $= 1$
Yscl $= 1$

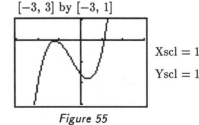   Xscl $= 1$
Yscl $= 1$

Figure 53   Figure 55

55 From the graph of f there are two zeros. They are -1.2 and 1.1. The zero at -1.2 has even multiplicity and the zero at 1.1 has odd multiplicity. Since f has degree 3, the zero at -1.2 must have multiplicity 2 and the zero at 1.1 has multiplicity 1.

$\boxed{57}$ From the graph of $A(t) = -\frac{1}{2400}t^3 + \frac{1}{20}t^2 + \frac{7}{6}t + 340$, we see that $A = 400$ when $t \approx 27.1$. Thus, the carbon dioxide concentration will be 400 in $1980 + 27.1 = 2007.1$, or, during the year 2007.

[0, 60] by [0, 600]   [0.5, 12.5] by [−30, 50]

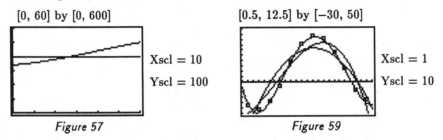

Xscl = 10   Xscl = 1
Yscl = 100   Yscl = 10

Figure 57   *Figure 59*

$\boxed{59}$ (a) Graphing the data and the functions show that the best fit is $h(x)$.

(b) Since the temperature changes sign between April and May and between October and November, an average temperature of $0°F$ occurs when $4 \le x \le 5$ and $10 \le x \le 11$.

(c) Finding the zeros of h between 1 and 12 gives us $x \approx 4.02, 10.53$.

$\boxed{61}$ Let $r = 6$ and $k = 0.7$. Graph $Y_1 = \frac{4k}{3}\pi r^3 - \pi x^2 r + \frac{1}{3}\pi x^3 = \frac{604.8\pi}{3} - 6\pi x^2 + \frac{1}{3}\pi x^3$ and determine the positive zeros. There are two zeros located at $x \approx 7.64, 15.47$. Since the sphere floats, it will not sink deeper than twice the radius, which is 12 centimeters. Thus, the pine sphere will sink approximately 7.64 centimeters into the water.

[−20, 20] by [−800, 800]   [−20, 20] by [−1000, 1000]

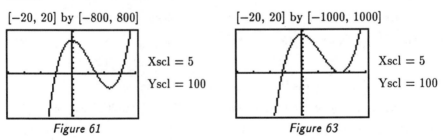

Xscl = 5   Xscl = 5
Yscl = 100   Yscl = 100

Figure 61   *Figure 63*

$\boxed{63}$ Let $r = 6$ and $k = 1$. Graph $Y_1 = \frac{4k}{3}\pi r^3 - \pi x^2 r + \frac{1}{3}\pi x^3 = 288\pi - 6\pi x^2 + \frac{1}{3}\pi x^3$ and determine the positive zero. There is a zero at $x = 12$. This means that the entire sphere is just submerged. The sphere has the same density as water and neither sinks nor floats, much like a balloon filled with water.

4.4 Exercises

Note: It is helpful to remember that if $a + bi$ is a zero, then $\left[x^2 - 2ax + (a^2 + b^2)\right]$

is the associated quadratic factor for the following exercises. (page 268 in the text)

$\boxed{1}$ Since $3 + 2i$ is a root, so is $3 - 2i$. As in the preceding note, $a = 3$ and $b = -2$, and hence, $-2a = -6$ and $a^2 + b^2 = 13$. Thus, the polynomial is of the form $\left[x - (3 + 2i)\right]\left[x - (3 - 2i)\right] = x^2 - 6x + 13$.

$\boxed{3}$ Since 2 is a zero, $x - 2$ is a factor. The associated quadratic factor for $-2 \pm 5i$ is $x^2 - 2(-2)x + (-2)^2 + (5)^2$, or, equivalently, $x^2 + 4x + 29$. Hence, the polynomial is of the form $(x - 2)(x^2 + 4x + 29)$.

$\boxed{5}$ Since -1 and 0 are zeros, $x + 1$ and x are factors. The associated quadratic factor for $3 \pm i$ is $x^2 - 2(3)x + (3)^2 + (1)^2$, or, equivalently, $x^2 - 6x + 10$. Hence, the polynomial is of the form $(x)(x + 1)(x^2 - 6x + 10)$.

$\boxed{9}$ If $-2i$ is a zero, then $2i$ is a zero. Thus, $(x - 2i)(x - (-2i)) = x^2 - 4i^2 = x^2 + 4$ is the associated factor. Hence, the polynomial is of the form $(x)(x^2 + 4)(x^2 - 2x + 2)$.

$\boxed{11}$ The constant term, 6, has positive integer divisors 1, 2, 3, 6. The leading coefficient, 1, has integer divisors ± 1. By the theorem on rational zeros of a polynomial, any rational root will be of the form

$$\frac{\text{positive divisors of 6}}{\text{divisors of 1}} = \frac{1,\ 2,\ 3,\ 6}{\pm 1} \longrightarrow \pm 1,\ \pm 2,\ \pm 3,\ \pm 6.$$

Note that this quotient gives us the same possible rational roots as the one on page 269:

$$\frac{\text{factors of } a_0}{\text{factors of } a_n} = \frac{\pm 1,\ \pm 2,\ \pm 3,\ \pm 6}{\pm 1} \longrightarrow \pm 1,\ \pm 2,\ \pm 3,\ \pm 6.$$

Since ± 1 are always potential solutions, they are as good as any other values to begin with. If the coefficients of the polynomial ($x^3 + 3x^2 - 4x + 6$ in this exercise) sum to 0, then 1 is a zero. In this case, the sum is $1 + 3 + (-4) + 6 = 6 \neq 0$. Thus, 1 is eliminated and we will try -1. If $f(x)$ is the polynomial, then -1 will be a zero if the coefficients of $f(-x)$ sum to 0. In this case, $f(-x) = -x^3 + 3x^2 - 10x + 6$ and the sum of the coefficients is $(-1) + 3 + (-10) + 6 = -2 \neq 0$. Thus, -1 is not a zero. We now use synthetic division to show that ± 2, ± 3, and ± 6 are not zeros of $f(x)$.

[15] The constant term, -8, has positive integer divisors 1, 2, 4, 8. The leading coefficient has integer divisors ± 1. Hence, any rational root will be of the form

$$\frac{\text{positive divisors of } -8}{\text{divisors of } 1} = \frac{1,\, 2,\, 4,\, 8}{\pm 1} \longrightarrow \pm 1,\ \pm 2,\ \pm 4,\ \pm 8.$$

We now need to try to find one solution of the equation. Once we have one solution, the resulting polynomial will be a quadratic and then we can use the quadratic formula to find the other 2 solutions. As in the solution to Exercise 11, we will try ± 1 first. After determining that 1 is not a solution, we try -1. In this case, $f(-x) = -x^3 - x^2 + 10x - 8$ and the sum of the coefficients is $(-1) + (-1) + 10 + (-8) = 0$. Thus, -1 is a zero and we will use synthetic division to find the remaining polynomial.

$$\begin{array}{r|rrrr}
-1 & 1 & -1 & -10 & -8 \\
   &   & -1 &   2 & 8 \\
\hline
   & 1 & -2 & -8 & 0
\end{array}$$

The remaining polynomial is $(x^2 - 2x - 8) = (x - 4)(x + 2)$.

Hence, the solutions are -2, -1, and 4.

[17] As previously noted, we first check ± 1. Neither of these values are solutions so we list the possible rational roots. The values listed below are obtained by taking quotients of each number in the numerator with ± 1, and then each number in the numerator with ± 2, while discarding any repeat choices.

$$\frac{1,\, 2,\, 3,\, 5,\, 6,\, 10,\, 15,\, 30}{\pm 1,\ \pm 2} \longrightarrow$$
$$\pm 1,\ \pm 2,\ \pm 3,\ \pm 5,\ \pm 6,\ \pm 10,\ \pm 15,\ \pm 30,\ \pm \tfrac{1}{2},\ \pm \tfrac{3}{2},\ \pm \tfrac{5}{2},\ \pm \tfrac{15}{2}$$

Trying 2 we obtain

$$\begin{array}{r|rrrr}
2 & 2 & -3 & -17 & 30 \\
  &   & 4 &   2 & -30 \\
\hline
  & 2 & 1 & -15 & 0
\end{array}$$

The remaining polynomial is $(2x^2 + x - 15) = (2x - 5)(x + 3)$.

Hence, the solutions are -3, 2, and $\tfrac{5}{2}$.

[19] $\dfrac{1,\ 2,\ 4,\ 7,\ 8,\ 14,\ 28,\ 56}{\pm 1} \longrightarrow \pm 1,\ \pm 2,\ \pm 4,\ \pm 7,\ \pm 8,\ \pm 14,\ \pm 28,\ \pm 56.$ Again, ± 1

are not solutions. Trying 4 and then -7 (using a slightly different format) we obtain

$$
\begin{array}{r|rrrrr}
4 & 1 & 3 & -30 & -6 & 56 \\
  &   & 4 & 28 & -8 & -56 \\
\hline
-7 & 1 & 7 & -2 & -14 & 0 \\
   &   & -7 & 0 & 14 & \\
\hline
   & 1 & 0 & -2 & 0 &
\end{array}
$$

The remaining polynomial is $x^2 - 2$. Its solutions are $\pm\sqrt{2}$.

Hence, the solutions are -7, $\pm\sqrt{2}$, and 4.

[21] We first factor out x^2, leaving us with the equation $x^2(6x^3 + 19x^2 + x - 6) = 0$. The x^2 factor indicates that the number 0 is a zero of multiplicity two. We can now concentrate on solving the equation $6x^3 + 19x^2 + x - 6 = 0$.

$$\dfrac{1,\ 2,\ 3,\ 6}{\pm 1,\ \pm 2,\ \pm 3,\ \pm 6} \longrightarrow \pm 1,\ \pm 2,\ \pm 3,\ \pm 6,\ \pm\tfrac{1}{2},\ \pm\tfrac{3}{2},\ \pm\tfrac{1}{3},\ \pm\tfrac{2}{3},\ \pm\tfrac{1}{6}$$

$$
\begin{array}{r|rrrr}
-3 & 6 & 19 & 1 & -6 \\
   &   & -18 & -3 & 6 \\
\hline
   & 6 & 1 & -2 & 0
\end{array}
$$

$6x^2 + x - 2 = (3x+2)(2x-1)$ $\bigstar\ -3,\ -\tfrac{2}{3},\ \tfrac{1}{2}$

[23] $\dfrac{1,\ 3,\ 9,\ 27}{\pm 1,\ \pm 2,\ \pm 4,\ \pm 8} \longrightarrow$

$\pm 1,\ \pm 3,\ \pm 9,\ \pm 27,\ \pm\tfrac{1}{2},\ \pm\tfrac{1}{4},\ \pm\tfrac{1}{8},\ \pm\tfrac{3}{2},\ \pm\tfrac{3}{4},\ \pm\tfrac{3}{8},\ \pm\tfrac{9}{2},\ \pm\tfrac{9}{4},\ \pm\tfrac{9}{8},\ \pm\tfrac{27}{2},\ \pm\tfrac{27}{4},\ \pm\tfrac{27}{8}$

This is a tough one because only $-\tfrac{3}{4}$ is a solution.

$$
\begin{array}{r|rrrr}
-\tfrac{3}{4} & 8 & 18 & 45 & 27 \\
              &   & -6 & -9 & -27 \\
\hline
              & 8 & 12 & 36 & 0
\end{array}
$$

$8x^2 + 12x + 36 = 0$ { divide by 4 } $\Rightarrow 2x^2 + 3x + 9 = 0 \Rightarrow$

$x = \dfrac{-3 \pm \sqrt{9-72}}{4} = -\tfrac{3}{4} \pm \tfrac{1}{4}\sqrt{63}\,i = -\tfrac{3}{4} \pm \tfrac{3}{4}\sqrt{7}\,i$ $\bigstar\ -\tfrac{3}{4},\ -\tfrac{3}{4} \pm \tfrac{3}{4}\sqrt{7}\,i$

[25] $f(x) = 6x^5 - 23x^4 + 24x^3 + x^2 - 12x + 4$ has zeros at $-\tfrac{2}{3}, \tfrac{1}{2}, 1$ (mult. 2), and 2.

Thus, $f(x) = 6(x + \tfrac{2}{3})(x - \tfrac{1}{2})(x-1)^2(x-2) = (3x+2)(2x-1)(x-1)^2(x-2)$.

27 From the graph, we see that $f(x) = 2x^3 - 25.4x^2 + 3.02x + 24.75$ has zeros of approximately -0.9, 1.1, and 12.5. Since the leading coefficient of f is 2, we have $f(x) = 2(x + 0.9)(x - 1.1)(x - 12.5)$.

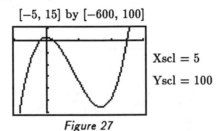

$[-5, 15]$ by $[-600, 100]$

Xscl $= 5$

Yscl $= 100$

Figure 27

33 (a) $V(x) = x(20 - 2x)(30 - 2x) = 1000 \Rightarrow 4x^3 - 100x^2 + 600x - 1000 = 0 \Rightarrow$
$4(x - 5)\left[x - (10 - 5\sqrt{2})\right]\left[x - (10 + 5\sqrt{2})\right] = 0$. The allowable range from Exercise 33 of Section 4.1 was $0 < x < 10$, so discard $10 + 5\sqrt{2}$.

The two boxes having volume 1000 in^3 have dimensions

[A] $5 \times 10 \times 20$ and   [B] $(10 - 5\sqrt{2}) \times (10\sqrt{2}) \times (10 + 10\sqrt{2})$.

(b) The surface area function is

$S(x) = (20 - 2x)(30 - 2x) + 2(x)(20 - 2x) + 2(x)(30 - 2x) = -4x^2 + 600$.

$S(5) = 500$ and $S(10 - 5\sqrt{2}) = 400\sqrt{2} \approx 565.7$ so box [A] has less surface area.

35 (a) Let x denote one of the sides of the triangle and $x + 1$ its hypotenuse.

Using the Pythagorean theorem, the third side y is given by

$$x^2 + y^2 = (x + 1)^2 \Rightarrow y^2 = (x^2 + 2x + 1) - x^2 \Rightarrow y = \sqrt{2x + 1}.$$

Hence, the sides of the triangle are x, $\sqrt{2x + 1}$, and $x + 1$.

$A = \frac{1}{2}bh \Rightarrow 30 = \frac{1}{2}x\sqrt{2x + 1} \Rightarrow 60 = x\sqrt{2x + 1} \Rightarrow 60^2 = x^2(2x + 1) \Rightarrow$
$$3600 = 2x^3 + x^2 \Rightarrow 2x^3 + x^2 - 3600 = 0.$$

(b) There is one sign change in $f(x) = 2x^3 + x^2 - 3600$.

By Descartes' rule of signs there is one positive real root.

Synthetically dividing 13 into f, we obtain

$$
\begin{array}{r|rrrr}
13 & 2 & 1 & 0 & -3600 \\
   &   & 26 & 351 & 4563 \\
\hline
   & 2 & 27 & 351 & 963
\end{array}
$$

The numbers in the third row are nonnegative,
so 13 is an upper bound for the zeros of f.

(c) $2x^3 + x^2 - 3600 = 0 \Leftrightarrow (x - 12)(2x^2 + 25x + 300) = 0$. The solutions of $2x^2 + 25x + 300 = 0$ are $x = -\frac{25}{4} \pm \frac{5}{4}\sqrt{71}\, i$, and hence, $x = 12$ is the only real solution. The legs of the triangle are 12 ft and 5 ft, and the hypotenuse is 13 ft.

$\boxed{37}$ (a) $\text{Volume}_{\text{total}} = \text{Volume}_{\text{cube}} + \text{Volume}_{\text{roof}}$

$$= x^3 + \tfrac{1}{2}bhx = x^3 + \tfrac{1}{2}(x)(6-x)(x) = x^3 + \tfrac{1}{2}x^2(6-x).$$

(b) Volume $= 80 \Rightarrow x^3 + \tfrac{1}{2}x^2(6-x) = 80 \Rightarrow x^3 + 3x^2 - \tfrac{1}{2}x^3 = 80 \Rightarrow \tfrac{1}{2}x^3 + 3x^2 = 80 \Rightarrow$

$x^3 + 6x^2 - 160 = 0 \Rightarrow (x-4)(x^2 + 10x + 40) = 0.$ The length of the side is 4 ft.

$\boxed{39}$ $x^5 + 1.1x^4 - 3.21x^3 - 2.835x^2 + 2.7x + 0.62 = -1 \Leftrightarrow$

$x^5 + 1.1x^4 - 3.21x^3 - 2.835x^2 + 2.7x + 1.62 = 0.$

The graph of $y = x^5 + 1.1x^4 - 3.21x^3 - 2.835x^2 + 2.7x + 1.62$ intersects the x-axis three times. The zeros at -1.5 and 1.2 have even multiplicity (since the graph is tangent to the x-axis at these points) and the zero at -0.5 has odd multiplicity (since the graph crosses the x-axis at this point). Since the equation has degree 5, the only possibility is that the zeros at -1.5 and 1.2 have multiplicity 2 and the zero at -0.5 has multiplicity 1. Thus, the equation has no nonreal solutions.

$[-4.5, 4.5]$ by $[-3, 3]$

$[-4.5, 4.5]$ by $[-3, 3]$

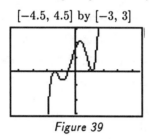

Xscl $= 1$

Yscl $= 1$

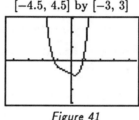

Xscl $= 1$

Yscl $= 1$

Figure 39

Figure 41

$\boxed{41}$ Graph $y = x^4 + 1.4x^3 + 0.44x^2 - 0.56x - 0.96$.

From the graph, zeros are located at -1.2 and 0.8. Using synthetic division,

$$\frac{x^4 + 1.4x^3 + 0.44x^2 - 0.56x - 0.96}{x + 1.2} = x^3 + 0.2x^2 + 0.2x - 0.8 \text{ and}$$

$$\frac{x^3 + 0.2x^2 + 0.2x - 0.8}{x - 0.8} = x^2 + x + 1. \text{ The zeros of } x^2 + x + 1 \text{ are } -\tfrac{1}{2} \pm \tfrac{\sqrt{3}}{2}i.$$

Thus, the solutions to the equation are $-1.2, 0.8, -\tfrac{1}{2} \pm \tfrac{\sqrt{3}}{2}i.$

$\boxed{43}$ From the graph, we see that $D(h) = 0.4$ when $h \approx 10{,}200$.

Thus, the density of the atmosphere is 0.4 kg/m^3 at $10{,}200$ m.

$[0, 30{,}000]$ by $[0, 1.2]$

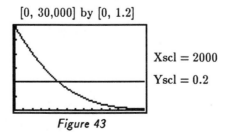

Xscl $= 2000$

Yscl $= 0.2$

Figure 43

Note: We will use the guidelines listed in the text on page 279.

Below is a summary of these guidelines.

$$\text{Let } f(x) = \frac{a_n x^n + a_{n-1} x^{n-1} + \cdots + a_1 x + a_0}{b_k x^k + b_{k-1} x^{k-1} + \cdots + b_1 x + b_0}, \text{ where } a_n \neq 0 \text{ and } b_k \neq 0.$$

(1) Find the x-intercepts { the zeros of the numerator }.

(2) Find the vertical asymptotes { the zeros of the denominator }.

(3) Find the y-intercept { the ratio of constant terms, a_0/b_0 }.

(4) Find the horizontal or oblique asymptote.

   (a) If $n < k$, then $y = 0$ { the x-axis } is the horizontal asymptote.

   (b) If $n = k$,

      then $y = a_n/b_k$ { the ratio of leading coefficients } is the horizontal asymptote.

   (c) If $n > k$, then the asymptote is found by long division. The graph of f is

      asymptotic to $y = q(x)$, where $q(x)$ is the quotient of the division process.

(5) Find the intersection points of the function and the asymptote found in step 3. This

   will help us decide *how* the function approaches the asymptote.

(6) Sketch the graph by regions, where the regions are determined by the vertical

   asymptotes. We may use the sign of a particular function value to help us determine

   if the function is positive or negative. We will not plot any points since the purpose

   in this section is to determine the general shape of the graph and understand the

   general principles involved with asymptotes.

$\boxed{1}$ $f(x) = 4/x$ •

(a) (1) There are no x-intercepts since the numerator is never equal to zero.

(2) There is a vertical asymptote at $x = 0$ {the y-axis}.

(3) There is no y-intercept since the function is undefined for $x = 0$.

(4) The degree of the numerator is 0, which is less than the degree of the denominator, 1, so the horizontal asymptote is $y = 0$ {the x-axis}.

(5) Setting the function equal to the value of the asymptote found in step 4 gives us $4/x = 0$, which has no solutions.

(6) There is one vertical asymptote, and it separates the plane into 2 regions. For the region $x < 0$, $f(x) = 4/x < 0$—that is, the y-values are negative. This indicates that the graph is under the y-axis as in *Figure 1*. For the region $x > 0$, $f(x) > 0$, and the graph is above the x-axis.

(b) The domain D is the set of all nonzero real numbers—that is, $\mathbb{R} - \{0\}$. The range R is equal to the same set of numbers.

(c) The function is decreasing on $(-\infty, 0)$ and also on $(0, \infty)$.

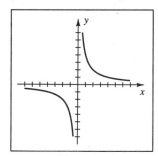

Figure 1

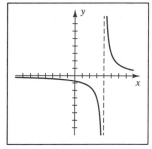

Figure 3

$\boxed{3}$ $f(x) = \dfrac{3}{x-4} = 3\left(\dfrac{1}{x-4}\right)$ •

Rather than use the guidelines, we will think of this graph in terms of the shifting and stretching properties presented in Section 3.5. Consider the graph of $g(x) = 1/x$, which is similar to the graph in Exercise 1. Now $h(x) = 1/(x-4)$ is just $g(x-4)$, which shifts the graph of g to the right by 4 units, making $x = 4$ the vertical asymptote. The effect of the 3 in the numerator is to vertically stretch the graph of h by a factor of 3.

⑤ $f(x) = \frac{-3x}{x+2}$ •

(1) $-3x = 0 \Rightarrow x = 0$, the <u>only</u> x-intercept.

(2) $x + 2 = 0 \Rightarrow x = -2$, the only vertical asymptote.

(3) $f(0) = \frac{0}{2} = 0 \Rightarrow (0, 0)$ is the y-intercept. { We already knew this from step 1. }

(4) Degree of numerator $= 1 =$ degree of denominator $\Rightarrow$

$$y = \frac{-3}{1} \{\text{ ratio of leading coefficients }\} = -3 \text{ is the horizontal asymptote.}$$

(5) Function $=$ asymptote $\Rightarrow f(x) = -3 \Rightarrow \frac{-3x}{x+2} = -3 \Rightarrow -3x = -3(x+2) \Rightarrow$

$-3x = -3x - 6 \Rightarrow 0 = -6$. This is a contradiction and indicates that there are <u>no</u> intersection points on the horizontal asymptote. Remember, the function can not intersect a vertical asymptote, but can intersect the horizontal asymptote.

(6) For the region $x < -2$, there are no x-intercepts and the graph must be below the horizontal asymptote $y = -3$. { If it was above $y = -3$, it would have to intersect the x-axis at some point. }

For the region $x > -2$, there is an x-intercept at 0 and the graph is above the horizontal asymptote. A common mistake is to confuse the x-axis with the horizontal asymptote. Remember, as $x \to \infty$ or $x \to -\infty$, $f(x)$ will get close to the horizontal or oblique asymptote, not the x-axis { unless the x-axis is the horizontal asymptote }.

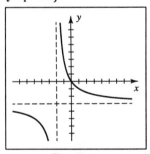

Figure 5

$\boxed{7}$ $f(x) = \dfrac{x-2}{x^2-x-6} = \dfrac{x-2}{(x+2)(x-3)}$ •

(1) $x - 2 = 0 \Rightarrow x = 2$, the only x-intercept

(2) $(x+2)(x-3) = 0 \Rightarrow x = -2$ and $x = 3$, the vertical asymptotes

(3) $f(0) = \dfrac{-2}{-6} = \dfrac{1}{3} \Rightarrow (0, \frac{1}{3})$ is the y-intercept

(4) Degree of numerator $= 1 < 2 =$ degree of denominator $\Rightarrow$

$y = 0$ is the horizontal asymptote.

(5) We have already solved $f(x) = 0$ in step 1.

Hence, we know that the function will cross the horizontal asymptote at $(2, 0)$.

(6) For $x < -2$, we have no information and will examine $f(-5)$ { -5 is to the left of -2 }. Using the factored form of f and only the signs of the values, we obtain $f(-5) = \dfrac{(-)}{(-)(-)} = \{$ some negative number $\} < 0$ since the combination of 3 negative signs will be negative. Hence, the graph will be below the horizontal asymptote.

For $-2 < x < 3$, we have an x-intercept at 2 and need to know what the function does on each side of 2. Choosing 0 and 2.5 for test points, we check the sign of $f(0)$ and $f(2.5)$ as above. $f(0) = \dfrac{(-)}{(+)(-)} > 0$ and $f(2.5) = \dfrac{(+)}{(+)(-)} < 0$. Hence, the function will change signs from positive to negative as it passes through 2.

For $x > 3$, $f(5) = \dfrac{(+)}{(+)(+)} > 0$ and the function is above the horizontal asymptote.

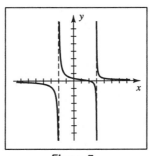

Figure 7

$\boxed{9}$ $f(x) = \dfrac{-4}{(x-2)^2}$ • Note that the function is always negative since it is the quotient of a negative and a positive, provided $x \neq 2$.

$\boxed{11}$ $f(x) = \dfrac{x-3}{x^2-1} = \dfrac{x-3}{(x+1)(x-1)}$ •

(6) For $-1 < x < 1$, we have the y-intercept at 3. Since the function cannot intersect the x-axis {the only x-intercept is at 3}, the function remains positive and we have $f(x) \to \infty$ as $x \to 1^-$ and $f(x) \to \infty$ as $x \to -1^+$.

For $x > 1$, we have the x-intercept at 3 as a point to work with. The function will change sign at 3 since 3 is a zero of odd multiplicity {1}. If the function was going to merely touch the point (3, 0) and "turn around", 3 would have to be a zero of even multiplicity. Hence, it suffices to find only one sign. We choose 2 as our test point. $f(2) = \dfrac{(-)}{(+)(+)} < 0$. The function values are negative for $1 < x < 3$ and positive for $x > 3$. Remember, eventually $f(x)$ will be extremely close to 0 as $x \to \infty$. Make sure your sketch reflects that fact.

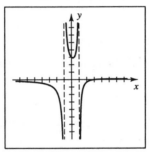

Figure 11

⬚13 $f(x) = \dfrac{2x^2 - 2x - 4}{x^2 + x - 12} = \dfrac{2(x+1)(x-2)}{(x+4)(x-3)}$ • As a generalization, you should always factor the function first. For steps 1–6 of the guidelines, it is convenient to use the original function for steps 3, 4, and 5, and to use the factored form for steps 1, 2, and 6. After this problem, we will concentrate on steps 3 { when appropriate }, 5, and 6. You should be able to pick up the information for steps 1–4 by merely looking at both forms of the function { except for the oblique asymptote case }.

(1) $2(x+1)(x-2) = 0 \Rightarrow x = -1, 2$; the x-intercepts

(2) $(x+4)(x-3) = 0 \Rightarrow x = -4, 3$; the vertical asymptotes

(3) $f(0) = \frac{-4}{-12} = \frac{1}{3} \Rightarrow (0, \frac{1}{3})$ is the y-intercept

(4) Degree of numerator $= 2 =$ degree of denominator $\Rightarrow$

$$y = \tfrac{2}{1} = 2 \text{ is the horizontal asymptote.}$$

(5) $\dfrac{2x^2 - 2x - 4}{x^2 + x - 12} = 2 \Rightarrow 2x^2 - 2x - 4 = 2(x^2 + x - 12) \Rightarrow$

$2x^2 - 2x - 4 = 2x^2 + 2x - 24 \Rightarrow 20 = 4x \Rightarrow x = 5$. Hence, the function will

intersect the horizontal asymptote <u>only</u> at the point $(5, 2)$.

(6) For $x < -4$, there are no x-intercepts so the function is above the horizontal asymptote.

For $-4 < x < 3$, the function must go through $(-1, 0)$, $(0, \frac{1}{3})$, and $(2, 0)$. It must go "down" by both vertical asymptotes because if it went "up", it would have to cross the horizontal asymptote. But we know this only occurs at the point $(5, 2)$ and not in this region.

For $x > 3$, we know that the graph goes up as we get close to 3 because if it went down, there would have to be an x-intercept, but there isn't one. The function goes through $(5, 2)$, but then turns around { before touching the x-axis } and gets very close to 2 as x increases.

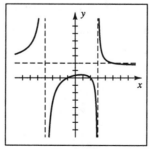

Figure 13

Note: We will let $I(x, y)$ denote the intersection point found in step 5.

$\boxed{15}$ $f(x) = \dfrac{-x^2 - x + 6}{x^2 + 3x - 4} = \dfrac{-1(x+3)(x-2)}{(x+4)(x-1)}$ •

(5) $\dfrac{-x^2 - x + 6}{x^2 + 3x - 4} = -1 \Rightarrow -x^2 - x + 6 = -x^2 - 3x + 4 \Rightarrow$

$$2x = -2 \Rightarrow x = -1, \, I = (-1, \, -1)$$

(6) For $x < -4$, the function is below $y = -1$ since there are no x-intercepts.

For $-4 < x < 1$, the function passes through $(-3, \, 0)$, $(-1, \, -1)$, and $(0, \, -\frac{3}{2})$. These points are known from finding information in steps 1–5.

For $x > 1$, the function passes through $(2, \, 0)$ and is always above the horizontal asymptote since it only intersects the horizontal asymptote at $(-1, \, -1)$.

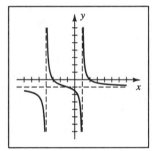

Figure 15

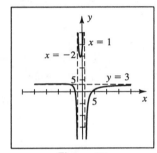

Figure 17

$\boxed{17}$ $f(x) = \dfrac{3x^2 - 3x - 36}{x^2 + x - 2} = \dfrac{3(x+3)(x-4)}{(x+2)(x-1)}$ •

(5) $\dfrac{3x^2 - 3x - 36}{x^2 + x - 2} = 3 \Rightarrow 3x^2 - 3x - 36 = 3x^2 + 3x - 6 \Rightarrow$

$$-30 = 6x \Rightarrow x = -5, \, I = (-5, \, 3)$$

(6) For $x < -2$, the function passes through $(-3, \, 0)$ and $(-5, \, 3)$ and stays above the horizontal asymptote, $y = 3$, as $x \to -\infty$.

For $-2 < x < 1$, we have the y-intercept at $(0, \, 18)$. Since the graph doesn't intersect the horizontal asymptote in this region, the function goes up as it gets close to each vertical asymptote.

For $x > 1$, we have the x-intercept $(4, \, 0)$ and the function stays below the horizontal asymptote.

$\boxed{19}$ $f(x) = \dfrac{-2x^2 + 10x - 12}{x^2 + x} = \dfrac{-2(x-2)(x-3)}{(x+1)(x)}$ •

(2) Since $(x+1)(x) = 0 \Rightarrow x = -1$ and 0,

we have the y-axis as a vertical asymptote, and hence, there is no y-intercept.

(5) $\dfrac{-2x^2 + 10x - 12}{x^2 + x} = -2 \Rightarrow -2x^2 + 10x - 12 = -2x^2 - 2x \Rightarrow$

$12x = 12 \Rightarrow x = 1,\ I = (1, -2)$

(6) For $-1 < x < 0$, $f(-\frac{1}{2}) = \dfrac{(-)(-)(-)}{(+)(-)} > 0$ and the function is above the x-axis.

For $x > 0$, the function goes through $(1, -2)$, $(2, 0)$, and $(3, 0)$ and gradually gets

closer to $y = -3$ as $x \to \infty$.

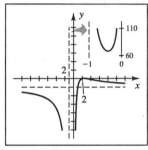

Figure 19

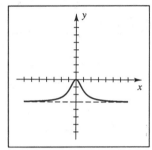

Figure 23

$\boxed{23}$ $f(x) = \dfrac{-3x^2}{x^2 + 1}$ • f is an even function

(2) There are no vertical asymptotes since $x^2 + 1 \neq 0$ for any real number.

(5) $\dfrac{-3x^2}{x^2 + 1} = -3 \Rightarrow -3x^2 = -3x^2 - 3 \Rightarrow 0 = -3 \Rightarrow$

there are no intersection points of the horizontal asymptote and the function.

(6) We have the x-intercept at $(0, 0)$ and the function must get close to the

horizontal asymptote, $y = -3$, as $x \to \infty$ or $x \to -\infty$.

[25] $f(x) = \dfrac{x^2 - x - 6}{x + 1} = \dfrac{(x + 2)(x - 3)}{x + 1}$ •

(4) We could use long division to find the oblique asymptote, but since the denominator is of the from $x - c$, we will use synthetic division.

$$
\begin{array}{r|rrr}
-1 & 1 & -1 & -6 \\
   &   & -1 & 2 \\
\hline
   & 1 & -2 & -4
\end{array}
$$

The third row indicates that $f(x) = x - 2 - \dfrac{4}{x + 1}$. The expression $\dfrac{4}{x + 1} \to 0$ as $x \to \pm\infty$, so $y = x - 2$ is an oblique asymptote for f.

(5) $\dfrac{x^2 - x - 6}{x + 1} = x - 2 \Rightarrow x^2 - x - 6 = (x - 2)(x + 1) \Rightarrow$

$x^2 - x - 6 = x^2 - x - 2 \Rightarrow -6 = -2 \Rightarrow$ there are no intersection points of the oblique asymptote and the function.

(6) For $x < -1$, we have the x-intercept at $(-2,\ 0)$ and the function is above the oblique asymptote.

For $x > -1$, we have the points $(0,\ -6)$ and $(3,\ 0)$ and the function is below the oblique asymptote.

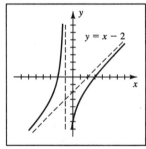 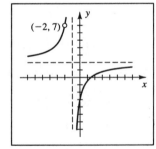

Figure 25   Figure 29

[29] $f(x) = \dfrac{2x^2 + x - 6}{x^2 + 3x + 2} = \dfrac{(x + 2)(2x - 3)}{(x + 2)(x + 1)} = \dfrac{2x - 3}{x + 1}$ for $x \neq -2$. We must reduce the function first and then *sketch the reduced function*. After sketching the reduced function, we need to remember to put a hole in the graph where the original function was undefined. In this case, the reduced function is $f(x) = \dfrac{2x - 3}{x + 1}$. We sketch this as we did the other rational functions. To determine the value of y when $x = -2$, substitute -2 into $\dfrac{2x - 3}{x + 1}$ to get 7. There is a hole in the graph at $(-2,\ 7)$.

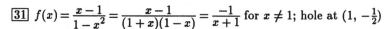

$\boxed{31}$ $f(x) = \dfrac{x-1}{1-x^2} = \dfrac{x-1}{(1+x)(1-x)} = \dfrac{-1}{x+1}$ for $x \neq 1$; hole at $(1, -\frac{1}{2})$

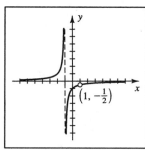

Figure 31   Figure 35

$\boxed{35}$ $f(x) = \dfrac{x^2+4x+4}{x^2+3x+2} = \dfrac{(x+2)^2}{(x+1)(x+2)} = \dfrac{x+2}{x+1}$ for $x \neq -2$.

Note that the hole is on the x-axis at $(-2, 0)$.

$\boxed{37}$ The vertical asymptote at $x = 4$ implies that $x - 4$ is a factor in the denominator. Since 3 is an x-intercept, $x - 3$ is a factor in the numerator. The ratio of leading coefficients must be -1 because the horizontal asymptote is $y = -1$, and hence, we must have a factor of -1 in the numerator. Therefore, f has the form

$$f(x) = \frac{-1(x-3)}{x-4} = \frac{3-x}{x-4}.$$

$\boxed{39}$ The hole at $x = 2$ indicates that $x - 2$ is a factor in the numerator and in the denominator. If we let a denote an arbitrary coefficient, then f has the form

$$f(x) = \frac{a(x+1)(x-2)}{(x-1)(x+3)(x-2)}.$$

To determine the value of a, we use the fact that $f(0) = -2$.

$f(0) = -2$ and $f(0) = \dfrac{a(1)}{(-1)(3)} = \dfrac{a}{-3} \Rightarrow \dfrac{a}{-3} = -2 \Rightarrow a = 6$.

Therefore, f has the form $f(x) = \dfrac{6(x+1)(x-2)}{(x-1)(x+3)(x-2)} = \dfrac{6x^2-6x-12}{x^3-7x+6}$.

$\boxed{41}$ (a) The radius of the outside cylinder is $(r + 0.5)$ ft and its height is $(h + 1)$ ft. Since the volume is 16π ft^3, we have $16\pi = \pi(r+0.5)^2(h+1) \Rightarrow$

$$h + 1 = \frac{16\pi}{\pi(r+0.5)^2} \Rightarrow h = \frac{16}{(r+0.5)^2} - 1.$$

(b) $V(r) = \pi r^2 h = \pi r^2 \left[\dfrac{16}{(r+0.5)^2} - 1 \right]$

(c) r and h must both be positive. $h > 0 \Rightarrow \dfrac{16}{(r+0.5)^2} - 1 > 0 \Rightarrow$

$16 > (r+0.5)^2 \Rightarrow |r+0.5| < 4 \Rightarrow -4.5 < r < 3.5$. The last inequality combined with $r > 0$ means that the excluded values are $r \leq 0$ and $r \geq 3.5$.

$\boxed{43}$ (a) Since 5 gallons of water flow into the tank each minute, $V(t) = 50 + 5t$. Since each additional gallon of water contains 0.1 lb of salt, $A(t) = 5(0.1)t = 0.5t$.

(b) $c(t) = \dfrac{A(t)}{V(t)} = \dfrac{0.5t}{50 + 5t} = \dfrac{t}{10t + 100}$ lb/gal

(c) As $t \to \infty$, $c(t) \to 0.1$ lb. of salt per gal.

$\boxed{45}$ (a) $R > S \Rightarrow \dfrac{4500\,S}{S + 500} > S \Rightarrow \dfrac{4500\,S}{S + 500} - S > 0 \Rightarrow \dfrac{4500\,S}{S + 500} - \dfrac{S(S + 500)}{S + 500} > 0 \Rightarrow$

$\dfrac{4500\,S - S^2 - 500\,S}{S + 500} > 0 \Rightarrow \dfrac{4000\,S - S^2}{S + 500} > 0 \Rightarrow \dfrac{S^2 - 4000\,S}{S + 500} < 0$ { multiply both

sides by -1 and reverse the direction of the inequality } $\Rightarrow \dfrac{S(S - 4000)}{S + 500} < 0$.

From *Diagram 45* and using the fact that $S > 0$,

we conclude that $R > S$ when $0 < S < 4000$.

Resulting sign:	$\ominus$	$\oplus$	$\ominus$	$\oplus$
$S - 4000$:	−	−	−	+
S:	−	−	+	+
$S + 500$:	−	+	+	+
S values:	−500	0	4000	

Diagram 45

(b) The greatest possible number of offspring that survive to maturity is 4500, the horizontal asymptote value. 90% of 4500 is 4050. $R = 4050 \Rightarrow$

$4050 = \dfrac{4500\,S}{S + 500} \Rightarrow 4050S + 2{,}025{,}000 = 4500S \Rightarrow 2{,}025{,}000 = 450S \Rightarrow S = 4500$.

(c) 80% of 4500 is 3600. $R = 3600 \Rightarrow$

$3600 = \dfrac{4500\,S}{S + 500} \Rightarrow 3600S + 1{,}800{,}000 = 4500S \Rightarrow 1{,}800{,}000 = 900S \Rightarrow S = 2000$.

(d) A 125% increase $\left\{ \dfrac{4500 - 2000}{2000} \times 100 \right\}$ in the number S of spawners produces only

a 12.5% increase $\left\{ \dfrac{4050 - 3600}{3600} \times 100 \right\}$ in the number R of offspring surviving to

maturity.

47 Assign $20x^2 + 80x + 72$ to Y_1, $10x^2 + 40x + 41$ to Y_2, and Y_1/Y_2 to Y_3. Zoom-in around $(-2, -8)$ to confirm that this a low point and that there is not a vertical asymptote at $x = -2$. To determine the vertical asymptotes, graph Y_2 only {turn off Y_3}, and look for its zeros. If these values are not zeros of the numerator, then they are the values of the vertical asymptotes. No vertical asymptotes in this case.

[−9, 3] by [−9, 3]

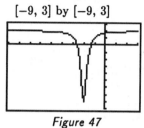

Xscl = 1

Yscl = 1

Figure 47

[0.7, 1.3] by [0.8, 1.2]

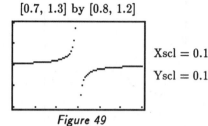

Xscl = 0.1

Yscl = 0.1

Figure 49

49 $f(x) = \dfrac{(x-1)^2}{(x - 0.999)^2}$ • Note that the standard viewing rectangle gives the horizontal line $y = 1$. *Figure 49* was obtained by using Dot mode. If Connected mode is used, the calculator will draw a near-vertical line at $x = 0.999$. An equation of the vertical asymptote is $x = 0.999$.

51 (a) The graph of g is the horizontal line $y = 1$ with holes at $x = 0$, ± 1, ± 2, ± 3.

The TI-85 shows a small break in the line to indicate a hole.

(b) The graph of h is the graph of p with holes at $x = 0$, ± 1, ± 2, ± 3.

4.6 Exercises

Note: For Exercises 1–12, we will put each parabola equation in one of the forms listed on page 293—either

$$(x - h)^2 = 4p(y - k) \quad \text{or} \quad (y - k)^2 = 4p(x - h).$$

Once in one of those forms, the information concerning the vertex, focus, and directrix is easily obtainable and illustrated in the chart in the text. Let V, F, and l denote the vertex, focus, and directrix, respectively.

3 $2y^2 = -3x \Rightarrow (y-0)^2 = -\frac{3}{2}(x-0) \Rightarrow 4p = -\frac{3}{2} \Rightarrow p = -\frac{3}{8}$. We know that the parabola opens either right or left since the variable "y" is squared. Since p is negative, we know that the parabola opens left and that the focus is $\frac{3}{8}$ unit to the left of the vertex. The directrix is $\frac{3}{8}$ unit to the right of the vertex.

$$V(0, 0); \ F(-\tfrac{3}{8}, 0); \ l: \ x = \tfrac{3}{8}$$

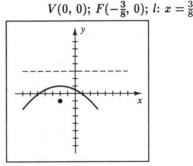

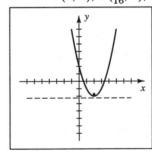

Figure 3	Figure 5

5 $(x+2)^2 = -8(y-1) \Rightarrow 4p = -8 \Rightarrow p = -2$. The $(x+2)$ and $(y-1)$ factors indicate that we need to shift the vertex, of the parabola having equation $x^2 = -8y$, 2 units left and 1 unit up—that is, move it from $(0, 0)$ to $(-2, 1)$.

$$V(-2, 1); \ F(-2, -1); \ l: \ y = 3$$

7 $(y-2)^2 = \frac{1}{4}(x-3) \Rightarrow 4p = \frac{1}{4} \Rightarrow p = \frac{1}{16}$. "$y$" squared and p positive imply that the parabola opens to the right and the focus is to the right of the vertex. The $(x-3)$ and $(y-2)$ factors indicate that we need to shift the vertex 3 units right and 2 units up from $(0, 0)$ to $(3, 2)$.

$$V(3, 2); \ F(\tfrac{49}{16}, 2); \ l: \ x = \tfrac{47}{16}$$

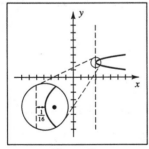

Figure 7	Figure 9

9 For this exercise, we need to "complete the square" in order to get the equation in proper form. The term we need to add is $\left[\frac{1}{2}(\text{coefficient of } x)\right]^2$. In this case, that value is $\left[\frac{1}{2}(-4)\right]^2 = 4$. Notice that we add and subtract the value 4 from the same side of the equation as opposed to adding 4 to both sides of the equation.

$y = x^2 - 4x + 2 = (x^2 - 4x + \underline{4}) + 2 - \underline{4} = (x-2)^2 - 2 \Rightarrow (y+2) = 1(x-2)^2 \Rightarrow$ $4p = 1 \Rightarrow p = \frac{1}{4}$.

$$V(2, -2); \ F(2, -\tfrac{7}{4}); \ l: \ y = -\tfrac{9}{4}$$

[13] Since the vertex is at $(1, 0)$ and the parabola has a horizontal axis, the standard equation has the form $y^2 = 4p(x-1)$. The distance from the focus $F(6, 0)$ to the vertex $V(1, 0)$ is $6-1=5$, which is the value of p. Hence, an equation of the parabola is $y^2 = 20(x-1)$.

[15] $V(-2, 3) \Rightarrow (x+2)^2 = 4p(y-3)$.
$$x = 2, \ y = 2 \Rightarrow 16 = 4p(-1) \Rightarrow p = -4. \quad (x+2)^2 = -16(y-3).$$

[17] The distance from the directrix to the focus is $2-(-2) = 4$ units. The vertex $V(0, 0)$ is halfway between the directrix and the focus—that is, 2 units from either one. Since the focus is 2 units to the right of the vertex, p is 2. Using one of the forms of an equation of a parabola with vertex at (h, k), we have $(y-0)^2 = 4p(x-0)$, or, equivalently, $y^2 = 8x$.

[19] $F(6, 4)$ and l: $y = -2 \Rightarrow p = 3$ and $V(6, 1)$.
$$(x-6)^2 = 4p(y-1) \Rightarrow (x-6)^2 = 12(y-1).$$

[21] $V(3, -5)$ and l: $x = 2 \Rightarrow p = 1$. $(y+5)^2 = 4p(x-3) \Rightarrow (y+5)^2 = 4(x-3)$.

[23] $V(-1, 0)$ and $F(-4, 0) \Rightarrow p = -3$. $(y-0)^2 = 4p(x+1) \Rightarrow y^2 = -12(x+1)$.

[25] The vertex at the origin and symmetric to the y-axis imply that the equation is of the form $y = ax^2$. Substituting $x = 2$ and $y = -3$ into that equation yields
$$-3 = a \cdot 4 \Rightarrow a = -\tfrac{3}{4}. \text{ Thus, an equation is } y = -\tfrac{3}{4}x^2, \text{ or } 3x^2 = -4y.$$

[27] The vertex at $(-3, 5)$ and axis parallel to the x-axis imply that the equation is of the form $(y-5)^2 = 4p(x+3)$. Substituting $x = 5$ and $y = 9$ into that equation yields $16 = 4p \cdot 8 \Rightarrow p = \tfrac{1}{2}$. Thus, an equation is $(y-5)^2 = 2(x+3)$.

[31] Refer to the definition of a parabola. The point $P(-6, 3)$ is the fixed point (focus) and the line l: $x = -2$ is the fixed line (directrix). The vertex is halfway between the focus and the directrix—that is, at $V(-4, 3)$. An equation is of the form $(y-3)^2 = 4p(x+4)$. The distance from the vertex to the focus is $p = -6-(-4) = -2$. Thus, an equation is $(y-3)^2 = -8(x+4)$.

Note: To find an equation for a lower or upper half, we need to solve for y (use $-$ or $+$ respectively). For the left or right half, solve for x (use $-$ or $+$ respectively).

[33] $(y+1)^2 = x+3 \Rightarrow y+1 = \pm\sqrt{x+3} \Rightarrow y = -\sqrt{x+3} - 1$

[35] $(x+1)^2 = y-4 \Rightarrow x+1 = \pm\sqrt{y-4} \Rightarrow x = \sqrt{y-4} - 1$

[37] A cross section of the mirror is a parabola with $V(0, 0)$ and passing through $P(4, 1)$. The incoming light will collect at the focus F. A general equation of this form of a parabola is $y = ax^2$. Substituting $x = 4$ and $y = 1$ gives us $1 = a(4)^2 \Rightarrow a = \tfrac{1}{16}$. $p = 1/(4a) = 1/(\tfrac{1}{4}) = 4$. The light will collect 4 inches from the center of the mirror.

[39] If we set up a coordinate system with a parabola opening upward and the vertex at the origin, then the phrase "3 feet across at the opening and 1 foot deep" implies that the points $(\pm\frac{3}{2}, 1)$ are on the parabola.

$$y = ax^2 \Rightarrow 1 = a(\tfrac{3}{2})^2 \Rightarrow a = \tfrac{4}{9}. \quad p = 1/(4a) = 1/(\tfrac{16}{9}) = \tfrac{9}{16} \text{ ft.}$$

[41] $p = 5 \Rightarrow a = 1/(4p) = 1/(4 \cdot 5) = \tfrac{1}{20}.$

$y = ax^2 \; \{y = 2 \underline{\text{ft}} = 24 \text{ inches}\} \Rightarrow 24 = \tfrac{1}{20}x^2 \Rightarrow x^2 = 480 \Rightarrow x = \sqrt{480}.$

The width is twice the value of x. Width $= 2\sqrt{480} \approx 43.82$ in.

[43] (a) Let the parabola have the equation $x^2 = 4py$. Since the point (r, h) is on the parabola, we can substitute r for x and h for y, giving us $r^2 = 4ph$. Solving for p we have $p = \dfrac{r^2}{4h}$.

(b) $p = 10$ and $h = 5 \Rightarrow r^2 = 4(10)(5) \Rightarrow r = 10\sqrt{2}.$

[45] With $a = 125$ and $p = 50$, $S = \dfrac{8\pi p^2}{3}\left[\left(1 + \dfrac{a^2}{4p^2}\right)^{3/2} - 1\right] \approx 64{,}968 \text{ ft}^2.$

[47] Depending on the type of calculator or software used, we may need to solve for y in terms of x. $x = -y^2 + 2y + 5 \Rightarrow y^2 - 2y + (x - 5) = 0$. This is a quadratic equation in y. Using the quadratic formula to solve for y yields

$$y = \dfrac{-(-2) \pm \sqrt{(-2)^2 - 4(1)(x - 5)}}{2(1)} = 1 \pm \sqrt{6 - x}.$$

[−11, 10] by [−7, 7]

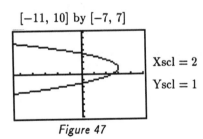

Xscl = 2
Yscl = 1

Figure 47

[−2, 4] by [−3, 3]

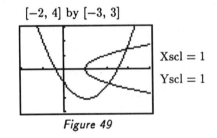

Xscl = 1
Yscl = 1

Figure 49

[49] $x = y^2 + 1 \Rightarrow y = \pm\sqrt{x - 1}$. From the graph, we can see that there are 2 points of intersection. Their coordinates are approximately $(2.08, -1.04)$ and $(2.92, 1.38)$.

4.7 Exercises

Note: Let C, V, F, and M denote the center, the vertices, the foci, and the end points of the minor axis, respectively. Let c denote the distance from the center of the ellipse to a focus.

1. $\frac{x^2}{9} + \frac{y^2}{4} = 1$ • The x-intercepts are at $(\pm\sqrt{9}, 0)$, or, equivalently, $(\pm 3, 0)$. The y-intercepts are at $(\pm\sqrt{4}, 0) = (\pm 2, 0)$. The major axis {the longer of the two axes} is the horizontal axis and has length $2(3) = 6$. The minor axis {the shorter} is the vertical axis and has length $2(2) = 4$. To find the foci, it is helpful to remember the relationship

$$\left[\tfrac{1}{2}(\text{minor axis})\right]^2 + \left[c\right]^2 = \left[\tfrac{1}{2}(\text{major axis})\right]^2.$$

Using the values from above we have $\left[\tfrac{1}{2}(4)\right]^2 + c^2 = \left[\tfrac{1}{2}(6)\right]^2 \Rightarrow 4 + c^2 = 9 \Rightarrow$ $c = \pm\sqrt{5}.$

$V(\pm 3, 0)$; $F(\pm\sqrt{5}, 0)$; $M(0, \pm 2)$ See *Figure 1*.

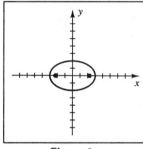

Figure 1

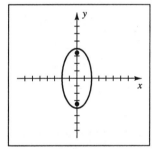

Figure 5

5. We first divide by 16 to obtain the "1" on the right side of the equation. $4x^2 + y^2 = 16 \Rightarrow \frac{x^2}{4} + \frac{y^2}{16} = 1$. Since the 16 in the denominator of the term with the variable y is larger than the 4 in the denominator of the term with the variable x, the vertices and foci are on the y-axis and the major axis is the vertical axis. $4 + c^2 = 16 \Rightarrow c^2 = 16 - 4 \Rightarrow c = \pm 2\sqrt{3}.$

$V(0, \pm 4)$; $F(0, \pm 2\sqrt{3})$; $M(\pm 2, 0)$

$\boxed{7}$ $4x^2 + 25y^2 = 1 \Rightarrow \dfrac{x^2}{\frac{1}{4}} + \dfrac{y^2}{\frac{1}{25}} = 1$. $\frac{1}{25} + c^2 = \frac{1}{4} \Rightarrow c^2 = \frac{1}{4} - \frac{1}{25} = \frac{21}{100} \Rightarrow c = \pm\frac{1}{10}\sqrt{21}$.

$V(\pm\frac{1}{2},\, 0)$; $F(\pm\frac{1}{10}\sqrt{21},\, 0)$; $M(0,\, \pm\frac{1}{5})$

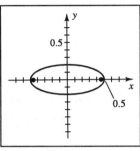

Figure 7

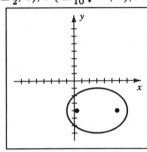

Figure 9

$\boxed{9}$ $\dfrac{(x-3)^2}{16} + \dfrac{(y+4)^2}{9} = 1$ •   The effect of the factors $(x-3)$ and $(y+4)$ is to shift the center of the ellipse from $(0,\,0)$ to $(3,\,-4)$. Since the larger denominator, 16, is in the term with x, the major axis will be horizontal. The end points are 4 units in either direction of the center. Their coordinates are the points $(3 \pm 4,\,-4)$, or, equivalently, $(7,\,-4)$ and $(-1,\,-4)$. The minor axis will be vertical with end points $(3,\,-4 \pm 3)$, or, equivalently, $(3,\,-1)$ and $(3,\,-7)$. $9 + c^2 = 16 \Rightarrow c^2 = 16 - 9 \Rightarrow$ $c = \pm\sqrt{7}$. Remember that c is the distance *from the center* to a focus. Hence, the coordinates of the foci are $(3 \pm \sqrt{7},\,-4)$.

$C(3,\,-4)$; $V(3 \pm 4,\,-4)$; $F(3 \pm \sqrt{7},\,-4)$; $M(3,\,-4 \pm 3)$

$\boxed{11}$ $4x^2 + 9y^2 - 32x - 36y + 64 = 0 \Rightarrow$

$(4x^2 - 32x) + (9y^2 - 36y) = -64 \Rightarrow$

{ first group the x terms and the y terms }

$4(x^2 - 8x) + 9(y^2 - 4y) = -64 \Rightarrow$

{ factor out the coefficients of x^2 and y^2 }

$4(x^2 - 8x + \underline{\;16\;}) + 9(y^2 - 4y + \underline{\;4\;}) = -64 + \underline{\;64\;} + \underline{\;36\;} \Rightarrow$

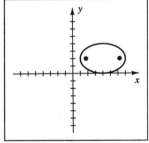

Figure 11

{ Complete the squares—remember that we have added $\underline{\;4\;}(16)$ and $\underline{\;9\;}(4)$, not 16 and 4. Thus, we add 64 and 36 to the right side of the equation. } $4(x-4)^2 + 9(y-2)^2 = 36 \Rightarrow \dfrac{(x-4)^2}{9} + \dfrac{(y-2)^2}{4} = 1$.

$c^2 = 9 - 4 \Rightarrow c = \pm\sqrt{5}$.

$C(4,\,2)$; $V(4 \pm 3,\,2)$; $F(4 \pm \sqrt{5},\,2)$; $M(4,\,2 \pm 2)$

Note: Let b denote the distance equal to $\frac{1}{2}$(minor axis length) and

let a denote the distance equal to $\frac{1}{2}$(major axis length).

15 $a = 2$ and $b = 6 \Rightarrow \frac{x^2}{a^2} + \frac{y^2}{b^2} = 1$ is $\frac{x^2}{4} + \frac{y^2}{36} = 1$.

17 The center of the ellipse is $(-2, 1)$. $a = 5$ and $b = 2$ give us $\frac{(x+2)^2}{25} + \frac{(y-1)^2}{4} = 1$.

19 Since the vertices are at $(\pm 8, 0)$, $\frac{1}{2}$(major axis) $= 8$. Since the foci are at $(\pm 5, 0)$,

$c = 5$. Using the relationship $\left[\frac{1}{2}\text{(minor axis)}\right]^2 + \left[c\right]^2 = \left[\frac{1}{2}\text{(major axis)}\right]^2$, we have

$\left[\frac{1}{2}\text{(minor axis)}\right]^2 + 5^2 = 8^2 \Rightarrow \left[\frac{1}{2}\text{(minor axis)}\right]^2 = 64 - 25 = 39$.

An equation is $\frac{x^2}{64} + \frac{y^2}{39} = 1$.

21 If the length of the minor axis is 3, then $b = \frac{3}{2}$.

An equation is $\frac{x^2}{(\frac{3}{2})^2} + \frac{y^2}{5^2} = 1$, or, equivalently, $\frac{4x^2}{9} + \frac{y^2}{25} = 1$.

23 With the vertices at $(0, \pm 6)$, an equation of the ellipse is $\frac{x^2}{b^2} + \frac{y^2}{6^2} = 1$. Substituting

$x = 3$ and $y = 2$ and solving for b^2 yields $\frac{9}{b^2} + \frac{4}{36} = 1 \Rightarrow \frac{9}{b^2} = \frac{8}{9} \Rightarrow b^2 = \frac{81}{8}$.

An equation is $\frac{x^2}{\frac{81}{8}} + \frac{y^2}{36} = 1$, or, equivalently, $\frac{8x^2}{81} + \frac{y^2}{36} = 1$.

25 With vertices $V(0, \pm 4)$, an equation of the ellipse is $\frac{x^2}{b^2} + \frac{y^2}{16} = 1$.

Remember the formula for the eccentricity:

$e = \text{eccentricity} = \frac{c}{a} = \frac{\text{distance from center to a focus}}{\text{distance from center to a vertex}}$.

Hence, $e = \frac{c}{a} = \frac{3}{4}$ and $a = 4 \Rightarrow c = 3$. Thus, $b^2 + c^2 = a^2 \Rightarrow$

$b^2 = a^2 - c^2 = 4^2 - 3^2 = 16 - 9 = 7$. An equation is $\frac{x^2}{7} + \frac{y^2}{16} = 1$.

29 Remember to divide the lengths of the major and minor axes by 2.

$\frac{x^2}{(\frac{1}{2} \cdot 8)^2} + \frac{y^2}{(\frac{1}{2} \cdot 5)^2} = 1 \Rightarrow \frac{x^2}{16} + \frac{y^2}{\frac{25}{4}} = 1 \Rightarrow \frac{x^2}{16} + \frac{4y^2}{25} = 1$.

31 Refer to the definition of an ellipse. As in the discussion on page 300, we will let the

positive constant, k, equal $2a$. $k = 2a = 10 \Rightarrow a = 5$. $F(3, 0)$ and $F'(-3, 0) \Rightarrow c = 3$.

$b^2 = a^2 - c^2 = 25 - 9 = 16$. An equation is $\frac{x^2}{25} + \frac{y^2}{16} = 1$.

33 $k = 2a = 34 \Rightarrow a = 17$. $F(0, 15)$ and $F'(0, -15) \Rightarrow c = 15$.

$b^2 = a^2 - c^2 = 289 - 225 = 64$. An equation is $\frac{x^2}{64} + \frac{y^2}{289} = 1$.

35 $y = 11\sqrt{1 - \frac{x^2}{49}} \Rightarrow \frac{y}{11} = \sqrt{1 - \frac{x^2}{49}} \Rightarrow \frac{x^2}{49} + \frac{y^2}{121} = 1$.

Since $y \geq 0$ in the original equation, its graph is the upper half of the ellipse.

$\boxed{37}$ $x = -\frac{1}{3}\sqrt{9-y^2} \Rightarrow -3x = \sqrt{9-y^2} \Rightarrow 9x^2 = 9 - y^2 \Rightarrow x^2 + \frac{y^2}{9} = 1.$

Since $x \le 0$ in the original equation, its graph is the left half of the ellipse.

$\boxed{39}$ $x = 1 + 2\sqrt{1 - \frac{(y+2)^2}{9}} \Rightarrow \frac{x-1}{2} = \sqrt{1 - \frac{(y+2)^2}{9}} \Rightarrow \frac{(x-1)^2}{4} + \frac{(y+2)^2}{9} = 1.$

Since $x \ge 1$ in the original equation, its graph is the right half of the ellipse.

$\boxed{41}$ $y = 2 - 7\sqrt{1 - \frac{(x+1)^2}{9}} \Rightarrow \frac{y-2}{-7} = \sqrt{1 - \frac{(x+1)^2}{9}} \Rightarrow \frac{(x+1)^2}{9} + \frac{(y-2)^2}{49} = 1.$

Since $y \le 2$ in the original equation, its graph is the lower half of the ellipse.

$\boxed{43}$ Model this problem as an ellipse with $V(\pm 15, 0)$ and $M(0, \pm 10)$.

Substituting $x = 6$ into $\frac{x^2}{15^2} + \frac{y^2}{10^2} = 1$ yields $\frac{y^2}{100} = \frac{189}{225} \Rightarrow y^2 = 84.$

The desired height is $\sqrt{84} = 2\sqrt{21} \approx 9.165$ ft.

$\boxed{45}$ $e = \frac{c}{a} = 0.017 \Rightarrow c = 0.017a = 0.017(93,000,000) = 1,581,000.$ The maximum and

minimum distances are $a + c = 94,581,000$ miles and $a - c = 91,419,000$ miles.

$\boxed{47}$ (a) Let c denote the distance from the center of the hemi-ellipsoid to F. Hence,

$(\frac{1}{2}k)^2 + c^2 = h^2 \Rightarrow c^2 = h^2 - \frac{1}{4}k^2 \Rightarrow c = \sqrt{h^2 - \frac{1}{4}k^2}.$ $d = d(V, F) = h - c \Rightarrow$

$d = h - \sqrt{h^2 - \frac{1}{4}k^2}$ and $d' = d(V, F') = h + c \Rightarrow d' = h + \sqrt{h^2 - \frac{1}{4}k^2}.$

(b) From part (a), $d' = h + c \Rightarrow c = d' - h = 32 - 17 = 15.$ $c = \sqrt{h^2 - \frac{1}{4}k^2} \Rightarrow$

$15 = \sqrt{17^2 - \frac{1}{4}k^2} \Rightarrow 225 = 289 - \frac{1}{4}k^2 \Rightarrow \frac{1}{4}k^2 = 64 \Rightarrow k^2 = 256 \Rightarrow k = 16$ cm.

$d = h - c = 17 - 15 = 2 \Rightarrow F$ should be located 2 cm from V.

$\boxed{49}$ $c^2 = (\frac{1}{2} \cdot 50)^2 - 15^2 = 625 - 225 = 400 \Rightarrow c = 20.$

Their feet should be $25 - 20 = 5$ ft from the vertices.

$\boxed{51}$ First determine an equation of the ellipse for the orbit of Earth. $e = \frac{c}{a} \Rightarrow c = ae =$

$0.093 \times 149.6 = 13.9128.$ $b^2 = a^2 - c^2 = 149.6^2 - 13.9128^2 \Rightarrow b \approx 148.95 \approx 149.0.$

An equation for the orbit of Earth is $\dfrac{x^2}{149.6^2} + \dfrac{y^2}{149.0^2} = 1.$

Graph $Y_1 = 149\sqrt{1 - (x^2/149.6^2)}$ and $Y_2 = -Y_1.$

The sun is at $(\pm 13.9128, 0)$. Plot the point $(13.9128, 0)$ for the sun.

[−300, 300] by [−200, 200]   [−3, 3] by [−2, 2]

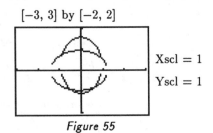

Xscl = 100   Xscl = 1

Yscl = 100   Yscl = 1

Figure 51   *Figure 55*

$\boxed{55}$ $\dfrac{(x+0.1)^2}{1.7} + \dfrac{y^2}{0.9} = 1 \Rightarrow y = \pm\sqrt{0.9[1-(x+0.1)^2/1.7]}$.

$\dfrac{x^2}{0.9} + \dfrac{(y-0.25)^2}{1.8} = 1 \Rightarrow y = 0.25 \pm \sqrt{1.8(1-x^2/0.9)}$.

See *Figure 55* on the preceding page. From the graph, the points of intersection are approximately $(-0.88, 0.76)$, $(-0.48, -0.91)$, $(0.58, -0.81)$, and $(0.92, 0.59)$.

4.8 Exercises

Note: Let C, V, F, and W denote the center, the vertices, the foci, and the end points of the conjugate axis, respectively. Let c denote the distance from the center of the hyperbola to a focus.

$\boxed{1}$ $\dfrac{x^2}{9} - \dfrac{y^2}{4} = 1$ • The hyperbola will have a right branch and a left branch since the term containing x is positive. The vertices will be on the horizontal transverse axis, $\pm\sqrt{9} = \pm 3$ units from the center. The end points of the vertical conjugate axis are $(0, \pm\sqrt{4}) = (0, \pm 2)$. The asymptotes have equations

$$y = \pm\left[\frac{\frac{1}{2}(\text{vertical axis length})}{\frac{1}{2}(\text{horizontal axis length})}\right](x).$$

Note that the terms are "vertical" and "horizontal" and not transverse and conjugate since the latter can be either vertical or horizontal. In this case, we have $y = \pm\dfrac{\frac{1}{2}(4)}{\frac{1}{2}(6)}(x) = \pm\frac{2}{3}x$. The positive sign corresponds to the asymptote with positive slope and the negative sign corresponds to the asymptote with negative slope. To find the foci, it is helpful to remember the relationship

$$\left[\tfrac{1}{2}(\text{transverse axis})\right]^2 + \left[\tfrac{1}{2}(\text{conjugate axis})\right]^2 = \left[c\right]^2.$$

Using the values from above we have $\left[\frac{1}{2}(6)\right]^2 + \left[\frac{1}{2}(4)\right]^2 = c^2 \Rightarrow 9+4 = c^2 \Rightarrow c = \pm\sqrt{13}$.

$V(\pm 3, 0)$; $F(\pm\sqrt{13}, 0)$; $W(0, \pm 2)$; $y = \pm\frac{2}{3}x$

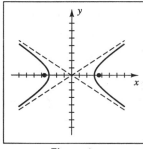

Figure 1

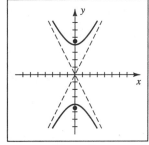

Figure 7

$\boxed{7}$ $y^2 - 4x^2 = 16 \Rightarrow \dfrac{y^2}{16} - \dfrac{x^2}{4} = 1$ •   The hyperbola will have an upper branch and a

lower branch since the term containing y is positive. The vertices will be
$\pm\sqrt{16} = \pm 4$ units from the center on the vertical transverse axis. The end points
of the horizontal conjugate axis are $(\pm\sqrt{4},\ 0) = (\pm 2,\ 0)$. The asymptotes have

equations $y = \pm\dfrac{\frac{1}{2}(8)}{\frac{1}{2}(4)}(x) = \pm 2x$. Using the foci relationship given in Exercise 1, we

have $\left[\frac{1}{2}(8)\right]^2 + \left[\frac{1}{2}(4)\right]^2 = c^2 \Rightarrow 16 + 4 = c^2 \Rightarrow c = \pm 2\sqrt{5}$.

    $V(0,\ \pm 4);\ F(0,\ \pm 2\sqrt{5});\ W(\pm 2,\ 0);\ y = \pm 2x$ (See *Figure 7* on the preceding page.)

$\boxed{9}$ $16x^2 - 36y^2 = 1 \Rightarrow \dfrac{x^2}{\frac{1}{16}} - \dfrac{y^2}{\frac{1}{36}} = 1.$ $c^2 = \frac{1}{16} + \frac{1}{36} \Rightarrow c = \pm\frac{1}{12}\sqrt{13}.$

                                $V(\pm\frac{1}{4},\ 0);\ F(\pm\frac{1}{12}\sqrt{13},\ 0);\ W(0,\ \pm\frac{1}{6});\ y = \pm\frac{2}{3}x$

Note that the branches of the hyperbola almost coincide with the asymptotes.

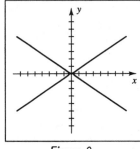
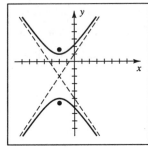

           Figure 9   *Figure 11*

$\boxed{11}$ $\dfrac{(y+2)^2}{9} - \dfrac{(x+2)^2}{4} = 1$ •   The effect of the factors $(x+2)$ and $(y+2)$ is to shift

the center of the hyperbola from $(0,\ 0)$ to $(-2,\ -2)$. Since the term involving y is
positive, the transverse axis will be vertical. The vertices are 3 units in either
direction of the center. Their coordinates are $(-2,\ -2\pm 3)$ or equivalently, $(-2,\ 1)$
and $(-2,\ -5)$. The conjugate axis will be horizontal with end points $(-2\pm 2,\ -2)$,
or, equivalently, $(0,\ -2)$ and $(-4,\ -2)$. Remember that c is the distance *from the
center* to a focus. $c^2 = 9 + 4 \Rightarrow c = \pm\sqrt{13}$. Hence, the coordinates of the foci are
$(-2,\ -2\pm\sqrt{13})$. If the center of the hyperbola was at the origin, we would have
asymptote equations $y = \pm\frac{3}{2}x$. Since the center of the hyperbola has been shifted by
the factors $(y+2)$ and $(x+2)$, we can shift the asymptote equations using the same
factors. Hence, these equations are $(y+2) = \pm\frac{3}{2}(x+2)$.

                    $C(-2,\ -2);\ V(-2,\ -2\pm 3);\ F(-2,\ -2\pm\sqrt{13});\ W(-2\pm 2,\ -2)$

13 $144x^2 - 25y^2 + 864x - 100y - 2404 = 0 \Rightarrow$

$144(x^2 + 6x + \underline{9}) - 25(y^2 + 4y + \underline{4}) =$

$$2404 + \underline{1296} - \underline{100} \Rightarrow$$

$144(x+3)^2 - 25(y+2)^2 = 3600 \Rightarrow \dfrac{(x+3)^2}{25} - \dfrac{(y+2)^2}{144} = 1.$

$c^2 = 25 + 144 \Rightarrow c = \pm 13.$

$C(-3, -2); V(-3 \pm 5, -2); F(-3 \pm 13, -2);$

$W(-3, -2 \pm 12); (y+2) = \pm \frac{12}{5}(x+3)$

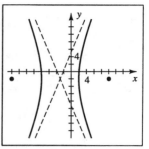

Figure 13

Note: Let b denote the distance equal to $\frac{1}{2}$(conjugate axis length) and

let a denote the distance equal to $\frac{1}{2}$(transverse axis length).

17 $a = 3$ and $c = 5 \Rightarrow b^2 = c^2 - a^2 = 16.$ $\dfrac{x^2}{a^2} - \dfrac{y^2}{b^2} = 1$ is then $\dfrac{x^2}{9} - \dfrac{y^2}{16} = 1.$

19 The center of the hyperbola is $(-2, -3)$. $a = 1$ and $c = 2 \Rightarrow b^2 = 2^2 - 1^2 = 3$ and an

equation is $\dfrac{(y+3)^2}{1^2} - \dfrac{(x+2)^2}{3} = 1$, or, equivalently, $(y+3)^2 - \dfrac{(x+2)^2}{3} = 1.$

21 $F(0, \pm 4) \Rightarrow c = 4.$ $V(0, \pm 1) \Rightarrow a = 1.$ $a^2 + b^2 = c^2 \Rightarrow b^2 = 4^2 - 1^2 = 15.$

Since the vertices are on the y-axis, the "1^2" is associated with the y^2 term.

An equation is $\dfrac{y^2}{1} - \dfrac{x^2}{15} = 1.$

23 $F(\pm 5, 0)$ and $V(\pm 3, 0) \Rightarrow W(0, \pm 4).$ An equation is $\dfrac{x^2}{9} - \dfrac{y^2}{16} = 1.$

25 Conjugate axis of length 4 implies that $b = 2.$ $F(0, \pm 5) \Rightarrow c = 5.$

$$a^2 + b^2 = c^2 \Rightarrow a^2 = 5^2 - 2^2 = 21.$$ An equation is $\dfrac{y^2}{21} - \dfrac{x^2}{4} = 1.$

27 Since the asymptote equations are $y = \pm 2x$ and we know that the point $(3, 0)$ is on

the hyperbola, we conclude that the upper right corner of the rectangle formed by the

transverse and conjugate axes has coordinates $(3, 6)$ {substitute $x = 3$ in $y = 2x$ to

obtain the 6}. Thus we have end points of the conjugate axis at $(0, \pm 6).$

An equation is $\dfrac{x^2}{9} - \dfrac{y^2}{36} = 1.$

29 $a = 5, b = 2(5) = 10.$ $\dfrac{x^2}{5^2} - \dfrac{y^2}{10^2} = 1 \Rightarrow \dfrac{x^2}{25} - \dfrac{y^2}{100} = 1.$

31 Since the transverse axis is vertical, the y^2 term will be positive.

An equation is $\dfrac{y^2}{(\frac{1}{2} \cdot 10)^2} - \dfrac{x^2}{(\frac{1}{2} \cdot 14)^2} = 1$, or $\dfrac{y^2}{25} - \dfrac{x^2}{49} = 1.$

35 Refer to the definition of a hyperbola. As in the discussion on page 312,

we will let the positive constant, k, equal $2a.$ $k = 2a = 16 \Rightarrow a = 8.$

$F(0, 10)$ and $F'(0, -10) \Rightarrow c = 10.$ $b^2 = c^2 - a^2 = 100 - 64 = 36.$

An equation is $\dfrac{y^2}{64} - \dfrac{x^2}{36} = 1.$

$\boxed{37}$ $x = \frac{5}{4}\sqrt{y^2 + 16} \Rightarrow \frac{4}{5}x = \sqrt{y^2 + 16} \Rightarrow \frac{16}{25}x^2 = y^2 + 16 \Rightarrow \frac{16}{25}x^2 - y^2 = 16 \Rightarrow \frac{x^2}{25} - \frac{y^2}{16} = 1.$

Since $x > 0$ in the original equation, its graph is the right branch of the hyperbola.

$\boxed{39}$ $y = \frac{3}{7}\sqrt{x^2 + 49} \Rightarrow \frac{7}{3}y = \sqrt{x^2 + 49} \Rightarrow \frac{49}{9}y^2 = x^2 + 49 \Rightarrow \frac{y^2}{9} - \frac{x^2}{49} = 1.$

Since $y > 0$ in the original equation, its graph is the upper branch of the hyperbola.

$\boxed{41}$ $y = -\frac{9}{4}\sqrt{x^2 - 16} \Rightarrow -\frac{4}{9}y = \sqrt{x^2 - 16} \Rightarrow \frac{16}{81}y^2 = x^2 - 16 \Rightarrow \frac{x^2}{16} - \frac{y^2}{81} = 1.$

Since $y \le 0$ in the original equation,

its graph is the lower halves of the branches of the hyperbola.

$\boxed{43}$ $x = -\frac{2}{3}\sqrt{y^2 - 36} \Rightarrow -\frac{3}{2}x = \sqrt{y^2 - 36} \Rightarrow \frac{9}{4}x^2 = y^2 - 36 \Rightarrow \frac{y^2}{36} - \frac{x^2}{16} = 1.$

Since $x \le 0$ in the original equation,

its graph is the left halves of the branches of the hyperbola.

$\boxed{47}$ The path is a hyperbola with $V(\pm 3, 0)$ and $W(0, \pm\frac{3}{2})$.

An equation is $\dfrac{x^2}{(3)^2} - \dfrac{y^2}{\left(\frac{3}{2}\right)^2} = 1$ or equivalently, $x^2 - 4y^2 = 9$.

If only the right branch is considered, then $x = \sqrt{9 + 4y^2}$ is an equation of the path.

$\boxed{49}$ Set up a coordinate system like the one in Example 6. Let the origin be located on the shoreline halfway between A and B and let $P(x, y)$ denote the coordinates of the ship. The coordinates of A and B (which can be thought of as the foci of the hyperbola) are $(-100, 0)$ and $(100, 0)$, respectively. Hence, $c = 100$. As in Exercise 35, the difference in distances is a constant—that is, $d(P, A) - d(P, B) = 2a = 160$ $\Rightarrow a = 80$. $b^2 = c^2 - a^2 = 100^2 - 80^2 \Rightarrow b = 60$. An equation of the hyperbola is

$\dfrac{x^2}{80^2} - \dfrac{y^2}{60^2} = 1$. Now, $y = 100 \Rightarrow \dfrac{x^2}{80^2} = 1 + \dfrac{100^2}{60^2} \Rightarrow x^2 = 80^2 \cdot \dfrac{13{,}600}{60^2} \Rightarrow$

$x = 80 \cdot \frac{10}{60}\sqrt{136} = \frac{80}{3}\sqrt{34}$. The ship's coordinates are $\left(\frac{80}{3}\sqrt{34},\ 100\right) \approx (155.5,\ 100)$.

$\boxed{51}$ $\dfrac{(y - 0.1)^2}{1.6} - \dfrac{(x + 0.2)^2}{0.5} = 1 \Rightarrow 0.5(y - 0.1)^2 - 1.6(x + 0.2)^2 = 0.8 \Rightarrow$

$$y = 0.1 \pm \sqrt{1.6 + 3.2(x + 0.2)^2}.$$

$\dfrac{(y - 0.5)^2}{2.7} - \dfrac{(x - 0.1)^2}{5.3} = 1 \Rightarrow 5.3(y - 0.5)^2 - 2.7(x - 0.1)^2 = 14.31 \Rightarrow$

$$y = 0.5 \pm \sqrt{\tfrac{1}{5.3}\left[14.31 + 2.7(x - 0.1)^2\right]}.$$

From the graph of *Figure 51* on the next page,

the point of intersection in the first quadrant is approximately $(0.741, 2.206)$.

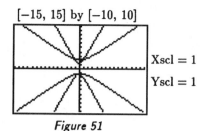

[-15, 15] by [-10, 10]

Xscl = 1
Yscl = 1

Figure 51

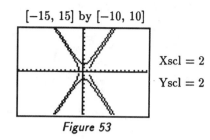

[-15, 15] by [-10, 10]

Xscl = 2
Yscl = 2

Figure 53

$\boxed{53}$ $\dfrac{(x-0.3)^2}{1.3} - \dfrac{y^2}{2.7} = 1 \Rightarrow y = \pm\sqrt{2.7[-1 + (x-0.3)^2/1.3]}$.

$\dfrac{y^2}{2.8} - \dfrac{(x-0.2)^2}{1.2} = 1 \Rightarrow y = \pm\sqrt{2.8[1 + (x-0.2)^2/1.2]}$.

The two graphs nearly intersect in the second and fourth quadrants,

but there are no points of intersection.

$\boxed{55}$ (a) The comet's path is hyperbolic with $a^2 = 26 \times 10^{14}$ and $b^2 = 18 \times 10^{14}$.

$c^2 = a^2 + b^2 = 26 \times 10^{14} + 18 \times 10^{14} = 44 \times 10^{14} \Rightarrow c \approx 6.63 \times 10^7$.

The coordinates of the sun are approximately $(6.63 \times 10^7, 0)$.

(b) The minimum distance between the comet and the sun will be

$$c - a = \sqrt{44 \times 10^{14}} - \sqrt{26 \times 10^{14}} = 1.53 \times 10^7 \text{ mi}.$$

Since r must be in meters, 1.53×10^7 mi $\times$ 1610 m/mi $\approx 2.47 \times 10^{10}$ m. At this

distance, v must be greater than $\sqrt{\dfrac{2k}{r}} \approx \sqrt{\dfrac{2(1.325 \times 10^{20})}{2.47 \times 10^{10}}} \approx 103{,}600$ m/sec.

Chapter 4 Review Exercises

$\boxed{3}$ $f(x) = -\frac{1}{4}(x+2)(x-1)^2(x-3)$ has zeros at -2, 1 (multiplicity 2), and 3.

$f(x) > 0$ if $-2 < x < 1$ or $1 < x < 3$, $f(x) < 0$ if $x < -2$ or $x > 3$.

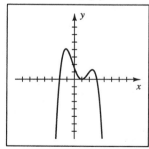

Figure 3

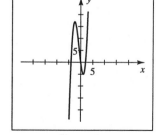

Figure 5

$\boxed{5}$ $f(x) = x^3 + 2x^2 - 8x = x(x^2 + 2x - 8) = x(x+4)(x-2)$.

$f(x) > 0$ if $-4 < x < 0$ or $x > 2$, $f(x) < 0$ if $x < -4$ or $0 < x < 2$.

[11] Synthetically dividing $f(x) = -4x^4 + 3x^3 - 5x^2 + 7x - 10$ by $x + 2$, we have

$$\begin{array}{r|rrrrr} -2 & -4 & 3 & -5 & 7 & -10 \\ & & 8 & -22 & 54 & -122 \\ \hline & -4 & 11 & -27 & 61 & -132 \end{array}$$

By the remainder theorem, $f(-2) = -132$.

[15] Since $-3 + 5i$ is a zero, so is $-3 - 5i$.

$$f(x) = a\big[x - (-3 + 5i)\big]\big[x - (-3 - 5i)\big](x + 1) = a(x^2 + 6x + 34)(x + 1).$$

$f(1) = a(41)(2)$ and $f(1) = 4 \Rightarrow 82a = 4 \Rightarrow a = \frac{2}{41}$.

Hence, $f(x) = \frac{2}{41}(x^2 + 6x + 34)(x + 1)$.

[17] $f(x) = x^5(x + 3)^2$

{ leading coefficient is 1 }

{ 0 is a zero of multiplicity 5 }

{ -3 is a zero of multiplicity 2 }

$= x^5(x^2 + 6x + 9)$

$= x^7 + 6x^6 + 9x^5$

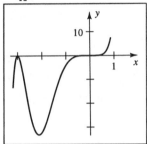

Figure 17

[20] $f(x) = x^6 + 2x^4 + x^2 = x^2(x^4 + 2x^2 + 1) = x^2(x^2 + 1)^2$

★ $0, \pm i$ (all have multiplicity 2)

[22] (a) Let $f(x) = x^5 - 4x^3 + 6x^2 + x + 4$. Since there are 2 sign changes in $f(x)$ and 3 sign changes in $f(-x) = -x^5 + 4x^3 + 6x^2 - x + 4$, there are either

2 positive and 3 negative solutions;

2 positive, 1 negative, and 2 nonreal complex;

3 negative and 2 nonreal complex;

or   1 negative and 4 nonreal complex solutions.

(b) Upper bound is 2, lower bound is -3

[25] $\dfrac{\text{positive divisors of 3}}{\text{divisors of 16}} = \dfrac{1, 3}{\pm 1, \pm 2, \pm 4, \pm 8, \pm 16} \longrightarrow$

$\pm 1, \pm 3, \pm \frac{1}{2}, \pm \frac{3}{2}, \pm \frac{1}{4}, \pm \frac{3}{4}, \pm \frac{1}{8}, \pm \frac{3}{8}, \pm \frac{1}{16}, \pm \frac{3}{16}$

After a few attempts, we try $-\frac{1}{2}$.

$$\begin{array}{r|rrrr} -\frac{1}{2} & 16 & -20 & -8 & 3 \\ & & -8 & 14 & -3 \\ \hline & 16 & -28 & 6 & 0 \end{array}$$

$16x^2 - 28x + 6 = 0 \Rightarrow 8x^2 - 14x + 3 = 0 \Rightarrow (4x - 1)(2x - 3) = 0 \Rightarrow x = \frac{1}{4}, \frac{3}{2}$.

Thus, the solutions of $16x^3 - 20x^2 - 8x + 3 = 0$ are $x = -\frac{1}{2}, \frac{1}{4}, \frac{3}{2}$.

$\boxed{29}$ $f(x) = \dfrac{3x^2}{16 - x^2} = \dfrac{3x^2}{(4 + x)(4 - x)}$. $f(x) = -3 \Rightarrow 3x^2 = 3x^2 - 48 \Rightarrow 0 = -48$,

a contradiction $\Rightarrow$ the function does not intersect the horizontal asymptote.

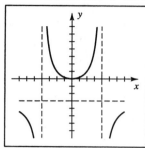

Figure 29

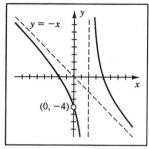

Figure 31

$\boxed{31}$ $f(x) = \dfrac{x^3 - 2x^2 - 8x}{-x^2 + 2x} = \dfrac{x(x^2 - 2x - 8)}{x(2 - x)} = \dfrac{(x - 4)(x + 2)}{2 - x}$ for $x \neq 0$.

Substituting $x = 0$ into $\dfrac{(x - 4)(x + 2)}{2 - x}$ gives us -4. Thus, we have a hole at $(0, -4)$.

$\boxed{32}$ $f(x) = \dfrac{x^2 - 2x + 1}{x^3 - x^2 + x - 1} = \dfrac{(x - 1)(x - 1)}{x^2(x - 1) + 1(x - 1)} = \dfrac{(x - 1)^2}{(x^2 + 1)(x - 1)} = \dfrac{x - 1}{x^2 + 1}$ for $x \neq 1$.

Substituting $x = 1$ into $\dfrac{x - 1}{x^2 + 1}$ gives us 0. Thus, we have a hole at $(1, 0)$.

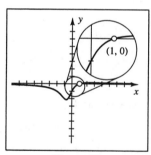

Figure 32

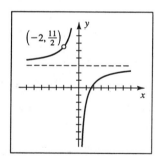

Figure 33

$\boxed{33}$ $f(x) = \dfrac{3x^2 + x - 10}{x^2 + 2x} = \dfrac{(3x - 5)(x + 2)}{x(x + 2)} = \dfrac{3x - 5}{x}$ if $x \neq -2$.

Substituting $x = -2$ into $\dfrac{3x - 5}{x}$ gives us $\dfrac{11}{2}$. Thus, we have a hole at $\left(-2, \dfrac{11}{2}\right)$.

$\boxed{34}$ $f(x) = \dfrac{-2x^2 - 8x - 6}{x^2 - 6x + 8} = \dfrac{-2(x^2 + 4x + 3)}{(x-2)(x-4)} = \dfrac{-2(x+1)(x+3)}{(x-2)(x-4)}$.

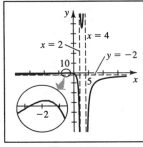
Figure 34

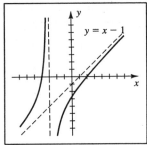
Figure 35

$\boxed{35}$ $f(x) = \dfrac{x^2 + 2x - 8}{x + 3} = \dfrac{(x+4)(x-2)}{x+3} = x - 1 - \dfrac{5}{x+3}$.

$y = x - 1$ is an oblique asymptote.

$\boxed{37}$ (a) $l = 10$, $x = 10$, and $y = 2 \Rightarrow 2 = 30{,}000c \Rightarrow c = \frac{1}{15{,}000}$

(b) $y \approx 0.9754 < 1$ if $x = 6.1$, and $y \approx 1.0006 > 1$ if $x = 6.2$

$\boxed{38}$ (a) The volume of the cylinder is $\pi r^2 x$, where r is the radius of the cylinder and

$x = \overline{AD} = \overline{BC}$ is the height. Edge AB has length $2\pi r$, since it is the circum-

ference of the lower edge of the cylinder. Also, $\overline{AB}^2 + \overline{BC}^2 = l^2 \Rightarrow$

$(2\pi r)^2 + x^2 = l^2 \Rightarrow 4\pi^2 r^2 = l^2 - x^2 \Rightarrow r^2 = \frac{1}{4\pi^2}(l^2 - x^2)$.

$$\text{Now, } V = \pi r^2 x = \pi \left[\frac{1}{4\pi^2}(l^2 - x^2) \right](x) = \frac{1}{4\pi} x(l^2 - x^2).$$

(b) If $x > 0$, $V > 0$ if $l^2 - x^2 > 0$ or $l > x$. Thus, when $0 < x < l$, $V > 0$.

$\boxed{39}$ $T = \frac{1}{20} t(t - 12)(t - 24) = 32 \Rightarrow t^3 - 36t^2 + 288t - 640 = 0$. Solving for t yields

$t = 4$ and $16 \pm 4\sqrt{6}$. Since $0 \le t \le 24$, $t = 4$ (10:00 A.M.) and

$t = 16 - 4\sqrt{6} \approx 6.2020$ (12:12 P.M.) are the times when the temperature was $32°$F.

$\boxed{40}$ $N(t) = -t^4 + 21t^2 + 100$ and $N(t) > 180 \Rightarrow t^4 - 21t^2 + 80 < 0 \Rightarrow (t^2 - 5)(t^2 - 16) < 0$.

The positive values of t satisfying this inequality are in the interval $(\sqrt{5}, 4)$.

$\boxed{41}$ (a) S is the value that is changing. As S gets large, the value of $R = \dfrac{kS^n}{S^n + a^n}$ will

approach the ratio of leading coefficients, $k/1$. Hence, an equation of the

horizontal asymptote is $R = k$.

(b) k is the maximum rate at which the liver can remove alcohol from the

bloodstream.

$\boxed{42}$ (a) $C(100) = \$2,000,000.00$ and

$C(90) \approx \$163,636.36$

(b) Notice the great difference in cost as x approaches 100.

$C(50)$ is only $\$19,607.84$ and $C(80)$ is $\$76,190.48$.

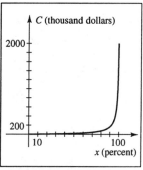

Figure 42

Note: Let the notation be the same as in §4.6–4.8.

$\boxed{43}$ $y^2 = 64x \Rightarrow x = \frac{1}{64}y^2 \Rightarrow a = \frac{1}{64}.$ $p = \dfrac{1}{4(\frac{1}{64})} = 16.$ $V(0, 0);$ $F(16, 0);$ $l : x = -16$

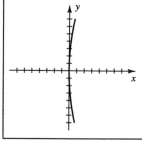

Figure 43

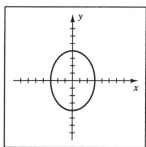

Figure 45

$\boxed{45}$ $9y^2 = 144 - 16x^2 \Rightarrow \dfrac{x^2}{9} + \dfrac{y^2}{16} = 1.$ $c^2 = 16 - 9 \Rightarrow c = \pm\sqrt{7}.$

$V(0, \pm 4);$ $F(0, \pm\sqrt{7});$ $M(\pm 3, 0)$

$\boxed{47}$ $x^2 - y^2 - 4 = 0 \Rightarrow \dfrac{x^2}{4} - \dfrac{y^2}{4} = 1.$ $c^2 = 4 + 4 \Rightarrow c = \pm 2\sqrt{2}.$

$V(\pm 2, 0);$ $F(\pm 2\sqrt{2}, 0);$ $W(0, \pm 2);$ $y = \pm x$

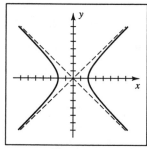

Figure 47

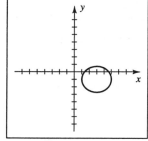

Figure 50

$\boxed{50}$ $3x^2 + 4y^2 - 18x + 8y + 19 = 0 \Rightarrow$

$3(x^2 - 6x + \underline{9}) + 4(y^2 + 2y + \underline{1}) = -19 + \underline{27} + \underline{4} \Rightarrow$

$3(x - 3)^2 + 4(y + 1)^2 = 12 \Rightarrow \dfrac{(x - 3)^2}{4} + \dfrac{(y + 1)^2}{3} = 1.$ $c^2 = 4 - 3 \Rightarrow c = \pm 1.$

$C(3, -1);$ $V(3 \pm 2, -1);$ $F(3 \pm 1, -1);$ $M(3, -1 \pm \sqrt{3})$

$\boxed{54}$ $4x^2 - y^2 - 40x - 8y + 88 = 0 \Rightarrow$

$4(x^2 - 10x + \underline{\;25\;}) - (y^2 + 8y + \underline{\;16\;}) = -88 + \underline{\;100\;} - \underline{\;16\;} \Rightarrow$

$4(x-5)^2 - (y+4)^2 = -4 \Rightarrow \dfrac{(y+4)^2}{4} - \dfrac{(x-5)^2}{1} = 1. \quad c^2 = 4 + 1 \Rightarrow c = \pm\sqrt{5}.$

$\qquad C(5, -4); \; V(5, -4 \pm 2); \; F(5, -4 \pm \sqrt{5}); \; W(5 \pm 1, -4); \; (y+4) = \pm 2(x-5)$

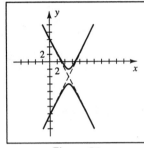

Figure 54

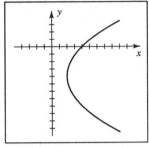

Figure 55

$\boxed{55}$ $y^2 - 8x + 8y + 32 = 0 \Rightarrow x = \frac{1}{8}y^2 + y + 4 \Rightarrow a = \frac{1}{8}. \quad p = \dfrac{1}{4(\frac{1}{8})} = 2.$

$\qquad\qquad\qquad\qquad\qquad\qquad V(2, -4); \; F(4, -4); \; l: x = 0$

$\boxed{59}$ The vertex is halfway between the x-intercepts, so it is of the form $V(-7, k)$.

$\qquad y = a(x + 10)(x + 4)$ and $x = 0, \; y = 80 \Rightarrow 80 = a(10)(4) \Rightarrow a = 2.$

$\qquad\qquad\qquad x = -7 \Rightarrow y = 2(3)(-3) = -18.$ Hence, $y = 2(x+7)^2 - 18.$

$\boxed{60}$ The vertex is $V(-4, k)$. $y = a(x+11)(x-3)$ and $x = 2, \; y = 39 \Rightarrow$

$\qquad 39 = a(13)(-1) \Rightarrow a = -3. \quad x = -4 \Rightarrow y = -3(7)(-7) = 147.$

$\qquad\qquad\qquad\qquad\qquad\qquad\qquad$ Hence, $y = -3(x+4)^2 + 147.$

$\boxed{62}$ $F(-4, 0)$ and $l: x = 4 \Rightarrow p = -4$ and $V(0, 0)$.

$\qquad\qquad$ An equation is $(y-0)^2 = \left[4(-4)\right](x - 0)$, or $y^2 = -16x.$

$\boxed{64}$ The general equation of a parabola that is symmetric to the x-axis and has its vertex

$\qquad$ at the origin is $x = ay^2$. Substituting $x = 5$ and $y = -1$ into that equation yields

$\qquad\qquad\qquad\qquad$ $a = 5$. An equation is $x = 5y^2.$

$\boxed{66}$ $F(\pm 10, 0)$ and $V(\pm 5, 0) \Rightarrow b^2 = 10^2 - 5^2 = 75.$

$\qquad\qquad\qquad$ An equation is $\dfrac{x^2}{5^2} - \dfrac{y^2}{75} = 1$ or $\dfrac{x^2}{25} - \dfrac{y^2}{75} = 1.$

$\boxed{68}$ $F(\pm 2, 0) \Rightarrow c^2 = 4.$ Now $\dfrac{x^2}{a^2} + \dfrac{y^2}{b^2} = 1$ can be written as $\dfrac{x^2}{a^2} + \dfrac{y^2}{a^2 - 4} = 1$ since

$\qquad b^2 = a^2 - c^2$. Substituting $x = 2$ and $y = \sqrt{2}$ into that equation yields

$\qquad \dfrac{4}{a^2} + \dfrac{2}{a^2 - 4} = 1 \Rightarrow 4a^2 - 16 + 2a^2 = a^4 - 4a^2 \Rightarrow a^4 - 10a^2 + 16 = 0 \Rightarrow$

$\qquad (a^2 - 2)(a^2 - 8) = 0 \Rightarrow a^2 = 2, \, 8.$ Since $a > c$, a^2 must be 8 and b^2 is equal to 4.

$\qquad\qquad\qquad\qquad$ An equation is $\dfrac{x^2}{8} + \dfrac{y^2}{4} = 1.$

69 $M(\pm 5, 0) \Rightarrow b = 5.$ $e = \frac{c}{a} = \frac{\sqrt{a^2 - b^2}}{a} = \frac{\sqrt{a^2 - 25}}{a} = \frac{2}{3} \Rightarrow \frac{2}{3}a = \sqrt{a^2 - 25} \Rightarrow$

$\frac{4}{9}a^2 = a^2 - 25 \Rightarrow \frac{5}{9}a^2 = 25 \Rightarrow a^2 = 45.$ An equation is $\frac{x^2}{25} + \frac{y^2}{45} = 1.$

72 The vertex of the square in the first quadrant has coordinates (x, x).

Since it is on the ellipse, $\frac{x^2}{a^2} + \frac{y^2}{b^2} = 1 \Rightarrow b^2 x^2 + a^2 x^2 = a^2 b^2 \Rightarrow x^2 = \frac{a^2 b^2}{a^2 + b^2}.$

x^2 is $\frac{1}{4}$ of the area of the square, hence $A = \frac{4a^2 b^2}{a^2 + b^2}.$

73 The focus is a distance of $p = 1/(4a) = 1/(4 \cdot \frac{1}{8}) = 2$ units from the origin.

An equation of the circle is $x^2 + (y - 2)^2 = 2^2 = 4.$

74 $P(x, y)$ is a distance of $(2 + d)$ from $(0, 0)$ and a distance of d from $(4, 0)$. The difference of these distances is $(2 + d) - d = 2$, a *positive constant*. By the definition of a hyperbola, $P(x, y)$ lies on the right branch of the hyperbola with foci $(0, 0)$ and $(4, 0)$. The center of the hyperbola is halfway between the foci, i.e., $(2, 0)$. The vertex is halfway from $(2, 0)$ to $(4, 0)$ since the distance from the circle to P equals the distance from P to $(4, 0)$.

Thus, the vertex is $(3, 0)$ and $a = 1.$

$b^2 = c^2 - a^2 = 2^2 - 1^2 = 3$ and

an equation of the right branch of the hyperbola is

$\frac{(x - 2)^2}{1} - \frac{y^2}{3} = 1, \ x \geq 3 \quad \text{or} \quad x = 2 + \sqrt{1 + \frac{y^2}{3}}.$

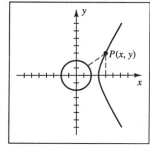

Figure 74

Chapter 4 Discussion Exercises

1 **For even-degreed polynomials:** the domain is ℝ and the number of x-intercepts ranges from *zero* to the degree of the polynomial; if the leading coefficient is positive, the range is of the form $[c, \infty)$ and the general shape has $y \to \infty$ as $|x| \to \infty$; if the leading coefficient is negative, the range is of the form $(-\infty, c]$ and the general shape has $y \to -\infty$ as $|x| \to \infty$.

For odd-degreed polynomials: the domain is ℝ, the range is ℝ, and the number of x-intercepts ranges from *one* to the degree of the polynomial; if the leading coefficient is positive, then $y \to \infty$ as $x \to \infty$ and $y \to -\infty$ as $x \to \infty$, if the leading coefficient is negative, then $y \to -\infty$ as $x \to \infty$ and $y \to \infty$ as $x \to -\infty$.

$\boxed{3}$ By long division, we obtain the quotient $2x^2 - 7x + 5$ with remainder -6. By synthetic division with $k = -3/2$, we obtain a bottom row of 4 -14 10 -6. The first three numbers are twice the coefficients of the quotient and the last number is the remainder. For the factor $ax + b$, we can use synthetic division with $k = -b/a$, and obtain a times the quotient and the remainder in the bottom row.

$\boxed{5}$ From your previous experience, you know that 2 points specify a first-degree polynomial (a line) and that 3 points specify a second-degree polynomial (a parabola). After working Discussion Exercise 4, you should guess that 4 points specify a third-degree polynomial, so a logical conclusion is that $n + 1$ points specify an n-degree polynomial.

$\boxed{7}$ If the common factor is never equal to zero for any real number, then it can be canceled and has no effect on its graph. Such a factor is $x^2 + 1$, and an example of a function is $f(x) = \dfrac{(x^2 + 1)(x - 1)}{(x^2 + 1)(x - 2)}$.

$\boxed{9}$ Let x, $x + 1$, and $x + 2$ denote three consecutive integers. Their product is $x(x + 1)(x + 2) = x^3 + 3x^2 + 3x + 1$. Only $x + 1$ or $x - 1$ could be factors, and it turns out that $x + 1$ is a factor three times. Thus, if you multiply three consecutive integers together and then add the second integer to that product, you obtain the cube of the second integer.

$\boxed{11}$ The circle goes through both foci and all 4 vertices of the auxiliary rectangle.

$\boxed{13}$ Refer to Figure 60 on text page 312. We begin with the first displayed equation on that page:

$$\left| d(P,\ F) - d(P,\ F') \right| = 2a$$

Using the distance formula to find $d(P,\ F)$ and $d(P,\ F')$, we obtain an equation of the hyperbola:

$$\left| \sqrt{(x-c)^2 + (y-0)^2} - \sqrt{(x+c)^2 + (y-0)^2} \right| = 2a$$

We can rewrite the preceding equation as (a similar argument follows if we use $-2a$ instead of $2a$)

$$\sqrt{(x-c)^2 + y^2} = 2a + \sqrt{(x+c)^2 + y^2}$$

Squaring both sides of the last equation and simplifying gives us

$$x^2 - 2cx + c^2 + y^2 = 4a^2 + 4a\sqrt{(x+c)^2 + y^2} + x^2 + 2cx + c^2 + y^2,$$

$$-4a^2 - 4cx = 4a\sqrt{(x+c)^2 + y^2}$$

$$-a^2 - cx = a\sqrt{(x+c)^2 + y^2}$$

Squaring both sides again and putting the variables on the left side yields

$$a^4 + 2a^2cx + c^2x^2 = a^2(x^2 + 2cx + c^2 + y^2)$$

$$a^4 + 2a^2cx + c^2x^2 = a^2x^2 + 2a^2cx + a^2c^2 + a^2y^2$$

$$c^2x^2 - a^2x^2 - a^2y^2 = a^2c^2 - a^4$$

$$x^2(c^2 - a^2) - a^2y^2 = a^2(c^2 - a^2)$$

Dividing both sides by $a^2(c^2 - a^2)$, we obtain

$$\frac{x^2}{a^2} - \frac{y^2}{c^2 - a^2} = 1 \ (*).$$

Recalling that $c > a$ (for the hyperbola) and therefore $c^2 - a^2 > 0$, we let

$$b = \sqrt{c^2 - a^2}, \quad \text{or} \quad b^2 = c^2 - a^2.$$

Substituting in $(*)$, we get the following equation general equation of a hyperbola:

$$\frac{x^2}{a^2} - \frac{y^2}{b^2} = 1$$

Chapter 5: Exponential and Logarithmic Functions

$\boxed{3}$ $3^{2x+3} = 3^{(x^2)} \Rightarrow 2x + 3 = x^2 \Rightarrow x^2 - 2x - 3 = 0 \Rightarrow (x-3)(x+1) = 0 \Rightarrow x = -1, 3$

$\boxed{5}$ We need to obtain the same base on each side of the equals sign, then we can apply

part (2) of the theorem about exponential functions being one-to-one.

$2^{-100x} = (0.5)^{x-4} \Rightarrow (2^{-1})^{100x} = \left(\frac{1}{2}\right)^{x-4} \Rightarrow \left(\frac{1}{2}\right)^{100x} = \left(\frac{1}{2}\right)^{x-4} \Rightarrow$

$$100x = x - 4 \Rightarrow 99x = -4 \Rightarrow x = -\frac{4}{99}$$

$\boxed{7}$ $4^{x-3} = 8^{4-x} \Rightarrow (2^2)^{x-3} = (2^3)^{4-x} \Rightarrow 2^{2x-6} = 2^{12-3x} \Rightarrow$

$$2x - 6 = 12 - 3x \Rightarrow 5x = 18 \Rightarrow x = \frac{18}{5}$$

$\boxed{9}$ (a) Let $g = f(x) = 2^x$ for reference purposes.

The graph of g goes through the points $(-1, \frac{1}{2})$, $(0, 1)$, and $(1, 2)$.

(b) $f(x) = -2^x$ •

Reflect g through the x-axis since f is just $-1 \cdot 2^x$. Do not confuse this function

with $(-2)^x$ — remember, the base is positive for exponential functions.

(c) $f(x) = 3 \cdot 2^x$ •   vertically stretch g by a factor of 3

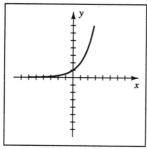
Figure 9(a)

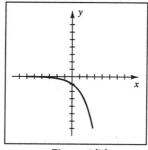
Figure 9(b)

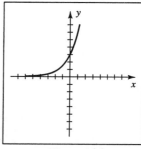
Figure 9(c)

(d) $f(x) = 2^{x+3}$ •   shift g left 3 units since f is $g(x+3)$

(e) $f(x) = 2^x + 3$ •   vertically shift g up 3 units

(f) $f(x) = 2^{x-3}$ •   shift g right 3 units since f is $g(x-3)$

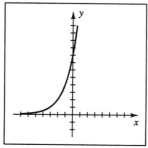
Figure 9(d)

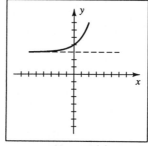
Figure 9(e)

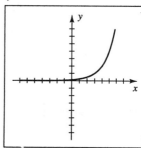
Figure 9(f)

(g) $f(x) = 2^x - 3$ • vertically shift g down 3 units

(h) $f(x) = 2^{-x}$ • reflect g through the y-axis since f is $g(-x)$

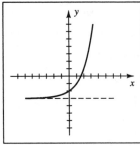 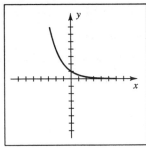

| Figure 9(g) | Figure 9(h) |

(i) $f(x) = (\frac{1}{2})^x$ • $(\frac{1}{2})^x = (2^{-1})^x = 2^{-x}$, same graph as in part (h)

(j) $f(x) = 2^{3-x}$ • $2^{3-x} = 2^{-(x-3)}$, shift g right 3 units and reflect through the line $x = 3$. Alternatively, $2^{3-x} = 2^3 2^{-x} = 8(\frac{1}{2})^x$, vertically stretch $y = (\frac{1}{2})^x$ (the graph in part (i))by a factor of 8.

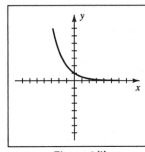 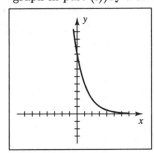

| Figure 9(i) | Figure 9(j) |

$\boxed{13}$ $f(x) = -(\frac{1}{2})^x + 4$ • reflect $y = (\frac{1}{2})^x$ through the x-axis and shift up 4 units

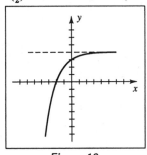 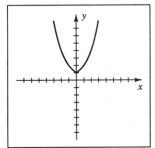

| Figure 13 | Figure 15 |

$\boxed{15}$ $f(x) = 2^{|x|} = \begin{cases} 2^x & \text{if } x \geq 0 \\ 2^{-x} & \text{if } x < 0 \end{cases} = \begin{cases} 2^x & \text{if } x \geq 0 \\ (\frac{1}{2})^x & \text{if } x < 0 \end{cases}$

Use the portion of $y = 2^x$ with $x \geq 0$ and reflect it through the y-axis since f is even.

Note: For Exercises 17, 18, and 5 of the review exercises, refer to Example 5 in the text

for the basic graph of $y = a^{-x^2} = (a^{-1})^{(x^2)} = \left(\frac{1}{a}\right)^{(x^2)}$, where $a > 1$.

$\boxed{17}$ $f(x) = 3^{1-x^2} = 3^1 3^{-x^2} = 3\left(\frac{1}{3}\right)^{(x^2)}$ •   stretch $y = \left(\frac{1}{3}\right)^{(x^2)}$ by a factor of 3

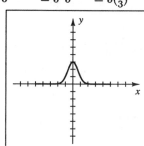

Figure 17

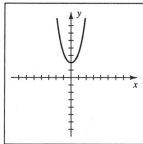

Figure 19

$\boxed{19}$ $f(x) = 3^x + 3^{-x}$ •   Adding the functions $g(x) = 3^x$ and $h(x) = 3^{-x} = \left(\frac{1}{3}\right)^x$ together, we see that the y-intercept will be $(0, 2)$. If $x > 0$, f looks like $y = 3^x$ since 3^x dominates 3^{-x} (3^{-x} gets close to 0 and 3^x grows very large). If $x < 0$, f looks like $y = 3^{-x}$ since 3^{-x} dominates 3^x.

$\boxed{23}$ (a) 8:00 A.M. corresponds to $t = 1$ and $f(1) = 600\sqrt{3} \approx 1039$.

10:00 A.M. corresponds to $t = 3$ and $f(3) = 600(3\sqrt{3}) = 1800\sqrt{3} \approx 3118$.

11:00 A.M. corresponds to $t = 4$ and $f(4) = 600(9) = 5400$.

(b) The graph of f is an increasing exponential that passes through $(0, 600)$ and the

points in part (a).

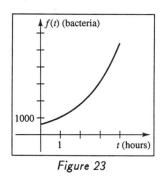

Figure 23

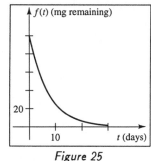

Figure 25

$\boxed{25}$ (a) $f(5) = 100(2)^{-1} = 50$ mg; $f(10) = 100(2)^{-2} = 25$ mg;

$$f(12.5) = 100(2)^{-2.5} = \frac{100}{4\sqrt{2}} = \frac{25}{2}\sqrt{2} \approx 17.7 \text{ mg}$$

(b) The end points are $(0, 100)$ and $(30, 1.5625)$.

$\boxed{27}$ A half-life of 1600 years means that when $t = 1600$, the amount remaining, $q(t)$, will be one-half the original amount—that is, $\frac{1}{2}q_0$. $q(t) = \frac{1}{2}q_0$ when $t = 1600 \Rightarrow$ $\frac{1}{2}q_0 = q_0 2^{k(1600)} \Rightarrow 2^{-1} = 2^{1600k} \Rightarrow 1600k = -1 \Rightarrow k = -\frac{1}{1600}$.

29. Using $A = P\left(1 + \frac{r}{n}\right)^{nt}$, we have $P = 1000$, $r = 0.12$, and $n = 12$.

   Consider A to be a function of t, that is, $A(t) = 1000\left(1 + \frac{0.12}{12}\right)^{12t} = 1000(1.01)^{12t}$.

   (a) $A\left(\frac{1}{12}\right) = \$1010.00$ (d) $A(20) \approx \$10{,}892.55$

31. $C = 10{,}000 \Rightarrow V(t) = 7800(0.85)^{t-1}$

   (a) $V(1) = \$7800$ (b) $V(4) \approx \$4790.18$, or \$4790   (c) $V(7) \approx \$2941.77$, or \$2942

33. $t = 1996 - 1626 = 370$; $A = \$24(1 + 0.06/4)^{4 \cdot 370} = \$89{,}115{,}607{,}202.91$.

   That's right—\$89 billion!

35. (a) Examine the pattern formed by the value y in the year n.

year (n)	value (y)
0	y_0
1	$(1-a)y_0 = y_1$
2	$(1-a)y_1 = (1-a)\left[(1-a)y_0\right] = (1-a)^2 y_0 = y_2$
3	$(1-a)y_2 = (1-a)\left[(1-a)^2 y_0\right] = (1-a)^3 y_0 = y_3$

   (b) $s = (1-a)^T y_0 \Rightarrow (1-a)^T = s/y_0 \Rightarrow 1 - a = \sqrt[T]{s/y_0} \Rightarrow a = 1 - \sqrt[T]{s/y_0}$

37. (a) $r = 0.12$, $t = 30$, $L = 90{,}000 \Rightarrow k \approx 35.95$, $M \approx 925.75$

   (b) (360 payments) $\times \$925.75 - \$90{,}000 = \$243{,}270$

43. Part (b) may be interpreted as doubling an investment at 8.5%.

   (a) If $y = (1.085)^x$ and $x = 40$, then $y \approx 26.13$.   (b) If $y = 2$, then $x \approx 8.50$.

[0, 60] by [0, 40]

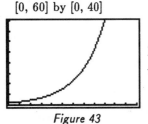

Xscl = 5

Yscl = 5

[−3, 3] by [−2, 2]

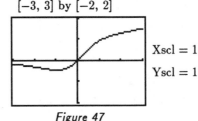

Xscl = 1

Yscl = 1

Figure 43   *Figure 47*

47. (a) f is not one-to-one since

   the horizontal line $y = -0.1$ intersects the graph of f more than once.

   (b) The only zero of f is $x = 0$.

51 *Figure 51* is a graph of $N(t) = 1000(0.9)^t$.

By tracing and zooming, we can determine that $N = 500$ when $t \approx 6.58$ yr.

[0, 10] by [0, 1000]

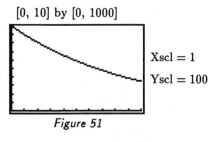

Xscl = 1
Yscl = 100

Figure 51

[0, 7.5] by [0, 5]

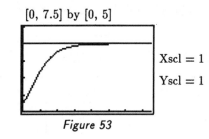

Xscl = 1
Yscl = 1

Figure 53

53 Graph $y = 4(0.125)^{(0.25^x)}$. The line $y = k = 4$ is a horizontal asymptote for the Gompertz function. The maximum number of sales of the product approaches k.

55 From the graph, we determine that $A = 100{,}000$ when $n \approx 32.8$.

[0, 40] by [0, 200,000]

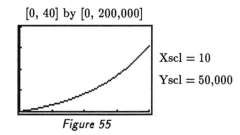

Xscl = 10
Yscl = 50,000

Figure 55

57 (a) Let $x = 0$ correspond to 1910, $x = 20$ to 1930, ... , and $x = 85$ to 1995.

Graph the data together with the functions

$f(x) = 0.809(1.094)^x$ and $g(x) = 0.375x^2 - 18.4x + 88.1$.

[−10, 90] by [−200, 1500]

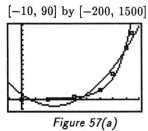

Xscl = 10
Yscl = 100

Figure 57(a)

[−10, 90] by [−200, 1500]

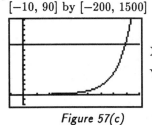

Xscl = 10
Yscl = 100

Figure 57(c)

(b) The exponential function f best models the data.

(c) Graph $Y_1 = f(x)$ and $Y_2 = 1000$. The graphs intersect at $x \approx 79$, or in 1989.

5.2 Exercises

Note: Examine Figure 13 in this section to reinforce the idea that $y = e^x$ is just a special case of $y = a^x$ with $a > 1$.

3   (a) $f(x) = e^{x+4}$ •   shift $y = e^x$ left 4 units

(b) $f(x) = e^x + 4$ •   shift $y = e^x$ up 4 units

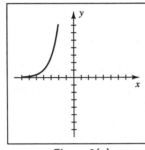

Figure 3(a)

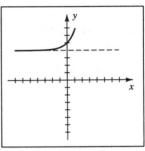

Figure 3(b)

5   $A = Pe^{rt} = 1000e^{(0.0825)(5)} \approx \$1510.59$

7   $100{,}000 = Pe^{(0.11)(18)} \Rightarrow P = \dfrac{100{,}000}{e^{1.98}} \approx \$13{,}806.92$

9   $13{,}464 = 1000e^{(r)(20)} \Rightarrow e^{20r} = 13.464.$ Using a trial and error approach on a scientific calculator or tracing and zooming or an intersect feature on a graphing calculator, we determine that $e^x \approx 13.464$ if $x \approx 2.6$. Thus, $20r = 2.6$ and $r = 0.13$ or 13%.

11   $e^{(x^2)} = e^{7x-12} \Rightarrow x^2 = 7x - 12 \Rightarrow x^2 - 7x + 12 = 0 \Rightarrow (x-3)(x-4) = 0 \Rightarrow x = 3,\ 4$

13   $xe^x + e^x = 0 \Rightarrow e^x(x+1) = 0 \Rightarrow x = -1$ { Note that $e^x \neq 0.$ }

15   $x^3(4e^{4x}) + 3x^2 e^{4x} = 0 \Rightarrow x^2 e^{4x}(4x+3) = 0 \Rightarrow x = -\frac{3}{4},\ 0.$ Note that $e^{4x} \neq 0.$

17   $\dfrac{(e^x + e^{-x})(e^x + e^{-x}) - (e^x - e^{-x})(e^x - e^{-x})}{(e^x + e^{-x})^2} =$

$$\frac{(e^{2x} + 2 + e^{-2x}) - (e^{2x} - 2 + e^{-2x})}{(e^x + e^{-x})^2} = \frac{4}{(e^x + e^{-x})^2}$$

19   $W(t) = W_0 e^{kt}$; $t = 30 \Rightarrow W(30) = 68e^{(0.2)(30)} \approx 27{,}433$ mg, or 27.43 grams

21   The year 2010 corresponds to $t = 2010 - 1980 = 30$. Using the law of growth formula on page 343 with $q_0 = 227$ and $r = 0.007$, we have $N(t) = 227e^{0.007t}$.

Thus, $N(30) = 227e^{(0.007)(30)} \approx 280.0$ million.

23   $N(10) = N_0 e^{-2}.$ The percentage of the original number still alive after 10 years is

$$100 \times \left(\frac{N(10)}{N_0} \right) = 100e^{-2} \approx 13.5\%.$$

25 The year 2010 corresponds to $t = 2010 - 1978 = 32$.

$$N(32) = 5000e^{(0.0036)(32)} = 5000e^{(0.1152)} \approx 5610.$$

27 $h = 40,000 \Rightarrow p = 29e^{(-0.000034)(40,000)} = 29e^{(-1.36)} \approx 7.44$ in.

29 $x = 1 \Rightarrow y = 79.041 + 6.39 - e^{2.268} \approx 75.77$ cm.

$$x = 1 \Rightarrow R = 6.39 + 0.993e^{2.268} \approx 15.98 \text{ cm/yr.}$$

31 $2010 - 1971 = 39 \Rightarrow t = 39$ years. Using the continuously compounded interest

formula with $P = 1.60$ and $r = 0.05$, we have $A = 1.60e^{(0.05)(39)} \approx \$11.25$ per hour.

33 (a) Note here that the amount of money invested is not of interest and that we are

only concerned with the percent of growth.

$$\left(1 + \tfrac{0.07}{4}\right)^{4 \cdot 1} \approx 1.0719. \quad (1.0719 - 1) \times 100\% = 7.19\%$$

(b) $e^{(0.07)(1)} \approx 1.0725.$ $(1.0725 - 1) \times 100\% = 7.25\%.$ The results indicate that we

would receive an extra 0.06% in interest by investing our money in an account

that is compounded continuously rather than quarterly. This is only an extra 6

cents on a \$100 investment, but \$600 extra on a \$1,000,000 investment (actually

\$649.15 if the computations are carried beyond 0.01%).

35 It may be of interest to compare this graph with the graph of $y = (1.085)^x$ in

Exercise 43 of §5.1. Both are compounding functions with $r = 8.5\%$.

Note that $e^{0.085x} = (e^{0.085})^x \approx (1.0887)^x > (1.085)^x$ for $x > 0$.

(a) If $y = e^{0.085x}$ and $x = 40$, then $y \approx 29.96.$ (b) If $y = 2$, then $x \approx 8.15.$

[0, 60] by [0, 40]

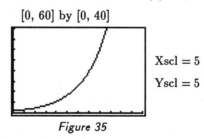

Xscl = 5

Yscl = 5

Figure 35

37 (a) As $x \to \infty$, $e^{-x} \to 0$ and f will resemble $\tfrac{1}{2}e^x$.

As $x \to -\infty$, $e^x \to 0$ and f will resemble $-\tfrac{1}{2}e^x$.

(b) At $x = 0$, $f(x) = 0$, and g will have a vertical asymptote since g is undefined (division by 0). As $x \to \infty$, $f(x) \to \infty$, and since the reciprocal of a large positive number is a small positive number, we have $g(x) \to 0$. As $x \to -\infty$, $f(x) \to -\infty$, and since the reciprocal of a large negative number is a small negative number, we have $g(x) \to 0$.

$[-7.5, 7.5]$ by $[-5, 5]$

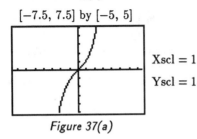

Xscl $= 1$

Yscl $= 1$

Figure 37(a)

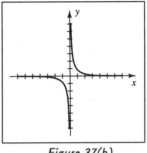

Figure 37(b)

39 (a) $f(x) = \dfrac{e^x - e^{-x}}{e^x + e^{-x}} = \dfrac{e^x - 1/e^x}{e^x + 1/e^x} \cdot \dfrac{e^x}{e^x} = \dfrac{e^{2x} - 1}{e^{2x} + 1}$.   At $x = 0$, $f(x) = 0$.   As $x \to \infty$,

$f(x) \to 1$ since the numerator and denominator are nearly the same number. As $x \to -\infty$, $f(x) \to -1$.

(b) At $x = 0$, we will have a vertical asymptote. As $x \to \infty$, $f(x) \to 1$, and since g is the reciprocal of f, $g(x) \to 1$. Similarly, as $x \to -\infty$, $g(x) \to -1$.

$[-4.5, 4.5]$ by $[-3, 3]$

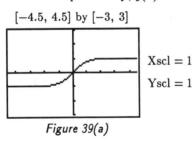

Xscl $= 1$

Yscl $= 1$

Figure 39(a)

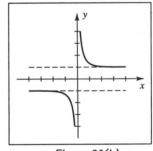

Figure 39(b)

41 The approximate coordinates of the points where the graphs of f and g intersect are $(-1.04, -0.92)$, $(2.11, 2.44)$, and $(8.51, 70.42)$. The region near the origin in *Figure 41(a)* is enhanced in *Figure 41(b)*. Thus, the solutions are $x \approx -1.04$, 2.11, and 8.51.

$[-3, 11]$ by $[-10, 80]$ $[-2.26, 3.34]$ by $[-7.14, 8.57]$

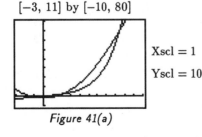

Xscl $= 1$

Yscl $= 10$

Figure 41(a)

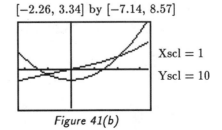

Xscl $= 1$

Yscl $= 10$

Figure 41(b)

45 From the graph, we see that f has zeros at $x \approx 0.11$, 0.79, and 1.13.

[−2, 2.5] by [−1, 2]

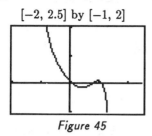

Xscl = 1
Yscl = 1

[0, 200] by [0, 8]

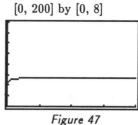

Xscl = 50
Yscl = 1

Figure 45   Figure 47

47 From the graph, there is a horizontal asymptote of $y \approx 2.71$.

f is approaching the value of e asymptotically.

49 $e^{-x} = x$ when $x \approx 0.567$.

[−4.5, 4.5] by [−3, 3]

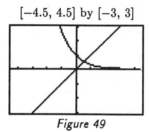

Xscl = 1
Yscl = 1

[−5.5, 5] by [−2, 5]

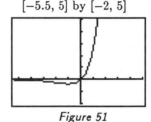

Xscl = 1
Yscl = 1

Figure 49   Figure 51

51 f is increasing on $[-1, \infty)$ and f is decreasing on $(-\infty, -1]$.

53 (a) When $y = 0$ and $z = 0$, the equation becomes $C = \dfrac{2Q}{\pi v a b} e^{-h^2/(2b^2)}$. As h increases, the exponent becomes a larger *negative* value, and hence the concentration C decreases.

(b) When $z = 0$, the equation becomes $C = \dfrac{2Q}{\pi v a b} e^{-y^2/(2a^2)} e^{-h^2/(2b^2)}$.

As y increases, the concentration C decreases.

55 (a) Chose two arbitrary points that appear to lie on the curve such as $(0, 1.225)$ and $(10,000, 0.414)$. $f(0) = Ce^0 = C = 1.225$ and $f(10,000) = 1.225e^{10,000k} = 0.414$. To solve the last equation, graph $Y_1 = 1.225e^{10,000x}$ and $Y_2 = 0.414$. The graphs intersect at $x \approx -0.0001085$. Thus, $f(x) = 1.225e^{-0.0001085x}$.

(b) $f(3000) \approx 0.885$ and $f(9000) \approx 0.461$.

[−1000, 10,100] by [0, 1.5]

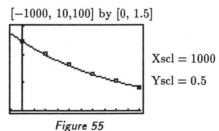

Xscl = 1000
Yscl = 0.5

Figure 55

Note: Exercises 1–4 are designed to familiarize the reader with the definition of $\log_a$ in this section. It is very important that you can generalize your understanding of this definition to the following case:

$$\log_{\text{base}}(\text{argument}) = \text{exponent} \quad \textit{is equivalent to} \quad (\text{base})^{\text{exponent}} = \text{argument}$$

Later in this section, you will also want to be able to use the following two special cases with ease:

$$\log(\text{argument}) = \text{exponent} \quad \textit{is equivalent to} \quad (10)^{\text{exponent}} = \text{argument}$$

$$\ln(\text{argument}) = \text{exponent} \quad \textit{is equivalent to} \quad (e)^{\text{exponent}} = \text{argument}$$

1 (e) In this case, the *base* is 5, the *exponent* is $7t$, and the *argument* is $\frac{a+b}{a}$. Thus,

$$5^{7t} = \frac{a+b}{a} \quad \text{is equivalent to} \quad \log_5 \frac{a+b}{a} = 7t.$$

3 (e) In this case, the *base* is 2, the *argument* is m, and the *exponent* is $3x + 4$. Thus,

$$\log_2 m = 3x + 4 \quad \text{is equivalent to} \quad 2^{3x+4} = m.$$

7 In order to solve for t, we must isolate the expression containing t—in this case, that expression is the exponential a^{Ct}. $A = Ba^{Ct} + D \Rightarrow A - D = Ba^{Ct} \Rightarrow$

$$\frac{A-D}{B} = a^{Ct} \Rightarrow Ct = \log_a\left(\frac{A-D}{B}\right) \Rightarrow t = \frac{1}{C}\log_a\left(\frac{A-D}{B}\right).$$

The confusing step to most students in the above solution is $\frac{A-D}{B} = a^{Ct} \Rightarrow Ct = \log_a\left(\frac{A-D}{B}\right).$ This is similar to $y = a^x \Rightarrow \log_a y = x$, except x and y are more complicated expressions.

9 (a) Changing $10^5 = 100{,}000$ to logarithmic form gives us $\log_{10} 100{,}000 = 5$.

Since this is a common logarithm, we denote it as $\log 100{,}000 = 5$.

(e) Changing $e^{2t} = 3 - x$ to logarithmic form gives us $\log_e(3 - x) = 2t$.

Since this is a natural logarithm, we denote it as $\ln(3 - x) = 2t$.

11 (b) Remember that $\log x = 20t$ is the same as $\log_{10} x = 20t$.

Changing to exponential form, we have $10^{20t} = x$.

(d) Remember that $\ln w = 4 + 3x$ is the same as $\log_e w = 4 + 3x$.

Changing to exponential form, we have $e^{4+3x} = w$.

[13] (c) Remember that you cannot take the logarithm, any base, of a negative number.

Hence, $\log_4(-2)$ is undefined.

(g) We will change the form of $\frac{1}{16}$ so that it can be written as an exponential expression with the same base as the logarithm—in this case, that base is 4.

$$\log_4 \tfrac{1}{16} = \log_4 4^{-2} = -2$$

[15] Parts (a)–(f) are direct applications of the properties in the chart on page 356.

For part (g), we use a property of exponents that will enable us to use the property $e^{\ln x} = x$. (g) $e^{2+\ln 3} = e^2 e^{\ln 3} = e^2(3) = 3e^2$

[17] $\log_4 x = \log_4(8-x) \Rightarrow x = 8 - x$ {since the logarithm function is one-to-one} $\Rightarrow 2x = 8 \Rightarrow x = 4$. We must check to make sure that all proposed solutions do not make any of the original expressions undefined. Checking $x = 4$ in the original equation gives us $\log_4 4 = \log_4(8-4)$, which is a true statement, so $x = 4$ is a valid solution.

[19] $\log_5(x-2) = \log_5(3x+7) \Rightarrow x - 2 = 3x + 7 \Rightarrow 2x = -9 \Rightarrow x = -\frac{9}{2}$. The value $x = -\frac{9}{2}$ is extraneous since it makes either of the original logarithm expressions undefined. Hence, there is no solution.

[21] $\log x^2 = \log(-3x-2) \Rightarrow x^2 = -3x - 2 \Rightarrow x^2 + 3x + 2 = 0 \Rightarrow (x+1)(x+2) = 0 \Rightarrow$
$x = -1, -2$. Checking -1 and -2, we find that both are valid solutions.

[23] $\log_3(x-4) = 2 \Rightarrow x - 4 = 3^2 \Rightarrow x = 13$

[25] $\log_9 x = \frac{3}{2} \Rightarrow x = 9^{3/2} = (9^{1/2})^3$ {remember, root first, power second} $= 3^3 = 27$

[27] $\ln x^2 = -2 \Rightarrow x^2 = e^{-2} = \frac{1}{e^2} \Rightarrow x = \pm\frac{1}{e}$

[29] $e^{2\ln x} = 9 \Rightarrow (e^{\ln x})^2 = 9 \Rightarrow x^2 = 9 \Rightarrow x = \pm 3$; -3 is extraneous

[31] (a) $f(x) = \log_4 x$ • This graph has a vertical asymptote of $x = 0$ and goes through $(\frac{1}{4}, -1)$, $(1, 0)$, and $(4, 1)$. For reference purposes, call this $F(x)$.

(b) $f(x) = -\log_4 x$ • reflect F through the x-axis

(c) $f(x) = 2\log_4 x$ • vertically stretch F by a factor of 2

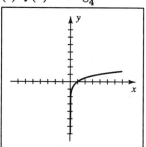

Figure 31(a)

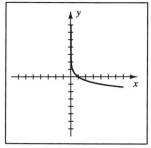

Figure 31(b)

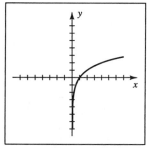

Figure 31(c)

(d) $f(x) = \log_4 (x + 2)$ •   shift F left 2 units

(e) $f(x) = (\log_4 x) + 2$ •   shift F up 2 units

(f) $f(x) = \log_4 (x - 2)$ •   shift F right 2 units

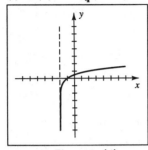
Figure 31(d)

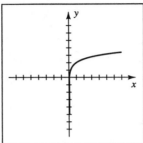
Figure 31(e)

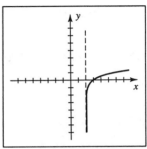
Figure 31(f)

(g) $f(x) = (\log_4 x) - 2$ •   shift F down 2 units

(h) $f(x) = \log_4 |x|$ •   include the reflection of F through the y-axis since x may

be positive or negative, but $\log_4 |x|$ will give the same result

(i) $f(x) = \log_4 (-x)$ • x must be negative so that $-x$ is positive,

reflect F through the y-axis

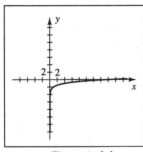

Figure 31(g)

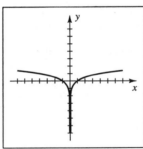
Figure 31(h)

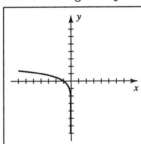
Figure 31(i)

(j) $f(x) = \log_4(3 - x) = \log_4[-(x - 3)]$ • Shift F right 3 units and reflect through the line $x = 3$. It may be helpful to determine the domain of this function. We know that $3 - x$ must be positive for the function to be defined. Thus, $3 - x > 0 \Rightarrow 3 > x$, or, equivalently, $x < 3$.

(k) $f(x) = |\log_4 x|$ •

reflect points with negative y-coordinates through the x-axis

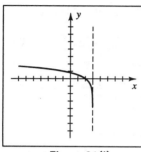

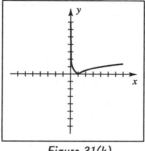

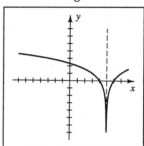

Figure 31(j)   Figure 31(k)   Figure 35

35 $f(x) = \log_2 |x - 5|$ • shift $y = \log_2 |x|$ right 5 units

37 This is the basic "logarithm with base 2" graph, call it $F(x) = \log_2 x$. ★ $f(x) = \log_2 x$

39 shift F down 1 unit   ★ $f(x) = \log_2 x - 1$

41 the reflection of F through the y-axis   ★ $f(x) = \log_2(-x)$

43 F appears to be stretched. Since the graph goes through (2, 2) instead of (2, 1) and (4, 4) instead of (4, 2), we might guess that the y-coordinates of F are doubled, that is, $f(x) = 2\log_2 x$. $f(x) \neq \log_2 x^2$ since the domain of f is $(0, \infty)$ and the domain of $g(x) = \log_2 x^2$ is $\mathbb{R} - \{0\}$.   ★ $f(x) = 2\log_2 x$

45 (a) $\log x = 3.6274 \Rightarrow x = 10^{3.6274} \approx 4240.333$, or 4240 to three significant figures

    (f) $\ln x = -1.6 \Rightarrow x = e^{-1.6} \approx 0.2019$, or 0.202

47 $q = q_0(2)^{-t/1600} \Rightarrow \dfrac{q}{q_0} = 2^{-t/1600}$ { change to logarithm form } $\Rightarrow$

$$-\frac{t}{1600} = \log_2\left(\frac{q}{q_0}\right) \text{ \{ multiply by } -1600 \text{ \}} \Rightarrow t = -1600\log_2\left(\frac{q}{q_0}\right)$$

49 $I = 20e^{-Rt/L} \Rightarrow \dfrac{I}{20} = e^{-Rt/L} \Rightarrow \ln\left(\dfrac{I}{20}\right) = -\dfrac{Rt}{L} \Rightarrow t = -\dfrac{L}{R}\ln\left(\dfrac{I}{20}\right)$

51 $I = 10^a I_0 \Rightarrow R = \log\left(\dfrac{I}{I_0}\right) = \log\left(\dfrac{10^a I_0}{I_0}\right) = \log 10^a = a$.

Hence, for $10^2 I_0$, $10^4 I_0$, and $10^5 I_0$, the answers are: (a) 2   (b) 4   (c) 5

53 We will find a general formula for α first.

$$I = 10^a I_0 \Rightarrow \alpha = 10\log\left(\frac{I}{I_0}\right) = 10\log\left(\frac{10^a I_0}{I_0}\right) = 10\left(\log 10^a\right) = 10(a) = 10a.$$

Hence, for $10^1 I_0$, $10^3 I_0$, and $10^4 I_0$, the answers are: (a) 10   (b) 30   (c) 40

55 1980 corresponds to $t = 0$ and $N(0) = 227$ million. $2 \cdot 227 = 227e^{0.007t} \Rightarrow$

$$2 = e^{0.007t} \Rightarrow \ln 2 = 0.007t \Rightarrow t \approx 99, \text{ which corresponds to the year 2079.}$$

Alternatively, using the doubling time formula, $t = (\ln 2)/r = (\ln 2)/0.007 \approx 99$.

57 (a) $\ln W = \ln 2.4 + (1.84)h$ { change to exponential form } $\Rightarrow W = e^{[\ln 2.4 + (1.84)h]} \Rightarrow$

$$W = e^{\ln 2.4}e^{1.84h} \text{ { since } e^x e^y = e^{x+y} \text{ } } \Rightarrow W = 2.4e^{1.84h}$$

(b) $h = 1.5 \Rightarrow W = 2.4e^{(1.84)(1.5)} = 2.4e^{2.76} \approx 37.92$ kg

59 (a) $10 = 14.7e^{-0.0000385h} \Rightarrow \frac{10}{14.7} = e^{-0.0000385h} \Rightarrow \ln\left(\frac{10}{14.7}\right) = -0.0000385h \Rightarrow$

$$h = -\frac{1}{0.0000385}\ln\left(\frac{10}{14.7}\right) \approx 10{,}007 \text{ ft.}$$

(b) At sea level, $h = 0$, and $p(0) = 14.7$. Setting $p(h)$ equal to $\frac{1}{2}(14.7)$,

and solving as in part (a), we have $h = -\frac{1}{0.0000385}\ln\left(\frac{1}{2}\right) \approx 18{,}004$ ft.

61 (a) $t = 0 \Rightarrow W = 2600(1 - 0.51)^3 = 2600(0.49)^3 \approx 305.9$ kg

(b) (1) From the graph, if $W = 1800$, t appears to be about 20.

(2) Solving the equation for t, we have $1800 = 2600\left(1 - 0.51e^{-0.075t}\right)^3 \Rightarrow$

$$\frac{1800}{2600} = \left(1 - 0.51e^{-0.075t}\right)^3 \Rightarrow \sqrt[3]{\frac{9}{13}} = 1 - 0.51e^{-0.075t} \Rightarrow$$

$$e^{-0.075t} = \left(1 - \sqrt[3]{\frac{9}{13}}\right)\left(\frac{100}{51}\right) \text{ { call this } } A \text{ } } \Rightarrow (-0.075)t = \ln A \Rightarrow t \approx 19.8 \text{ yr.}$$

63 $D = 2 \Rightarrow 5.5e^{-0.1x} = 2 \Rightarrow e^{-0.1x} = \frac{4}{11} \Rightarrow -0.1x = \ln\frac{4}{11} \Rightarrow x = -10\ln\frac{4}{11}$ mi ≈ 10.1 mi

65 Since the half-life is eight days, $A(t) = \frac{1}{2}A_0$ when $t = 8$.

Thus, $\frac{1}{2}A_0 = A_0 a^{-8} \Rightarrow a^{-8} = \frac{1}{2} \Rightarrow \frac{1}{a^8} = \frac{1}{2} \Rightarrow$

$$a^8 = 2 \Rightarrow a = 2^{1/8} \text{ { take the eighth root } } \approx 1.09.$$

67 (a) Since $\log P$ is an increasing function, increasing the population increases the

walking speed. Pedestrians have faster average walking speeds in large cities.

(b) $S = 5 \Rightarrow 5 = 0.05 + 0.86\log P \Rightarrow 4.95 = 0.86\log P \Rightarrow$

$$\frac{4.95}{0.86} = \log P \Rightarrow P = 10^{4.95/0.86} \approx 570{,}000$$

71 *Figure 71* shows a graph of $Y_1 = x\ln x$ and $Y_2 = 1$. By tracing and zooming or using

an intersect feature, we determine that $x\ln x = 1$ when $x \approx 1.763$. Alternatively, we

could graph $Y_1 = x\ln x - 1$, and find the zero of that graph.

[0, 4] by [−1, 1.67]

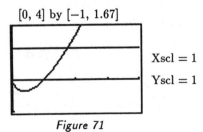

Xscl $= 1$

Yscl $= 1$

Figure 71

73 The domain of $f(x) = 2.2 \log (x + 2)$ is $x > -2$. The domain of $g(x) = \ln x$ is $x > 0$. From *Figure 73*, we determine that f intersects g at about 14.90. Thus, $f(x) \geq g(x)$ on approximately (0, 14.90), not (−2, 14.90) since g is not defined if $x \leq 0$. In general, the larger the base of the logarithm, the slower its graph will rise. In this case, we have base 10 and base $e \approx 2.72$, so we know that g will eventually intersect f even if we don't see this intersection in our first window.

[−2, 16] by [−4, 8]

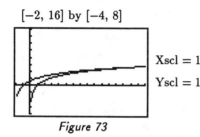

Xscl = 1
Yscl = 1

[3, 5] by [0, 1]

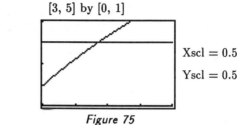

Xscl = 0.5
Yscl = 0.5

Figure 73   *Figure 75*

75 (a) $R = 2.07 \ln \frac{242}{78} - 2.04 \approx 0.3037 \approx 30\%$.

    (b) Graph $Y_1 = 2.07 \ln x - 2.04$ and $Y_2 = 0.75$.

                       From the graph, $R \approx 0.75$ when $x \approx 3.85$.

5.4 Exercises

3   $\log_a \frac{x^3 w}{y^2 z^4} = \log_a x^3 w - \log_a y^2 z^4 = \log_a x^3 + \log_a w - (\log_a y^2 + \log_a z^4) =$

$$3 \log_a x + \log_a w - 2 \log_a y - 4 \log_a z$$

The most common mistake is to not have the minus sign in front of $4 \log_a z$.

This error results from not having the parentheses in the correct place.

5   $\log \frac{\sqrt[3]{z}}{x \sqrt{y}} = \log \sqrt[3]{z} - \log x \sqrt{y} = \log z^{1/3} - \log x - \log y^{1/2} = \frac{1}{3} \log z - \log x - \frac{1}{2} \log y$

7   $\ln \sqrt[4]{\frac{x^7}{y^5 z}} = \ln x^{7/4} - \ln y^{5/4} z^{1/4} = \ln x^{7/4} - \ln y^{5/4} - \ln z^{1/4} = \frac{7}{4} \ln x - \frac{5}{4} \ln y - \frac{1}{4} \ln z$

As a generalization for exercises similar to those in 1–8, if the exponents on the variables are positive, then the sign in front of the individual logarithms will be positive if the variable was originally in the numerator and negative if the variable was originally in the denominator.

11   $2 \log_a x + \frac{1}{3} \log_a (x - 2) - 5 \log_a (2x + 3)$

     $= \log_a x^2 + \log_a (x - 2)^{1/3} - \log_a (2x + 3)^5$ {logarithm law (3)}

     $= \log_a x^2 \sqrt[3]{x - 2} - \log_a (2x + 3)^5$ {logarithm law (1)}

     $= \log_a \frac{x^2 \sqrt[3]{x - 2}}{(2x + 3)^5}$ {logarithm law (2)}

$\boxed{13}$ $\log{(x^3 y^2)} - 2\log{x}\sqrt[3]{y} - 3\log{\left(\frac{x}{y}\right)} = \log{(x^3 y^2)} - \left[\log{\left(x\sqrt[3]{y}\right)^2} + \log{\left(\frac{x}{y}\right)^3}\right]$

$$= \log{(x^3 y^2)} - \left[\log{\left(x^2 y^{2/3} \cdot (x^3/y^3)\right)}\right]$$

$$= \log{\frac{x^3 y^2}{x^5 y^{-7/3}}} = \log{\frac{y^{13/3}}{x^2}}$$

$\boxed{15}$ $\ln{y^3} + \frac{1}{3}\ln{(x^3 y^6)} - 5\ln{y} = \ln{y^3} + \ln{(xy^2)} - \ln{y^5} = \ln{\left[(xy^5)/y^5\right]} = \ln{x}$

$\boxed{17}$ $\log_6{(2x-3)} = \log_6{12} - \log_6{3} \Rightarrow \log_6{(2x-3)} = \log_6{\frac{12}{3}} \Rightarrow 2x - 3 = 4 \Rightarrow x = \frac{7}{2}$

$\boxed{19}$ $2\log_3{x} = 3\log_3{5} \Rightarrow \log_3{x^2} = \log_3{5^3} \Rightarrow x^2 = 125 \Rightarrow x = \pm 5\sqrt{5};$

$-5\sqrt{5}$ is extraneous since it would make $\log_3{x}$ undefined

$\boxed{21}$ $\log{x} - \log{(x+1)} = 3\log{4} \Rightarrow \log{\frac{x}{x+1}} = \log{64} \Rightarrow$

$\frac{x}{x+1} = 64 \Rightarrow x = 64x + 64 \Rightarrow x = -\frac{64}{63}; \; -\frac{64}{63}$ is extraneous, no solution

$\boxed{23}$ $\ln{(-4-x)} + \ln{3} = \ln{(2-x)} \Rightarrow \ln{(-12-3x)} = \ln{(2-x)} \Rightarrow -12 - 3x = 2 - x \Rightarrow$

$2x = -14 \Rightarrow x = -7.$ Remember, the solution of a logarithmic equation may be negative—you must examine what happens to the original logarithm expressions. In this case, we have $\ln{3} + \ln{3} = \ln{9}$, which is true.

$\boxed{25}$ $\log_2{(x+7)} + \log_2{x} = 3 \Rightarrow \log_2{(x^2 + 7x)} = 3 \Rightarrow x^2 + 7x = 2^3 \Rightarrow x^2 + 7x - 8 = 0 \Rightarrow$

$(x+8)(x-1) = 0 \Rightarrow x = -8, 1; \; -8$ is extraneous

$\boxed{27}$ $\log_3{(x+3)} + \log_3{(x+5)} = 1 \Rightarrow \log_3{\left[(x+3)(x+5)\right]} = 1 \Rightarrow \log_3{(x^2 + 8x + 15)} = 1 \Rightarrow$

$x^2 + 8x + 15 = 3 \Rightarrow x^2 + 8x + 12 = 0 \Rightarrow (x+2)(x+6) = 0 \Rightarrow x = -6, -2;$

-6 is extraneous

$\boxed{29}$ $\log{(x+3)} = 1 - \log{(x-2)} \Rightarrow \log{(x+3)} + \log{(x-2)} = 1 \Rightarrow$

$\log{\left[(x+3)(x-2)\right]} = 1 \Rightarrow x^2 + x - 6 = 10^1 \Rightarrow$

$x^2 + x - 16 = 0 \Rightarrow x = \frac{-1 + \sqrt{65}}{2} \approx 3.53; \; \frac{-1 - \sqrt{65}}{2} \approx -4.53$ is extraneous

$\boxed{31}$ $\ln{x} = 1 - \ln{(x+2)} \Rightarrow \ln{x} + \ln{(x+2)} = 1 \Rightarrow \ln{\left[x(x+2)\right]} = 1 \Rightarrow x^2 + 2x = e^1 \Rightarrow$

$x^2 + 2x - e = 0 \Rightarrow x = \frac{-2 \pm \sqrt{4 + 4e}}{2} = \frac{-2 \pm 2\sqrt{1+e}}{2} = -1 \pm \sqrt{1+e}.$

$x = -1 + \sqrt{1+e} \approx 0.93$ is a valid solution,

but $x = -1 - \sqrt{1+e} \approx -2.93$ is extraneous.

$\boxed{33}$ $f(x) = \log_3 (3x) = \log_3 3 + \log_3 x = \log_3 x + 1$ • shift $y = \log_3 x$ up 1 unit

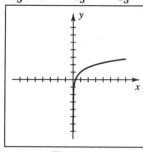

Figure 33

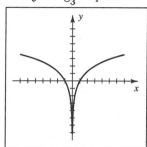

Figure 37

$\boxed{37}$ $f(x) = \log_3 (x^2) = 2\log_3 x$ • Vertically stretch $y = \log_3 x$ by a factor of 2 and include its reflection through the y-axis since the domain of the original function, $f(x) = \log_3 (x^2)$, is $\mathbb{R} - \{0\}$. Keep in mind that the laws of logarithms are established for positive real numbers, so that when we make the step $\log_3 (x^2) = 2\log_3 x$, it is only for positive x.

$\boxed{41}$ $f(x) = \log_2 \sqrt{x} = \log_2 x^{1/2} = \frac{1}{2}\log_2 x$ •

vertically compress $y = \log_2 x$ by a factor of $1/(1/2) = 2$

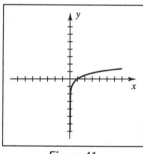

Figure 41

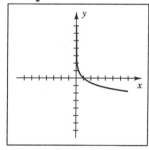

Figure 43

$\boxed{43}$ $f(x) = \log_3 \left(\frac{1}{x}\right) = \log_3 x^{-1} = -\log_3 x$ • reflect $y = \log_3 x$ through the x-axis

$\boxed{45}$ The values of $F(x) = \log_2 x$ are doubled and the reflection of F through the y-axis is included. The domain of f is $\mathbb{R} - \{0\}$ and $f(x) = \log_2 x^2$.

$\boxed{47}$ $F(x) = \log_2 x$ is shifted up 3 units since $(1, 0)$ on F is $(1, 3)$ on the graph.

Hence, $f(x) = 3 + \log_2 x = \log_2 2^3 + \log_2 x = \log_2 (8x)$.

$\boxed{49}$ $\log y = \log b - k \log x \Rightarrow \log y = \log b - \log x^k \Rightarrow \log y = \log \dfrac{b}{x^k} \Rightarrow y = \dfrac{b}{x^k}$

$\boxed{53}$ (a) $R(x_0) = a\log \left(\frac{x_0}{x_0}\right) = a\log 1 = a \cdot 0 = 0$

(b) $R(2x) = a\log \left(\frac{2x}{x_0}\right) = a\left[\log 2 + \log \left(\frac{x}{x_0}\right)\right] = a\log 2 + a\log \left(\frac{x}{x_0}\right) = R(x) + a\log 2$

$\boxed{55}$ $\ln I_0 - \ln I = kx \Rightarrow \ln \dfrac{I_0}{I} = kx \Rightarrow x = \dfrac{1}{k}\ln \dfrac{I_0}{I} = \dfrac{1}{0.39}\ln 1.12 \approx 0.29$ cm.

57 From the graph, the coordinates of the points of intersection are approximately

(1.01, 0.48) and (2.4, 0.86). $f(x) \geq g(x)$ on the intervals $(0, 1.01]$ and $[2.4, \infty)$.

[0, 6] by [−1, 3]   [0, 8] by [−1.67, 3.67]

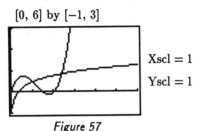   Xscl = 1   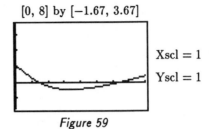   Xscl = 1
                              Yscl = 1   Yscl = 1

Figure 57   Figure 59

59 Graph $y = e^{-x} - 2\log(1 + x^2) + 0.5x$ and estimate any x-intercepts. From the graph,

we see that the roots of the equation are approximately $x \approx 1.41, 6.59$.

63 Graph $Y_1 = x\log x - \log x$ and $Y_2 = 5$. The graphs intersect at $x \approx 6.94$.

[−5, 10] by [−2, 8]   [0, 150] by [0, 100]

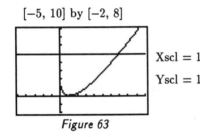   Xscl = 1   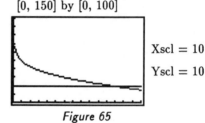   Xscl = 10
                              Yscl = 1   Yscl = 10

Figure 63   Figure 65

65 Let $d = x$. Graph $Y_1 = I_0 - 20\log x - kx = 70 - 20\log x - 0.076x$ and $Y_2 = 20$.

At the point of intersection, $x \approx 115.3$. The distance is approximately 115 meters.

5.5 Exercises

1   (a) $5^x = 8 \Rightarrow \log 5^x = \log 8 \Rightarrow x\log 5 = \log 8 \Rightarrow x = \dfrac{\log 8}{\log 5} \approx 1.29$

(b) $5^x = 8 \Rightarrow x = \log_5 8 = \dfrac{\log 8}{\log 5} \approx 1.29$

3   (a) $3^{4-x} = 5 \Rightarrow \log(3^{4-x}) = \log 5 \Rightarrow (4 - x)\log 3 = \log 5 \Rightarrow$

$4 - x = \dfrac{\log 5}{\log 3} \Rightarrow x = 4 - \dfrac{\log 5}{\log 3} \approx 2.54$. *Note:* The answer could also

be written as $4 - \dfrac{\log 5}{\log 3} = \dfrac{4\log 3 - \log 5}{\log 3} = \dfrac{\log 81 - \log 5}{\log 3} = \dfrac{\log\frac{81}{5}}{\log 3}$.

(b) $3^{4-x} = 5 \Rightarrow 4 - x = \log_3 5 \Rightarrow x = 4 - \dfrac{\log 5}{\log 3} \approx 2.54$.

7   $\log_9 0.2 = \dfrac{\log 0.2}{\log 9}\left\{ \text{or, equivalently, } \dfrac{\ln 0.2}{\ln 9} \right\} \approx -0.7325$.

Note that either log or ln can be used here. You should get comfortable using both.

9   $\dfrac{\log_5 16}{\log_5 4} = \log_4 16 = \log_4 4^2 = 2$

$\boxed{13}$ The steps are similar to those in Example 4. $2^{2x-3} = 5^{x-2} \Rightarrow$

$\log\left(2^{2x-3}\right) = \log\left(5^{x-2}\right) \Rightarrow (2x-3)\log 2 = (x-2)\log 5 \Rightarrow$

$2x\log 2 - 3\log 2 = x\log 5 - 2\log 5 \Rightarrow 2x\log 2 - x\log 5 = 3\log 2 - 2\log 5 \Rightarrow$

$$x(2\log 2 - \log 5) = \log 2^3 - \log 5^2 \Rightarrow x = \frac{\log 8 - \log 25}{\log 4 - \log 5} \Rightarrow x = \frac{\log \frac{8}{25}}{\log \frac{4}{5}} \approx 5.11$$

$\boxed{17}$ $\log x = 1 - \log(x-3) \Rightarrow \log x + \log(x-3) = 1 \Rightarrow \log(x^2 - 3x) = 1 \Rightarrow$

$$x^2 - 3x = 10^1 \Rightarrow (x-5)(x+2) = 0 \Rightarrow x = 5, -2; \; -2 \text{ is extraneous}$$

$\boxed{19}$ $\log(x^2+4) - \log(x+2) = 2 + \log(x-2) \Rightarrow \log\left(\dfrac{x^2+4}{x+2}\right) - \log(x-2) = 2 \Rightarrow$

$\log\left(\dfrac{x^2+4}{x^2-4}\right) = 2 \Rightarrow \dfrac{x^2+4}{x^2-4} = 10^2 \Rightarrow x^2 + 4 = 100x^2 - 400 \Rightarrow 404 = 99x^2 \Rightarrow$

$$x = \pm\sqrt{\frac{404}{99}} = \pm\frac{2}{3}\sqrt{\frac{101}{11}} \approx \pm 2.02; \; -\frac{2}{3}\sqrt{\frac{101}{11}} \text{ is extraneous}$$

$\boxed{21}$ See Example 5 for more detail concerning this type of exercise.

$5^x + 125(5^{-x}) = 30 \; \{\text{multiply by } 5^x\} \Rightarrow$

$(5^x)^2 - 30(5^x) + 125 = 0 \; \{\text{recognize as a quadratic in } 5^x \text{ and factor}\} \Rightarrow$

$$(5^x - 5)(5^x - 25) = 0 \Rightarrow 5^x = 5, 25 \Rightarrow 5^x = 5^1, 5^2 \Rightarrow x = 1, 2$$

$\boxed{23}$ $4^x - 3(4^{-x}) = 8 \; \{\text{multiply by } 4^x\} \Rightarrow$

$(4^x)^2 - 8(4^x) - 3 = 0 \; \{\text{recognize as a quadratic in } 4^x\} \Rightarrow$

$4^x = \dfrac{8 \pm \sqrt{76}}{2} = \dfrac{8 \pm 2\sqrt{19}}{2} = 4 \pm \sqrt{19}.$

Since since 4^x is positive and $4 - \sqrt{19}$ is negative, $4 - \sqrt{19}$ is discarded.

Continuing, $4^x = 4 + \sqrt{19} \Rightarrow x = \log_4(4 + \sqrt{19}) = \dfrac{\log(4 + \sqrt{19})}{\log 4} \; \{\text{use the change of}$

base formula to approximate $\} \approx 1.53$

$\boxed{25}$ $\log(x^2) = (\log x)^2 \Rightarrow 2\log x = (\log x)^2 \Rightarrow (\log x)^2 - 2\log x = 0 \Rightarrow$

$$(\log x)(\log x - 2) = 0 \Rightarrow \log x = 0, 2 \Rightarrow x = 10^0, 10^2 \Rightarrow x = 1 \text{ or } 100$$

$\boxed{27}$ Don't confuse $\log(\log x)$ with $(\log x)(\log x)$. The first expression is

the log of the log of x, whereas the second expression is the log of x times itself.

$$\log(\log x) = 2 \Rightarrow \log x = 10^2 = 100 \Rightarrow x = 10^{100}$$

$\boxed{29}$ $x^{\sqrt{\log x}} = 10^8 \; \{\text{take the log of both sides}\} \Rightarrow \log\left(x^{\sqrt{\log x}}\right) = \log 10^8 \Rightarrow$

$\sqrt{\log x}\,(\log x) = 8 \Rightarrow (\log x)^{1/2}(\log x)^1 = 8 \Rightarrow (\log x)^{3/2} = 8 \Rightarrow$

$$\left[(\log x)^{3/2}\right]^{2/3} = (8)^{2/3} \Rightarrow \log x = (\sqrt[3]{8})^2 = 4 \Rightarrow x = 10{,}000$$

Note: For Exercises 31–38 and 39–40 of the Chapter Review Exercises, let D denote the domain of the function determined by the original equation, and R its range. These are then the range and domain, respectively, of the equation listed in the answer.

$\boxed{31}$ $y = \dfrac{10^x + 10^{-x}}{2}$ $\{D = \mathbb{R}, R = [1, \infty)\} \Rightarrow$

$2y = 10^x + 10^{-x}$ $\left\{\text{since } 10^{-x} = \frac{1}{10^x}, \text{ multiply by } 10^x \text{ to eliminate denominator}\right\} \Rightarrow$

$10^{2x} - 2y\,10^x + 1 = 0$ $\{\text{treat as a quadratic in } 10^x\} \Rightarrow$

$$10^x = \frac{2y \pm \sqrt{4y^2 - 4}}{2} = y \pm \sqrt{y^2 - 1} \Rightarrow x = \log\left(y \pm \sqrt{y^2 - 1}\right)$$

$\boxed{33}$ $y = \dfrac{10^x - 10^{-x}}{10^x + 10^{-x}}$ $\{D = \mathbb{R}, R = (-1, 1)\} \Rightarrow y\,10^x + y\,10^{-x} = 10^x - 10^{-x} \Rightarrow$

$y\,10^{2x} + y = 10^{2x} - 1 \Rightarrow (y - 1)\,10^{2x} = -1 - y \Rightarrow$

$$10^{2x} = \frac{-1 - y}{y - 1} \Rightarrow 2x = \log\left(\frac{1+y}{1-y}\right) \Rightarrow x = \tfrac{1}{2}\log\left(\frac{1+y}{1-y}\right)$$

$\boxed{35}$ $y = \dfrac{e^x - e^{-x}}{2}$ $\{D = R = \mathbb{R}\} \Rightarrow 2y = e^x - e^{-x} \Rightarrow$

$e^{2x} - 2y\,e^x - 1 = 0 \Rightarrow e^x = \dfrac{2y \pm \sqrt{4y^2 + 4}}{2} = y \pm \sqrt{y^2 + 1};$

$$\sqrt{y^2 + 1} > y, \text{ so } y - \sqrt{y^2 + 1} < 0, \text{ but } e^x > 0 \text{ and thus, } x = \ln\left(y + \sqrt{y^2 + 1}\right)$$

$\boxed{37}$ $y = \dfrac{e^x + e^{-x}}{e^x - e^{-x}}$ $\{D = \mathbb{R} - \{0\}, R = (-\infty, -1) \cup (1, \infty)\} \Rightarrow$

$ye^x - ye^{-x} = e^x + e^{-x} \Rightarrow ye^{2x} - y = e^{2x} + 1 \Rightarrow$

$$(y - 1)e^{2x} = y + 1 \Rightarrow e^{2x} = \frac{y+1}{y-1} \Rightarrow 2x = \ln\left(\frac{y+1}{y-1}\right) \Rightarrow x = \tfrac{1}{2}\ln\left(\frac{y+1}{y-1}\right)$$

$\boxed{39}$ $f(x) = \log_2(x + 3)$ • $x = 0 \Rightarrow y\text{-intercept} = \log_2 3 = \dfrac{\log 3}{\log 2} \approx 1.5850$

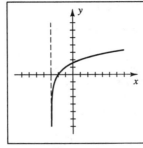

Figure 39

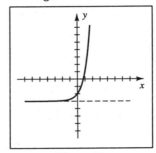

Figure 41

$\boxed{41}$ $f(x) = 4^x - 3$ • $y = 0 \Rightarrow 4^x = 3 \Rightarrow x\text{-intercept} = \log_4 3 = \dfrac{\log 3}{\log 4} \approx 0.7925$

43 (a) vinegar: $pH \approx -\log(6.3 \times 10^{-3}) = -(\log 6.3 + \log 10^{-3}) = -(\log 6.3 - 3) =$

$$3 - \log 6.3 \approx 2.2$$

(b) carrots: $pH \approx -\log(1.0 \times 10^{-5}) = 5 - \log 1.0 = 5$

(c) sea water: $pH \approx -\log(5.0 \times 10^{-9}) = 9 - \log 5.0 \approx 8.3$

45 $[H^+] < 10^{-7} \Rightarrow \log[H^+] < \log 10^{-7}$ { since log is an increasing function } $\Rightarrow$

$\log[H^+] < -7 \Rightarrow -\log[H^+] > -(-7) \Rightarrow pH > 7$ for basic solutions;

similarly, $pH < 7$ for acidic solutions.

47 Solving $A = P(1 + \frac{r}{n})^{nt}$ for t with $A = 2P$, $r = 0.06$, and $n = 12$ yields

$2P = P(1 + \frac{0.06}{12})^{12t}$ { divide by P } $\Rightarrow 2 = (1.005)^{12t}$ { take the ln of both sides } $\Rightarrow$

$\ln 2 = \ln(1.005)^{12t} \Rightarrow \ln 2 = 12t \ln(1.005) \Rightarrow$

$$t = \frac{\ln 2}{12 \ln(1.005)} \approx 11.58 \text{ yr, or about 11 years and 7 months.}$$

49 50% of the light reaching a depth of 13 meters corresponds to the equation

$\frac{1}{2}I_0 = I_0 c^{13}$. Solving for c, we have $c^{13} = \frac{1}{2} \Rightarrow c = \sqrt[13]{\frac{1}{2}} = 2^{-1/13}$.

Now letting $I = 0.01 I_0$, $c = 2^{-1/13}$, and using the formula from Example 6,

$$x = \frac{\log(I/I_0)}{\log c} = \frac{\log[(0.01 I_0)/I_0]}{\log 2^{-1/13}} = \frac{\log 10^{-2}}{-\frac{1}{13}\log 2} = \frac{26}{\log 2} \approx 86.4 \text{ m.}$$

51 (a) A is an *increasing* exponential that contains $(0, 0)$,

$(5, \approx 41)$, and $(10, \approx 65)$.

(b) $A = 50 \Rightarrow 50 = 100[1 - (0.9)^t] \Rightarrow$

$1 - (0.9)^t = 0.5 \Rightarrow (0.9)^t = 0.5 \Rightarrow$

$t = \log_{0.9}(0.5) = \frac{\log 0.5}{\log 0.9} \approx 6.58$ min.

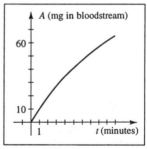

Figure 51

53 (a) $F = F_0(1 - m)^t \Rightarrow (1 - m)^t = \frac{F}{F_0} \Rightarrow \log(1 - m)^t = \log\left(\frac{F}{F_0}\right) \Rightarrow$

$$t \log(1 - m) = \log\left(\frac{F}{F_0}\right) \Rightarrow t = \frac{\log(F/F_0)}{\log(1 - m)}$$

(b) Using part (a) with $F = \frac{1}{2}F_0$ and $m = 0.00005$,

$$t = \frac{\log(F/F_0)}{\log(1 - m)} = \frac{\log(\frac{1}{2}F_0/F_0)}{\log(1 - 0.00005)} = \frac{\log\frac{1}{2}}{\log 0.99995} \approx 13,863 \text{ generations.}$$

[55] (a) $t = 10 \Rightarrow h = \dfrac{120}{1 + 200e^{-2}} \approx 4.28$ ft

(b) $h = 50 \Rightarrow 50 = \dfrac{120}{1 + 200e^{-0.2t}} \Rightarrow 1 + 200e^{-0.2t} = \dfrac{120}{50} \Rightarrow 200e^{-0.2t} = \dfrac{12}{5} - 1 \Rightarrow$

$e^{-0.2t} = \dfrac{7}{5} \cdot \dfrac{1}{200} \Rightarrow e^{-0.2t} = 0.007 \Rightarrow -0.2t = \ln 0.007 \Rightarrow t = \dfrac{\ln 0.007}{-0.2} \approx 24.8$ yr

[57] $\dfrac{v_0}{v_1} = \left(\dfrac{h_0}{h_1}\right)^P \Rightarrow \ln \dfrac{v_0}{v_1} = P \ln \dfrac{h_0}{h_1} \Rightarrow P = \dfrac{\ln(v_0/v_1)}{\ln(h_0/h_1)} = \dfrac{\ln(25/6)}{\ln(200/35)} \approx 0.82$

[59] When $x = 0$, $y = c2^0 = c = 4$. Thus, $y = 4(2)^{kx}$. Similarly, $x = 1 \Rightarrow$ $y = 4(2)^k = 3.249 \Rightarrow k = \log_2\left(\dfrac{3.249}{4}\right) \approx -0.300$. Thus, $y = 4(2)^{-0.3x}$. Checking the remaining two points, we see that $x = 2 \Rightarrow y \approx 2.639$ and $x = 3 \Rightarrow y \approx 2.144$. The points all lie on the graph of $y = 4(2)^{-0.3x}$ to within three-decimal-place accuracy.

[61] When $x = 0$, $y = c \log 10 = c = 1.5$. Thus, $y = 1.5 \log(kx + 10)$. Similarly, $x = 1 \Rightarrow$ $y = 1.5 \log(k + 10) = 1.619 \Rightarrow k + 10 = 10^{1.619/1.5} \Rightarrow k = 10^{1.619/1.5} - 10 \approx 2.004$. Thus, $y = 1.5 \log(2.004x + 10)$. Checking the remaining two points, we see that $x = 2 \Rightarrow y \approx 1.720$, and $x = 3 \Rightarrow y \approx 1.807$. The points do not all lie on the graph of $y = c2^{kx}$ to within three-decimal-place accuracy.

[63] $h(5.3) = \log_4 5.3 - 2 \log_8 (1.2 \times 5.3) = \dfrac{\ln 5.3}{\ln 4} - \dfrac{2 \ln 6.36}{\ln 8} \approx -0.5764$

[65] From the graph of $y = x - \ln(0.3x) - 3 \log_3 x$,

we see that there are *no* x-intercepts, and hence, no roots of the equation.

[−1, 17] by [−1, 11]

[−1, 17] by [−1, 11]

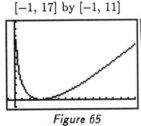

Xscl = 1
Yscl = 1

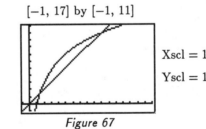

Xscl = 1
Yscl = 1

Figure 65

Figure 67

[67] The graphs of $f(x) = x$ and $g(x) = 3 \log_2 x$ intersect at approximately $(1.37, 1.37)$ and $(9.94, 9.94)$. Hence, the solutions of the equation $f(x) = g(x)$ are 1.37 and 9.94.

69 From the graph, we see that the graphs of f and g intersect at three points. Their coordinates are approximately $(-0.32, 0.50)$, $(1.52, -1.33)$, and $(6.84, -6.65)$. The region near the origin in *Figure 69(a)* is enhanced in *Figure 69(b)*. Thus, $f(x) > g(x)$ on $(-\infty, -0.32)$ and $(1.52, 6.84)$.

$[-5, 10]$ by $[-8, 2]$

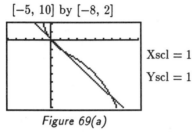

Xscl $= 1$

Yscl $= 1$

$[-1.53, 2.26]$ by $[-2.92, 1.05]$

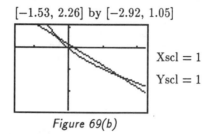

Xscl $= 1$

Yscl $= 1$

Figure 69(a)   *Figure 69(b)*

71 (1) The graph of $n(t) = 85e^{t/3}$ is increasing rapidly and soon becomes greater than 100. It is doubtful that the average score would improve dramatically without any review.

$[0, 5]$ by $[0, 200]$

$[0, 5]$ by $[0, 100]$

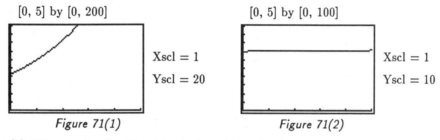

Xscl $= 1$

Yscl $= 20$

Xscl $= 1$

Yscl $= 10$

Figure 71(1)   *Figure 71(2)*

(2) The graph of $n(t) = 70 + \ln(t+1)$ is also increasing. It is incorrect because $n(0) = 70 \neq 85$. Also, one would not expect the average score to improve without any review.

(3) The graph of $n(t) = 86 - e^t$ decreases rapidly to zero. The *average* score probably would not be zero after 5 weeks.

$[0, 5]$ by $[0, 100]$

$[0, 5]$ by $[0, 100]$

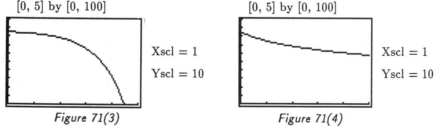

Xscl $= 1$

Yscl $= 10$

Xscl $= 1$

Yscl $= 10$

Figure 71(3)   *Figure 71(4)*

(4) The graph of $n(t) = 85 - 15\ln(t+1)$ is decreasing. During the first weeks it decreases most rapidly and then starts to level off. Of the four functions, this function seems to best model the situation.

[73] (a) The ozone level is decreasing by 11% per year. The fraction of ozone remaining x years after April 1992 is given by the function $f(x) = (0.89)^x$. We must approximate x when $f(x) = 0.5$. Using a table, $f(x) = 0.5$ when $x \approx 6$. Thus, in 1998 the ozone level would be 50% of its normal amount.

x	1	2	3	4	5	6	7
$f(x)$	0.89	0.79	0.70	0.63	0.56	0.50	0.44

(b) $(0.89)^t = 0.5 \Rightarrow \ln(0.89)^t = \ln 0.5 \Rightarrow t \ln 0.89 = \ln 0.5 \Rightarrow$

$$t = \frac{\ln 0.5}{\ln 0.89} \approx 5.948, \text{ or in 1998.}$$

Chapter 5 Review Exercises

[3] $f(x) = \left(\frac{3}{2}\right)^{-x} = \left(\frac{2}{3}\right)^x$ • goes through $\left(-1, \frac{3}{2}\right)$, $(0, 1)$, and $\left(1, \frac{2}{3}\right)$

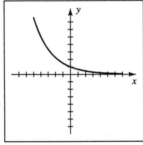

Figure 3

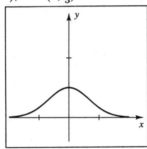

Figure 5

[5] $f(x) = 3^{-x^2} = (3^{-1})^{(x^2)} = \left(\frac{1}{3}\right)^{(x^2)}$ • see the note in §5.1 before Exercise 17

[7] $f(x) = e^{x/2} = (e^{1/2})^x \approx (1.65)^x$ • goes through $(-1, 1/\sqrt{e})$, $(0, 1)$, and $(1, \sqrt{e})$;

or approximately $(-1, 0.61)$, $(0, 1)$, and $(1, 1.65)$

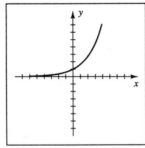

Figure 7

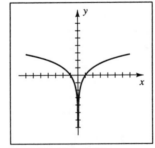

Figure 13

[13] $f(x) = \log_4(x^2) = 2\log_4 x$ • stretch $y = \log_4 x$ by a factor of 2 and include its reflection through the y-axis since the domain of the original function is $\mathbb{R} - \{0\}$

$\boxed{17}$ (a) $\log_2 \frac{1}{16} = \log_2 2^{-4} = -4$ (b) $\log_\pi 1 = 0$ (c) $\ln e = 1$

(d) $6^{\log_6 4} = 4$ (e) $\log 1{,}000{,}000 = \log 10^6 = 6$

(f) $10^{3\log 2} = 10^{\log 2^3} = 2^3 = 8$ (g) $\log_4 2 = \log_4 4^{1/2} = \frac{1}{2}$

$\boxed{19}$ $2^{3x-1} = \frac{1}{2} \Rightarrow 2^{3x-1} = 2^{-1} \Rightarrow 3x - 1 = -1 \Rightarrow 3x = 0 \Rightarrow x = 0$

$\boxed{22}$ $\log_4 (x+1) = 2 + \log_4 (3x-2) \Rightarrow \log_4 (x+1) - \log_4 (3x-2) = 2 \Rightarrow$

$$\log_4 \left(\frac{x+1}{3x-2}\right) = 2 \Rightarrow \frac{x+1}{3x-2} = 16 \Rightarrow x + 1 = 48x - 32 \Rightarrow x = \frac{33}{47}$$

$\boxed{23}$ $2\ln(x+3) - \ln(x+1) = 3\ln 2 \Rightarrow \ln\frac{(x+3)^2}{x+1} = \ln 2^3 \Rightarrow (x+3)^2 = 8(x+1) \Rightarrow$

$$x^2 + 6x + 9 = 8x + 8 \Rightarrow x^2 - 2x + 1 = 0 \Rightarrow (x-1)^2 = 0 \Rightarrow x = 1$$

$\boxed{26}$ $3^{(x^2)} = 7 \Rightarrow x^2 = \log_3 7 \Rightarrow x^2 = \frac{\log 7}{\log 3} \Rightarrow x = \pm\sqrt{\frac{\log 7}{\log 3}}$

$\boxed{27}$ $2^{5x+3} = 3^{2x+1} \Rightarrow \log(2^{5x+3}) = \log(3^{2x+1}) \Rightarrow (5x+3)\log 2 = (2x+1)\log 3 \Rightarrow$

$5x \log 2 + 3\log 2 = 2x \log 3 + \log 3 \Rightarrow 5x \log 2 - 2x \log 3 = \log 3 - 3 \log 2 \Rightarrow$

$$x(5\log 2 - 2\log 3) = \log 3 - \log 2^3 \Rightarrow x = \frac{\log 3 - \log 8}{\log 32 - \log 9} = \frac{\log\frac{3}{8}}{\log\frac{32}{9}}$$

$\boxed{29}$ $\log_4 x = \sqrt[3]{\log_4 x} \Rightarrow \log_4 x = (\log_4 x)^{1/3} \Rightarrow (\log_4 x)^3 = \log_4 x \Rightarrow$

$(\log_4 x)^3 - \log_4 x = 0 \Rightarrow \log_4 x \left[(\log_4 x)^2 - 1\right] = 0 \Rightarrow$

$$\log_4 x = 0 \text{ or } \log_4 x = \pm 1 \Rightarrow x = 1 \text{ or } x = 4, \frac{1}{4}$$

$\boxed{31}$ $10^{2\log x} = 5 \Rightarrow 10^{\log x^2} = 5 \Rightarrow x^2 = 5 \Rightarrow x = \pm\sqrt{5}; \ -\sqrt{5}$ is extraneous

$\boxed{34}$ $e^x + 2 = 8e^{-x} \Rightarrow e^{2x} + 2e^x - 8 = 0 \Rightarrow (e^x+4)(e^x-2) = 0 \Rightarrow e^x = -4, 2 \Rightarrow$

$$x = \ln 2 \text{ since } e^x \neq -4$$

$\boxed{35}$ (a) $\log x^2 = \log(6-x) \Rightarrow x^2 = 6 - x \Rightarrow x^2 + x - 6 = 0 \Rightarrow (x+3)(x-2) = 0 \Rightarrow$

$$x = -3, 2$$

(b) $2\log x = \log(6-x) \Rightarrow \log x^2 = \log(6-x)$, which is the equation in part (a).

This equation has the same solutions provided they are in the domain.

But -3 is extraneous, so 2 is the only solution.

$\boxed{36}$ (a) $\ln(e^x)^2 = 16 \Rightarrow 2\ln e^x = 16 \Rightarrow 2x = 16 \Rightarrow x = 8$

(b) $\ln e^{(x^2)} = 16 \Rightarrow x^2 = 16 \Rightarrow x = \pm 4$

$\boxed{38}$ $\log(x^2/y^3) + 4\log y - 6\log\sqrt{xy} = \log\left(\frac{x^2}{y^3}\right) + \log y^4 - \log(x^3 y^3)$

$$= \log\left(\frac{x^2 y}{x^3 y^3}\right) = \log\left(\frac{1}{xy^2}\right) = -\log(xy^2)$$

39 $y = \dfrac{1}{10^x + 10^{-x}} \left\{ D = \mathbb{R},\ R = (0, \frac{1}{2}] \right\} \Rightarrow y\, 10^x + y\, 10^{-x} = 1 \Rightarrow$

$y\, 10^{2x} + y = 10^x \Rightarrow y\, 10^{2x} - 10^x + y = 0 \Rightarrow 10^x = \dfrac{1 \pm \sqrt{1 - 4y^2}}{2y} \Rightarrow$

$$x = \log\left(\dfrac{1 \pm \sqrt{1 - 4y^2}}{2y}\right)$$

43 (a) For $y = \log_2(x + 1)$, $D = (-1, \infty)$ and $R = \mathbb{R}$.

(b) $y = \log_2(x + 1) \Rightarrow x = \log_2(y + 1) \Rightarrow 2^x = y + 1 \Rightarrow y = 2^x - 1$,

$$D = \mathbb{R},\ R = (-1, \infty)$$

45 (a) $Q(0) = 2(3^0) = 2$ {in thousands}, or 2000

(b) $Q(\frac{10}{60}) = 2000(3^{1/6}) \approx 2401;\ Q(\frac{30}{60}) = 2000(3^{1/2}) \approx 3464;\ Q(1) = 2(3) = 6000$

47 (a) $N = 64(0.5)^{t/8} = 64\left[(0.5)^{1/8}\right]^t \approx 64(0.917)^t$

(b) $N = \frac{1}{2}N_0 \Rightarrow \frac{1}{2}N_0 = N_0(0.5)^{t/8} \Rightarrow$

$(\frac{1}{2})^1 = (\frac{1}{2})^{t/8} \Rightarrow 1 = t/8 \Rightarrow t = 8$ days

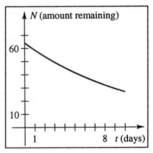

Figure 47

49 (a) Using $A = Pe^{rt}$ with $A = \$35{,}000$, $P = \$10{,}000$, and $r = 11\%$, we have

$35{,}000 = 10{,}000e^{0.11t} \Rightarrow e^{0.11t} = 3.5 \Rightarrow 0.11t = \ln 3.5 \Rightarrow t = \frac{1}{0.11}\ln 3.5 \approx 11.39$ yr.

(b) $2 \cdot 10{,}000 = 10{,}000e^{0.11t} \Rightarrow e^{0.11t} = 2 \Rightarrow 0.11t = \ln 2 \Rightarrow t \approx 6.30$ yr

Alternatively, using the doubling time formula, $t = (\ln 2)/r = (\ln 2)/0.11 \approx 6.30$.

51 (a) $\alpha = 10\log\left(\dfrac{I}{I_0}\right) \Rightarrow \dfrac{\alpha}{10} = \log\left(\dfrac{I}{I_0}\right) \Rightarrow 10^{\alpha/10} = \dfrac{I}{I_0} \Rightarrow I = I_0 10^{\alpha/10}$

(b) Let $I(\alpha)$ be the intensity corresponding to α decibels.

$I(\alpha + 1) = I_0 10^{(\alpha + 1)/10} = I_0 10^{\alpha/10} 10^{1/10} = I(\alpha)\, 10^{1/10} \approx 1.26\, I(\alpha)$,

which represents a 26% increase in $I(\alpha)$.

53 $R = 2.3\log(A + 3000) - 5.1 \Rightarrow R + 5.1 = 2.3\log(A + 3000) \Rightarrow$

$\dfrac{R + 5.1}{2.3} = \log(A + 3000) \Rightarrow$

$A + 3000 = 10^{(R + 5.1)/2.3}$ {change to exponential form} $\Rightarrow A = 10^{(R + 5.1)/2.3} - 3000$

55 $R = 4 \Rightarrow 2.3\log(A + 14{,}000) - 6.6 = 4 \Rightarrow \log(A + 14{,}000) = \frac{10.6}{2.3} \Rightarrow$

$$A = 10^{106/23} - 14{,}000 \approx 26{,}615.9 \text{ mi}^2.$$

57 Substituting $v = 0$ and $m = m_1 + m_2$ in $v = -a \ln m + b$ yields

$0 = -a \ln(m_1 + m_2) + b$. Thus, $b = a \ln(m_1 + m_2)$. At burnout, $m = m_1$,

and hence, $v = -a \ln m_1 + b = -a \ln m_1 + a \ln(m_1 + m_2)$

$$= a[\ln(m_1 + m_2) - \ln m_1] \Rightarrow v = a \ln\left(\frac{m_1 + m_2}{m_1}\right)$$

59 (a) $\log E = 11.4 + (1.5)R \Rightarrow E = 10^{11.4 + 1.5R}$ { merely change the form }

(b) $R = 8.4 \Rightarrow E = 10^{11.4 + 1.5(8.4)} = 10^{24}$ ergs

62 $I = \frac{V}{R}(1 - e^{-Rt/L}) \Rightarrow \frac{RI}{V} = 1 - e^{-Rt/L} \Rightarrow$

$$e^{-Rt/L} = \left(\frac{V - RI}{V}\right) \Rightarrow -\frac{Rt}{L} = \ln\left(\frac{V - RI}{V}\right) \Rightarrow t = -\frac{L}{R} \ln\left(\frac{V - RI}{V}\right)$$

63 (a) $x = 4\% \Rightarrow T = -8310 \ln(0.04) \approx 26{,}749$ yr.

(b) $T = 10{,}000 \Rightarrow 10{,}000 = -8310 \ln x \Rightarrow -\frac{10{,}000}{8310} = \ln x \Rightarrow$

$$x = e^{-1000/831} \approx 0.30, \text{ or } 30\%.$$

65 $N(t) = \frac{1}{2}N_0 \Rightarrow \frac{1}{2}N_0 = N_0(0.805)^t \Rightarrow 2^{-1} = (0.805)^t \Rightarrow \ln(2^{-1}) = \ln(0.805)^t \Rightarrow$

$$-\ln 2 = t \ln(0.805) \Rightarrow t = -\frac{\ln 2}{\ln(0.805)} \approx 3.196 \text{ millennia, or } 3196 \text{ yr}$$

Chapter 5 Discussion Exercises

1 (a) The y-intercept is a, so it increases as a does. The graph flattens out as a increases.

(b) Graph $Y_1 = \frac{a}{2}(e^{x/a} + e^{-x/a}) + (30 - a)$ on $[-20, 20]$ by $[30, 32]$ and check for $Y_1 < 32$ at $x = 20$. In this case it turns out that $a = 101$ is the smallest integer value that satisfies the conditions, so an equation is

$$y = \frac{101}{2}(e^{x/101} + e^{-x/101}) - 71.$$

3 (a) $x^y = y^x \Rightarrow \ln(x^y) = \ln(y^x) \Rightarrow y \ln x = x \ln y \Rightarrow \frac{\ln x}{x} = \frac{\ln y}{y}$

(b) Once you find two values of $(\ln x)/x$ that are the same (such as 0.36652), you know that the corresponding x-values, x_1 and x_2, satisfy the relationship $(x_1)^{x_2} \approx (x_2)^{x_1}$. In particular, when $(\ln x)/x \approx 0.36652$, we find that $x_1 \approx 2.50$ and $x_2 \approx 2.97$. Note that $2.50^{2.97} \approx 2.97^{2.50} \approx 15.20$.

(c) Note that $f(e) = \frac{1}{e}$. Any horizontal line $y = k$, with $0 < k < \frac{1}{e}$, will intersect the graph at the two points $\left(x_1, \frac{\ln x_1}{x_1}\right)$ and $\left(x_2, \frac{\ln x_2}{x_2}\right)$, where $1 < x_1 < e$ and $x_2 > e$.

⑤ Logarithm law 3 states that it is valid only for positive real numbers, so $y = \log_3(x^2)$ is equivalent to $y = 2\log_3 x$ only for $x > 0$. The domain of $y = \log_3(x^2)$ is $\mathbb{R} - \{0\}$, whereas the domain of $y = 2\log_3 x$ is $x > 0$.

⑦ There are 4 points of intersection. Listed in order of difficulty to find we have: $(-0.9999011, 0.00999001)$, $(-0.0001, 0.01)$, $(100, 0.01105111)$, and $(36,102.844, 4.6928 \times 10^{13})$. Exponential function values (with base > 1) are greater than polynomial function values (with leading term positive) for very large values of x.

⑨ $60,000 = 40,000b^5 \Rightarrow b = \sqrt[5]{1.5} \approx 1.0844717712$, or 8.44717712%. Mentally—it would take $70/8.5 \approx 8^+$ years to double and there would be about $40/8 = 5$ doubling periods, $2^5 = 32$ and $32 \cdot \$40,000 = \$1,280,000$.

$$\text{Actual} = \$40,000(1 + 0.0844717712)^{40} = \$1,025,156.25.$$

⑪ On the TI-82, enter the sum of the days, $\{0, 5168, 6728, 8136, 8407, 8735\}$, and the averages, $\{1003.16, 2002.25, 3004.46, 4003.33, 5023.55, 6010.00\}$, in L_1 and L_2, respectively. Use ExpReg under STAT CALC to obtain $y = ab^x$, where $a = 898.7637295$ and $b = 1.00019458$. Plot the data along with the exponential function and the line $y = 10,000$. The functions intersect at approximately 12,383. This value corresponds to October 10, 2006. *Note:* The TI-83 has a convenient "day between dates" function for problems of this nature.

Using the first and last milestone figures and the continuously compounded interest formula gives us $A = Pe^{rt} \Leftrightarrow 6010 = 1003.16e^{r(8735)} \Rightarrow r \approx 0.000204953604$ (daily) or about 7.48%, or 7% annually. *Note:* The Dow was 100.25 on 1/12/1906. You may want to include this information and examine the differences it makes in any type of prediction.

A discussion of practical considerations should lead to mention of crashes, corrections, and the validity of any model over too long of a period of time.

Chapter 6: The Trigonometric Functions

Note: Exercises 1 and 3: The answers listed are the smallest (in magnitude) two positive
coterminal angles and two negative coterminal angles.

$\boxed{1}$ (a) $120° + 1(360°) = 480°$, $\qquad$ $120° + 2(360°) = 840°$;

$\qquad$ $120° - 1(360°) = -240°$, $\qquad$ $120° - 2(360°) = -600°$

$\quad$ (b) $135° + 1(360°) = 495°$, $\qquad$ $135° + 2(360°) = 855°$;

$\qquad$ $135° - 1(360°) = -225°$, $\qquad$ $135° - 2(360°) = -585°$

$\quad$ (c) $-30° + 1(360°) = 330°$, $\qquad$ $-30° + 2(360°) = 690°$;

$\qquad$ $-30° - 1(360°) = -390°$, $\qquad$ $-30° - 2(360°) = -750°$

$\boxed{3}$ (a) $620° - 1(360°) = 260°$, $\qquad$ $620° + 1(360°) = 980°$;

$\qquad$ $620° - 2(360°) = -100°$, $\qquad$ $620° - 3(360°) = -460°$

$\quad$ (b) $\frac{5\pi}{6} + 1(2\pi) = \frac{5\pi}{6} + \frac{12\pi}{6} = \frac{17\pi}{6}$, $\qquad$ $\frac{5\pi}{6} + 2(2\pi) = \frac{5\pi}{6} + \frac{24\pi}{6} = \frac{29\pi}{6}$;

$\qquad$ $\frac{5\pi}{6} - 1(2\pi) = \frac{5\pi}{6} - \frac{12\pi}{6} = -\frac{7\pi}{6}$, $\qquad$ $\frac{5\pi}{6} - 2(2\pi) = \frac{5\pi}{6} - \frac{24\pi}{6} = -\frac{19\pi}{6}$

$\quad$ (c) $-\frac{\pi}{4} + 1(2\pi) = -\frac{\pi}{4} + \frac{8\pi}{4} = \frac{7\pi}{4}$, $\qquad$ $-\frac{\pi}{4} + 2(2\pi) = -\frac{\pi}{4} + \frac{16\pi}{4} = \frac{15\pi}{4}$;

$\qquad$ $-\frac{\pi}{4} - 1(2\pi) = -\frac{\pi}{4} - \frac{8\pi}{4} = -\frac{9\pi}{4}$, $\qquad$ $-\frac{\pi}{4} - 2(2\pi) = -\frac{\pi}{4} - \frac{16\pi}{4} = -\frac{17\pi}{4}$

$\boxed{5}$ (a) $90° - 5°17'34'' = 84°42'26''$ $\qquad$ (b) $90° - 32.5° = 57.5°$

$\boxed{7}$ (a) $180° - 48°51'37'' = 131°8'23''$ $\qquad$ (b) $180° - 136.42° = 43.58°$

Note: Multiply each degree measure by $\frac{\pi}{180}$ to obtain the listed radian measure.

$\boxed{9}$ (a) $150° \cdot \frac{\pi}{180} = \frac{5 \cdot 30\pi}{6 \cdot 30} = \frac{5\pi}{6}$ $\qquad\qquad$ (b) $-60° \cdot \frac{\pi}{180} = -\frac{60\pi}{3 \cdot 60} = -\frac{\pi}{3}$

$\quad$ (c) $225° \cdot \frac{\pi}{180} = \frac{5 \cdot 45\pi}{4 \cdot 45} = \frac{5\pi}{4}$

Note: Multiply each radian measure by $\frac{180}{\pi}$ to obtain the listed degree measure.

$\boxed{15}$ (a) $-\frac{7\pi}{2} \cdot \left(\frac{180}{\pi}\right)° = -\left(\frac{7 \cdot 90 \cdot 2\pi}{2\pi}\right)° = -630°$ $\quad$ (b) $7\pi \cdot \left(\frac{180}{\pi}\right)° = (7 \cdot 180)° = 1260°$

$\quad$ (c) $\frac{\pi}{9} \cdot \left(\frac{180}{\pi}\right)° = \left(\frac{20 \cdot 9\pi}{9\pi}\right)° = 20°$

$\boxed{17}$ *Note:* Some calculators can easily change radians to degrees, minutes, and seconds by
pressing a few keys. For **TI-82/83** users, enter

$$\boxed{2} \;\boxed{\times}\; \boxed{180} \;\boxed{\div}\; \boxed{\text{2nd}} \;\boxed{\pi}\; \boxed{\text{2nd}} \;\boxed{\text{ANGLE}}\; \boxed{4} \;\boxed{\text{ENTER}}.$$

Check your calculator manual for this feature.

We first convert 2 radians to degrees. $2 \cdot \left(\frac{180}{\pi}\right)° \approx 114.59156° = 114° + 0.59156°$.

We now use the decimal portion, $0.59156°$, and convert it to minutes.

Since $60' = 1°$, we have $0.59156° = 0.59156(60') = 35.4936'$.

Using the decimal portion, $0.4936'$, we convert it to seconds.

Since $60'' = 1'$, we have $0.4936' = 0.4936(60'') \approx 30''$. $\qquad$ $\therefore$ 2 radians $\approx 114°35'30''$

[21] Since $1' = (\frac{1}{60})^\circ$, $41' = (\frac{41}{60})^\circ$. Thus, $37^\circ41' = \left(37 + \frac{41}{60}\right)^\circ \approx 37.6833^\circ$.

[23] Since $1'' = (\frac{1}{3600})^\circ$, $27'' = (\frac{27}{3600})^\circ$. Thus, $115^\circ26'27'' = \left(115 + \frac{26}{60} + \frac{27}{3600}\right)^\circ \approx 115.4408^\circ$.

[25] We have 63° and a portion of one more degree. Since $1^\circ = 60'$,

$0.169^\circ = 0.169\,(60') = 10.14'$. We now have $10'$ and a portion of one more minute.

Since $1' = 60''$, $0.14' = 0.14\,(60'') = 8.4''$. $\therefore\ 63.169^\circ \approx 63^\circ10'8''$

[29] We will use the formula for the length of a circular arc.

$$s = r\theta \Rightarrow r = \frac{s}{\theta} = \frac{10}{4} = 2.5 \text{ cm}.$$

[31] (a) $s = r\theta = 8 \cdot (45 \cdot \frac{\pi}{180}) = 8 \cdot \frac{\pi}{4} = 2\pi \approx 6.28$ cm

(b) $A = \frac{1}{2}r^2\theta = \frac{1}{2}(8)^2(\frac{\pi}{4}) = 8\pi \approx 25.13$ cm^2

[33] (a) Remember that θ is measured in radians. $s = r\theta \Rightarrow \theta = \frac{s}{r} = \frac{7}{4} = 1.75$ radians.

Converting to degrees, we have $\frac{7}{4} \cdot (\frac{180}{\pi})^\circ = (\frac{315}{\pi})^\circ \approx 100.27^\circ$.

(b) $A = \frac{1}{2}r^2\theta = \frac{1}{2}(4)^2(\frac{7}{4}) = 14$ cm^2

[35] (a) A measure of 50° is equivalent to $(50 \cdot \frac{\pi}{180})$ radians. The radius is one-half of the

diameter. Thus, $s = r\theta = (\frac{1}{2} \cdot 16)(50 \cdot \frac{\pi}{180}) = 8 \cdot \frac{5\pi}{18} = \frac{20\pi}{9} \approx 6.98$ m.

(b) $A = \frac{1}{2}r^2\theta = \frac{1}{2}(8)^2(\frac{5\pi}{18}) = \frac{80\pi}{9} \approx 27.93$ m^2

[37] radius $= \frac{1}{2} \cdot 8000$ miles $= 4000$ miles

(a) $s = r\theta = 4000\,(60 \cdot \frac{\pi}{180}) = \frac{4000\pi}{3} \approx 4189$ miles

(b) $s = r\theta = 4000\,(45 \cdot \frac{\pi}{180}) = 1000\pi \approx 3142$ miles

(c) $s = r\theta = 4000\,(30 \cdot \frac{\pi}{180}) = \frac{2000\pi}{3} \approx 2094$ miles

(d) $s = r\theta = 4000\,(10 \cdot \frac{\pi}{180}) = \frac{2000\pi}{9} \approx 698$ miles

(e) $s = r\theta = 4000\,(1 \cdot \frac{\pi}{180}) = \frac{200\pi}{9} \approx 70$ miles

[39] $\theta = \frac{s}{r} = \frac{500}{4000} = \frac{1}{8}$ radian; $(\frac{1}{8})(\frac{180}{\pi})^\circ = (\frac{45}{2\pi})^\circ \approx 7^\circ10'$

[41] 23 hours, 56 minutes, and 4 seconds $= 23(60)^2 + 56(60) + 4 = 86,164$ sec.

Since the earth turns through 2π radians in 86,164 seconds,

it rotates through $\frac{2\pi}{86,164} \approx 7.29 \times 10^{-5}$ radians in one second.

[43] (a) $\left(40 \, \frac{\text{revolutions}}{\text{minute}}\right)\left(2\pi \, \frac{\text{radians}}{\text{revolution}}\right) = 80\pi \, \frac{\text{radians}}{\text{minute}}$.

Note: Remember to write out and "cancel" the units if you are unsure about

what units your answer is measured in.

(b) The distance that a point on the circumference travels is

$$s = r\theta = (5 \text{ in.}) \cdot 80\pi = 400\pi \text{ in.}$$

Hence, its linear speed is

$$400\pi \text{ in./min} = \frac{100\pi}{3} \text{ ft/min} \; \{\, 400\pi \cdot \tfrac{1}{12} \,\} \approx 104.72 \text{ ft/min}.$$

45 (a) As in Exercise 43, we take the number of

revolutions per minute times the number of radians per revolution and

obtain $(33\frac{1}{3})(2\pi) = \frac{200\pi}{3}$ and $45(2\pi) = 90\pi$.

(b) $s = r\theta = (\frac{1}{2} \cdot 12)(\frac{200\pi}{3}) = 400\pi$ in. Linear speed $= 400\pi$ in./min $= \frac{100\pi}{3}$ ft/min.

$s = r\theta = (\frac{1}{2} \cdot 7)(90\pi) = 315\pi$ in. Linear speed $= 315\pi$ in./min $= \frac{105\pi}{4}$ ft/min.

47 (a) The distance that the cargo is lifted is equal to the arc length that the cable is

moved through. $s = r\theta = (\frac{1}{2} \cdot 3)(\frac{7\pi}{4}) = \frac{21\pi}{8} \approx 8.25$ ft.

(b) $s = r\theta \Rightarrow d = (\frac{1}{2} \cdot 3)\theta \Rightarrow \theta = (\frac{2}{3}d)$ radians. For example, to lift the cargo 6 feet,

the winch must rotate $\frac{2}{3} \cdot 6 = 4$ radians, or about 229°.

49 $\text{Area}_{small} = \frac{1}{2}r^2\theta = \frac{1}{2}(\frac{1}{2} \cdot 18)^2 \cdot (\frac{2\pi}{6}) = \frac{27\pi}{2}$. $\text{Area}_{large} = \frac{1}{2}(\frac{1}{2} \cdot 26)^2 \cdot (\frac{2\pi}{8}) = \frac{169\pi}{8}$.

$\text{Cost}_{small} = \text{Area}_{small} \div \text{Cost} = \frac{27\pi}{2} \div 2 \approx 21.21$ in.²/dollar.

$\text{Cost}_{large} = \text{Area}_{large} \div \text{Cost} = \frac{169\pi}{8} \div 3 \approx 22.12$ in.²/dollar.

The large slice provides slightly more pizza per dollar.

51 $\frac{40 \text{ miles}}{\text{hour}} = \frac{40 \text{ miles}}{\text{hour}} \cdot \frac{1 \text{ hour}}{3600 \text{ seconds}} \cdot \frac{5280 \text{ feet}}{\text{mile}} \cdot \frac{12 \text{ inches}}{\text{foot}} = \frac{704 \text{ inches}}{\text{second}}$.

The circumference of the wheel is $2\pi(14)$ inches. The back sprocket then rotates

$\frac{704 \text{ inches}}{\text{second}} \cdot \frac{1 \text{ revolution}}{28\pi \text{ inches}} = \frac{704 \text{ revolutions}}{28\pi \text{ second}}$ or $\frac{704}{28\pi} \cdot 2\pi = \frac{352}{7}$ radians per second.

The front sprocket's angular speed is given by { from Exercise 50 }

$\theta_1 = \frac{r_2\theta_2}{r_1} = \frac{2 \cdot \frac{352}{7}}{5} = \frac{704}{35} \approx 20.114$ radians per second

or 3.2 revolutions per second or 192.08 revolutions per minute.

6.2 Exercises

Note: Answers are in the order *sin, cos, tan, cot, sec, csc* for any exercises that require
the values of the six trigonometric functions.

1 Using the definition of the trigonometric functions in terms of a unit circle with
$P(-\frac{15}{17}, \frac{8}{17})$, we have

$$\sin t = y = \frac{8}{17}, \qquad \cos t = x = -\frac{15}{17}, \qquad \tan t = \frac{y}{x} = \frac{8/17}{-15/17} = -\frac{8}{15},$$

$$\csc t = \frac{1}{y} = \frac{1}{8/17} = \frac{17}{8}, \qquad \sec t = \frac{1}{x} = \frac{1}{-15/17} = -\frac{17}{15}, \qquad \cot t = \frac{x}{y} = \frac{-15/17}{8/17} = -\frac{15}{8}.$$

$\boxed{5}$ See *Figure 5.* $P(t) = (\frac{3}{5}, \frac{4}{5})$.

(a) $(t + \pi)$ will be $\frac{1}{2}$ revolution from (t) in the counterclockwise direction.

$$\text{Hence, } P(t + \pi) = (-\tfrac{3}{5}, -\tfrac{4}{5}).$$

(b) $(t - \pi)$ will be $\frac{1}{2}$ revolution from (t) in the clockwise direction.

$$\text{Hence, } P(t - \pi) = (-\tfrac{3}{5}, -\tfrac{4}{5}). \text{ Observe that } P(t + \pi) = P(t - \pi) \text{ for any } t.$$

(c) $(-t)$ is the angle measure in the opposite direction of (t).

$$\text{Hence, } P(-t) = (\tfrac{3}{5}, -\tfrac{4}{5}).$$

(d) $(-t - \pi)$ will be $\frac{1}{2}$ revolution from $(-t)$ in the clockwise direction.

$$\text{Hence, } P(-t - \pi) = (-\tfrac{3}{5}, \tfrac{4}{5}).$$

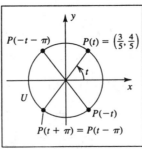

Figure 5

$\boxed{9}$ (a) The point P on the unit circle U that corresponds to $t = 2\pi$ has coordinates $(1, 0)$. Thus, we choose $x = 1$ and $y = 0$ and use the definition of the trigonometric functions in terms of a unit circle.

$\sin t = y \Rightarrow \sin 2\pi = 0.$ $\qquad\qquad$ $\cos t = x \Rightarrow \cos 2\pi = 1.$

$\tan t = \frac{y}{x} \Rightarrow \tan 2\pi = \frac{0}{1} = 0.$ $\qquad$ $\cot t = \frac{x}{y} \Rightarrow \cot 2\pi = \frac{1}{0}$ and is undefined.

$\sec t = \frac{1}{x} \Rightarrow \sec 2\pi = \frac{1}{1} = 1.$ $\qquad$ $\csc t = \frac{1}{y} \Rightarrow \csc 2\pi = \frac{1}{0}$ and is undefined.

(b) $t = -3\pi$ is coterminal with $t = \pi$. The point P on the unit circle U that corresponds to $t = -3\pi$ has coordinates $(-1, 0)$. Thus, we choose $x = -1$ and $y = 0$.

$\sin t = y \Rightarrow \sin (-3\pi) = 0.$ $\qquad\qquad$ $\cos t = x \Rightarrow \cos (-3\pi) = -1.$

$\tan t = \frac{y}{x} \Rightarrow \tan (-3\pi) = \frac{0}{-1} = 0.$ $\qquad$ $\cot t = \frac{x}{y} \Rightarrow \cot (-3\pi) = \frac{-1}{0}$ and is und.

$\sec t = \frac{1}{x} \Rightarrow \sec (-3\pi) = \frac{1}{-1} = -1.$ $\qquad$ $\csc t = \frac{1}{y} \Rightarrow \csc (-3\pi) = \frac{1}{0}$ and is und.

$\boxed{11}$ (a) The point P on the unit circle U that corresponds to $t = \frac{3\pi}{2}$ has coordinates $(0, -1)$. Thus, we choose $x = 0$ and $y = -1$.

$\sin t = y \Rightarrow \sin \frac{3\pi}{2} = -1.$ $\qquad$ $\cos t = x \Rightarrow \cos \frac{3\pi}{2} = 0.$

$\tan t = \frac{y}{x} \Rightarrow \tan \frac{3\pi}{2} = \frac{-1}{0}$ and is und. $\quad$ $\cot t = \frac{x}{y} \Rightarrow \cot \frac{3\pi}{2} = \frac{0}{-1} = 0.$

$\sec t = \frac{1}{x} \Rightarrow \sec \frac{3\pi}{2} = \frac{1}{0}$ and is und. $\quad$ $\csc t = \frac{1}{y} \Rightarrow \csc \frac{3\pi}{2} = \frac{1}{-1} = -1.$

(b) $t = -\frac{7\pi}{2}$ is coterminal with $t = \frac{\pi}{2}$. The point P on the unit circle U that corresponds to $t = -\frac{7\pi}{2}$ has coordinates $(0, 1)$. Thus, we choose $x = 0$ and $y = 1$.

$\sin t = y \Rightarrow \sin\left(-\frac{7\pi}{2}\right) = 1.$ $\qquad$ $\cos t = x \Rightarrow \cos\left(-\frac{7\pi}{2}\right) = 0.$

$\tan t = \frac{y}{x} \Rightarrow \tan\left(-\frac{7\pi}{2}\right) = \frac{1}{0}$ and is und. $\quad$ $\cot t = \frac{x}{y} \Rightarrow \cot\left(-\frac{7\pi}{2}\right) = \frac{0}{1} = 0.$

$\sec t = \frac{1}{x} \Rightarrow \sec\left(-\frac{7\pi}{2}\right) = \frac{1}{0}$ and is und. $\quad$ $\csc t = \frac{1}{y} \Rightarrow \csc\left(-\frac{7\pi}{2}\right) = \frac{1}{1} = 1.$

$\boxed{13}$ (a) $t = \frac{9\pi}{4}$ is coterminal with $t = \frac{\pi}{4}$. The point P on the unit circle U that corresponds to $t = \frac{9\pi}{4}$ has coordinates $\left(\frac{\sqrt{2}}{2}, \frac{\sqrt{2}}{2}\right)$. Thus, we choose $x = \frac{\sqrt{2}}{2}$ and $y = \frac{\sqrt{2}}{2}$.

$\sin t = y \Rightarrow \sin \frac{9\pi}{4} = \frac{\sqrt{2}}{2}.$ $\qquad$ $\cos t = x \Rightarrow \cos \frac{9\pi}{4} = \frac{\sqrt{2}}{2}.$

$\tan t = \frac{y}{x} \Rightarrow \tan \frac{9\pi}{4} = \frac{\sqrt{2}/2}{\sqrt{2}/2} = 1.$

We now use the reciprocal relationships to find the values of the 3 remaining trigonometric functions.

$\cot t = \frac{1}{\tan t} = \frac{1}{1} = 1.$ $\qquad$ $\sec t = \frac{1}{\cos t} = \frac{1}{\sqrt{2}/2} = \frac{2}{\sqrt{2}} = \frac{2^1}{2^{1/2}} = 2^{1/2} = \sqrt{2}.$

$\csc t = \frac{1}{\sin t} = \frac{1}{\sqrt{2}/2} = \sqrt{2}.$

(b) $t = -\frac{5\pi}{4}$ is coterminal with $t = \frac{3\pi}{4}$. The point P on the unit circle U that corresponds to $t = -\frac{5\pi}{4}$ has coordinates $\left(-\frac{\sqrt{2}}{2}, \frac{\sqrt{2}}{2}\right)$. Thus, we choose $x = -\frac{\sqrt{2}}{2}$ and $y = \frac{\sqrt{2}}{2}$.

$\sin t = y \Rightarrow \sin\left(-\frac{5\pi}{4}\right) = \frac{\sqrt{2}}{2}.$ $\qquad$ $\cos t = x \Rightarrow \cos\left(-\frac{5\pi}{4}\right) = -\frac{\sqrt{2}}{2}.$

$\tan t = \frac{y}{x} \Rightarrow \tan\left(-\frac{5\pi}{4}\right) = \frac{\sqrt{2}/2}{-\sqrt{2}/2} = -1.$ $\quad$ $\cot t = \frac{1}{\tan t} = \frac{1}{-1} = -1.$

$\sec t = \frac{1}{\cos t} = \frac{1}{-\sqrt{2}/2} = -\sqrt{2}.$ $\qquad$ $\csc t = \frac{1}{\sin t} = \frac{1}{\sqrt{2}/2} = \sqrt{2}.$

15 (a) The point P on the unit circle U that corresponds to $t = \frac{5\pi}{4}$ has coordinates $\left(-\frac{\sqrt{2}}{2}, -\frac{\sqrt{2}}{2} \right)$. Thus, we choose $x = -\frac{\sqrt{2}}{2}$ and $y = -\frac{\sqrt{2}}{2}$.

$$\sin t = y \Rightarrow \sin \frac{5\pi}{4} = -\frac{\sqrt{2}}{2}. \qquad \cos t = x \Rightarrow \cos \frac{5\pi}{4} = -\frac{\sqrt{2}}{2}.$$

$$\tan t = \frac{y}{x} \Rightarrow \tan \frac{5\pi}{4} = \frac{-\sqrt{2}/2}{-\sqrt{2}/2} = 1. \qquad \cot t = \frac{1}{\tan t} = \frac{1}{1} = 1.$$

$$\sec t = \frac{1}{\cos t} = \frac{1}{-\sqrt{2}/2} = -\sqrt{2}. \qquad \csc t = \frac{1}{\sin t} = \frac{1}{-\sqrt{2}/2} = -\sqrt{2}.$$

(b) $t = -\frac{\pi}{4}$ is coterminal with $t = \frac{7\pi}{4}$. The point P on the unit circle U that corresponds to $t = -\frac{\pi}{4}$ has coordinates $\left(\frac{\sqrt{2}}{2}, -\frac{\sqrt{2}}{2} \right)$. Thus, we choose $x = \frac{\sqrt{2}}{2}$ and $y = -\frac{\sqrt{2}}{2}$.

$$\sin t = y \Rightarrow \sin \left(-\frac{\pi}{4} \right) = -\frac{\sqrt{2}}{2}. \qquad \cos t = x \Rightarrow \cos \left(-\frac{\pi}{4} \right) = \frac{\sqrt{2}}{2}.$$

$$\tan t = \frac{y}{x} \Rightarrow \tan \left(-\frac{\pi}{4} \right) = \frac{-\sqrt{2}/2}{\sqrt{2}/2} = -1. \qquad \cot t = \frac{1}{\tan t} = \frac{1}{-1} = -1.$$

$$\sec t = \frac{1}{\cos t} = \frac{1}{\sqrt{2}/2} = \sqrt{2}. \qquad \csc t = \frac{1}{\sin t} = \frac{1}{-\sqrt{2}/2} = -\sqrt{2}.$$

17 (a) $\cos t > 0 \ \{ x > 0 \}$ implies that the point P corresponding to t is in quadrant I or quadrant IV. $\sin t < 0 \ \{ y < 0 \}$ implies that the point P corresponding to t is in quadrant III or quadrant IV. P must be in quadrant IV to satisfy both conditions.

(b) $\sin t < 0 \Rightarrow P$ is in QIII or QIV. $\cot t > 0 \Rightarrow P$ is in QI or QIII.

Hence, P is in QIII.

(c) $\csc t > 0 \Rightarrow P$ is in QI or QII. $\sec t < 0 \Rightarrow P$ is in QII or QIII.

Hence, P is in QII.

(d) $\sec t < 0 \Rightarrow P$ is in QII or QIII. $\tan t > 0 \Rightarrow P$ is in QI or QIII.

Hence, P is in QIII.

19 $\cot t = \frac{\cos t}{\sin t} \ \{ \text{cotangent identity} \} = \frac{\sqrt{1 - \sin^2 t}}{\sin t} \ \{ \sin^2 t + \cos^2 t = 1 \}$ Note that the " $\pm$ " notation is not needed in front of the square root since all of the trigonometric functions of an acute angle are positive.

21 $\sec t = \frac{1}{\cos t} \ \{ \text{reciprocal identity} \} = \frac{1}{\sqrt{1 - \sin^2 t}}$

23 One solution is $\sin t = \sqrt{1 - \cos^2 t} = \sqrt{1 - \frac{1}{\sec^2 t}} = \frac{\sqrt{\sec^2 t - 1}}{\sec t}$.

Alternatively, $\sin t = \frac{\sin t / \cos t}{1 / \cos t} = \frac{\tan t}{\sec t} = \frac{\sqrt{\sec^2 t - 1}}{\sec t} \ \{ 1 + \tan^2 t = \sec^2 t \}$.

Note: Steps to determine 2 function values using only the fundamental identities are shown. The other 3 function values are just the reciprocals of those given and are listed in the answer.

25 $\tan t = -\frac{3}{4}$ and $\sin t > 0 \Rightarrow t$ is in QII.

$1 + \tan^2 t = \sec^2 t \Rightarrow \sec t = \pm \sqrt{1 + \tan^2 t}$ { Use the "$-$" since the secant is negative in the second quadrant. } $= -\sqrt{1 + \tan^2 t} = -\sqrt{1 + \frac{9}{16}} = -\frac{5}{4}$.

$\tan t = \frac{\sin t}{\cos t} \Rightarrow -\frac{3}{4} = \frac{\sin t}{-4/5} \Rightarrow \sin t = (-\frac{3}{4})(-\frac{4}{5}) = \frac{3}{5}$. $\bigstar \ \frac{3}{5}, -\frac{4}{5}, -\frac{3}{4}, -\frac{4}{3}, -\frac{5}{4}, \frac{5}{3}$

27 $\sin t = -\frac{5}{13}$ and $\sec t > 0 \Rightarrow t$ is in QIV.

$\sin^2 t + \cos^2 t = 1 \Rightarrow \cos^2 t = 1 - \sin^2 t \Rightarrow \cos t = \pm \sqrt{1 - \sin^2 t} =$

{ Use the "$+$" since the cosine is positive in the fourth quadrant. } $= \sqrt{1 - \frac{25}{169}} = \frac{12}{13}$.

$\tan t = \frac{\sin t}{\cos t} = \frac{-5/13}{12/13} = -\frac{5}{12}$. $\bigstar \ -\frac{5}{13}, \frac{12}{13}, -\frac{5}{12}, -\frac{12}{5}, \frac{13}{12}, -\frac{13}{5}$

29 $\cos t = -\frac{1}{3}$ and $\sin t < 0 \Rightarrow t$ is in QIII. $\sin t = -\sqrt{1 - \cos^2 t} = -\sqrt{1 - \frac{1}{9}} = -\frac{\sqrt{8}}{3}$.

$\tan t = \frac{\sin t}{\cos t} = \frac{-\sqrt{8}/3}{-1/3} = \sqrt{8}$, or $2\sqrt{2}$. $\bigstar \ -\frac{\sqrt{8}}{3}, -\frac{1}{3}, \sqrt{8}, \frac{1}{\sqrt{8}}, -3, -\frac{3}{\sqrt{8}}$

31 $\sec t = -4$ and $\csc t > 0 \Rightarrow t$ is in QII. $\tan t = -\sqrt{\sec^2 t - 1} = -\sqrt{16 - 1} = -\sqrt{15}$.

$\tan t = \frac{\sin t}{\cos t} \Rightarrow -\sqrt{15} = \frac{\sin t}{-1/4} \Rightarrow \sin t = \frac{\sqrt{15}}{4}$.

$\bigstar \ \frac{\sqrt{15}}{4}, -\frac{1}{4}, -\sqrt{15}, -\frac{1}{\sqrt{15}}, -4, \frac{4}{\sqrt{15}}$

33 (a) Since $1 + \tan^2 4\beta = \sec^2 4\beta$, $\tan^2 4\beta - \sec^2 4\beta = -1$.

(b) $4 \tan^2 \beta - 4 \sec^2 \beta = 4(\tan^2 \beta - \sec^2 \beta) = 4(-1) = -4$

35 (a) $5 \sin^2 \theta + 5 \cos^2 \theta = 5(\sin^2 \theta + \cos^2 \theta) = 5(1) = 5$

(b) $5 \sin^2 (\theta/4) + 5 \cos^2 (\theta/4) = 5 \left[\sin^2 (\theta/4) + \cos^2 (\theta/4) \right] = 5(1) = 5$

37 $\dfrac{\sin^3 t + \cos^3 t}{\sin t + \cos t} = \dfrac{(\sin t + \cos t)(\sin^2 t - \sin t \cos t + \cos^2 t)}{\sin t + \cos t}$ { factor, sum of cubes }

$= \sin^2 t - \sin t \cos t + \cos^2 t$ { cancel $(\sin t + \cos t)$ }

$= (\sin^2 t + \cos^2 t) - \sin t \cos t$ { group terms }

$= 1 - \sin t \cos t$ { Pythagorean identity }

39 $\dfrac{2 - \tan t}{2 \csc t - \sec t} = \dfrac{2 - \dfrac{\sin t}{\cos t}}{2 \cdot \dfrac{1}{\sin t} - \dfrac{1}{\cos t}}$ { tangent and reciprocal identities }

$= \dfrac{\dfrac{2 \cos t - \sin t}{\cos t}}{\dfrac{2 \cos t - \sin t}{\sin t \cos t}}$ { combine terms } $= \dfrac{\dfrac{1}{1}}{\dfrac{1}{\sin t}}$ { cancel like terms } $= \sin t$

$\boxed{41}$ $\cos t \sec t = \cos t\,(1/\cos t)$ { reciprocal identity } $= 1$

$\boxed{43}$ $\sin t \sec t = \sin t\,(1/\cos t) = \sin t/\cos t = \tan t$ { tangent identity }

$\boxed{45}$ $\dfrac{\csc t}{\sec t} = \dfrac{1/\sin t}{1/\cos t}$ { reciprocal identities } $= \dfrac{\cos t}{\sin t} = \cot t$ { cotangent identity }

$\boxed{47}$ $(1 + \cos 2t)(1 - \cos 2t) = 1 - \cos^2 2t = \sin^2 2t$ { Pythagorean identity }

$\boxed{49}$ $\cos^2 t\,(\sec^2 t - 1) \; = \cos^2 t\,(\tan^2 t)$ { Pythagorean identity }

$$= \cos^2 t \cdot \frac{\sin^2 t}{\cos^2 t} \text{ \{ tangent identity \} } = \sin^2 t$$

$\boxed{51}$ $\dfrac{\sin(t/2)}{\csc(t/2)} + \dfrac{\cos(t/2)}{\sec(t/2)} = \dfrac{\sin(t/2)}{1/\sin(t/2)} + \dfrac{\cos(t/2)}{1/\cos(t/2)} = \sin^2(t/2) + \cos^2(t/2) = 1$

$\boxed{53}$ $(1 + \sin t)(1 - \sin t) = 1 - \sin^2 t$ { multiply as a difference of squares } $= \cos^2 t = \dfrac{1}{\sec^2 t}$

$\boxed{55}$ $\sec t - \cos t = \dfrac{1}{\cos t} - \cos t = \dfrac{1 - \cos^2 t}{\cos t} = \dfrac{\sin^2 t}{\cos t} = \dfrac{\sin t}{\cos t} \cdot \sin t = \tan t \sin t$

$\boxed{57}$ $(\cot t + \csc t)(\tan t - \sin t)$

$$= \cot t \tan t - \cot t \sin t + \csc t \tan t - \csc t \sin t \text{ \{ multiply binomials \} }$$

$$= \frac{1}{\tan t} \tan t - \frac{\cos t}{\sin t} \sin t + \frac{1}{\sin t} \frac{\sin t}{\cos t} - \frac{1}{\sin t} \sin t$$

{ reciprocal, cotangent, and tangent identities }

$$= 1 - \cos t + \frac{1}{\cos t} - 1 \text{ \{ cancel terms \} }$$

$$= -\cos t + \sec t = \sec t - \cos t$$

$\boxed{59}$ $\sec^2 3t\,\csc^2 3t \;\; = (1 + \tan^2 3t)(1 + \cot^2 3t)$ { Pythagorean identities }

$$= 1 + \tan^2 3t + \cot^2 3t + 1 \qquad \text{\{ multiply binomials \}}$$

$$= (1 + \tan^2 3t) + (\cot^2 3t + 1) \quad \text{\{ group terms \}}$$

$$= \sec^2 3t + \csc^2 3t \qquad\qquad \text{\{ Pythagorean identities \}}$$

$\boxed{61}$ $\log \csc t \;\; = \log\!\left(\dfrac{1}{\sin t}\right)$ { reciprocal identity }

$$= \log 1 - \log \sin t \quad \text{\{ property of logarithms \}}$$

$$= 0 - \log \sin t \qquad \text{\{ } \log 1 = 0 \text{ \}}$$

$$= -\log \sin t$$

$\boxed{63}$ $\sqrt{\sec^2 t - 1} = \sqrt{\tan^2 t}$ { Pythagorean identity } $= |\tan t| \; \left\{ \sqrt{x^2} = |x| \right\} =$

$$-\tan t \text{ since } \tan t < 0 \text{ if } \pi/2 < t < \pi.$$

$\boxed{65}$ $\sqrt{1 + \tan^2 t} = \sqrt{\sec^2 t}$ { Pythagorean identity } $= |\sec t| \; \left\{ \sqrt{x^2} = |x| \right\} =$

$$\sec t \text{ since } \sec t > 0 \text{ if } 3\pi/2 < t < 2\pi.$$

$\boxed{67}$ $\sqrt{\sin^2(t/2)} = |\sin(t/2)| \; \left\{ \sqrt{x^2} = |x| \right\} =$

$$-\sin(t/2) \text{ since } \sin(t/2) < 0 \text{ if } 2\pi < t < 4\pi \;\{\, \pi < t/2 < 2\pi \,\}.$$

boxed{69} (a) From (1, 0), move counterclockwise on the unit circle to the point at the tick marked 4. The projection of this point on the y-axis,

approximately -0.7 or -0.8, is the value of $\sin 4$.

(b) From (1, 0), move clockwise 1.2 units to about 5.1.

As in part (a), $\sin(-1.2)$ is about -0.9.

(c) Draw the horizontal line $y = 0.5$.

This line intersects the circle at about 0.5 and 2.6.

boxed{71} (a) From (1, 0), move counterclockwise on the unit circle to the point at the tic marked 4. The projection of this point on the x-axis,

approximately -0.6 or -0.7, is the value of $\cos 4$.

(b) Proceeding as in part (a), $\cos(-1.2)$ is about 0.4.

(c) Draw the vertical line $x = -0.6$.

This line intersects the circle at about 2.2 and 4.1.

boxed{73} (a) Note that midnight occurs when $t = -6$.

Time	Temp.	Humidity	Time	Temp.	Humidity
12 A.M.	60	60	12 P.M.	60	60
3 A.M.	52	74	3 P.M.	68	46
6 A.M.	48	80	6 P.M.	72	40
9 A.M.	52	74	9 P.M.	68	46

(b) Since $T(t) = -12\cos\left(\frac{\pi}{12}t\right) + 60$, its maximum is $60 + 12 = 72\,°\text{F}$ at $t = 12$ or 6:00 P.M., and its minimum is $60 - 12 = 48\,°\text{F}$ at $t = 0$ or 6:00 A.M. Since $H(t) = 20\cos\left(\frac{\pi}{12}t\right) + 60$, its maximum is $60 + 20 = 80\%$ at $t = 0$ or 6:00 A.M., and its minimum is $60 - 20 = 40\%$ at $t = 12$ or 6:00 P.M.

(c) When the temperature increases, the relative humidity decreases and vice versa. As the temperature cools, the air can hold less moisture and the relative humidity increases. Because of this phenomenon, fog often occurs during the evening hours rather than in the middle of the day.

6.3 Exercises

boxed{1} (a) $\sin(-90°) = -\sin 90°$ { since $\sin(-t) = -\sin t$ } $= -1$

(b) $\cos\left(-\frac{3\pi}{4}\right) = \cos\frac{3\pi}{4}$ { since $\cos(-t) = \cos t$ } $= -\frac{\sqrt{2}}{2}$

(c) $\tan(-45°) = -\tan 45°$ { since $\tan(-t) = -\tan t$ } $= -1$

3 (a) $\cot\left(-\frac{3\pi}{4}\right) = -\cot\frac{3\pi}{4}$ { since $\cot(-t) = -\cot t$ } $= -(-1) = 1$

(b) $\sec(-180°) = \sec 180°$ { since $\sec(-t) = \sec t$ } $= -1$

(c) $\csc\left(-\frac{3\pi}{2}\right) = -\csc\frac{3\pi}{2}$ { since $\csc(-t) = -\csc t$ } $= -(-1) = 1$

5 $\sin(-t)\sec(-t) = (-\sin t)\sec t$ { formulas for negatives }

$= (-\sin t)(1/\cos t)$ { reciprocal identity }

$= -\tan t$ { tangent identity }

7 $\dfrac{\cot(-t)}{\csc(-t)} = \dfrac{-\cot t}{-\csc t}$ { formulas for negatives }

$= \dfrac{\cos t/\sin t}{1/\sin t}$ { cotangent identity and reciprocal identity }

$= \cos t$ { simplify }

9 $\dfrac{1}{\cos(-t)} - \tan(-t)\sin(-t) = \dfrac{1}{\cos t} - (-\tan t)(-\sin t)$ { formulas for negatives }

$= \dfrac{1}{\cos t} - \dfrac{\sin t}{\cos t}\sin t$ { tangent identity }

$= \dfrac{1 - \sin^2 t}{\cos t}$ { combine terms }

$= \dfrac{\cos^2 t}{\cos t}$ { Pythagorean identity }

$= \cos t$ { cancel $\cos t$ }

11 (a) Using Figure 22 in the text, we see that as t gets close to 0 through numbers greater than 0 (from the *right* of 0), $\sin t$ approaches 0.

(b) As t approaches $-\frac{\pi}{2}$ through numbers less than $-\frac{\pi}{2}$ (from the *left* of $-\frac{\pi}{2}$), $\sin t$ approaches -1.

13 (a) Using Figure 24 in the text, we see that as t gets close to $\frac{\pi}{4}$ through numbers greater than $\frac{\pi}{4}$ (from the *right* of $\frac{\pi}{4}$), $\cos t$ approaches $\frac{\sqrt{2}}{2}$. Note that the value $\frac{\sqrt{2}}{2}$ is in the table containing specific values of the cosine function.

(b) As t approaches π through numbers less than π (from the *left* of π), $\cos t$ approaches -1.

15 (a) Using Figure 26 in the text, we see that as t gets close to $\frac{\pi}{4}$ through numbers greater than $\frac{\pi}{4}$ (from the *right* of $\frac{\pi}{4}$), $\tan t$ approaches 1.

(b) As t approaches $\frac{\pi}{2}$ through numbers greater than $\frac{\pi}{2}$ (from the *right* of $\frac{\pi}{2}$), $\tan t$ is approaching the vertical asymptote at $t = \frac{\pi}{2}$. $\tan t$ is *decreasing* without bound, and we use the notation $\underline{\tan t \to -\infty}$ to denote this.

17 (a) Using Figure 29 in the text, we see that as t gets close to $-\frac{\pi}{4}$ through numbers less than $-\frac{\pi}{4}$ (from the *left* of $-\frac{\pi}{4}$), $\cot t$ approaches -1.

(b) As t approaches 0 through numbers greater than 0 (from the *right* of 0), $\cot t$ is approaching the vertical asymptote at $t = 0$ {the y-axis}. $\cot t$ is *increasing without bound*, and we use the notation $\underline{\cot t \to \infty}$ to denote this.

19 (a) Using Figure 28 in the text, we see that as t gets close to $\frac{\pi}{2}$ through numbers less than $\frac{\pi}{2}$ (from the *left* of $\frac{\pi}{2}$), $\sec t \to \infty$.

(b) As t approaches $\frac{\pi}{4}$ through numbers greater than $\frac{\pi}{4}$ (from the *right* of $\frac{\pi}{4}$), $\sec t$ approaches $\sqrt{2}$. Recall that $\cos \frac{\pi}{4} = \frac{1}{\sqrt{2}}$ and that $\sec t = \frac{1}{\cos t}$.

21 (a) Using Figure 27 in the text, we see that as t gets close to 0 through numbers less than 0 (from the *left* of 0), $\csc t \to -\infty$.

(b) As t approaches $\frac{\pi}{2}$ through numbers greater than $\frac{\pi}{2}$ (from the *right* of $\frac{\pi}{2}$), $\csc t$ approaches 1.

23 Refer to Figure 21 and the accompanying table. We see that $\sin \frac{3\pi}{2} = -1$.

Since the period of the sine is 2π, the second value in $[0, 4\pi]$ is $\frac{3\pi}{2} + 2\pi = \frac{7\pi}{2}$.

29 Refer to Figure 23 and the accompanying table.

We see that $\cos \frac{3\pi}{4} = \cos \frac{5\pi}{4} = -\frac{\sqrt{2}}{2}$. Since the period of the cosine is 2π,

other values in $[0, 4\pi]$ are $\frac{3\pi}{4} + 2\pi = \frac{11\pi}{4}$ and $\frac{5\pi}{4} + 2\pi = \frac{13\pi}{4}$.

31 Refer to Figure 26. In the interval $(-\frac{\pi}{2}, \frac{\pi}{2})$, $\tan t = 1$ only if $t = \frac{\pi}{4}$. Since the period of the tangent is π, the desired value in the interval $(\frac{\pi}{2}, \frac{3\pi}{2})$ is $\frac{\pi}{4} + \pi = \frac{5\pi}{4}$.

33 $y = \sin t$; $[-2\pi, 2\pi]$; $a = \sqrt{2}/2$ •   Refer to Figure 21.

$\sin t = \sqrt{2}/2 \Rightarrow t = \frac{\pi}{4}$ and $\frac{3\pi}{4}$. Also, $\frac{\pi}{4} - 2\pi = -\frac{7\pi}{4}$ and $\frac{3\pi}{4} - 2\pi = -\frac{5\pi}{4}$.

$\sin t > \sqrt{2}/2$ when the graph is *above* the horizontal line $y = \sqrt{2}/2$.

$\sin t < \sqrt{2}/2$ when the graph is *below* the horizontal line $y = \sqrt{2}/2$.

★ (a) $-\frac{7\pi}{4}, -\frac{5\pi}{4}, \frac{\pi}{4}, \frac{3\pi}{4}$ (b) $-\frac{7\pi}{4} < t < -\frac{5\pi}{4}, \frac{\pi}{4} < t < \frac{3\pi}{4}$

(c) $-2\pi \le t < -\frac{7\pi}{4}, -\frac{5\pi}{4} < t < \frac{\pi}{4}$, and $\frac{3\pi}{4} < t \le 2\pi$

35 $y = \cos t$; $[-2\pi, 2\pi]$; $a = -\sqrt{2}/2$ •   Refer to Figure 23.

$\cos t = -\sqrt{2}/2 \Rightarrow t = \frac{3\pi}{4}$ and $\frac{5\pi}{4}$. Also, $\frac{3\pi}{4} - 2\pi = -\frac{5\pi}{4}$ and $\frac{5\pi}{4} - 2\pi = -\frac{3\pi}{4}$.

$\cos t > -\sqrt{2}/2$ when the graph is *above* the horizontal line $y = -\sqrt{2}/2$.

$\cos t < -\sqrt{2}/2$ when the graph is *below* the horizontal line $y = -\sqrt{2}/2$.

★ (a) $-\frac{5\pi}{4}, -\frac{3\pi}{4}, \frac{3\pi}{4}, \frac{5\pi}{4}$ (b) $-2\pi \le t < -\frac{5\pi}{4}, -\frac{3\pi}{4} < t < \frac{3\pi}{4}$, and $\frac{5\pi}{4} < t \le 2\pi$

(c) $-\frac{5\pi}{4} < t < -\frac{3\pi}{4}$ and $\frac{3\pi}{4} < t < \frac{5\pi}{4}$

37 $y = 2 + \sin t$　•　Shift $y = \sin t$ up 2 units.

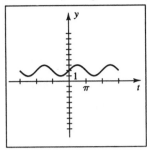

Figure 37

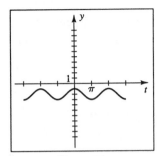

Figure 39

39 $y = \cos t - 2$　•　Shift $y = \cos t$ down 2 units.

41 $y = 1 + \tan t$　•　Shift $y = \tan t$ up 1 unit.

Since $1 + \tan\left(-\frac{\pi}{4}\right) = 1 + (-1) = 0$, and the period of the tangent is π,

there are t-intercepts at $t = -\frac{\pi}{4} + \pi n$.

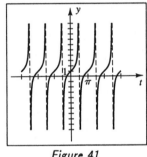

Figure 41

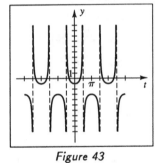

Figure 43

43 $y = \sec t - 2$　•

Shift $y = \sec t$ down 2 units, t-intercepts are at $t = \frac{\pi}{3} + 2\pi n, \frac{5\pi}{3} + 2\pi n$.

45 (a) As we move from left to right, the function increases { goes up } on the

intervals $[-2\pi, -\frac{3\pi}{2})$, $(-\frac{3\pi}{2}, -\pi]$, $[0, \frac{\pi}{2})$, $(\frac{\pi}{2}, \pi]$.

(b) As we move from left to right, the function decreases { goes down } on the

intervals $[-\pi, -\frac{\pi}{2})$, $(-\frac{\pi}{2}, 0]$, $[\pi, \frac{3\pi}{2})$, $(\frac{3\pi}{2}, 2\pi]$.

47 (a) The tangent function increases on *all* intervals on which it is defined.

Between -2π and 2π,

these intervals are $[-2\pi, -\frac{3\pi}{2})$, $(-\frac{3\pi}{2}, -\frac{\pi}{2})$, $(-\frac{\pi}{2}, \frac{\pi}{2})$, $(\frac{\pi}{2}, \frac{3\pi}{2})$, and $(\frac{3\pi}{2}, 2\pi]$.

(b) The tangent function is *never* decreasing on any interval for which it is defined.

49 This is good advice.

51 Graph $y = \sin(t^2)$ and $y = 0.5$ on the same coordinate plane. From the graph, we see that $\sin(t^2)$ assumes the value of 0.5 at $x \approx \pm 0.72, \pm 1.62, \pm 2.61, \pm 2.98$.

$[-\pi, \pi]$ by $[-2.09, 2.09]$

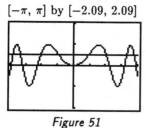

Xscl $= \pi/4$
Yscl $= 1$

Figure 51

$[-2\pi, 2\pi]$ by $[-5.19, 3.19]$

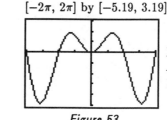

Xscl $= \pi/2$
Yscl $= 1$

Figure 53

53 We see that the graph of $y = t \sin t$ assumes a maximum value of approximately

1.82 at $t \approx \pm 2.03$, and a minimum value of -4.81 at $t \approx \pm 4.91$.

55 As $t \to 0^+$, $f(t) = \dfrac{1 - \cos t}{t} \to 0$.

$[-1, 1]$ by $[-0.67, 0.67]$

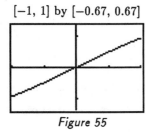

Xscl $= 0.5$
Yscl $= 0.5$

Figure 55

$[-2, 2]$ by $[-1.33, 1.33]$

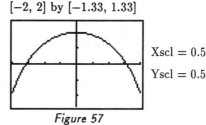

Xscl $= 0.5$
Yscl $= 0.5$

Figure 57

57 As $t \to 0^+$, $f(t) = t \cot t \to 1$.

59 As $t \to 0^+$, $f(t) = \dfrac{\tan t}{t} \to 1$.

$[-1.5, 1.5]$ by $[0, 3]$

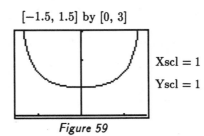

Xscl $= 1$
Yscl $= 1$

Figure 59

1 For the point $P(x, y)$, we let r denote the distance from the origin to P. Applying the theorem on trigonometric functions as ratios with $x = 4$, $y = -3$, and $r = \sqrt{x^2 + y^2} = \sqrt{4^2 + (-3)^2} = 5$, we obtain the following:

$$\sin\theta = \frac{y}{r} = \frac{-3}{5} = -\frac{3}{5} \qquad\qquad \cos\theta = \frac{x}{r} = \frac{4}{5}$$

$$\tan\theta = \frac{y}{x} = \frac{-3}{4} = -\frac{3}{4} \qquad\qquad \cot\theta = \frac{x}{y} = \frac{4}{-3} = -\frac{4}{3}$$

$$\sec\theta = \frac{r}{x} = \frac{5}{4} \qquad\qquad\qquad \csc\theta = \frac{r}{y} = \frac{5}{-3} = -\frac{5}{3}$$

3 $x = -2$ and $y = -5 \Rightarrow r = \sqrt{x^2 + y^2} = \sqrt{(-2)^2 + (-5)^2} = \sqrt{29}$.

$$\sin\theta = \frac{y}{r} = \frac{-5}{\sqrt{29}}, \text{ or } -\frac{5}{29}\sqrt{29} \qquad \cos\theta = \frac{x}{r} = \frac{-2}{\sqrt{29}}, \text{ or } -\frac{2}{29}\sqrt{29}$$

$$\tan\theta = \frac{y}{x} = \frac{-5}{-2} = \frac{5}{2} \qquad\qquad\qquad \cot\theta = \frac{x}{y} = \frac{-2}{-5} = \frac{2}{5}$$

$$\sec\theta = \frac{r}{x} = \frac{\sqrt{29}}{-2}, \text{ or } -\frac{1}{2}\sqrt{29} \qquad \csc\theta = \frac{r}{y} = \frac{\sqrt{29}}{-5} = -\frac{1}{5}\sqrt{29}$$

Note: In the following exercises, we will only find the values of x, y, and r. The above definitions can then be used to find the values of the trigonometric functions of θ. These values are listed in the usual order in the answer.

5 Since the terminal side of θ is in QII, choose x to be negative.

If $x = -1$, then $y = 4$ and $(-1, 4)$ is a point on the terminal side of θ.

$x = -1$ and $y = 4 \Rightarrow r = \sqrt{(-1)^2 + 4^2} = \sqrt{17}$. $\bigstar \ \dfrac{4}{\sqrt{17}}, \ -\dfrac{1}{\sqrt{17}}, \ -4, \ -\dfrac{1}{4}, \ -\sqrt{17}, \ \dfrac{\sqrt{17}}{4}$

7 Remember that $y = mx$ is an equation of a line that passes through the origin and has slope m. Thus, an equation of the line is $y = \frac{4}{3}x$.

If $x = 3$, then $y = 4$ and $(3, 4)$ is a point on the terminal side of θ.

$x = 3$ and $y = 4 \Rightarrow r = \sqrt{3^2 + 4^2} = 5$. $\bigstar \ \dfrac{4}{5}, \dfrac{3}{5}, \dfrac{4}{3}, \dfrac{3}{4}, \dfrac{5}{3}, \dfrac{5}{4}$

9 $2y - 7x + 2 = 0 \Leftrightarrow y = \frac{7}{2}x - 1$. Thus, the slope of the given line is $\frac{7}{2}$.

An equation of the line through the origin with that slope is $y = \frac{7}{2}x$.

If $x = -2$, then $y = -7$ and $(-2, -7)$ is a point on the terminal side of θ.

$x = -2$ and $y = -7 \Rightarrow r = \sqrt{(-2)^2 + (-7)^2} = \sqrt{53}$.

$$\bigstar \ -\frac{7}{\sqrt{53}}, \ -\frac{2}{\sqrt{53}}, \ \frac{7}{2}, \frac{2}{7}, \ -\frac{\sqrt{53}}{2}, \ -\frac{\sqrt{53}}{7}$$

11 *Note:* U denotes *undefined.*

(a) For $\theta = 90°$, choose $x = 0$ and $y = 1$. $r = 1$.   ★ 1, 0, U, 0, U, 1

(b) For $\theta = 0°$, choose $x = 1$ and $y = 0$. $r = 1$.   ★ 0, 1, 0, U, 1, U

(c) For $\theta = \frac{7\pi}{2}$, choose $x = 0$ and $y = -1$. $r = 1$.   ★ -1, 0, U, 0, U, -1

(d) For $\theta = 3\pi$, choose $x = -1$ and $y = 0$. $r = 1$.   ★ 0, -1, 0, U, -1, U

Note: It may help to sketch a triangle as shown for Exercises 13 and 17.

13 Use the Pythagorean theorem to find the remaining side.

$$(\text{adj})^2 + (\text{opp})^2 = (\text{hyp})^2 \Rightarrow (\text{adj})^2 + 3^2 = 5^2 \Rightarrow \text{adj} = \sqrt{25 - 9} = 4.$$

We now apply the "trigonometric functions of an acute angle of a right triangle."

$$\sin\theta = \frac{\text{opp}}{\text{hyp}} = \frac{3}{5} \qquad \cos\theta = \frac{\text{adj}}{\text{hyp}} = \frac{4}{5} \qquad \tan\theta = \frac{\text{opp}}{\text{adj}} = \frac{3}{4}$$

$$\csc\theta = \frac{\text{hyp}}{\text{opp}} = \frac{5}{3} \qquad \sec\theta = \frac{\text{hyp}}{\text{adj}} = \frac{5}{4} \qquad \cot\theta = \frac{\text{adj}}{\text{opp}} = \frac{4}{3}$$

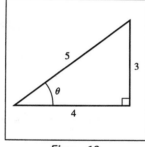

Figure 13

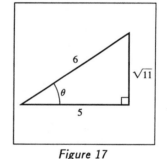

Figure 17

15 $12^2 + 5^2 = (\text{hyp})^2 \Rightarrow \text{hyp} = \sqrt{144 + 25} = 13.$ ★ $\frac{5}{13}, \frac{12}{13}, \frac{5}{12}, \frac{12}{5}, \frac{13}{12}, \frac{13}{5}$

17 $5^2 + (\text{opp})^2 = 6^2 \Rightarrow \text{opp} = \sqrt{36 - 25} = \sqrt{11}.$ ★ $\frac{\sqrt{11}}{6}, \frac{5}{6}, \frac{\sqrt{11}}{5}, \frac{5}{\sqrt{11}}, \frac{6}{5}, \frac{6}{\sqrt{11}}$

19 Using the theorem on trigonometric functions,

$$\sin\theta = \frac{\text{opp}}{\text{hyp}} = \frac{4}{5}, \quad \cos\theta = \frac{\text{adj}}{\text{hyp}} = \frac{3}{5}, \quad \text{and} \quad \tan\theta = \frac{\text{opp}}{\text{adj}} = \frac{4}{3}.$$

We now use the reciprocal identities to find the values of the other trigonometric functions:

$$\cot\theta = \frac{1}{\tan\theta} = \frac{3}{4}, \quad \sec\theta = \frac{1}{\cos\theta} = \frac{5}{3}, \quad \text{and} \quad \csc\theta = \frac{1}{\sin\theta} = \frac{5}{4}.$$

21 Using the Pythagorean theorem, $(\text{adj})^2 + (\text{opp})^2 = (\text{hyp})^2 \Rightarrow$

$$(\text{adj})^2 = (\text{hyp})^2 - (\text{opp})^2 \Rightarrow \text{adj} = \sqrt{(\text{hyp})^2 - (\text{opp})^2} = \sqrt{5^2 - 2^2} = \sqrt{21}.$$

★ $\frac{2}{5}, \frac{\sqrt{21}}{5}, \frac{2}{\sqrt{21}}, \frac{\sqrt{21}}{2}, \frac{5}{\sqrt{21}}, \frac{5}{2}$

[23] Using the Pythagorean theorem, hyp $= \sqrt{(\text{adj})^2 + (\text{opp})^2} = \sqrt{a^2 + b^2}$.

$$\bigstar \quad \frac{a}{\sqrt{a^2 + b^2}}, \ \frac{b}{\sqrt{a^2 + b^2}}, \ \frac{a}{b}, \ \frac{b}{a}, \ \frac{\sqrt{a^2 + b^2}}{b}, \ \frac{\sqrt{a^2 + b^2}}{a}$$

[27] Since we want to find the value of the hypotenuse x, we need to use a trigonometric function which relates x to two given parts of the triangle—in this case, the angle $30°$ and the opposite side of length 4. The sine function relates the opposite side and the hypotenuse. Hence, $\sin 30° = \frac{4}{x} \Rightarrow \frac{1}{2} = \frac{4}{x} \Rightarrow x = 8$. The tangent function relates the opposite and the adjacent side. Hence, $\tan 30° = \frac{4}{y} \Rightarrow \frac{\sqrt{3}}{3} = \frac{4}{y} \Rightarrow y = 4\sqrt{3}$.

[31] $\sin 60° = \frac{x}{8} \Rightarrow \frac{\sqrt{3}}{2} = \frac{x}{8} \Rightarrow x = 4\sqrt{3}$ and $\cos 60° = \frac{y}{8} \Rightarrow \frac{1}{2} = \frac{y}{8} \Rightarrow y = 4$.

[33] Let h denote the height of the tree.

$$\tan \theta = \frac{\text{opp}}{\text{adj}} \Rightarrow \tan 60° = \frac{h}{200} \Rightarrow h = 200 \tan 60° = 200\sqrt{3} \approx 346.4 \text{ ft.}$$

[35] Let d be the distance that the stone was moved. The side opposite $9°$ is 30 feet and the hypotenuse is d. Thus, $\sin 9° = \frac{30}{d} \Rightarrow d = \frac{30}{\sin 9°} \approx 192 \text{ ft.}$

[37] $\sin \theta = \frac{1.22\lambda}{D} \Rightarrow D = \frac{1.22\lambda}{\sin \theta} = \frac{1.22 \times 550 \times 10^{-9}}{\sin 0.00003769°} \approx 1.02 \text{ meters}$

6.5 Exercises

Note: Let θ_C denote the coterminal angle of θ such that $0° \le \theta_C < 360°$ { or $0 \le \theta_C < 2\pi$ }.

The following formulas are then used in the solutions { on text page 433 }.

(1) If θ_C is in QI, then $\theta_R = \theta_C$.

(2) If θ_C is in QII, then $\theta_R = 180° - \theta_C$ { or $\pi - \theta_C$ }.

(3) If θ_C is in QIII, then $\theta_R = \theta_C - 180°$ { or $\theta_C - \pi$ }.

(4) If θ_C is in QIV, then $\theta_R = 360° - \theta_C$ { or $2\pi - \theta_C$ }.

[1] (a) Since $240°$ is in QIII, $\theta_R = 240° - 180° = 60°$.

(b) Since $340°$ is in QIV, $\theta_R = 360° - 340° = 20°$.

(c) $\theta_C = -202° + 1(360°) = 158° \in$ QII. $\theta_R = 180° - 158° = 22°$.

(d) $\theta_C = -660° + 2(360°) = 60° \in$ QI. $\theta_R = 60°$.

[3] (a) Since $\frac{3\pi}{4}$ is in QII, $\theta_R = \pi - \frac{3\pi}{4} = \frac{\pi}{4}$.

(b) Since $\frac{4\pi}{3}$ is in QIII, $\theta_R = \frac{4\pi}{3} - \pi = \frac{\pi}{3}$.

(c) $\theta_C = -\frac{\pi}{6} + 1(2\pi) = \frac{11\pi}{6} \in$ QIV. $\theta_R = 2\pi - \frac{11\pi}{6} = \frac{\pi}{6}$.

(d) $\theta_C = \frac{9\pi}{4} - 1(2\pi) = \frac{\pi}{4} \in$ QI. $\theta_R = \frac{\pi}{4}$.

$\boxed{5}$ (a) Since $\frac{\pi}{2} < 3 < \pi$, θ is in QII and $\theta_R = \pi - 3 \approx 0.14$, or 8.1°.

  (b) $\theta_C = -2 + 1(2\pi) = 2\pi - 2 \approx 4.28$.

   Since $\pi < 4.28 < \frac{3\pi}{2}$, θ_C is in QIII and $\theta_R = (2\pi - 2) - \pi = \pi - 2 \approx 1.14$, or 65.4°.

  (c) Since $\frac{3\pi}{2} < 5.5 < 2\pi$, θ is in QIV and $\theta_R = 2\pi - 5.5 \approx 0.78$, or 44.9°.

  (d) The number of revolutions formed by θ is $\frac{100}{2\pi} \approx 15.92$, so

   $\theta_C = 100 - 15(2\pi) = 100 - 30\pi \approx 5.75$. Since $\frac{3\pi}{2} < 5.75 < 2\pi$,

   θ_C is in QIV and $\theta_R = 2\pi - (100 - 30\pi) = 32\pi - 100 \approx 0.53$, or 30.4°.

   Alternatively, if your calculator is capable of computing trigonometric functions

   of large values, then computing $\sin^{-1}(\sin 100) \approx -0.53 \Rightarrow \theta_R = 0.53$, or 30.4°.

Note: For the following problems, we use the theorem on reference angles before

  evaluating.

$\boxed{7}$ (a) $\sin\frac{2\pi}{3} = \sin\frac{\pi}{3}$ { since the sine is positive in QII

   and $\frac{\pi}{3}$ is the reference angle for $\frac{2\pi}{3}$ } $= \frac{\sqrt{3}}{2}$

  (b) $\sin\left(-\frac{5\pi}{4}\right) = \sin\frac{3\pi}{4}$ { since $\frac{3\pi}{4}$ is coterminal with $-\frac{5\pi}{4}$ } $=$

   $\sin\frac{\pi}{4}$ { since the sine is positive in QII and $\frac{\pi}{4}$ is the reference angle for $\frac{3\pi}{4}$ } $= \frac{\sqrt{2}}{2}$

$\boxed{9}$ (a) $\cos 150° = -\cos 30°$ { since the cosine is negative in QII } $= -\frac{\sqrt{3}}{2}$

  (b) $\cos(-60°) = \cos 300° = \cos 60°$ { since the cosine is positive in QIV } $= \frac{1}{2}$

$\boxed{11}$ (a) $\tan\frac{5\pi}{6} = -\tan\frac{\pi}{6}$ { since the tangent is negative in QII } $= -\frac{\sqrt{3}}{3}$

  (b) $\tan\left(-\frac{\pi}{3}\right) = \tan\frac{5\pi}{3} = -\tan\frac{\pi}{3}$ { since the tangent is negative in QIV } $= -\sqrt{3}$

$\boxed{13}$ (a) $\cot 120° = -\cot 60°$ { since the cotangent is negative in QII } $= -\frac{\sqrt{3}}{3}$

  (b) $\cot(-150°) = \cot 210° = \cot 30°$ { since the cotangent is positive in QIII } $= \sqrt{3}$

$\boxed{15}$ (a) $\sec\frac{2\pi}{3} = -\sec\frac{\pi}{3}$ { since the secant is negative in QII } $= -2$

  (b) $\sec\left(-\frac{\pi}{6}\right) = \sec\frac{11\pi}{6} = \sec\frac{\pi}{6}$ { since the secant is positive in QIV } $= \frac{2}{\sqrt{3}}$

$\boxed{17}$ (a) $\csc 240° = -\csc 60°$ { since the cosecant is negative in QIII } $= -\frac{2}{\sqrt{3}}$

  (b) $\csc(-330°) = \csc 30° = 2$

$\boxed{19}$ (a) Using the *degree* mode on a calculator, $\sin 73°20' \approx 0.958$.

  (b) Using the *radian* mode on a calculator, $\cos 0.68 \approx 0.778$.

$\boxed{23}$ (a) First compute $\cos 67°50'$, obtaining 0.3773. Now use the reciprocal key,

   usually labeled as either $\boxed{1/x}$ or $\boxed{x^{-1}}$, to obtain $\sec 67°50' \approx 2.650$.

  (b) As in part (a), since $\sin 0.32 \approx 0.3146$, we have $\csc 0.32 \approx 3.179$.

$\boxed{25}$ (a) Using the degree mode, calculate $\cos^{-1}(0.8620)$ to obtain 30.46° to

   the nearest one-hundredth of a degree.

25 (b) Using the answer from part (a), subtract 30 and multiply that result by 60

to obtain $30°27'$ to the nearest minute.

29 (a) $\sin\theta = 0.4217 \Rightarrow \theta = \sin^{-1}(0.4217) \approx 24.94°$ (b) $24.94° \approx 24°57'$

31 (a) After entering 4.246, use $\boxed{1/x}$ and then $\boxed{\text{INV}}\,\boxed{\text{COS}}$, or, equivalently, $\boxed{\text{COS}^{-1}}$. Similarly, for the cosecant function use $\boxed{1/x}$ and then $\boxed{\text{INV}}\,\boxed{\text{SIN}}$ and for the cotangent function use $\boxed{1/x}$ and then $\boxed{\text{INV}}\,\boxed{\text{TAN}}$. $\sec\theta = 4.246 \Rightarrow \cos\theta = \frac{1}{4.246} \Rightarrow \theta = \cos^{-1}\left(\frac{1}{4.246}\right) \approx 76.38°.$

(b) $76.38° \approx 76°23'$

35 (a) Use the degree mode. $\sin\theta = -0.5640 \Rightarrow \theta = \sin^{-1}(-0.5640) \approx -34.3° \Rightarrow$

$\theta_R \approx 34.3°$. Since the sine is negative in QIII and QIV, we use θ_R in those quadrants. $180° + 34.3° = \underline{214.3°}$ and $360° - 34.3° = \underline{325.7°}$

(b) $\cos\theta = 0.7490 \Rightarrow \theta = \cos^{-1}(0.7490) \approx 41.5°$. $\theta_R \approx 41.5°$, QI: $41.5°$, QIV: $318.5°$

(c) $\tan\theta = 2.798 \Rightarrow \theta = \tan^{-1}(2.798) \approx 70.3°$. $\theta_R \approx 70.3°$, QI: $70.3°$, QIII: $250.3°$

(d) $\cot\theta = -0.9601 \Rightarrow \tan\theta = -\frac{1}{0.9601} \Rightarrow \theta = \tan^{-1}\left(-\frac{1}{0.9601}\right) \approx -46.2°.$

$\theta_R \approx 46.2°$, QII: $133.8°$, QIV: $313.8°$

(e) $\sec\theta = -1.116 \Rightarrow \cos\theta = -\frac{1}{1.116} \Rightarrow \theta = \cos^{-1}\left(-\frac{1}{1.116}\right) \approx 153.6°.$

$\theta_R \approx 180° - 153.6° = 26.4°$, QII: $153.6°$, QIII: $206.4°$

(f) $\csc\theta = 1.485 \Rightarrow \sin\theta = \frac{1}{1.485} \Rightarrow \theta = \sin^{-1}\left(\frac{1}{1.485}\right) \approx 42.3°.$

$\theta_R \approx 42.3°$, QI: $42.3°$, QII: $137.7°$

37 (a) Use the radian mode. $\sin\theta = 0.4195 \Rightarrow \theta = \sin^{-1}(0.4195) \approx 0.43.$

$\theta_R \approx 0.43$ is one answer. Since the sine is positive in QI and QII, we also use the reference angle for θ in quadrant II. QII: $\pi - 0.43 \approx 2.71$

(b) $\cos\theta = -0.1207 \Rightarrow \theta = \cos^{-1}(-0.1207) \approx 1.69$ is one answer. Since 1.69 is in QII, $\theta_R \approx \pi - 1.69 \approx 1.45$. The cosine is also negative in QIII. QIII: $\pi + 1.45 \approx 4.59$

(c) $\tan\theta = -3.2504 \Rightarrow \theta = \tan^{-1}(-3.2504) \approx -1.27 \Rightarrow \theta_R \approx 1.27.$

QII: $\pi - 1.27 \approx 1.87$, QIV: $2\pi - 1.27 \approx 5.01$

(d) $\cot\theta = 2.6815 \Rightarrow \tan\theta = \frac{1}{2.6815} \Rightarrow \theta = \tan^{-1}\left(\frac{1}{2.6815}\right) \approx 0.36 \Rightarrow$

$\theta_R \approx 0.36$ is one answer. QIII: $\pi + 0.36 \approx 3.50$

(e) $\sec\theta = 1.7452 \Rightarrow \cos\theta = \frac{1}{1.7452} \Rightarrow \theta = \cos^{-1}\left(\frac{1}{1.7452}\right) \approx 0.96 \Rightarrow$

$\theta_R \approx 0.96$ is one answer. QIV: $2\pi - 0.96 \approx 5.32$

(f) $\csc\theta = -4.8521 \Rightarrow \sin\theta = -\frac{1}{4.8521} \Rightarrow \theta = \sin^{-1}\left(-\frac{1}{4.8521}\right) \approx -0.21 \Rightarrow \theta_R \approx 0.21.$

QIII: $\pi + 0.21 \approx 3.35$, QIV: $2\pi - 0.21 \approx 6.07$

39 $\ln I_0 - \ln I = kx \sec\theta \Rightarrow \ln\frac{I_0}{I} = kx\sec\theta$ { property of logarithms } $\Rightarrow$

$\qquad x = \frac{1}{k\sec\theta}\ln\frac{I_0}{I}$ { solve for x }

$\qquad = \frac{1}{1.88\sec 12°}\ln 1.72$ { substitute and approximate } ≈ 0.28 cm.

41 (a) Since $\cos\theta$ and $\sin\phi$ are both less than or equal to 1, $R = R_0\cos\theta\sin\phi$, the solar radiation R will equal its maximum R_0 when $\cos\theta = \sin\phi = 1$. This occurs when $\theta = 0°$ and $\phi = 90°$, and corresponds to when the sun is just rising in the east.

   (b) The sun located in the southeast corresponds to $\phi = 45°$.

$$\text{percentage of } R_0 = \frac{\text{amount of } R_0}{R_0} = \frac{R_0\cos 60°\sin 45°}{R_0} = \frac{1}{2}\cdot\frac{\sqrt{2}}{2} = \frac{\sqrt{2}}{4} \approx 35\%.$$

43 $\sin\theta = \frac{b}{c} \Rightarrow \sin 60° = \frac{b}{18} \Rightarrow b = 18\sin 60° = 18\cdot\frac{\sqrt{3}}{2} = 9\sqrt{3} \approx 15.6$.

$\qquad \cos\theta = \frac{a}{c} \Rightarrow \cos 60° = \frac{a}{18} \Rightarrow a = 18\cos 60° = 18\cdot\frac{1}{2} = 9$.

$\qquad\qquad\qquad\qquad\qquad\qquad\qquad$ The hand is located at $(9, 9\sqrt{3})$.

6.6 Exercises

Note: Exercises 1 & 3: We will refer to $y = \sin x$ as just $\sin x$ ($y = \cos x$ as $\cos x$, etc.).
For the form $y = a\sin bx$, the amplitude is $|a|$ and the period is $\frac{2\pi}{|b|}$.

These are merely listed in the answer along with the values of the x-intercepts.

Let n denote any integer.

1 (a) $y = 4\sin x$ • Vertically stretch $\sin x$ by a factor of 4. The x-intercepts are not affected by a vertical stretch or compression.   ★ 4, 2π, x-int. @ πn

   (b) $y = \sin 4x$ • Horizontally compress $\sin x$ by a factor of 4. The x-intercepts are affected by a horizontal stretch or compression by the same factor—that is, a horizontal compression by a factor of k will move the x-intercepts of $\sin x$ from πn to $\frac{\pi}{k}n$, and a horizontal stretch by a factor of k will move the x-intercepts of $\sin x$ from πn to $k\pi n$.   ★ 1, $\frac{\pi}{2}$, x-int. @ $\frac{\pi}{4}n$

   (c) $y = \frac{1}{4}\sin x$ • Vertically compress $\sin x$ by a factor of 4.   ★ $\frac{1}{4}$, 2π, x-int. @ πn

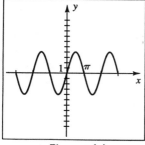

Figure 1(a)

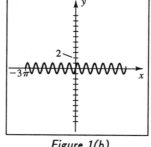

Figure 1(b)

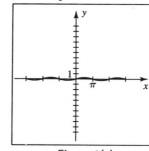

Figure 1(c)

(d) $y = \sin\frac{1}{4}x$ • Horizontally stretch $\sin x$ by a factor of 4. ★ 1, 8π, x-int. @ $4\pi n$

(e) $y = 2\sin\frac{1}{4}x$ • Vertically stretch the graph in part (d) by a factor of 2.

★ 2, 8π, x-int. @ $4\pi n$

(f) $y = \frac{1}{2}\sin 4x$ • Vertically compress the graph in part (b) by a factor of 2.

★ $\frac{1}{2}$, $\frac{\pi}{2}$, x-int. @ $\frac{\pi}{4}n$

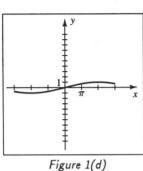
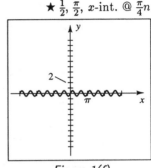

Figure 1(d)   Figure 1(e)   Figure 1(f)

(g) $y = -4\sin x$ • Reflect the graph in part (a) through the x-axis.

★ 4, 2π, x-int. @ πn

(h) $y = \sin(-4x) = -\sin 4x$ using a formula for negatives. Reflect the graph in part

(b) through the x-axis. ★ 1, $\frac{\pi}{2}$, x-int. @ $\frac{\pi}{4}n$

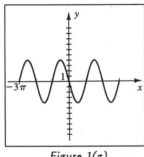
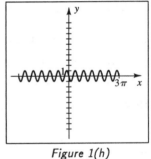

Figure 1(g)   Figure 1(h)

3 (a) $y = 3\cos x$ • Vertically stretch $\cos x$ by a factor of 3. Figures for Exercise 3 are on the next page.

★ 3, 2π, x-int. @ $\frac{\pi}{2} + \pi n$

(b) $y = \cos 3x$ • Horizontally compress $\cos x$ by a factor of 3. The x-intercepts are affected by a horizontal stretch or compression by the same factor—that is, horizontal compression by a factor of k will move the x-intercepts of $\cos x$ from $\frac{\pi}{2} + \pi n$ to $\frac{\pi}{2k} + \frac{\pi}{k}n$, and a horizontal stretch by a factor of k will move the x-intercepts of $\cos x$ from $\frac{\pi}{2} + \pi n$ to $\frac{k\pi}{2} + k\pi n$. ★ 1, $\frac{2\pi}{3}$, x-int. @ $\frac{\pi}{6} + \frac{\pi}{3}n$

(c) $y = \frac{1}{3}\cos x$ • Vertically compress $\cos x$ by a factor of 3.

★ $\frac{1}{3}$, 2π, x-int. @ $\frac{\pi}{2} + \pi n$

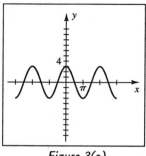

Figure 3(a)

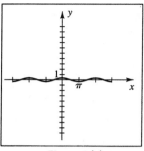

Figure 3(b)

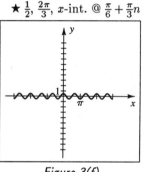

Figure 3(c)

(d) $y = \cos\frac{1}{3}x$ • Horizontally stretch $\cos x$ by a factor of 3.

★ 1, 6π, x-int. @ $\frac{3\pi}{2} + 3\pi n$

(e) $y = 2\cos\frac{1}{3}x$ • Vertically stretch the graph in part (d) by a factor of 2.

★ 2, 6π, x-int. @ $\frac{3\pi}{2} + 3\pi n$

(f) $y = \frac{1}{2}\cos 3x$ • Vertically compress the graph in part (b) by a factor of 2.

★ $\frac{1}{2}$, $\frac{2\pi}{3}$, x-int. @ $\frac{\pi}{6} + \frac{\pi}{3}n$

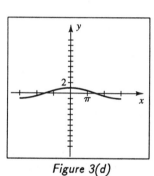

Figure 3(d)

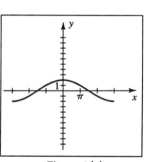

Figure 3(e)

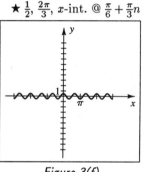

Figure 3(f)

(g) $y = -3\cos x$ • Reflect the graph in part (a) through the x-axis.

★ 3, 2π, x-int. @ $\frac{\pi}{2} + \pi n$

(h) $y = \cos(-3x) = \cos 3x$ using a formula for negatives. This is the same as the graph in part (b).

★ 1, $\frac{2\pi}{3}$, x-int. @ $\frac{\pi}{6} + \frac{\pi}{3}n$

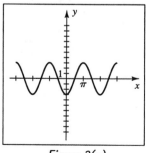

Figure 3(g)

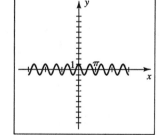

Figure 3(h)

Note: We will write $y = a\sin(bx + c)$ in the form $y = a\sin\left[b\left(x + \frac{c}{b}\right)\right]$. From this form we have the amplitude, $|a|$, the period, $\frac{2\pi}{|b|}$, and the phase shift, $-\frac{c}{b}$. We will also list the interval that corresponds to $[0, 2\pi]$ for the sine functions and to $[-\frac{\pi}{2}, \frac{3\pi}{2}]$ for the cosine functions—this interval gives us one wave (between zeros) of the graph of the function. The work to determine those intervals is shown in each exercise.

⑤ $y = \sin\left(x - \frac{\pi}{2}\right)$ • $0 \le x - \frac{\pi}{2} \le 2\pi \Rightarrow \frac{\pi}{2} \le x \le \frac{5\pi}{2}$

Phase shift $= -\left(-\frac{\pi}{2}\right) = \frac{\pi}{2}$.

★ $1, 2\pi, \frac{\pi}{2}, [\frac{\pi}{2}, \frac{5\pi}{2}]$

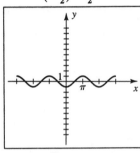

Figure 5

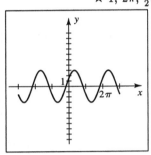

Figure 7

⑦ $y = 3\sin\left(x + \frac{\pi}{6}\right)$ • $0 \le x + \frac{\pi}{6} \le 2\pi \Rightarrow -\frac{\pi}{6} \le x \le \frac{11\pi}{6}$

Phase shift $= -\left(\frac{\pi}{6}\right) = -\frac{\pi}{6}$.

★ $3, 2\pi, -\frac{\pi}{6}, [-\frac{\pi}{6}, \frac{11\pi}{6}]$

⑨ $y = \cos\left(x + \frac{\pi}{2}\right)$ • $-\frac{\pi}{2} \le x + \frac{\pi}{2} \le \frac{3\pi}{2} \Rightarrow -\pi \le x \le \pi$

Phase shift $= -\left(\frac{\pi}{2}\right) = -\frac{\pi}{2}$.

★ $1, 2\pi, -\frac{\pi}{2}, [-\pi, \pi]$

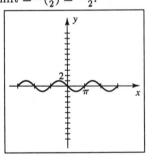

Figure 9

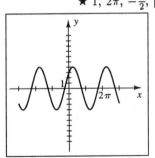

Figure 11

⑪ $y = 4\cos\left(x - \frac{\pi}{4}\right)$ • $-\frac{\pi}{2} \le x - \frac{\pi}{4} \le \frac{3\pi}{2} \Rightarrow -\frac{\pi}{4} \le x \le \frac{7\pi}{4}$

Phase shift $= -\left(-\frac{\pi}{4}\right) = \frac{\pi}{4}$.

★ $4, 2\pi, \frac{\pi}{4}, [-\frac{\pi}{4}, \frac{7\pi}{4}]$

$\boxed{13}$ $y = \sin(2x - \pi) + 1 = \sin\left[2(x - \frac{\pi}{2})\right] + 1$. • Period $= \frac{2\pi}{|2|} = \pi$. The normal range of the sine, -1 to 1, is affected by the "+1" at the end of the equation. It shifts the graph up 1 unit and the resulting range is 0 to 2. It may be easiest to graph $y = \sin\left[2(x - \frac{\pi}{2})\right]$ {1 period of a sine wave with end points at $\frac{\pi}{2}$ and $\frac{3\pi}{2}$} and then make a vertical shift of 1 unit up to complete the graph of $y = \sin\left[2(x - \frac{\pi}{2})\right] + 1$.

$0 \le 2x - \pi \le 2\pi \Rightarrow \pi \le 2x \le 3\pi \Rightarrow \frac{\pi}{2} \le x \le \frac{3\pi}{2}$ ★ $1, \pi, \frac{\pi}{2}, [\frac{\pi}{2}, \frac{3\pi}{2}]$

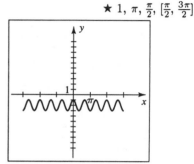

Figure 13   Figure 15

$\boxed{15}$ $y = -\cos(3x + \pi) - 2 = -\cos\left[3(x + \frac{\pi}{3})\right] - 2$ • The "$-$" in front of cos has the effect of reflecting the graph of $y = \cos(3x + \pi)$ through the x-axis. The "-2" at the end of the equation lowers the range 2 units to -3 to -1. Period $= \frac{2\pi}{|3|} = \frac{2\pi}{3}$, phase shift $= -(\frac{\pi}{3}) = -\frac{\pi}{3}$. $-\frac{\pi}{2} \le 3x + \pi \le \frac{3\pi}{2} \Rightarrow -\frac{3\pi}{2} \le 3x \le \frac{\pi}{2} \Rightarrow -\frac{\pi}{2} \le x \le \frac{\pi}{6}$

★ $1, \frac{2\pi}{3}, -\frac{\pi}{3}, [-\frac{\pi}{2}, \frac{\pi}{6}]$

$\boxed{17}$ $y = -2\sin(3x - \pi) = -2\sin\left[3(x - \frac{\pi}{3})\right]$. • Amplitude $= |-2| = 2$. The negative before the "2" has the effect of reflecting the graph of $y = 2\sin(3x - \pi)$ through the x-axis. Period $= \frac{2\pi}{|3|} = \frac{2\pi}{3}$, phase shift $= -(-\frac{\pi}{3}) = \frac{\pi}{3}$, $0 \le 3x - \pi \le 2\pi \Rightarrow$

$\pi \le 3x \le 3\pi \Rightarrow \frac{\pi}{3} \le x \le \pi$.   ★ $2, \frac{2\pi}{3}, \frac{\pi}{3}, [\frac{\pi}{3}, \pi]$

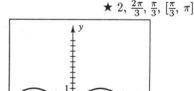

Figure 17   Figure 19

$\boxed{19}$ $y = \sin(\frac{1}{2}x - \frac{\pi}{3}) = \sin\left[\frac{1}{2}(x - \frac{2\pi}{3})\right]$. • Period $= \frac{2\pi}{|1/2|} = 4\pi$.

$0 \le \frac{1}{2}x - \frac{\pi}{3} \le 2\pi \Rightarrow \frac{\pi}{3} \le \frac{1}{2}x \le \frac{7\pi}{3} \Rightarrow \frac{2\pi}{3} \le x \le \frac{14\pi}{3}$ ★ $1, 4\pi, \frac{2\pi}{3}, [\frac{2\pi}{3}, \frac{14\pi}{3}]$

$\boxed{21}$ $y = 6\sin \pi x$ • Period $= \dfrac{2\pi}{|\pi|} = 2.$ $0 \le \pi x \le 2\pi \Rightarrow 0 \le x \le 2$ ★ 6, 2, 0, [0, 2]

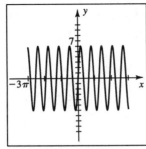

Figure 21

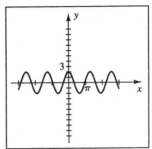

Figure 23

$\boxed{23}$ $y = 2\cos \dfrac{\pi}{2}x$ • Period $= \dfrac{2\pi}{|\pi/2|} = 4.$ $-\dfrac{\pi}{2} \le \dfrac{\pi}{2}x \le \dfrac{3\pi}{2} \Rightarrow -1 \le x \le 3$

★ 2, 4, 0, [−1, 3]

$\boxed{25}$ $y = \dfrac{1}{2}\sin 2\pi x$ • $0 \le 2\pi x \le 2\pi \Rightarrow 0 \le x \le 1$ ★ $\dfrac{1}{2}$, 1, 0, [0, 1]

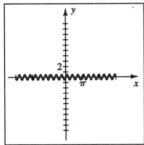

Figure 25

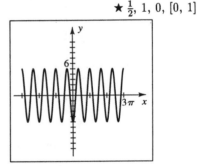

Figure 27

$\boxed{27}$ $y = 5\sin\left(3x - \dfrac{\pi}{2}\right) = 5\sin\left[3\left(x - \dfrac{\pi}{6}\right)\right].$ •   ★ 5, $\dfrac{2\pi}{3}$, $\dfrac{\pi}{6}$, $[\dfrac{\pi}{6}, \dfrac{5\pi}{6}]$

$0 \le 3x - \dfrac{\pi}{2} \le 2\pi \Rightarrow \dfrac{\pi}{2} \le 3x \le \dfrac{5\pi}{2} \Rightarrow \dfrac{\pi}{6} \le x \le \dfrac{5\pi}{6}.$ To graph this function, draw one period of a sine wave with end points at $\dfrac{\pi}{6}$ and $\dfrac{5\pi}{6}$ and an amplitude of 5.

$\boxed{29}$ $y = 3\cos\left(\dfrac{1}{2}x - \dfrac{\pi}{4}\right) = 3\cos\left[\dfrac{1}{2}\left(x - \dfrac{\pi}{2}\right)\right].$ •

$-\dfrac{\pi}{2} \le \dfrac{1}{2}x - \dfrac{\pi}{4} \le \dfrac{3\pi}{2} \Rightarrow -\dfrac{\pi}{4} \le \dfrac{1}{2}x \le \dfrac{7\pi}{4} \Rightarrow -\dfrac{\pi}{2} \le x \le \dfrac{7\pi}{2}$ ★ 3, 4π, $\dfrac{\pi}{2}$, $[-\dfrac{\pi}{2}, \dfrac{7\pi}{2}]$

Figure 29

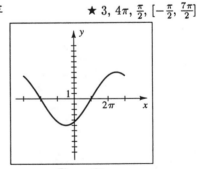

Figure 31

$\boxed{31}$ $y = -5\cos\left(\dfrac{1}{3}x + \dfrac{\pi}{6}\right) = -5\cos\left[\dfrac{1}{3}\left(x + \dfrac{\pi}{2}\right)\right].$ •

$-\dfrac{\pi}{2} \le \dfrac{1}{3}x + \dfrac{\pi}{6} \le \dfrac{3\pi}{2} \Rightarrow -\dfrac{2\pi}{3} \le \dfrac{1}{3}x \le \dfrac{4\pi}{3} \Rightarrow -2\pi \le x \le 4\pi$ ★ 5, 6π, $-\dfrac{\pi}{2}$, $[-2\pi, 4\pi]$

$\boxed{33}$ $y = 3\cos(\pi x + 4\pi) = 3\cos[\pi(x+4)]$. •

$-\frac{\pi}{2} \le \pi x + 4\pi \le \frac{3\pi}{2} \Rightarrow -\frac{9\pi}{2} \le \pi x \le -\frac{5\pi}{2} \Rightarrow -\frac{9}{2} \le x \le -\frac{5}{2}$ ★ $3, 2, -4, [-\frac{9}{2}, -\frac{5}{2}]$

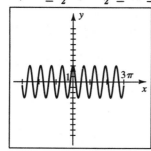

Figure 33

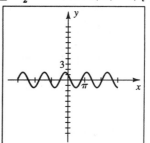

Figure 35

$\boxed{35}$ $y = -\sqrt{2}\sin\left(\frac{\pi}{2}x - \frac{\pi}{4}\right) = -\sqrt{2}\sin\left[\frac{\pi}{2}(x - \frac{1}{2})\right]$. •

$0 \le \frac{\pi}{2}x - \frac{\pi}{4} \le 2\pi \Rightarrow \frac{\pi}{4} \le \frac{\pi}{2}x \le \frac{9\pi}{4} \Rightarrow \frac{1}{2} \le x \le \frac{9}{2}$ ★ $\sqrt{2}, 4, \frac{1}{2}, [\frac{1}{2}, \frac{9}{2}]$

$\boxed{37}$ $y = -2\sin(2x - \pi) + 3 = -2\sin[2(x - \frac{\pi}{2})] + 3$. •

$0 \le 2x - \pi \le 2\pi \Rightarrow \pi \le 2x \le 3\pi \Rightarrow \frac{\pi}{2} \le x \le \frac{3\pi}{2}$. The amplitude of 2 makes the normal sine range of -1 to 1 change to -2 to 2. The "$+3$" at the end of the equation raises the graph up 3 units and the range is 1 to 5.   ★ $2, \pi, \frac{\pi}{2}, [\frac{\pi}{2}, \frac{3\pi}{2}]$

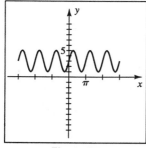

Figure 37

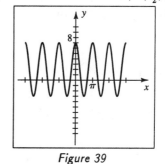

Figure 39

$\boxed{39}$ $y = 5\cos(2x + 2\pi) + 2 = 5\cos[2(x + \pi)] + 2$. •

$-\frac{\pi}{2} \le 2x + 2\pi \le \frac{3\pi}{2} \Rightarrow -\frac{5\pi}{2} \le 2x \le -\frac{\pi}{2} \Rightarrow -\frac{5\pi}{4} \le x \le -\frac{\pi}{4}$ ★ $5, \pi, -\pi, [-\frac{5\pi}{4}, -\frac{\pi}{4}]$

$\boxed{41}$ (a) The amplitude a is 4 and the period { from $-\pi$ to π } is 2π.

The phase shift is the first negative zero that occurs before a maximum, $-\pi$.

(b) Period $= \frac{2\pi}{b} \Rightarrow 2\pi = \frac{2\pi}{b} \Rightarrow b = 1$. Phase shift $= -\frac{c}{b} \Rightarrow -\pi = -\frac{c}{1} \Rightarrow c = \pi$.

Hence, $y = a\sin(bx + c) = 4\sin(x + \pi)$.

$\boxed{43}$ (a) The amplitude a is 2 and the period { from -3 to 1 } is 4.

The phase shift is the first negative zero that occurs before a maximum, -3.

(b) Period $= \frac{2\pi}{b} \Rightarrow 4 = \frac{2\pi}{b} \Rightarrow b = \frac{\pi}{2}$. Phase shift $= -\frac{c}{b} \Rightarrow -3 = -\frac{c}{\pi/2} \Rightarrow c = \frac{3\pi}{2}$.

Hence, $y = a\sin(bx + c) = 2\sin\left(\frac{\pi}{2}x + \frac{3\pi}{2}\right)$.

45 In the first second, there are 2 complete cycles.

Hence 1 cycle is completed in $\frac{1}{2}$ second and thus, the period is $\frac{1}{2}$.

Also, the period is $\frac{2\pi}{b}$. Equating these expressions yields $\frac{2\pi}{b} = \frac{1}{2} \Rightarrow b = 4\pi$.

47 We first note that $\frac{1}{2}$ period takes place in $\frac{1}{4}$ second and thus 1 period in $\frac{1}{2}$ second.

As in Exercise 45, $\frac{2\pi}{b} = \frac{1}{2} \Rightarrow b = 4\pi$. Since the maximum flow rate is 8 liters/minute, the amplitude is 8. $a = 8$ and $b = 4\pi \Rightarrow y = 8\sin 4\pi t$.

49 $f(t) = \frac{1}{2}\cos\left[\frac{\pi}{6}\left(t - \frac{11}{2}\right)\right]$, amplitude $= \frac{1}{2}$, period $= \frac{2\pi}{\pi/6} = 12$, phase shift $= \frac{11}{2}$

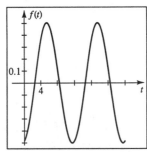

Figure 49

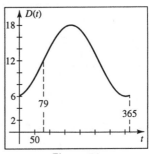

Figure 51

51 $D(t) = 6\sin\left[\frac{2\pi}{365}(t - 79)\right] + 12$, amplitude $= 6$, period $= \frac{2\pi}{2\pi/365} = 365$,

phase shift $= 79$, range $= \underline{12 - 6}$ to $\underline{12 + 6}$ or 6 to 18

53 The temperature is 20°F at 9:00 A.M. ($t = 0$). It increases to a high of 35°F at 3:00 P.M. ($t = 6$) and then decreases to 20°F at 9:00 P.M. ($t = 12$). It continues to decrease to a low of 5°F at 3:00 A.M. ($t = 18$). It then rises to 20°F at 9:00 A.M. ($t = 24$).

$[0, 24]$ by $[0, 40]$

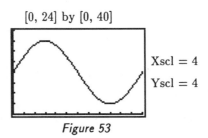

Xscl $= 4$
Yscl $= 4$

Figure 53

Note: Exer. 55–58: The period is 24 hours. Thus, $24 = \frac{2\pi}{b} \Rightarrow b = \frac{\pi}{12}$.

55 A high of 10°C and a low of −10°C imply that $d = \dfrac{\text{high} + \text{low}}{2} = \dfrac{10 + (-10)}{2} = 0$ and

$a = \text{high} - \text{average} = 10 - 0 = 10$. The average temperature of 0°C will occur 6 hours { one-half of 12 } after the low at 4 A.M., which corresponds to $t = 10$. Letting this correspond to the first zero of the sine function, we have

$f(t) = 10\sin\left[\frac{\pi}{12}(t-10)\right]+0 = 10\sin\left(\frac{\pi}{12}t - \frac{5\pi}{6}\right)$ with $a=10$, $b=\frac{\pi}{12}$, $c=-\frac{5\pi}{6}$, $d=0$.

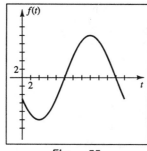

Figure 55

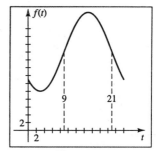

Figure 57

$\boxed{57}$ A high of 30°C and a low of 10°C imply that $d = \dfrac{30+10}{2} = 20$ and $a = 30-20 = 10$.

The average temperature of 20°C at 9 A.M. corresponds to $t=9$.

Letting this correspond to the first zero of the sine function, we have

$f(t) = 10\sin\left[\frac{\pi}{12}(t-9)\right]+20 = 10\sin\left(\frac{\pi}{12}t - \frac{3\pi}{4}\right)+20$ with

$a=10$, $b=\frac{\pi}{12}$, $c=-\frac{3\pi}{4}$, $d=20$.

$\boxed{59}$ (b) Since the period is 12 months, $12 = \frac{2\pi}{b} \Rightarrow b = \frac{\pi}{6}$. The maximum precipitation is 6.1 and the minimum is 0.2, so the sine wave is centered vertically at $d = \frac{6.1+0.2}{2} = 3.15$ and its amplitude is $a = \frac{6.1-0.2}{2} = 2.95$. Since the maximum precipitation occurs at $t=1$ (January), we must have $bt + c = \frac{\pi}{2} \Rightarrow \frac{\pi}{6}(1) + c = \frac{\pi}{2} \Rightarrow c = \frac{\pi}{3}$. Thus, $P(t) = a\sin(bt+c) + d = 2.95\sin\left(\frac{\pi}{6}t + \frac{\pi}{3}\right) + 3.15$.

$[0.5, 24.5]$ by $[-1, 8]$ $[0.5, 24.5]$ by $[0, 20]$

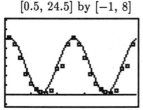

Xscl = 4
Yscl = 1

Figure 59

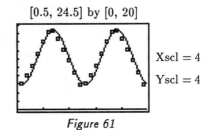

Xscl = 4
Yscl = 4

Figure 61

$\boxed{61}$ (b) Since the period is 12 months, $b = \frac{2\pi}{12} = \frac{\pi}{6}$. From the table, the maximum number of daylight hours is 18.72 and the minimum is 5.88. Thus, the sine wave is centered vertically at $d = \dfrac{18.72+5.88}{2} = 12.3$ and its amplitude is $a = \dfrac{18.72-5.88}{2} = 6.42$. Since the maximum daylight occurs at $t=7$ (July), we must have $bt + c = \frac{\pi}{2} \Rightarrow \frac{\pi}{6}(7) + c = \frac{\pi}{2} \Rightarrow c = -\frac{2\pi}{3}$.

Thus, $D(t) = a\sin(bt+c)+d = 6.42\sin\left(\frac{\pi}{6}t - \frac{2\pi}{3}\right) + 12.3$.

63 As $x \to 0^-$ or as $x \to 0^+$,

y oscillates between -1 and 1 and does not approach a unique value.

[-2, 2] by [-1.33, 1.33]

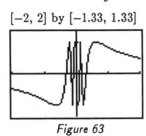

Xscl $= 0.5$
Yscl $= 0.5$

Figure 63

[-2, 2] by [-0.33, 2.33]

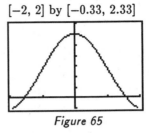

Xscl $= 0.5$
Yscl $= 0.5$

Figure 65

65 As $x \to 0^-$ or as $x \to 0^+$, y appears to approach 2.

67 From the graph, we see that there is a horizontal asymptote of $y = 4$.

[-20, 20] by [-1, 5]

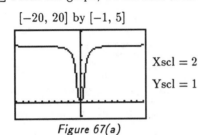

Xscl $= 2$
Yscl $= 1$

Figure 67(a)

[-1, 1] by [-0.67, 0.67]

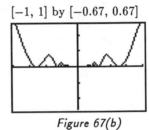

Xscl $= 0.25$
Yscl $= 0.25$

Figure 67(b)

69 Graph $Y_1 = \cos 3x$ and $Y_2 = \frac{1}{2}x - \sin x$.

From the graph, Y_1 intersects Y_2 at $x \approx -1.63$, -0.45, 0.61, 1.49, 2.42.

Thus, $\cos 3x \geq \frac{1}{2}x - \sin x$ on $[-\pi, -1.63] \cup [-0.45, 0.61] \cup [1.49, 2.42]$.

[-π, π] by [-2.09, 2.09]

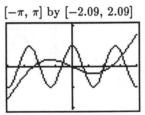

Xscl $= \pi/4$
Yscl $= 1$

Figure 69

6.7 Exercises

Note: If $y = a \tan(bx + c)$ or $y = a \cot(bx + c)$,

then the periods for the tangent and cotangent graphs are $\pi / |b|$.

If $y = a \sec(bx + c)$ or $y = a \csc(bx + c)$,

then the periods for the secant and cosecant graphs are $2\pi / |b|$.

$\boxed{1}$ $y = 4 \tan x$ • Vertically stretch $\tan x$ by a factor of 4. The *x-intercepts* of $\tan x$ and $\cot x$ are not affected by vertically stretching or compressing their graphs. The *vertical asymptotes* of $\tan x$, $\cot x$, $\sec x$, and $\csc x$ are not affected by vertically stretching or compressing their graphs. ★ π

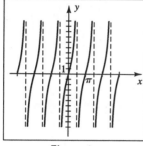

Figure 1

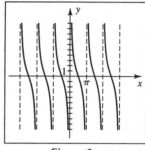

Figure 3

$\boxed{3}$ $y = 3 \cot x$ • Vertically stretch $\cot x$ by a factor of 3. ★ π

$\boxed{5}$ $y = 2 \csc x$ • Vertically stretch $\csc x$ by a factor of 2.

Note that there is now a minimum value of 2 at $x = \frac{\pi}{2}$.

The range of this function is $(-\infty, -2] \cup [2, \infty)$, or $|y| \geq 2$. ★ 2π

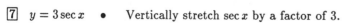

Figure 5

Figure 7

$\boxed{7}$ $y = 3 \sec x$ • Vertically stretch $\sec x$ by a factor of 3.

Note that there is now a minimum value of 3 at $x = 0$.

The range of this function is $(-\infty, -3] \cup [3, \infty)$ or $|y| \geq 3$. ★ 2π

Note: The vertical asymptotes of each function are denoted by VA @ $x =$. The work to determine two consecutive vertical asymptotes is shown for each exercise. For the tangent and secant functions, the region from $-\frac{\pi}{2}$ to $\frac{\pi}{2}$ is used. For the cotangent and cosecant functions, the region from 0 to π is used.

$\boxed{9}$ $y = \tan\left(x - \frac{\pi}{4}\right)$ • Shift $\tan x$ right $\frac{\pi}{4}$ units, VA @ $x = -\frac{\pi}{4} + \pi n$. Note that the asymptotes remain π units apart. $-\frac{\pi}{2} \le x - \frac{\pi}{4} \le \frac{\pi}{2} \Rightarrow -\frac{\pi}{4} \le x \le \frac{3\pi}{4}$ ★ π

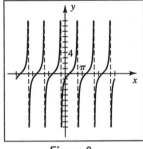

Figure 9

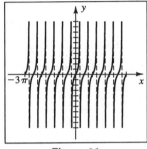

Figure 11

$\boxed{11}$ $y = \tan 2x$ • Horizontally compress $\tan x$ by a factor of 2, VA @ $x = -\frac{\pi}{4} + \frac{\pi}{2} n$. Note that the asymptotes are only $\frac{\pi}{2}$ units apart. $-\frac{\pi}{2} \le 2x \le \frac{\pi}{2} \Rightarrow -\frac{\pi}{4} \le x \le \frac{\pi}{4}$ ★ $\frac{\pi}{2}$

$\boxed{13}$ $y = \tan\frac{1}{4}x$ • Horizontally stretch $\tan x$ by a factor of 4, VA @ $x = -2\pi + 4\pi n$. Note that the asymptotes are 4π units apart. $-\frac{\pi}{2} \le \frac{1}{4}x \le \frac{\pi}{2} \Rightarrow -2\pi \le x \le 2\pi$ ★ 4π

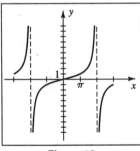

Figure 13

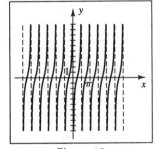

Figure 15

$\boxed{15}$ $y = 2\tan\left(2x + \frac{\pi}{2}\right) = 2\tan\left[2(x + \frac{\pi}{4})\right]$. •

The phase shift is $-\frac{\pi}{4}$, the period is $\frac{\pi}{2}$, and we have a vertical stretching factor of 2.

$-\frac{\pi}{2} \le 2x + \frac{\pi}{2} \le \frac{\pi}{2} \Rightarrow -\pi \le 2x \le 0 \Rightarrow -\frac{\pi}{2} \le x \le 0$, VA @ $x = \frac{\pi}{2} n$ ★ $\frac{\pi}{2}$

17 $y = -\frac{1}{4}\tan\left(\frac{1}{2}x + \frac{\pi}{3}\right) = -\frac{1}{4}\tan\left[\frac{1}{2}\left(x + \frac{2\pi}{3}\right)\right].$ • Note that the "−" in front of the $\frac{1}{4}$ reflects the graph through the x-axis. This changes the appearance of a tangent graph to that of a cotangent graph {increasing to decreasing}.

$-\frac{\pi}{2} \le \frac{1}{2}x + \frac{\pi}{3} \le \frac{\pi}{2} \Rightarrow -\frac{5\pi}{6} \le \frac{1}{2}x \le \frac{\pi}{6} \Rightarrow -\frac{5\pi}{3} \le x \le \frac{\pi}{3},\ VA @ x = -\frac{5\pi}{3} + 2\pi n$ ★ 2π

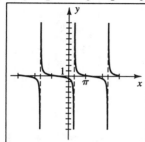

Figure 17

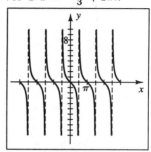

Figure 19

19 $y = \cot\left(x - \frac{\pi}{2}\right)$ • $0 \le x - \frac{\pi}{2} \le \pi \Rightarrow \frac{\pi}{2} \le x \le \frac{3\pi}{2},\ VA @ x = \frac{\pi}{2} + \pi n$ ★ π

21 $y = \cot 2x$ • $0 \le 2x \le \pi \Rightarrow 0 \le x \le \frac{\pi}{2},\ VA @ x = \frac{\pi}{2}n$ ★ $\frac{\pi}{2}$

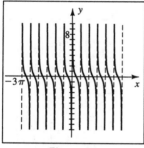

Figure 21

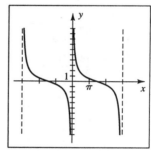

Figure 23

23 $y = \cot\frac{1}{3}x$ • $0 \le \frac{1}{3}x \le \pi \Rightarrow 0 \le x \le 3\pi,\ VA @ x = 3\pi n$ ★ 3π

25 $y = 2\cot\left(2x + \frac{\pi}{2}\right) = 2\cot\left[2\left(x + \frac{\pi}{4}\right)\right].$ •

$0 \le 2x + \frac{\pi}{2} \le \pi \Rightarrow -\frac{\pi}{2} \le 2x \le \frac{\pi}{2} \Rightarrow -\frac{\pi}{4} \le x \le \frac{\pi}{4},\ VA @ x = -\frac{\pi}{4} + \frac{\pi}{2}n$ ★ $\frac{\pi}{2}$

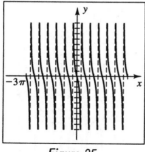

Figure 25

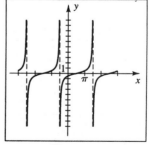

Figure 27

27 $y = -\frac{1}{2}\cot\left(\frac{1}{2}x + \frac{\pi}{4}\right) = -\frac{1}{2}\cot\left[\frac{1}{2}\left(x + \frac{\pi}{2}\right)\right].$ •

$0 \le \frac{1}{2}x + \frac{\pi}{4} \le \pi \Rightarrow -\frac{\pi}{4} \le \frac{1}{2}x \le \frac{3\pi}{4} \Rightarrow -\frac{\pi}{2} \le x \le \frac{3\pi}{2},\ VA @ x = -\frac{\pi}{2} + 2\pi n$ ★ 2π

29 $y = \sec\left(x - \frac{\pi}{2}\right)$ •   Note that this is the same graph as the graph of $y = \csc x$.

$-\frac{\pi}{2} \le x - \frac{\pi}{2} \le \frac{\pi}{2} \Rightarrow 0 \le x \le \pi$, $VA \ @ \ x = \pi n$ ★ 2π

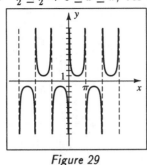

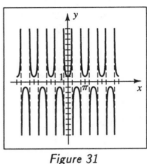

Figure 29   Figure 31

31 $y = \sec 2x$ •   Note that the asymptotes move closer together by a factor of 2.

$-\frac{\pi}{2} \le 2x \le \frac{\pi}{2} \Rightarrow -\frac{\pi}{4} \le x \le \frac{\pi}{4}$, $VA \ @ \ x = -\frac{\pi}{4} + \frac{\pi}{2}n$ ★ π

33 $y = \sec \frac{1}{3}x$ •   Note that the asymptotes move farther apart by a factor of 3,

just as they did with the graphs of tan and cot.

$-\frac{\pi}{2} \le \frac{1}{3}x \le \frac{\pi}{2} \Rightarrow -\frac{3\pi}{2} \le x \le \frac{3\pi}{2}$, $VA \ @ \ x = -\frac{3\pi}{2} + 3\pi n$ ★ 6π

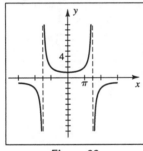

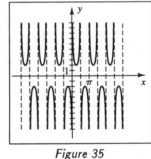

Figure 33   Figure 35

35 $y = 2\sec\left(2x - \frac{\pi}{2}\right) = 2\sec\left[2\left(x - \frac{\pi}{4}\right)\right]$.   •

$-\frac{\pi}{2} \le 2x - \frac{\pi}{2} \le \frac{\pi}{2} \Rightarrow 0 \le 2x \le \pi \Rightarrow 0 \le x \le \frac{\pi}{2}$, $VA \ @ \ x = \frac{\pi}{2}n$ ★ π

37 $y = -\frac{1}{3}\sec\left(\frac{1}{2}x + \frac{\pi}{4}\right) = -\frac{1}{3}\sec\left[\frac{1}{2}\left(x + \frac{\pi}{2}\right)\right]$.   •

$-\frac{\pi}{2} \le \frac{1}{2}x + \frac{\pi}{4} \le \frac{\pi}{2} \Rightarrow -\frac{3\pi}{4} \le \frac{1}{2}x \le \frac{\pi}{4} \Rightarrow -\frac{3\pi}{2} \le x \le \frac{\pi}{2}$, $VA \ @ \ x = -\frac{3\pi}{2} + 2\pi n$ ★ 4π

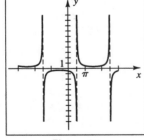

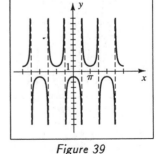

Figure 37   Figure 39

39 $y = \csc\left(x - \frac{\pi}{2}\right)$ • $0 \le x - \frac{\pi}{2} \le \pi \Rightarrow \frac{\pi}{2} \le x \le \frac{3\pi}{2}$, $VA \ @ \ x = \frac{\pi}{2} + \pi n$ ★ 2π

$\boxed{41}$ $y = \csc 2x$ • $0 \le 2x \le \pi \Rightarrow 0 \le x \le \frac{\pi}{2}$, $VA @ x = \frac{\pi}{2}n$ ★ π

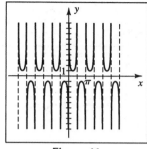   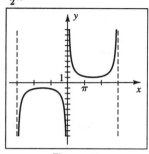

            Figure 41   *Figure 43*

$\boxed{43}$ $y = \csc \frac{1}{3}x$ • $0 \le \frac{1}{3}x \le \pi \Rightarrow 0 \le x \le 3\pi$, $VA @ x = 3\pi n$ ★ 6π

$\boxed{45}$ $y = 2\csc\left(2x + \frac{\pi}{2}\right) = 2\csc\left[2\left(x + \frac{\pi}{4}\right)\right]$.   •

    $0 \le 2x + \frac{\pi}{2} \le \pi \Rightarrow -\frac{\pi}{2} \le 2x \le \frac{\pi}{2} \Rightarrow -\frac{\pi}{4} \le x \le \frac{\pi}{4}$, $VA @ x = -\frac{\pi}{4} + \frac{\pi}{2}n$ ★ π

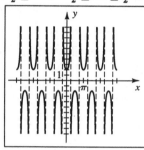   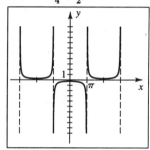

            Figure 45   *Figure 47*

$\boxed{47}$ $y = -\frac{1}{4}\csc\left(\frac{1}{2}x + \frac{\pi}{2}\right) = -\frac{1}{4}\csc\left[\frac{1}{2}(x + \pi)\right]$.   •

    $0 \le \frac{1}{2}x + \frac{\pi}{2} \le \pi \Rightarrow -\frac{\pi}{2} \le \frac{1}{2}x \le \frac{\pi}{2} \Rightarrow -\pi \le x \le \pi$, $VA @ x = -\pi + 2\pi n$ ★ 4π

$\boxed{49}$ $y = \tan \frac{\pi}{2}x$ •   Horizontally stretch $\tan x$ by a factor of $2/\pi$, $VA @ x = -1 + 2n$.

    $-\frac{\pi}{2} \le \frac{\pi}{2}x \le \frac{\pi}{2} \Rightarrow -1 \le x \le 1$ ★ 2

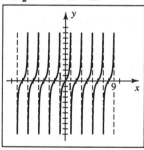   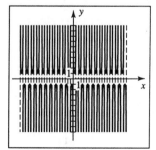

            Figure 49   *Figure 51*

$\boxed{51}$ $y = \csc 2\pi x$ • $0 \le 2\pi x \le \pi \Rightarrow 0 \le x \le \frac{1}{2}$, $VA @ x = \frac{1}{2}n$ ★ 1

$\boxed{53}$ Reflecting the graph of $y = \cot x$ about the x-axis, which is $y = -\cot x$, gives us the graph of $y = \tan\left(x + \frac{\pi}{2}\right)$. If we shift this graph to the left (or right), we will obtain the graph of $y = \tan x$. Thus, one equation is $y = -\cot\left(x + \frac{\pi}{2}\right)$.

$\boxed{55}$ $y = |\sin x|$ • Reflect the negative values of $y = \sin x$ through the x-axis. In general, when sketching the graph of $y = |f(x)|$, reflect the negative values of $f(x)$ through the x-axis. The absolute value does not affect the nonnegative values.

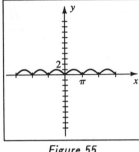

Figure 55

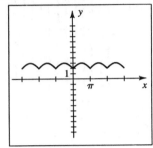

Figure 57

$\boxed{57}$ $y = |\sin x| + 2$ • Shift $y = |\sin x|$ up 2 units.

$\boxed{59}$ $y = -|\cos x| + 1$ • Similar to Exercise 55, we first reflect the negative values of $y = \cos x$ through the x-axis. The "−" in front of $|\cos x|$ has the effect of reflecting $y = |\cos x|$ through the x-axis. Finally, we shift that graph 1 unit up to obtain the graph of $y = -|\cos x| + 1$.

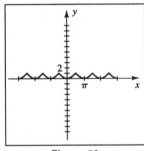

Figure 59

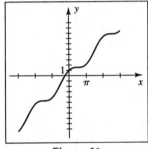

Figure 61

$\boxed{61}$ $y = x + \cos x$ • The value of $\cos x$ is between -1 and 1—adding this relatively small amount to the value of x has the effect of oscillating the graph about the line $y = x$.

$\boxed{63}$ $y = 2^{-x} \cos x$ • This graph is similar to the graph in Example 8.

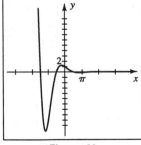

Figure 63

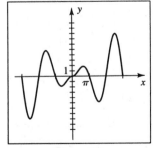

Figure 65

65 $y = |x| \sin x$ •   See *Figure 65* on the preceding page. The graph will coincide with the graph of $y = |x|$ if $\sin x = 1$—that is, if $x = \frac{\pi}{2} + 2\pi n$. The graph will coincide with the graph of $y = -|x|$ if $\sin x = -1$—that is, if $x = \frac{3\pi}{2} + 2\pi n$.

67 $f(x) = \tan(0.5x)$; $g(x) = \tan[0.5(x + \pi/2)]$ •   Since $g(x) = f(x + \pi/2)$, the graph of g can be obtained by shifting the graph of f left a distance of $\frac{\pi}{2}$.

$[-2\pi, 2\pi]$ by $[-4, 4]$

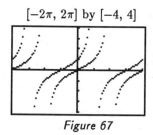

Xscl $= \pi/2$
Yscl $= 1$

Figure 67

$[-2\pi, 2\pi]$ by $[-4, 4]$

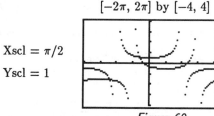

Xscl $= \pi/2$
Yscl $= 1$

Figure 69

69 $f(x) = 0.5 \sec 0.5x$; $g(x) = 0.5 \sec[0.5(x - \pi/2)] - 1$ •

Since $g(x) = f(x - \pi/2) - 1$, the graph of g can be obtained by shifting the graph of f horizontally to the right a distance of $\frac{\pi}{2}$ and vertically downward a distance of 1.

73 The damping factor of $y = e^{-x/4} \sin 4x$ is $e^{-x/4}$.

$[-2\pi, 2\pi]$ by $[-4.19, 4.19]$

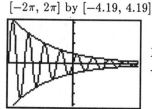

Xscl $= \pi/2$
Yscl $= 1$

Figure 73

$[-\pi, \pi]$ by $[-4, 4]$

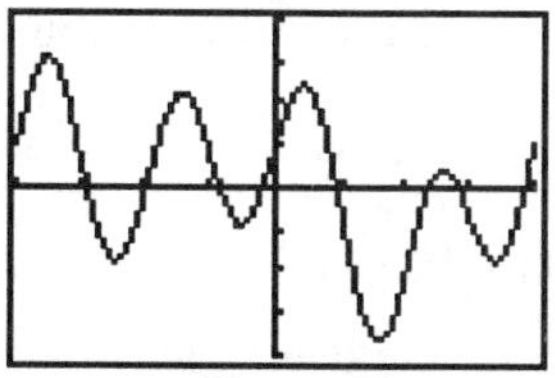

Xscl $= \pi/4$
Yscl $= 1$

Figure 75

75 From the graph, we see that the maximum occurs at the approximate coordinates $(-2.76, 3.09)$, and the minimum occurs at the approximate coordinates $(1.23, -3.68)$.

77 From the graph, we see that f is increasing and one-to-one between

     $a \approx -0.70$ and $b \approx 0.12$. Thus, the interval is approximately $[-0.70, 0.12]$.

$[-2, 2]$ by $[-1.33, 1.33]$

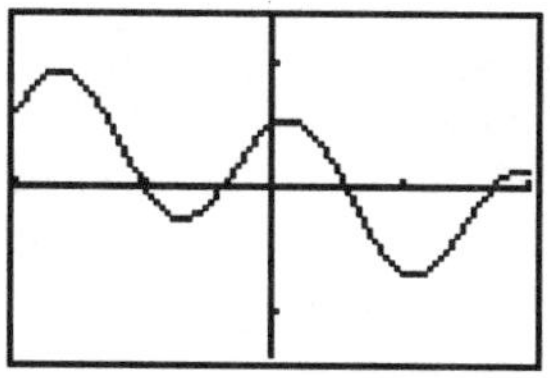

Xscl $= 1$
Yscl $= 1$

Figure 77

$[-\pi, \pi]$ by $[-2.09, 2.09]$

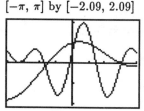

Xscl $= \pi/4$
Yscl $= 1$

Figure 79

79 Graph $Y_1 = \cos(2x - 1) + \sin 3x$ and $Y_2 = \sin\frac{1}{3}x + \cos x$.

From the graph, Y_1 intersects Y_2 at $x \approx -1.31,\ 0.11,\ 0.95,\ 2.39$.

     Thus, $\cos(2x - 1) + \sin 3x \geq \sin\frac{1}{3}x + \cos x$ on $[-\pi, -1.31] \cup [0.11, 0.95] \cup [2.39, \pi]$.

$\boxed{81}$ (a) $\theta = 0 \Rightarrow I = \frac{1}{2}I_0[1 + \cos(\pi \sin 0)] = \frac{1}{2}I_0[1 + \cos(0)] = \frac{1}{2}I_0(2) = I_0$.

(b) $\theta = \pi/3 \Rightarrow I = \frac{1}{2}I_0[1 + \cos(\pi \sin(\pi/3))] \approx 0.044I_0$.

(c) $\theta = \pi/7 \Rightarrow I = \frac{1}{2}I_0[1 + \cos(\pi \sin(\pi/7))] \approx 0.603I_0$.

$\boxed{83}$ (a) The damping factor of $S = A_0 e^{-\alpha z} \sin(kt - \alpha z)$ is $A_0 e^{-\alpha z}$.

(b) The phase shift at depth z_0 can be found by solving the equation $kt - \alpha z_0 = 0$ for
t. Doing so gives us $kt = \alpha z_0$, and hence, $t = \frac{\alpha}{k}z_0$.

(c) At the surface, $z = 0$. Hence, $S = A_0 \sin kt$ and the amplitude at the surface is
A_0. $\text{Amplitude}_{\text{wave}} = \frac{1}{2}\text{Amplitude}_{\text{surface}} \Rightarrow$
$$A_0 e^{-\alpha z} = \frac{1}{2}A_0 \Rightarrow e^{-\alpha z} = \frac{1}{2} \Rightarrow -\alpha z = \ln\frac{1}{2} \Rightarrow z = \frac{-\ln 2}{-\alpha} = \frac{\ln 2}{\alpha}.$$

6.8 Exercises

Note: The missing values are found in terms of the given values.

We could also use proportions to find the remaining parts.

$\boxed{1}$ Since α is given and $\gamma = 90°$, we can easily find β. $\beta = 90° - \alpha = 90° - 30° = 60°$.

To find a, we will relate it to the given parts, α and b, using the tangent function.

$$\tan\alpha = \frac{a}{b} \Rightarrow a = b\tan\alpha = 20\tan 30° = 20(\frac{1}{3}\sqrt{3}) = \frac{20}{3}\sqrt{3}.$$

$$\sec\alpha = \frac{c}{b} \Rightarrow c = b\sec\alpha = 20\sec 30° = 20(\frac{2}{3}\sqrt{3}) = \frac{40}{3}\sqrt{3}.$$

$\boxed{3}$ $\alpha = 90° - \beta = 90° - 45° = 45°$.

$\cos\beta = \frac{a}{c} \Rightarrow a = c\cos\beta = 30\cos 45° = 30(\frac{1}{2}\sqrt{2}) = 15\sqrt{2}$. $b = a$ in a $45°-45°-90°$ $\triangle$.

$\boxed{5}$ $\tan\alpha = \frac{a}{b} = \frac{5}{5} = 1 \Rightarrow \alpha = 45°$. $\beta = 90° - \alpha = 90° - 45° = 45°$. Using the
Pythagorean theorem, $a^2 + b^2 = c^2 \Rightarrow c = \sqrt{a^2 + b^2} = \sqrt{25 + 25} = \sqrt{50} = 5\sqrt{2}$.

$\boxed{7}$ $\cos\alpha = \frac{b}{c} = \frac{5\sqrt{3}}{10\sqrt{3}} = \frac{1}{2} \Rightarrow \alpha = 60°$. $\beta = 90° - \alpha = 90° - 60° = 30°$.

$$a = \sqrt{c^2 - b^2} = \sqrt{300 - 75} = \sqrt{225} = 15.$$

$\boxed{9}$ $\beta = 90° - \alpha = 90° - 37° = 53°$. $\tan\alpha = \frac{a}{b} \Rightarrow a = b\tan\alpha = 24\tan 37° \approx 18$.

$$\sec\alpha = \frac{c}{b} \Rightarrow c = b\sec\alpha = 24\sec 37° \approx 30.$$

$\boxed{11}$ $\alpha = 90° - \beta = 90° - 71°51' = 18°9'$. $\cot\beta = \frac{a}{b} \Rightarrow a = b\cot\beta = 240.0\cot 71°51' \approx 78.7$.

$$\csc\beta = \frac{c}{b} \Rightarrow c = b\csc\beta = 240.0\csc 71°51' \approx 252.6.$$

$\boxed{13}$ $\tan\alpha = \frac{a}{b} = \frac{25}{45} \Rightarrow \alpha = \tan^{-1}\frac{25}{45} \approx 29°$. $\beta = 90° - \alpha \approx 90° - 29° = 61°$.

$$c = \sqrt{a^2 + b^2} = \sqrt{25^2 + 45^2} = \sqrt{625 + 2025} = \sqrt{2650} \approx 51.$$

15 $\cos\alpha = \frac{b}{c} = \frac{2.1}{5.8} \Rightarrow \alpha = \cos^{-1}\frac{21}{58} \approx 69°.$ $\beta = 90° - \alpha \approx 90° - 69° = 21°.$

$$a = \sqrt{c^2 - b^2} = \sqrt{(5.8)^2 - (2.1)^2} = \sqrt{33.64 - 4.41} = \sqrt{29.23} \approx 5.4.$$

Note: Refer to Figures 74 and 75 in the text for the labeling of the sides and angles.

17 We need to find a relationship involving b, c, and α. We want angle α with its adjacent side b and hypotenuse c. The cosine or secant are the functions of α that involve b and c. We choose the cosine since it is easier to solve for b {b is in the numerator}. $\cos\alpha = \frac{b}{c} \Rightarrow b = c\cos\alpha.$

19 We want angle β with its adjacent side a and opposite side b. The tangent or cotangent are the functions of β that involve a and b. $\cot\beta = \frac{a}{b} \Rightarrow a = b\cot\beta.$

21 We want angle α with its opposite side a and hypotenuse c. The sine or cosecant are the functions of α that involve a and c. $\csc\alpha = \frac{c}{a} \Rightarrow c = a\csc\alpha.$

23 $a^2 + b^2 = c^2 \Rightarrow b^2 = c^2 - a^2 \Rightarrow b = \sqrt{c^2 - a^2}$

25 Let h denote the height of the kite and $x = h - 4$. $\sin 60° = \frac{x}{500} \Rightarrow$

$$x = 500\sin 60° = 500(\tfrac{1}{2}\sqrt{3}) = 250\sqrt{3}. \quad h = x + 4 = 250\sqrt{3} + 4 \approx 437 \text{ ft.}$$

27 $\sin 10° = \dfrac{5000}{x} \Rightarrow x = \dfrac{5000}{\sin 10°} \Rightarrow$

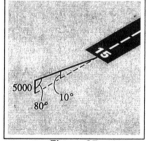

$x = 5000\csc 10° \approx 28{,}793.85$, or 28,800 ft.

Figure 27

31 The 10,000 feet would represent the hypotenuse in a triangle depicting this information. Let h denote the altitude. $\sin 75° = \dfrac{h}{10{,}000} \Rightarrow h \approx 9659$ ft.

33 (a) The bridge section is 75 feet long. Using the right triangle with the 75 foot section as its hypotenuse and $(d - 15)$ as the side opposite the 35° angle, we have

$$\sin 35° = \frac{d - 15}{75} \Rightarrow d = 75\sin 35° + 15 \approx 58 \text{ ft.}$$

(b) Let x be the horizontal distance from the end of a bridge section to a point directly underneath the end of the section. $\cos 35° = \frac{x}{75} \Rightarrow x = 75\cos 35°.$

The distance between the ends of the two sections is (total distance) −

(the 2 horizontal distances under the bridge sections) $= 150 - 2x \approx 27$ ft.

[37] Let D denote the position of the duck and t the number of seconds required for a direct hit. The duck will move $(7t)$ cm. and the bullet will travel $(25t)$ cm.

$$\sin \varphi = \frac{\overline{AD}}{\overline{OD}} = \frac{7t}{25t} \Rightarrow \sin \varphi = \frac{7}{25} \Rightarrow \varphi \approx 16.3^\circ.$$

[39] Let h denote the height of the tower.

$$\tan 21^\circ 20' 24'' = \tan 21.34^\circ = \frac{h}{5280} \Rightarrow h = 5280 \tan 21.34^\circ \approx 2063 \text{ ft.}$$

[41] The central angle of a section of the Pentagon has measure $\frac{360^\circ}{5} = 72^\circ$.

Bisecting that angle, we have an angle of 36° whose opposite side is $\frac{921}{2}$.

The height h is given by $\tan 36^\circ = \dfrac{\frac{921}{2}}{h} \Rightarrow h = \dfrac{921}{2 \tan 36^\circ}$.

$$\text{Area} = 5(\tfrac{1}{2}bh) = 5(\tfrac{1}{2})(921)\left(\frac{921}{2 \tan 36^\circ}\right) \approx 1{,}459{,}379 \text{ ft}^2.$$

[43] The diagonal of the base is $\sqrt{8^2 + 6^2} = 10$. $\tan \theta = \frac{4}{10} \Rightarrow \theta \approx 21.8^\circ$

[45] $\cot 53^\circ 30' = \frac{x}{h} \Rightarrow x = h \cot 53^\circ 30'$. $\cot 26^\circ 50' = \frac{x + 25}{h} \Rightarrow$

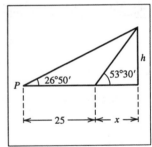

$x + 25 = h \cot 26^\circ 50' \Rightarrow x = h \cot 26^\circ 50' - 25.$

We now have two expressions for x.

We will set these equal to each other and solve for h.

Thus, $h \cot 53^\circ 30' = h \cot 26^\circ 50' - 25 \Rightarrow$

$25 = h \cot 26^\circ 50' - h \cot 53^\circ 30' \Rightarrow$

$h = \dfrac{25}{\cot 26^\circ 50' - \cot 53^\circ 30'} \approx 20.2 \text{ m.}$ *Figure 45*

[47] When the angle of elevation is $19^\circ 20'$, $\tan 19^\circ 20' = \dfrac{h_1}{110} \Rightarrow h_1 = 110 \tan 19^\circ 20'.$

When the angle of elevation is $31^\circ 50'$, $\tan 31^\circ 50' = \dfrac{h_2}{110} \Rightarrow h_2 = 110 \tan 31^\circ 50'.$

The change in elevation is $h_2 - h_1 \approx 68.29 - 38.59 = 29.7 \text{ km.}$

[49] The distance from the spacelab to the center of the earth is $(380 + r)$ miles.

$$\sin 65.8^\circ = \frac{r}{r + 380} \Rightarrow r \sin 65.8^\circ + 380 \sin 65.8^\circ = r \Rightarrow r - r \sin 65.8^\circ = 380 \sin 65.8^\circ \Rightarrow$$

$$r(1 - \sin 65.8^\circ) = 380 \sin 65.8^\circ \Rightarrow r = \frac{380 \sin 65.8^\circ}{1 - \sin 65.8^\circ} \approx 3944 \text{ mi.}$$

[51] Let d be the distance traveled. $\tan 42^\circ = \dfrac{10{,}000}{d} \Rightarrow d = 10{,}000 \cot 42^\circ.$

Converting to mi/hr, we have $\dfrac{10{,}000 \cot 42^\circ \text{ ft}}{1 \text{ minute}} \cdot \dfrac{60 \text{ minutes}}{1 \text{ hour}} \cdot \dfrac{1 \text{ mile}}{5280 \text{ ft}} \approx 126 \text{ mi/hr.}$

$\boxed{53}$ (a) As in Exercise 49, there is a right angle formed on the earth's surface.

Bisecting angle θ and forming a right triangle, we have

$$\cos\frac{\theta}{2} = \frac{R}{R+a} = \frac{4000}{26,300}. \text{ Thus, } \frac{\theta}{2} \approx 81.25° \Rightarrow \theta \approx 162.5°.$$

The percentage of the equator that is within signal range is $\frac{162.5°}{360°} \times 100 \approx 45\%$.

(b) Each satellite has a signal range of more than $120°$,

and thus all 3 will cover all points on the equator.

$\boxed{55}$ Let $x = h - c$. $\sin\alpha = \frac{x}{d} \Rightarrow x = d\sin\alpha$. $h = x + c = d\sin\alpha + c$.

$\boxed{57}$ Let x denote the distance from the base of the tower to the closer point.

$$\cot\beta = \frac{x}{h} \Rightarrow x = h\cot\beta. \ \cot\alpha = \frac{x+d}{h} \Rightarrow x + d = h\cot\alpha \Rightarrow x = h\cot\alpha - d.$$

Thus, $h\cot\beta = h\cot\alpha - d \Rightarrow d = h\cot\alpha - h\cot\beta \Rightarrow d = h(\cot\alpha - \cot\beta) \Rightarrow$

$$h = \frac{d}{\cot\alpha - \cot\beta}.$$

$\boxed{59}$ When the angle of elevation is α, $\tan\alpha = \frac{h_1}{d} \Rightarrow h_1 = d\tan\alpha$.

When the angle of elevation is β, $\tan\beta = \frac{h_2}{d} \Rightarrow h_2 = d\tan\beta$.

$$h = h_2 - h_1 = d\tan\beta - d\tan\alpha = d(\tan\beta - \tan\alpha).$$

$\boxed{61}$ The bearing from P to A is $90° - 20° = 70°$ east of north and is denoted by N70°E.

The bearing from P to B is $40°$ west of north and is denoted by N40°W.

The bearing from P to C is $90° - 75° = 15°$ west of south and is denoted by S15°W.

The bearing from P to D is $25°$ east of south and is denoted by S25°E.

$\boxed{63}$ (a) The first ship travels 2 hours @ 24 mi/hr for a
distance of 48 miles. The second ship travels
$1\frac{1}{2}$ hours @ 18 mi/hr for a distance of 27 miles.
The paths form a right triangle with legs of 48 miles
and 27 miles. The distance between the two ships is

$$\sqrt{27^2 + 48^2} \approx 55 \text{ miles.}$$

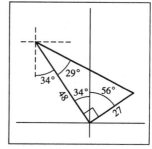

Figure 63

(b) The angle between the side of length 48 and the
hypotenuse is found by solving $\tan\alpha = \frac{27}{48}$ for α. Now
$\alpha \approx 29°$, so the second ship is approximately $29° + 34° = $ S63°E of the first ship.

65 30 minutes @ 360 mi/hr = 180 miles.   45 minutes @ 360 mi/hr = 270 miles. $227° - 137° = 90°$, and hence, the plane's flight forms a right triangle with legs 180 miles and 270 miles. The distance from A to the airplane is equal to the hypotenuse of the triangle—that is, $\sqrt{180^2 + 270^2} \approx 324.5$ mi.

67 Amplitude, 10 cm; period $= \frac{2\pi}{6\pi} = \frac{1}{3}$ sec; frequency $= \frac{6\pi}{2\pi} = 3$ oscillations/sec. The point is at the origin at $t = 0$. It moves upward with decreasing speed, reaching the point with coordinate 10 when $6\pi t = \frac{\pi}{2}$ or $t = \frac{1}{12}$.

It then reverses direction and moves downward, gaining speed until it reaches the origin when $6\pi t = \pi$ or $t = \frac{1}{6}$. It continues downward with decreasing speed, reaching the point with coordinate -10 when $6\pi t = \frac{3\pi}{2}$ or $t = \frac{1}{4}$.

It then reverses direction and moves upward with increasing speed, returning to the origin when $6\pi t = 2\pi$ or $t = \frac{1}{3}$ to complete one oscillation.

Another approach is to simply model this movement in terms of proportions of the sine curve. For one period, the sine increases for $\frac{1}{4}$ period, decreases for $\frac{1}{2}$ period, and increases for its last $\frac{1}{4}$ period.

69 Amplitude, 4 cm; period $= \frac{2\pi}{3\pi/2} = \frac{4}{3}$ sec; frequency $= \frac{3\pi/2}{2\pi} = \frac{3}{4}$ oscillation/sec. The point is at $d = 4$ when $t = 0$. It then decreases in height until $\frac{3\pi}{2}t = \pi$ or $t = \frac{2}{3}$ where it obtains a minimum of $d = -4$. It then reverses direction and increases to a height of $d = 4$ when $\frac{3\pi}{2}t = 2\pi$ or $t = \frac{4}{3}$ to complete one oscillation.

71 Period $= 3 \Rightarrow \frac{2\pi}{\omega} = 3 \Rightarrow \omega = \frac{2\pi}{3}$. Amplitude $= 5 \Rightarrow a = 5$. $d = 5\cos\frac{2\pi}{3}t$

73 (a) period $= 30 \Rightarrow \frac{2\pi}{\omega} = 30 \Rightarrow \omega = \frac{\pi}{15}$. When $t = 0$, the wave is at its highest

point, thus, we use the cosine function. $y = 25\cos\frac{\pi}{15}t$, where t is in minutes.

(b) 180 ft/sec = 10,800 ft/min.

10,800 ft/min for 30 minutes is a distance of 324,000 ft, or $61\frac{4}{11}$ miles.

$\boxed{1}$ $330° \cdot \frac{\pi}{180} = \frac{11 \cdot 30\pi}{6 \cdot 30} = \frac{11\pi}{6}$; $405° \cdot \frac{\pi}{180} = \frac{9 \cdot 45\pi}{4 \cdot 45} = \frac{9\pi}{4}$

$\boxed{2}$ $\frac{9\pi}{2} \cdot \left(\frac{180}{\pi}\right)° = \left(\frac{9 \cdot 90 \cdot 2\pi}{2\pi}\right)° = 810°$; $-\frac{2\pi}{3} \cdot \left(\frac{180}{\pi}\right)° = -\left(\frac{2 \cdot 60 \cdot 3\pi}{3\pi}\right)° = -120°$;

$\boxed{3}$ (a) $\theta = \frac{s}{r} = \frac{20 \text{ cm}}{2 \text{ m}} = \frac{20 \text{ cm}}{2(100) \text{ cm}} = 0.1 \text{ radian}$

   (b) $A = \frac{1}{2}r^2\theta = \frac{1}{2}(2)^2(0.1) = 0.2 \text{ m}^2$

$\boxed{4}$ (a) $s = r\theta = (15 \cdot \frac{1}{2})(70 \cdot \frac{\pi}{180}) = \frac{35\pi}{12} \approx 9.16 \text{ cm}$

   (b) $A = \frac{1}{2}r^2\theta = \frac{1}{2}(15 \cdot \frac{1}{2})^2(70 \cdot \frac{\pi}{180}) = \frac{175\pi}{16} \approx 34.4 \text{ cm}^2$

$\boxed{5}$ 7π is coterminal with π. $P(7\pi) = P(\pi) = (-1, 0)$.

   $-\frac{5\pi}{2}$ is coterminal with $-\frac{\pi}{2}$. $P(-\frac{5\pi}{2}) = P(-\frac{\pi}{2}) = (0, -1)$.

$\boxed{6}$ $P(t) = (-\frac{3}{5}, -\frac{4}{5})$ is in QIII. $P(t + \pi)$ is in QI and will have the same coordinates as

   $P(t)$, but with opposite (positive) signs. $P(t + 3\pi) = P(t + \pi) = P(t - \pi) = (\frac{3}{5}, \frac{4}{5})$.

$\boxed{7}$ (b) $\cot t = \frac{\cos t}{\sin t} = \frac{\csc t}{\sec t} \Rightarrow -\frac{3}{2} = \frac{\sqrt{13}/2}{\sec t} \Rightarrow \sec t = -\frac{\sqrt{13}}{3}$;

   the other values are just the reciprocals.

$\boxed{8}$ (a) $\sec t < 0 \Rightarrow P$ is in QII or QIII. $\sin t > 0 \Rightarrow P$ is in QI or QII.

   Hence, P is in QII.

   (b) $\cot t > 0 \Rightarrow P$ is in QI or QIII. $\csc t < 0 \Rightarrow P$ is in QIII or QIV.

   Hence, P is in QIII.

$\boxed{9}$ $1 + \tan^2 t = \sec^2 t \Rightarrow \tan^2 t = \sec^2 t - 1 \Rightarrow \tan t = \sqrt{\sec^2 t - 1}$

$\boxed{11}$ $\sin t (\csc t - \sin t) = \sin t \csc t - \sin^2 t$ { multiply terms }

$\qquad\qquad\qquad\quad = \sin t \cdot \frac{1}{\sin t} - \sin^2 t$ { reciprocal identity }

$\qquad\qquad\qquad\quad = 1 - \sin^2 t$ { simplify }

$\qquad\qquad\qquad\quad = \cos^2 t$ { Pythagorean identity }

$\boxed{13}$ $(\cos^2 t - 1)(\tan^2 t + 1) = (\cos^2 t - 1)(\sec^2 t)$ { Pythagorean identity }

$\qquad\qquad\qquad\qquad\quad = \cos^2 t \sec^2 t - \sec^2 t$ { multiply terms }

$\qquad\qquad\qquad\qquad\quad = 1 - \sec^2 t$ { reciprocal identity }

$\boxed{15}$ $\frac{1 + \tan^2 t}{\tan^2 t} = \frac{1}{\tan^2 t} + \frac{\tan^2 t}{\tan^2 t}$ { split up the fraction }

$\qquad\qquad\quad = \cot^2 t + 1$ { reciprocal identity, simplify }

$\qquad\qquad\quad = \csc^2 t$ { Pythagorean identity }

17   $\dfrac{\cot t - 1}{1 - \tan t} = \dfrac{\dfrac{\cos t}{\sin t} - 1}{1 - \dfrac{\sin t}{\cos t}}$ { put in terms of sines and cosines }

$\qquad = \dfrac{\dfrac{\cos t - \sin t}{\sin t}}{\dfrac{\cos t - \sin t}{\cos t}}$ $\left\{\begin{array}{l}\text{make the numerator and the} \\ \text{denominator each a single fraction}\end{array}\right\}$

$\qquad = \dfrac{(\cos t - \sin t)\cos t}{(\cos t - \sin t)\sin t}$ { simplify a complex fraction }

$\qquad = \dfrac{\cos t}{\sin t}$ { cancel like term }

$\qquad = \cot t$ { cotangent identity }

19   $\dfrac{\tan(-t) + \cot(-t)}{\tan t} = \dfrac{-\tan t - \cot t}{\tan t}$ { formulas for negatives }

$\qquad = -\dfrac{\tan t}{\tan t} - \dfrac{\cot t}{\tan t}$ { split up fraction }

$\qquad = -1 - \cot^2 t$ { simplify, reciprocal identity }

$\qquad = -(1 + \cot^2 t)$ { factor out -1 }

$\qquad = -\csc^2 t$ { Pythagorean identity }

21   (a) $x = 30$ and $y = -40 \Rightarrow r = \sqrt{30^2 + (-40)^2} = 50.$ ★ (a) $-\frac{4}{5}, \frac{3}{5}, -\frac{4}{3}, -\frac{3}{4}, \frac{5}{3}, -\frac{5}{4}$

(b) $2x + 3y + 6 = 0 \Leftrightarrow y = -\frac{2}{3}x - 2$, so the slope of the given line is $-\frac{2}{3}$.

The line through the origin with that slope is $y = -\frac{2}{3}x$.

If $x = -3$, then $y = 2$ and $(-3, 2)$ is a point on the terminal side of θ.

$\qquad\qquad x = -3$ and $y = 2 \Rightarrow r = \sqrt{(-3)^2 + 2^2} = \sqrt{13}.$

$\qquad\qquad$ ★ (b) $\dfrac{2}{\sqrt{13}}, -\dfrac{3}{\sqrt{13}}, -\dfrac{2}{3}, -\dfrac{3}{2}, -\dfrac{\sqrt{13}}{3}, \dfrac{\sqrt{13}}{2}$

(c) For $\theta = -90°$, choose $x = 0$ and $y = -1$. r is 1.   ★ (c) $-1, 0, U, 0, U, -1$

22   opp $= \sqrt{\text{hyp}^2 - \text{adj}^2} = \sqrt{7^2 - 4^2} = \sqrt{33}.$ ★ $\dfrac{\sqrt{33}}{7}, \dfrac{4}{7}, \dfrac{\sqrt{33}}{4}, \dfrac{4}{\sqrt{33}}, \dfrac{7}{4}, \dfrac{7}{\sqrt{33}}$

23   $\sin 60° = \frac{9}{x} \Rightarrow \frac{\sqrt{3}}{2} = \frac{9}{x} \Rightarrow x = 6\sqrt{3}$; $\tan 60° = \frac{9}{y} \Rightarrow \sqrt{3} = \frac{9}{y} \Rightarrow y = 3\sqrt{3}$

25   (a) $t = -\frac{9\pi}{8} \Rightarrow \theta_C = \frac{7\pi}{8}$ and $t_R = \pi - \frac{7\pi}{8} = \frac{\pi}{8}.$

(b) $\theta = 892° \Rightarrow \theta_C = 172°$ and $\theta_R = 180° - 172° = 8°.$

26   (b) For $\theta = -\frac{5\pi}{4}$, choose $x = -1$ and $y = 1$. $r = \sqrt{2}.$

$\qquad\qquad$ ★ (b) $\dfrac{\sqrt{2}}{2}, -\dfrac{\sqrt{2}}{2}, -1, -1, -\sqrt{2}, \sqrt{2}$

(d) For $\theta = \frac{11\pi}{6}$, choose $x = \sqrt{3}$ and $y = -1$. $r = 2.$

$\qquad\qquad$ ★ (d) $-\dfrac{1}{2}, \dfrac{\sqrt{3}}{2}, -\dfrac{\sqrt{3}}{3}, -\sqrt{3}, \dfrac{2}{\sqrt{3}}, -2$

27 (a) $\cos 225° = -\cos 45° = -\dfrac{\sqrt{2}}{2}$ (b) $\tan 150° = -\tan 30° = -\dfrac{\sqrt{3}}{3}$

(c) $\sin\left(-\dfrac{\pi}{6}\right) = -\sin\dfrac{\pi}{6} = -\dfrac{1}{2}$ (d) $\sec\dfrac{4\pi}{3} = -\sec\dfrac{\pi}{3} = -2$

(e) $\cot\dfrac{7\pi}{4} = -\cot\dfrac{\pi}{4} = -1$ (f) $\csc 300° = -\csc 60° = -\dfrac{2}{\sqrt{3}}$

28 $\sin\theta = -0.7604 \Rightarrow \theta = \sin^{-1}(-0.7604) \approx -49.5° \Rightarrow \theta_R \approx 49.5°$.

Since the sine is negative in QIII and QIV, and the secant is positive in QIV,

we want the fourth-quadrant angle having $\theta_R = 49.5°$. $360° - 49.5° = 310.5°$

29 $y = 5\cos x$ •   Vertically stretch $\cos x$ by a factor of 5.   ★ $5, 2\pi,$ x-int. @ $\dfrac{\pi}{2} + \pi n$

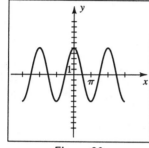

Figure 29

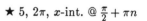
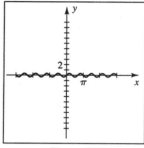

Figure 31

31 $y = \dfrac{1}{3}\sin 3x$ •   Horizontally compress $\sin x$ by a factor of 3 and vertically compress

that graph by a factor of 3.   ★ $\dfrac{1}{3}, \dfrac{2\pi}{3},$ x-int. @ $\dfrac{\pi}{3} n$

33 $y = -3\cos\dfrac{1}{2}x$ •   Horizontally stretch $\cos x$ by a factor of 2, vertically stretch that

graph by a factor of 3, and reflect that graph through the x-axis.

★ $3, 4\pi,$ x-int. @ $\pi + 2\pi n$

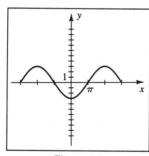

Figure 33

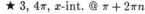
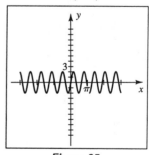

Figure 35

35 $y = 2\sin\pi x$ •   Horizontally compress $\sin x$ by a factor of π, and then vertically

stretch that graph by a factor of 2.   ★ $2, 2,$ x-int. @ n

Note: Let a denote the amplitude and p the period.

[37] (a) $a = |-1.43| = 1.43$. We have $\frac{3}{4}$ of a period from $(0, 0)$ to $(1.5, -1.43)$.

$$\tfrac{3}{4}p = 1.5 \Rightarrow p = 2.$$

(b) Since the period p is given by $\frac{2\pi}{b}$, we can solve for b, obtaining $b = \frac{2\pi}{p}$.

Hence, $b = \frac{2\pi}{p} = \frac{2\pi}{2} = \pi$ and consequently, $y = 1.43 \sin \pi x$.

[39] (a) Since the y-intercept is -3, $a = |-3| = 3$. The second positive x-intercept is π, so $\frac{3}{4}$ of a period occurs from $(0, -3)$ to $(\pi, 0)$. Thus, $\frac{3}{4}p = \pi \Rightarrow p = \frac{4\pi}{3}$.

(b) $b = \frac{2\pi}{p} = \frac{2\pi}{4\pi/3} = \frac{3}{2}$, $y = -3 \cos \frac{3}{2}x$.

[41] $y = 2 \sin \left(x - \frac{2\pi}{3} \right)$ • $0 \le x - \frac{2\pi}{3} \le 2\pi \Rightarrow \frac{2\pi}{3} \le x \le \frac{8\pi}{3}$.

There are x-intercepts at $x = \frac{2\pi}{3} + \pi n$.

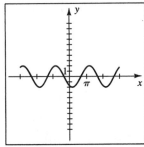

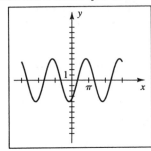

Figure 41   Figure 43

[43] $y = -4 \cos \left(x + \frac{\pi}{6} \right)$ • $-\frac{\pi}{2} \le x + \frac{\pi}{6} \le \frac{3\pi}{2} \Rightarrow -\frac{2\pi}{3} \le x \le \frac{4\pi}{3}$.

There are x-intercepts at $x = -\frac{2\pi}{3} + \pi n$.

[45] $y = 2 \tan \left(\frac{1}{2}x - \pi \right) = 2 \tan \left[\frac{1}{2}(x - 2\pi) \right]$. •

$-\frac{\pi}{2} \le \frac{1}{2}x - \pi \le \frac{\pi}{2} \Rightarrow \frac{\pi}{2} \le \frac{1}{2}x \le \frac{3\pi}{2} \Rightarrow \pi \le x \le 3\pi$, *VA* @ $x = \pi + 2\pi n$

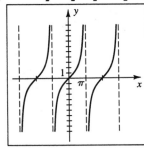

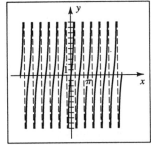

Figure 45   Figure 47

[47] $y = -4 \cot \left(2x - \frac{\pi}{2} \right) = -4 \cot \left[2 \left(x - \frac{\pi}{4} \right) \right]$. •

$0 \le 2x - \frac{\pi}{2} \le \pi \Rightarrow \frac{\pi}{2} \le 2x \le \frac{3\pi}{2} \Rightarrow \frac{\pi}{4} \le x \le \frac{3\pi}{4}$, *VA* @ $x = \frac{\pi}{4} + \frac{\pi}{2}n$

49 $y = \sec\left(\frac{1}{2}x + \pi\right) = \sec\left[\frac{1}{2}(x + 2\pi)\right].$ •

$-\frac{\pi}{2} \le \frac{1}{2}x + \pi \le \frac{\pi}{2} \Rightarrow -\frac{3\pi}{2} \le \frac{1}{2}x \le -\frac{\pi}{2} \Rightarrow -3\pi \le x \le -\pi,\ VA @ x = -3\pi + 2\pi n$

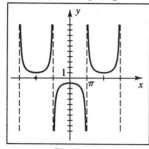

Figure 49

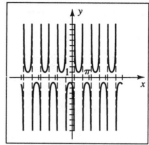

Figure 51

51 $y = \csc\left(2x - \frac{\pi}{4}\right) = \csc\left[2(x - \frac{\pi}{8})\right].$ •

$0 \le 2x - \frac{\pi}{4} \le \pi \Rightarrow \frac{\pi}{4} \le 2x \le \frac{5\pi}{4} \Rightarrow \frac{\pi}{8} \le x \le \frac{5\pi}{8},\ VA @ x = \frac{\pi}{8} + \frac{\pi}{2}n$

53 α and β are complementary, so $\alpha = 90° - \beta = 90° - 60° = 30°.$

$\cot\beta = \frac{a}{b} \Rightarrow a = b\cot\beta = 40\cot 60° = 40(\frac{1}{3}\sqrt{3}) \approx 23.$

$\csc\beta = \frac{c}{b} \Rightarrow c = b\csc\beta = 40\csc 60° = 40(\frac{2}{3}\sqrt{3}) \approx 46.$

55 $\tan\alpha = \frac{a}{b} = \frac{62}{25} \Rightarrow \alpha \approx 68°.\quad \beta = 90° - \alpha \approx 90° - 68° = 22°.$

$$c = \sqrt{a^2 + b^2} = \sqrt{62^2 + 25^2} = \sqrt{3844 + 625} = \sqrt{4469} \approx 67.$$

57 (a) $\left(\frac{545\text{ rev}}{1\text{ min}}\right)\left(\frac{2\pi\text{ rad}}{1\text{ rev}}\right)\left(\frac{1\text{ min}}{60\text{ sec}}\right) = \frac{109\pi}{6}$ rad/sec ≈ 57 rad/sec

(b) $d = 22.625$ ft $\Rightarrow C = \pi d = 22.625\pi$ ft.

$$\left(\frac{22.625\pi\text{ ft}}{1\text{ rev}}\right)\left(\frac{545\text{ rev}}{1\text{ min}}\right)\left(\frac{1\text{ mile}}{5280\text{ ft}}\right)\left(\frac{60\text{ min}}{1\text{ hour}}\right) \approx 440.2\text{ mi/hr}$$

59 $\Delta f = \frac{2fv}{c} \Rightarrow v = \frac{c(\Delta f)}{2f} = \frac{186{,}000 \times 10^8}{2 \times 10^{14}} = 0.093$ mi/sec

60 The angle φ has an adjacent side of $\frac{1}{2}(230\text{ m})$, or 115 m. $\tan\varphi = \frac{147}{115} \Rightarrow \varphi \approx 52°.$

62 The depth of the cone is 4 inches and its slant height is 5 inches. A right triangle is formed by a cross section of the cone with sides 4, 5, and r. Thus, $4^2 + r^2 = 5^2 \Rightarrow r = 3$ inches. The circumference of the rim of the cone is $2\pi r = 2\pi(3) = 6\pi.$ This circumference is the arc length from A to B on the circle.

Using the formula for arc length, $\theta = \frac{s}{r} = \frac{6\pi}{5}$ radians $= 216°.$

[64] (a) Let h denote the height of the building and x the distance between the two buildings. $\tan 59° = \dfrac{h - 50}{x}$ and $\tan 62° = \dfrac{h}{x}$. We now solve both of these equations for h, giving us $h = x \tan 59° + 50$ and $h = x \tan 62°$. Setting these expressions equal to each other and solving for x, we have

$x \tan 62° = x \tan 59° + 50 \Rightarrow x \tan 62° - x \tan 59° = 50 \Rightarrow$

$$x(\tan 62° - \tan 59°) = 50 \Rightarrow x = \frac{50}{\tan 62° - \tan 59°} \approx 231.0 \text{ ft.}$$

(b) From part (a), $h = x \tan 62° \approx 434.5$ ft.

[66] (a) Extend the two boundary lines for h { call these l_{top} and l_{bottom} } to the right until they intersect a line l extended down from the front edge of the building. Let x denote the distance from the intersection of the incline and l_{bottom} to l and y the distance on l from l_{top} to the lower left corner of the building.

$\cos \alpha = \dfrac{x}{d} \Rightarrow x = d \cos \alpha.$ $\sin \alpha = \dfrac{h + y}{d} \Rightarrow h + y = d \sin \alpha \Rightarrow y = d \sin \alpha - h.$

$\tan \theta = \dfrac{y + T}{x} \Rightarrow y + T = x \tan \theta \Rightarrow$

$$
\begin{aligned}
T = x \tan \theta - y &= (d \cos \alpha) \tan \theta - (d \sin \alpha - h) \\
&= d \cos \alpha \tan \theta - d \sin \alpha + h \\
&= h + (d \cos \alpha \tan \theta - d \sin \alpha) = h + d (\cos \alpha \tan \theta - \sin \alpha).
\end{aligned}
$$

(b) $T = 6 + 50 (\cos 15° \tan 31.4° - \sin 15°) \approx 6 + 50 (0.3308) \approx 22.54$ ft.

[68] (a) Let $x = \overline{PT}$ and $y = \overline{QT}$. Now $x^2 + d^2 = y^2$, $h = x \sin \alpha$, and $h = y \sin \beta$.

$$d^2 = y^2 - x^2 = \left(\frac{h}{\sin \beta}\right)^2 - \left(\frac{h}{\sin \alpha}\right)^2 = \frac{h^2}{\sin^2 \beta} - \frac{h^2}{\sin^2 \alpha} = \frac{h^2 \sin^2 \alpha - h^2 \sin^2 \beta}{\sin^2 \beta \sin^2 \alpha} =$$

$$\frac{h^2 (\sin^2 \alpha - \sin^2 \beta)}{\sin^2 \alpha \sin^2 \beta} \Rightarrow h^2 = \frac{d^2 \sin^2 \alpha \sin^2 \beta}{\sin^2 \alpha - \sin^2 \beta} \Rightarrow h = \frac{d \sin \alpha \sin \beta}{\sqrt{\sin^2 \alpha - \sin^2 \beta}}.$$

(b) $\alpha = 30°$, $\beta = 20°$, and $d = 10 \Rightarrow h = \dfrac{10 \sin 30° \sin 20°}{\sqrt{\sin^2 30° - \sin^2 20°}} \approx 4.69$ miles.

70 (a) The side opposite angle θ is one-half the length of the base, $\frac{1}{2}x$. $\sin\theta = \dfrac{\frac{1}{2}x}{a} \Rightarrow$

$x = 2a\sin\theta$. The area of each face is the area of a triangle, and S is the total

area of the four faces. The area of one face is $\frac{1}{2}$(base)(height) $= \frac{1}{2}xa$.

$$S = 4(\tfrac{1}{2}ax) = 2ax = 2a(2a\sin\theta) = 4a^2\sin\theta.$$

(b) $\cos\theta = \frac{y}{a} \Rightarrow y = a\cos\theta$. The volume of a pyramid is one-third times the area

of its base times its height. Hence,

$$V = \tfrac{1}{3}(\text{base area})(\text{height}) = \tfrac{1}{3}x^2y = \tfrac{1}{3}(2a\sin\theta)^2(a\cos\theta) = \tfrac{4}{3}a^3\sin^2\theta\,\cos\theta.$$

72 $y = 1 - 1\cos\left(\frac{1}{2}\pi x/10\right) = -\cos\left(\frac{\pi}{20}x\right) + 1$. To obtain the range for y, we will start with

the range for x, perform operations on these values, and try to obtain the expression

for y. For $0 \le x \le 10$, $0 \le \frac{\pi}{20}x \le \frac{\pi}{2}$ { multiply by $\frac{\pi}{20}$ }, $1 \ge \cos\left(\frac{\pi}{20}x\right) \ge 0$ { take the

cosine of all 3 parts }, $-1 \le -\cos\left(\frac{\pi}{20}x\right) \le 0$ { multiply by -1 }, $0 \le -\cos\left(\frac{\pi}{20}x\right) + 1 \le 1$

{ add 1 }, which is equivalent to $0 \le y \le 1$.

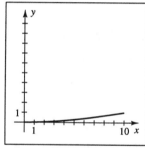

Figure 72

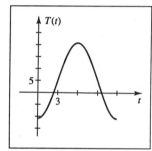

Figure 74

74 (a) $p = \frac{2\pi}{\pi/6} = 12$ months; range $= \underline{5 - 15.8}$ to $\underline{5 + 15.8}$, or, equivalently,

-10.8 to 20.8; phase shift $= 3$

(b) The highest temperature will occur when the argument of the sine is $\frac{\pi}{2}$ since this

is when the sine function has a maximum. $\frac{\pi}{6}(t - 3) = \frac{\pi}{2} \Rightarrow t - 3 = 3 \Rightarrow$

$t = 6$ months. This is July 1st and the temperature is $20.8\,°\text{C}$, or $69.44\,°\text{F}$.

76 (a) The cork is in simple harmonic motion. At $t = 0$, its height is

$s(0) = 12 + \cos 0 = 12 + 1 = 13$ ft. It decreases until $t = 1$, reaching a minimum

of $s(1) = 12 + \cos\pi = 12 + (-1) = 11$ ft. It then increases, reaching a maximum

of 13 ft at $t = 2$.

(b) From part (a), the cork is rising for $1 \le t \le 2$.

$\boxed{1}$ On the TI-82/83 with $a = 15$, there is an indication that there are 15 sine waves on each side of the y-axis, but the minimums and maximums do not get to -1 and 1, respectively. With $a = 30$, the number of sine waves is undetectable—there simply aren't enough pixels for any degree of clarity. With $a = 45$, there are 2 sine waves on each side of the y-axis—there should be 45!

$\boxed{3}$ The sum on the left-side of the equation can never be greater than 3—no solutions.

$\boxed{5}$ The graph of $y_1 = x$, $y_2 = \sin x$, and $y_3 = \tan x$ in the suggested viewing rectangle, $[-0.1, 0.1]$ by $[-0.1, 0.1]$, indicates that their values are very close to each other near $x = 0$—in fact, the graphs of the functions are indistinguishable. Creating a table of values on the order of 10^{-10} also shows that all three functions are nearly equal.

$\boxed{7}$ (a) S is at $(0, -1)$ on the rectangular coordinate system. Starting at S, subtract 1 from the 2 km to get to the circular portion of the track. Now consider $t = (1 + \frac{3\pi}{2})$ as the radian measurement (starting from S) of the circular track in Discussion Exercise 6. $\{\frac{3\pi}{2}$ to get to the bottom of the circle and 1 to make the second km.$\}$ $x = \cos t + 1 \approx 1.8415$ and $y = \sin t \approx -0.5403$.

   (b) The perimeter of the track is $(4 + 2\pi)$ km. $\frac{500}{4 + 2\pi} \approx 48.623066$ laps. Subtracting the 48 whole laps, we have a portion of a lap left, 0.623066 lap. Multiply by $4 + 2\pi$ to see how many km we are from S. $(4 + 2\pi)(0.623066) \approx 6.4071052$. To determine where this places us on the track, start at S and subtract 1 $\{$to get to $(1, -1)\}$, subtract π $\{$to get to $(1, 1)\}$, and subtract 2 $\{$to get to $(-1, 1)\}$. $6.4071052 - (3 + \pi) \approx 0.2655126$. Now consider $t = (0.2655126 + \frac{\pi}{2})$ as the radian measurement of the circular track in Discussion Exercise 6 (but with S at the bottom). $x \approx \cos t - 1 \approx -1.2624$ and $y \approx \sin t \approx 0.9650$.

Chapter 7: Analytic Trigonometry

$\boxed{1}$ $\csc\theta - \sin\theta = \dfrac{1}{\sin\theta} - \sin\theta = \dfrac{1-\sin^2\theta}{\sin\theta} = \dfrac{\cos^2\theta}{\sin\theta} = \dfrac{\cos\theta}{\sin\theta}\cdot\cos\theta = \cot\theta\,\cos\theta$

$\boxed{3}$ $\dfrac{\sec^2 u - 1}{\sec^2 u} = 1 - \dfrac{1}{\sec^2 u}\ \{\,\text{split up fraction}\,\} = 1 - \cos^2 u = \sin^2 u$

$\boxed{5}$ $\dfrac{\csc^2\theta}{1+\tan^2\theta} = \dfrac{\csc^2\theta}{\sec^2\theta} = \dfrac{1/\sin^2\theta}{1/\cos^2\theta} = \dfrac{\cos^2\theta}{\sin^2\theta} = \left(\dfrac{\cos\theta}{\sin\theta}\right)^2 = \cot^2\theta$

$\boxed{7}$ $\dfrac{1+\cos t}{\sin t} + \dfrac{\sin t}{1+\cos t} = \dfrac{(1+\cos t)^2 + \sin^2 t}{\sin t\,(1+\cos t)}$ $\qquad\{\,\text{combine fractions}\,\}$

$\qquad\qquad = \dfrac{1 + 2\cos t + \cos^2 t + \sin^2 t}{\sin t\,(1+\cos t)}$ $\{\,\text{expand}\,\}$

$\qquad\qquad = \dfrac{2 + 2\cos t}{\sin t\,(1+\cos t)}$ $\qquad\{\,\cos^2 t + \sin^2 t = 1\,\}$

$\qquad\qquad = \dfrac{2(1+\cos t)}{\sin t\,(1+\cos t)}$ $\qquad\{\,\text{factor out 2}\,\}$

$\qquad\qquad = 2\csc t$ $\qquad\{\,\text{cancel like term}\,\}$

$\boxed{9}$ $\dfrac{1}{1-\cos\gamma} + \dfrac{1}{1+\cos\gamma} = \dfrac{1+\cos\gamma + 1 - \cos\gamma}{1 - \cos^2\gamma} = \dfrac{2}{\sin^2\gamma} = 2\csc^2\gamma$

$\boxed{11}$ $(\sec u - \tan u)(\csc u + 1) = \left(\dfrac{1}{\cos u} - \dfrac{\sin u}{\cos u}\right)\left(\dfrac{1}{\sin u} + 1\right)$ $\{\,\text{change to sines and cosines}\,\}$

$\qquad\qquad = \left(\dfrac{1-\sin u}{\cos u}\right)\left(\dfrac{1+\sin u}{\sin u}\right)$ $\qquad\{\,\text{combine fractions}\,\}$

$\qquad\qquad = \dfrac{1 - \sin^2 u}{\cos u\,\sin u}$ $\qquad\{\,\text{multiply terms}\,\}$

$\qquad\qquad = \dfrac{\cos^2 u}{\cos u\,\sin u} = \dfrac{\cos u}{\sin u} = \cot u$

$\boxed{13}$ $\csc^4 t - \cot^4 t = (\csc^2 t)^2 - (\cot^2 t)^2$ $\qquad\{\,\text{recognize as the diff. of 2 squares}\,\}$

$\qquad\qquad = (\csc^2 t + \cot^2 t)(\csc^2 t - \cot^2 t)$ $\{\,\text{factor}\,\}$

$\qquad\qquad = (\csc^2 t + \cot^2 t)(1)$ $\qquad\{\,\text{Pythagorean id., }1 + \cot^2 t = \csc^2 t\,\}$

$\qquad\qquad = \csc^2 t + \cot^2 t$

15 The first step in the following verification is to multiply the denominator, $1 - \sin \beta$, by its conjugate, $1 + \sin \beta$. This procedure will give us the difference of two squares, $1 - \sin^2 \beta$, which is equal to $\cos^2 \beta$. This step is often helpful when simplifying trigonometric expressions because manipulation of the Pythagorean identities often allows us to reduce the resulting expression to a single term.

$$\frac{\cos \beta}{1 - \sin \beta} = \frac{\cos \beta}{1 - \sin \beta} \cdot \frac{1 + \sin \beta}{1 + \sin \beta} = \frac{\cos \beta \, (1 + \sin \beta)}{1 - \sin^2 \beta} = \frac{\cos \beta \, (1 + \sin \beta)}{\cos^2 \beta} = \frac{1 + \sin \beta}{\cos \beta} =$$

$$\frac{1}{\cos \beta} + \frac{\sin \beta}{\cos \beta} = \sec \beta + \tan \beta$$

17 $\dfrac{\tan^2 x}{\sec x + 1} = \dfrac{\sec^2 x - 1}{\sec x + 1} = \dfrac{(\sec x + 1)(\sec x - 1)}{\sec x + 1} = \sec x - 1 = \dfrac{1}{\cos x} - 1 = \dfrac{1 - \cos x}{\cos x}$

19 $\dfrac{\cot u - 1}{\cot u + 1} = \dfrac{\dfrac{1}{\tan u} - 1}{\dfrac{1}{\tan u} + 1} = \dfrac{\dfrac{1 - \tan u}{\tan u}}{\dfrac{1 + \tan u}{\tan u}} = \dfrac{1 - \tan u}{1 + \tan u}$

21 $\sin^4 r - \cos^4 r = (\sin^2 r - \cos^2 r)(\sin^2 r + \cos^2 r) = (\sin^2 r - \cos^2 r)(1) = \sin^2 r - \cos^2 r$

23 $\tan^4 k - \sec^4 k = (\tan^2 k - \sec^2 k)(\tan^2 k + \sec^2 k)$ { factor as the diff. of 2 squares }

$\qquad\qquad = (-1)(\sec^2 k - 1 + \sec^2 k)$ { Pythagorean identity }

$\qquad\qquad = (-1)(2 \sec^2 k - 1) = 1 - 2 \sec^2 k$

25 $(\sec t + \tan t)^2 = \left(\dfrac{1}{\cos t} + \dfrac{\sin t}{\cos t}\right)^2 = \left(\dfrac{1 + \sin t}{\cos t}\right)^2 = \dfrac{(1 + \sin t)^2}{\cos^2 t} =$

$$\frac{(1 + \sin t)^2}{1 - \sin^2 t} = \frac{(1 + \sin t)^2}{(1 + \sin t)(1 - \sin t)} = \frac{1 + \sin t}{1 - \sin t}$$

27 $(\sin^2 \theta + \cos^2 \theta)^3 = (1)^3 = 1$

29 $\dfrac{1 + \csc \beta}{\cot \beta + \cos \beta} = \dfrac{1 + \dfrac{1}{\sin \beta}}{\dfrac{\cos \beta}{\sin \beta} + \cos \beta} = \dfrac{\dfrac{\sin \beta + 1}{\sin \beta}}{\dfrac{\cos \beta + \cos \beta \sin \beta}{\sin \beta}} = \dfrac{\sin \beta + 1}{\cos \beta \, (1 + \sin \beta)} = \dfrac{1}{\cos \beta} = \sec \beta$

31 As demonstrated in the following verification, it may be beneficial to try to combine expressions rather than expand them. $(\csc t - \cot t)^4 (\csc t + \cot t)^4 =$

$$\left[(\csc t - \cot t)(\csc t + \cot t) \right]^4 = (\csc^2 t - \cot^2 t)^4 = (1)^4 = 1$$

33 $\text{RS} = \dfrac{\tan \alpha + \tan \beta}{1 - \tan \alpha \tan \beta} = \dfrac{\dfrac{\sin \alpha}{\cos \alpha} + \dfrac{\sin \beta}{\cos \beta}}{1 - \dfrac{\sin \alpha}{\cos \alpha} \cdot \dfrac{\sin \beta}{\cos \beta}} = \dfrac{\dfrac{\sin \alpha \cos \beta + \cos \alpha \sin \beta}{\cos \alpha \cos \beta}}{\dfrac{\cos \alpha \cos \beta - \sin \alpha \sin \beta}{\cos \alpha \cos \beta}} =$

$$\frac{\sin \alpha \cos \beta + \cos \alpha \sin \beta}{\cos \alpha \cos \beta - \sin \alpha \sin \beta} = \text{LS}$$

Note: We could obtain the RS by dividing

the numerator and denominator of the LS by $(\cos \alpha \, \cos \beta)$.

35 $\dfrac{\tan\alpha}{1+\sec\alpha} + \dfrac{1+\sec\alpha}{\tan\alpha} = \dfrac{\tan^2\alpha + (1+\sec\alpha)^2}{(1+\sec\alpha)\tan\alpha} = \dfrac{(\sec^2\alpha - 1) + (1+2\sec\alpha+\sec^2\alpha)}{(1+\sec\alpha)\tan\alpha} =$

$\dfrac{2\sec^2\alpha + 2\sec\alpha}{(1+\sec\alpha)\tan\alpha} = \dfrac{2\sec\alpha\,(\sec\alpha + 1)\cot\alpha}{1+\sec\alpha} = \dfrac{2}{\cos\alpha}\cdot\dfrac{\cos\alpha}{\sin\alpha} = \dfrac{2}{\sin\alpha} = 2\csc\alpha$

37 $\dfrac{1}{\tan\beta + \cot\beta} = \dfrac{1}{\dfrac{\sin\beta}{\cos\beta} + \dfrac{\cos\beta}{\sin\beta}} = \dfrac{1}{\dfrac{\sin^2\beta + \cos^2\beta}{\cos\beta\,\sin\beta}} = \dfrac{1}{\dfrac{1}{\cos\beta\,\sin\beta}} = \sin\beta\,\cos\beta$

39 $\sec\theta + \csc\theta - \cos\theta - \sin\theta = \dfrac{1}{\cos\theta} + \dfrac{1}{\sin\theta} - \cos\theta - \sin\theta$

$= \left(\dfrac{1}{\cos\theta} - \cos\theta\right) + \left(\dfrac{1}{\sin\theta} - \sin\theta\right)$

$= \dfrac{1-\cos^2\theta}{\cos\theta} + \dfrac{1-\sin^2\theta}{\sin\theta}$

$= \dfrac{\sin^2\theta}{\cos\theta} + \dfrac{\cos^2\theta}{\sin\theta}$

$= \left(\sin\theta\cdot\dfrac{\sin\theta}{\cos\theta}\right) + \left(\cos\theta\cdot\dfrac{\cos\theta}{\sin\theta}\right) = \sin\theta\,\tan\theta + \cos\theta\,\cot\theta$

41 $\text{RS} = \sec^4\phi - 4\tan^2\phi = (\sec^2\phi)^2 - 4\tan^2\phi = (1+\tan^2\phi)^2 - 4\tan^2\phi =$

$(1 + 2\tan^2\phi + \tan^4\phi) - 4\tan^2\phi = 1 - 2\tan^2\phi + \tan^4\phi = (1-\tan^2\phi)^2 = \text{LS}$

43 $\dfrac{\cot(-t) + \tan(-t)}{\cot t} = \dfrac{-\cot t - \tan t}{\cot t} = -\dfrac{\cot t}{\cot t} - \dfrac{\tan t}{\cot t} = -(1+\tan^2 t) = -\sec^2 t$

45 $\log 10^{\tan t} = \log_{10} 10^{\tan t} = \tan t$, since $\log_a a^x = x$

47 $\ln\cot x = \ln(\cot x) = \ln(\tan x)^{-1} = -\ln(\tan x)$ { since $\ln a^x = x\ln a$ } $= -\ln\tan x$

49 $\ln|\sec\theta + \tan\theta| = \ln\left|\dfrac{(\sec\theta + \tan\theta)(\sec\theta - \tan\theta)}{\sec\theta - \tan\theta}\right|$ { multiply by the conjugate }

$= \ln\left|\dfrac{\sec^2\theta - \tan^2\theta}{\sec\theta - \tan\theta}\right|$ { simplify }

$= \ln\left|\dfrac{1}{\sec\theta - \tan\theta}\right|$ { Pythagorean identity }

$= \ln\dfrac{|1|}{|\sec\theta - \tan\theta|}$ $\left\{\left|\dfrac{a}{b}\right| = \dfrac{|a|}{|b|}\right\}$

$= \ln|1| - \ln|\sec\theta - \tan\theta|$ $\left\{\ln\dfrac{a}{b} = \ln a - \ln b\right\}$

$= -\ln|\sec\theta - \tan\theta|$ { $\ln 1 = 0$ }

51 $\cos^2 t = 1 - \sin^2 t \Rightarrow \cos t = \pm\sqrt{1-\sin^2 t}$. Since the given equation is

$\cos t = +\sqrt{1-\sin^2 t}$, we may choose any t such that $\cos t < 0$.

Using $t = \pi$, $\text{LS} = \cos\pi = -1$. $\text{RS} = \sqrt{1-\sin^2\pi} = 1$. Since $-1 \neq 1$, $\text{LS} \neq \text{RS}$.

---

53 $\sqrt{\sin^2 t} = |\sin t| = \pm \sin t$. Hence, choose any t such that $\sin t < 0$.

      Using $t = \frac{3\pi}{2}$, LS $= \sqrt{(-1)^2} = 1$. RS $= \sin \frac{3\pi}{2} = -1$. Since $1 \neq -1$, LS $\neq$ RS.

55 $(\sin \theta + \cos \theta)^2 = \sin^2 \theta + 2 \sin \theta \cos \theta + \cos^2 \theta$. Since the right side of the given

      equation is only $\sin^2 \theta + \cos^2 \theta$, we may choose any θ such that $2 \sin \theta \cos \theta \neq 0$.

      Using $\theta = \frac{\pi}{4}$, LS $= (\frac{1}{2}\sqrt{2} + \frac{1}{2}\sqrt{2})^2 = (\sqrt{2})^2 = 2$.

                 RS $= (\frac{1}{2}\sqrt{2})^2 + (\frac{1}{2}\sqrt{2})^2 = \frac{1}{2} + \frac{1}{2} = 1$. Since $2 \neq 1$, LS $\neq$ RS.

57 $\cos(-t) = -\cos t$ •   Since $\cos(-t) = \cos t$, we may choose any t such that

      $\cos t \neq -\cos t$—that is, any t such that $\cos t \neq 0$. Using $t = \pi$, LS $= \cos(-\pi) = -1$.

                 RS $= -\cos \pi = -(-1) = 1$. Since $-1 \neq 1$, LS $\neq$ RS.

59 Don't confuse $\cos(\sec t) = 1$ with $\cos t \cdot \sec t = 1$. The former is true if $\sec t = 2\pi$ or

      an integer multiple of 2π. The latter is true for any value of t as long as $\sec t$ is

      defined. Choose any t such that $\sec t \neq 2\pi n$.

                 Using $t = \frac{\pi}{4}$, LS $= \cos(\sec \frac{\pi}{4}) = \cos \sqrt{2} \neq 1 =$ RS.

61 $\sin^2 t - 4 \sin t - 5 = 0 \Rightarrow (\sin t - 5)(\sin t + 1) = 0 \Rightarrow \sin t = 5$ or $\sin t = -1$.

      Since $\sin t$ cannot equal 5, we may choose any t such that $\sin t \neq -1$.

                 Using $t = \pi$, LS $= -5 \neq 0 =$ RS.

Note: Exer. 63–66: Use $\sqrt{a^2 - x^2} = a \cos \theta$ because

$$\sqrt{a^2 - x^2} = \sqrt{a^2 - a^2 \sin^2 \theta} = \sqrt{a^2(1 - \sin^2 \theta)} = \sqrt{a^2 \cos^2 \theta} = |a| \, |\cos \theta| = a \cos \theta$$

                 since $\cos \theta > 0$ if $-\frac{\pi}{2} < \theta < \frac{\pi}{2}$ and $a > 0$.

63 $(a^2 - x^2)^{3/2} = (\sqrt{a^2 - x^2})^3 = (a \cos \theta)^3$ { see above note } $= a^3 \cos^3 \theta$

65 $\dfrac{x^2}{\sqrt{a^2 - x^2}} = \dfrac{a^2 \sin^2 \theta}{a \cos \theta} = a \cdot \dfrac{\sin \theta}{\cos \theta} \cdot \sin \theta = a \tan \theta \sin \theta$

Note: Exer. 67–70: Use $\sqrt{a^2 + x^2} = a \sec \theta$ because

$$\sqrt{a^2 + x^2} = \sqrt{a^2 + a^2 \tan^2 \theta} = \sqrt{a^2(1 + \tan^2 \theta)} = \sqrt{a^2 \sec^2 \theta} = |a| \, |\sec \theta| =$$

                 $a \sec \theta$ since $\sec \theta > 0$ if $-\frac{\pi}{2} < \theta < \frac{\pi}{2}$ and $a > 0$.

67 $\sqrt{a^2 + x^2} = a \sec \theta$ { see above note }

69 $\dfrac{1}{x^2 + a^2} = \dfrac{1}{(\sqrt{a^2 + x^2})^2} = \dfrac{1}{(a \sec \theta)^2} = \dfrac{1}{a^2 \sec^2 \theta} = \dfrac{1}{a^2} \cos^2 \theta$

Note: Exer. 71–74: Use $\sqrt{x^2 - a^2} = a \tan \theta$ because

$$\sqrt{x^2 - a^2} = \sqrt{a^2 \sec^2 \theta - a^2} = \sqrt{a^2(\sec^2 \theta - 1)} = \sqrt{a^2 \tan^2 \theta} = |a| \, |\tan \theta| =$$

                 $a \tan \theta$ since $\tan \theta > 0$ if $0 < \theta < \frac{\pi}{2}$ and $a > 0$.

71 $\sqrt{x^2 - a^2} = a \tan \theta$ { see above note }

73 $x^3 \sqrt{x^2 - a^2} = (a^3 \sec^3 \theta)(a \tan \theta) = a^4 \sec^3 \theta \tan \theta$

[75] The graph of $f(x) = \dfrac{\sin^2 x - \sin^4 x}{(1 - \sec^2 x)\cos^4 x}$ appears to be that of the horizontal line

$y = g(x) = -1$. Verifying this identity, we have

$$\frac{\sin^2 x - \sin^4 x}{(1 - \sec^2 x)\cos^4 x} = \frac{\sin^2 x(1 - \sin^2 x)}{-\tan^2 x \, \cos^4 x} = \frac{\sin^2 x \, \cos^2 x}{-(\sin^2 x/\cos^2 x)\cos^4 x} = \frac{\sin^2 x \, \cos^2 x}{-\sin^2 x \, \cos^2 x} = -1.$$

[77] The graph of $f(x) = \sec x \, (\sin x \, \cos x + \cos^2 x) - \sin x$ appears to be that of

$y = g(x) = \cos x$. Verifying this identity, we have

$$\sec x \, (\sin x \, \cos x + \cos^2 x) - \sin x = \sec x \, \cos x \, (\sin x + \cos x) - \sin x$$

$$= (\sin x + \cos x) - \sin x = \cos x.$$

7.2 Exercises

[1] In $[0, 2\pi)$, $\sin x = -\dfrac{\sqrt{2}}{2}$ only if $x = \dfrac{5\pi}{4}, \dfrac{7\pi}{4}$. <u>All solutions</u> would include these angles

plus all angles coterminal with them. Hence, $x = \dfrac{5\pi}{4} + 2\pi n, \dfrac{7\pi}{4} + 2\pi n$.

[3] $\tan \theta = \sqrt{3} \Rightarrow \theta = \dfrac{\pi}{3} + \pi n$. *Note:* This solution could be written as $\dfrac{\pi}{3} + 2\pi n$ and

$\dfrac{4\pi}{3} + 2\pi n$. Since the period of the tangent (and the cotangent) is π, we use the

abbreviated form $\dfrac{\pi}{3} + \pi n$ to describe all solutions. We will use "πn" for solutions of

exercises involving the tangent and cotangent functions.

[7] $\sin x = \dfrac{\pi}{2}$ has no solution since $\dfrac{\pi}{2} > 1$, which is not in the range $[-1, 1]$. Your

calculator will give some kind of error message if you attempt to find a solution to

this problem or any similar type problem.

[9] $\cos \theta = \dfrac{1}{\sec \theta}$ is true for all values *for which the equation is defined.*

★ All θ except $\theta = \dfrac{\pi}{2} + \pi n$

[11] $2 \cos 2\theta - \sqrt{3} = 0 \Rightarrow \cos 2\theta = \dfrac{\sqrt{3}}{2}$ { 2θ is just an angle — so we solve this equation

for 2θ and then divide those solutions by 2 } $\Rightarrow$

$$2\theta = \frac{\pi}{6} + 2\pi n, \frac{11\pi}{6} + 2\pi n \Rightarrow \theta = \frac{\pi}{12} + \pi n, \frac{11\pi}{12} + \pi n$$

[13] $\sqrt{3} \tan \frac{1}{3}t = 1 \Rightarrow \tan \frac{1}{3}t = \dfrac{1}{\sqrt{3}} \Rightarrow \frac{1}{3}t = \frac{\pi}{6} + \pi n \Rightarrow t = \frac{\pi}{2} + 3\pi n$

[15] $\sin\left(\theta + \frac{\pi}{4}\right) = \frac{1}{2} \Rightarrow \theta + \frac{\pi}{4} = \frac{\pi}{6} + 2\pi n, \frac{5\pi}{6} + 2\pi n \Rightarrow \theta = -\frac{\pi}{12} + 2\pi n, \frac{7\pi}{12} + 2\pi n$

[17] $\sin\left(2x - \frac{\pi}{3}\right) = \frac{1}{2} \Rightarrow 2x - \frac{\pi}{3} = \frac{\pi}{6} + 2\pi n, \frac{5\pi}{6} + 2\pi n \Rightarrow 2x = \frac{\pi}{2} + 2\pi n, \frac{7\pi}{6} + 2\pi n \Rightarrow$

$$x = \frac{\pi}{4} + \pi n, \frac{7\pi}{12} + \pi n$$

[21] $\tan^2 x = 1 \Rightarrow \tan x = \pm 1 \Rightarrow x = \frac{\pi}{4} + \pi n, \frac{3\pi}{4} + \pi n$, or simply $\frac{\pi}{4} + \frac{\pi}{2}n$

[23] $(\cos \theta - 1)(\sin \theta + 1) = 0 \Rightarrow (\cos \theta - 1) = 0$ or $(\sin \theta + 1) = 0 \Rightarrow$

$$\cos \theta = 1 \text{ or } \sin \theta = -1 \Rightarrow \theta = 2\pi n \text{ or } \theta = \frac{3\pi}{2} + 2\pi n$$

$\boxed{25}$ $\sec^2\alpha - 4 = 0 \Rightarrow \sec^2\alpha = 4 \Rightarrow \sec\alpha = \pm 2 \Rightarrow$

$$\alpha = \tfrac{\pi}{3} + 2\pi n, \ \tfrac{5\pi}{3} + 2\pi n, \ \tfrac{2\pi}{3} + 2\pi n, \ \tfrac{4\pi}{3} + 2\pi n, \text{ or simply } \tfrac{\pi}{3} + \pi n, \ \tfrac{2\pi}{3} + \pi n$$

$\boxed{29}$ $\cot^2 x - 3 = 0 \Rightarrow \cot^2 x = 3 \Rightarrow \cot x = \pm\sqrt{3} \Rightarrow x = \tfrac{\pi}{6} + \pi n, \ \tfrac{5\pi}{6} + \pi n$

$\boxed{31}$ $(2\sin\theta + 1)(2\cos\theta + 3) = 0 \Rightarrow \sin\theta = -\tfrac{1}{2}$ or $\sin\theta = -\tfrac{3}{2} \Rightarrow$

$$\theta = \tfrac{7\pi}{6} + 2\pi n, \ \tfrac{11\pi}{6} + 2\pi n \ \{\sin\theta = -\tfrac{3}{2} \text{ has no solutions}\}$$

$\boxed{33}$ $\sin 2x\,(\csc 2x - 2) = 0 \Rightarrow 1 - 2\sin 2x = 0 \Rightarrow \sin 2x = \tfrac{1}{2} \Rightarrow$

$$2x = \tfrac{\pi}{6} + 2\pi n, \ \tfrac{5\pi}{6} + 2\pi n \Rightarrow x = \tfrac{\pi}{12} + \pi n, \ \tfrac{5\pi}{12} + \pi n$$

$\boxed{35}$ $\cos(\ln x) = 0 \Rightarrow \ln x = \tfrac{\pi}{2} + \pi n \Rightarrow x = e^{(\pi/2)\,+\,\pi n} \ \{\text{since } \ln x = y \Leftrightarrow x = e^y\}$

$\boxed{37}$ $\cos(2x - \tfrac{\pi}{4}) = 0 \Rightarrow 2x - \tfrac{\pi}{4} = \tfrac{\pi}{2} + \pi n \Rightarrow 2x = \tfrac{3\pi}{4} + \pi n \Rightarrow x = \tfrac{3\pi}{8} + \tfrac{\pi}{2}n.$

x will be in the interval $[0, 2\pi)$ if $n = 0, 1, 2,$ or 3. Thus, $x = \tfrac{3\pi}{8}, \tfrac{7\pi}{8}, \tfrac{11\pi}{8}, \tfrac{15\pi}{8}$.

$\boxed{39}$ $2 - 8\cos^2 t = 0 \Rightarrow \cos^2 t = \tfrac{1}{4} \Rightarrow \cos t = \pm\tfrac{1}{2} \Rightarrow t = \tfrac{\pi}{3}, \tfrac{2\pi}{3}, \tfrac{4\pi}{3}, \tfrac{5\pi}{3}$

$\boxed{41}$ $2\sin^2 u = 1 - \sin u \Rightarrow 2\sin^2 u + \sin u - 1 = 0 \Rightarrow (2\sin u - 1)(\sin u + 1) = 0 \Rightarrow$

$$\sin u = \tfrac{1}{2}, \ -1 \Rightarrow u = \tfrac{\pi}{6}, \tfrac{5\pi}{6}, \tfrac{3\pi}{2}$$

$\boxed{43}$ $\tan^2 x \sin x = \sin x \Rightarrow \tan^2 x \sin x - \sin x = 0 \Rightarrow \sin x(\tan^2 x - 1) = 0 \Rightarrow \sin x = 0$ or

$\tan x = \pm 1 \Rightarrow x = 0, \ \pi, \ \tfrac{\pi}{4}, \tfrac{3\pi}{4}, \tfrac{5\pi}{4}, \tfrac{7\pi}{4}$. *Note:* A common mistake is to divide both sides of the given equation by $\sin x$—doing so results in losing the solutions for $\sin x = 0$.

$\boxed{45}$ $2\cos^2\gamma + \cos\gamma = 0 \Rightarrow \cos\gamma(2\cos\gamma + 1) = 0 \Rightarrow \cos\gamma = 0, \ -\tfrac{1}{2} \Rightarrow \gamma = \tfrac{\pi}{2}, \tfrac{3\pi}{2}, \tfrac{2\pi}{3}, \tfrac{4\pi}{3}$

$\boxed{47}$ $\sin^2\theta + \sin\theta - 6 = 0 \Rightarrow (\sin\theta + 3)(\sin\theta - 2) = 0 \Rightarrow \sin\theta = -3, 2.$

There are *no solutions* for either equation.

$\boxed{49}$ $1 - \sin t = \sqrt{3}\cos t$ •   Square both sides to obtain an equation in either sin or cos.

$(1 - \sin t)^2 = (\sqrt{3}\cos t)^2 \Rightarrow 1 - 2\sin t + \sin^2 t = 3\cos^2 t \Rightarrow$

$\sin^2 t - 2\sin t + 1 = 3(1 - \sin^2 t) \Rightarrow 4\sin^2 t - 2\sin t - 2 = 0 \Rightarrow 2\sin^2 t - \sin t - 1 = 0 \Rightarrow$

$(2\sin t + 1)(\sin t - 1) = 0 \Rightarrow \sin t = -\tfrac{1}{2}, \ 1 \Rightarrow t = \tfrac{7\pi}{6}, \tfrac{11\pi}{6}, \tfrac{\pi}{2}$. Since each side of the equation was squared, the solutions must be checked in the original equation. Checking $\tfrac{7\pi}{6}$, we have $\text{LS} = 1 - \sin\tfrac{7\pi}{6} = 1 - (-\tfrac{1}{2}) = \tfrac{3}{2}$ and $\text{RS} = \sqrt{3}\cos\tfrac{7\pi}{6} = \sqrt{3}\left(-\tfrac{\sqrt{3}}{2}\right) = -\tfrac{3}{2}$. Since $\text{LS} \neq \text{RS}$, $\tfrac{7\pi}{6}$ is an extraneous solution. Checking $\tfrac{11\pi}{6}$, we have $\text{LS} = \tfrac{3}{2}$ and $\text{RS} = \tfrac{3}{2}$. Since $\text{LS} = \text{RS}$, $\tfrac{11\pi}{6}$ is a valid solution. Similarly, $\tfrac{\pi}{2}$ is a valid solution and our solution is $\tfrac{\pi}{2}$ and $\tfrac{11\pi}{6}$.

$\boxed{51}$ $\cos\alpha + \sin\alpha = 1 \Rightarrow \cos\alpha = 1 - \sin\alpha \ \{\text{square both sides}\} \Rightarrow$

$\cos^2\alpha = 1 - 2\sin\alpha + \sin^2\alpha$

$\{\text{change } \cos^2\alpha \text{ to } 1 - \sin^2\alpha \text{ to obtain an equation involving only } \sin\alpha\} \Rightarrow$

$1 - \sin^2\alpha = 1 - 2\sin\alpha + \sin^2\alpha \Rightarrow 2\sin^2\alpha - 2\sin\alpha = 0 \Rightarrow$

$2\sin\alpha(\sin\alpha - 1) = 0 \Rightarrow \sin\alpha = 0, 1 \Rightarrow \alpha = 0, \ \pi, \ \tfrac{\pi}{2}$. π is an extraneous solution.

$\boxed{53}$ $2\tan t - \sec^2 t = 0 \Rightarrow 2\tan t - (1+\tan^2 t) = 0 \Rightarrow \tan^2 t - 2\tan t + 1 = 0 \Rightarrow$

$$(\tan t - 1)^2 = 0 \Rightarrow \tan t = 1 \Rightarrow t = \frac{\pi}{4}, \frac{5\pi}{4}$$

$\boxed{55}$ $\cot\alpha + \tan\alpha = \csc\alpha \sec\alpha \Rightarrow \frac{\cos\alpha}{\sin\alpha} + \frac{\sin\alpha}{\cos\alpha} = \frac{1}{\sin\alpha \cos\alpha} \Rightarrow$

$\frac{\cos^2\alpha + \sin^2\alpha}{\sin\alpha \cos\alpha} = \frac{1}{\sin\alpha \cos\alpha}$. This is an identity and is true for *all numbers in* $[0, 2\pi)$

except 0, $\frac{\pi}{2}$, π, and $\frac{3\pi}{2}$ since these values make the original equation undefined.

$\boxed{57}$ $2\sin^3 x + \sin^2 x - 2\sin x - 1 = 0$ $\{$ factor by grouping since there are four terms $\} \Rightarrow$

$\sin^2 x(2\sin x + 1) - 1(2\sin x + 1) = 0 \Rightarrow$

$$(\sin^2 x - 1)(2\sin x + 1) = 0 \Rightarrow \sin x = \pm 1, -\frac{1}{2} \Rightarrow x = \frac{\pi}{2}, \frac{3\pi}{2}, \frac{7\pi}{6}, \frac{11\pi}{6}$$

$\boxed{59}$ $2\tan t \csc t + 2\csc t + \tan t + 1 = 0 \Rightarrow 2\csc t(\tan t + 1) + 1(\tan t + 1) \Rightarrow$

$(2\csc t + 1)(\tan t + 1) = 0 \Rightarrow \csc t = -\frac{1}{2}$ or $\tan t = -1 \Rightarrow t = \frac{3\pi}{4}, \frac{7\pi}{4}$ $\{$ since $\csc t \ne -\frac{1}{2}$ $\}$

$\boxed{61}$ $\sin^2 t - 4\sin t + 1 = 0 \Rightarrow \sin t = \frac{4 \pm \sqrt{12}}{2} = 2 \pm \sqrt{3}$.

$(2 + \sqrt{3}) > 1$ is not in the range of the sine, so $\sin t = 2 - \sqrt{3} \Rightarrow$

$$t = 15°30' \text{ or } 164°30' \ \{ \text{to the nearest ten minutes} \}$$

$\boxed{63}$ $\tan^2\theta + 3\tan\theta + 2 = 0 \Rightarrow (\tan\theta + 1)(\tan\theta + 2) = 0 \Rightarrow$

$$\tan\theta = -1, -2 \Rightarrow \theta = 135°, 315°, 116°30', 296°30'$$

$\boxed{65}$ $12\sin^2 u - 5\sin u - 2 = 0 \Rightarrow (3\sin u - 2)(4\sin u + 1) = 0 \Rightarrow \sin u = \frac{2}{3}, -\frac{1}{4} \Rightarrow$

$$u = 41°50', 138°10', 194°30', 345°30'$$

$\boxed{67}$ The top of the wave will be above the sea wall when its height is greater than 12.5.

$y > 12.5 \Rightarrow 25\cos\frac{\pi}{15}t > 12.5 \Rightarrow \cos\frac{\pi}{15}t > \frac{1}{2} \Rightarrow$

$\{$ To visualize this step, it may help to look at a unit circle and draw a vertical line

through 0.5 on the x-axis — $x > 0.5$ is the same as $\cos\frac{\pi}{15}t > 0.5$. $\}$

$-\frac{\pi}{3} < \frac{\pi}{15}t < \frac{\pi}{3} \Rightarrow -5 < t < 5$ $\{$ multiply by $\frac{15}{\pi}$ $\} \Rightarrow$

$y > 12.5$ for about $5 - (-5) = 10$ minutes of each 30-minute period.

$\boxed{69}$ (a)

[1, 25] by [0, 100]

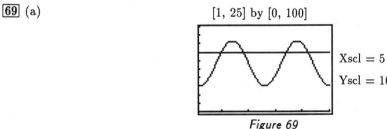

Xscl = 5
Yscl = 10

Figure 69

(b) July: $T(7) = 83°F$; October: $T(10) = 56.5°F$.

(c) Graph $Y_1 = 26.5\sin\left(\frac{\pi}{6}x - \frac{2\pi}{3}\right) + 56.5$ and $Y_2 = 69$. Their graphs intersect at $t \approx 4.94$, 9.06 on $[1, 13]$. The average high temperature is above $69°F$ approximately May through September.

(continued)

(d) A sine function is periodic and varies between a maximum and minimum value. Average monthly high temperatures are also seasonal with a 12-month period. Therefore, a sine function is a reasonable function to model these temperatures.

$\boxed{71}$ $I = \frac{1}{2}I_M$ and $D = 12 \Rightarrow \frac{1}{2}I_M = I_M \sin^3 \frac{\pi}{12}t \Rightarrow \sin^3 \frac{\pi}{12}t = \frac{1}{2} \Rightarrow \sin\frac{\pi}{12}t = \sqrt[3]{\frac{1}{2}} \Rightarrow$

$\frac{\pi}{12}t \approx 0.9169$ and 2.2247 { $\pi - 0.9169 \approx 2.2247$ is the reference angle for 0.9169 in

QII. } $\Rightarrow t \approx 3.50$ and $t \approx 8.50$

$\boxed{73}$ 75% of the maximum intensity $= 0.75\,I_M$

(a) $I > 0.75\,I_M \Rightarrow I_M \sin^3 \frac{\pi}{12}t > 0.75\,I_M \Rightarrow \sin^3 \frac{\pi}{12}t > \frac{3}{4} \Rightarrow \sin\frac{\pi}{12}t > \sqrt[3]{\frac{3}{4}} \Rightarrow$

$1.1398 < \frac{\pi}{12}t < 2.0018 \Rightarrow 4.3538 < t < 7.6462$, or approximately 3.29 hours.

(b) $I > 0.75\,I_M \Rightarrow I_M \sin^2 \frac{\pi}{12}t > 0.75\,I_M \Rightarrow \sin^2 \frac{\pi}{12}t > \frac{3}{4} \Rightarrow \sin\frac{\pi}{12}t > \frac{1}{2}\sqrt{3} \Rightarrow$

$\frac{\pi}{3} < \frac{\pi}{12}t < \frac{2\pi}{3} \Rightarrow 4 < t < 8$, or 4 hours.

$\boxed{75}$ (a) $N(t) = 1000\cos\frac{\pi}{5}t + 4000$, amplitude $= 1000$,

period $= \frac{2\pi}{\pi/5} = 10$ years

(b) $N > 4500 \Rightarrow 1000\cos\frac{\pi}{5}t + 4000 > 4500 \Rightarrow$

$\cos\frac{\pi}{5}t > \frac{1}{2} \Rightarrow$ { The cosine function is greater

than $\frac{1}{2}$ on $[0, \frac{\pi}{3})$ and $(\frac{5\pi}{3}, 2\pi]$. }

$0 \le \frac{\pi}{5}t < \frac{\pi}{3}$ and $\frac{5\pi}{3} < \frac{\pi}{5}t \le 2\pi \Rightarrow$

$0 \le t < \frac{5}{3}$ and $\frac{25}{3} < t \le 10$

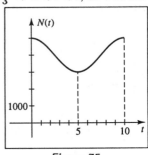

Figure 75

$\boxed{77}$ $\frac{1}{2} + \cos x = 0 \Rightarrow \cos x = -\frac{1}{2} \Rightarrow x = -\frac{4\pi}{3}, -\frac{2\pi}{3}, \frac{2\pi}{3},$ and $\frac{4\pi}{3}$ { for x in $[-2\pi, 2\pi]$ }

for A, B, C, and D, respectively. The corresponding y values are found by using

$y = \frac{1}{2}x + \sin x$ with each of the above values. The points are:

$A(-\frac{4\pi}{3}, -\frac{2\pi}{3} + \frac{1}{2}\sqrt{3})$, $B(-\frac{2\pi}{3}, -\frac{\pi}{3} - \frac{1}{2}\sqrt{3})$, $C(\frac{2\pi}{3}, \frac{\pi}{3} + \frac{1}{2}\sqrt{3})$, and $D(\frac{4\pi}{3}, \frac{2\pi}{3} - \frac{1}{2}\sqrt{3})$

$\boxed{79}$ $I(t) = k \Rightarrow 20\sin(60\pi t - 6\pi) = -10 \Rightarrow \sin(60\pi t - 6\pi) = -\frac{1}{2} \Rightarrow 60\pi t_1 - 6\pi = \frac{7\pi}{6} + 2\pi n$

or $60\pi t_2 - 6\pi = \frac{11\pi}{6} + 2\pi n$ { We use t_1 and t_2 to distinguish between angles

coterminal with $\frac{7\pi}{6}$ and those coterminal with $\frac{11\pi}{6}$. } $\Rightarrow 60t_1 = \frac{43}{6} + 2n$ or

$60t_2 = \frac{47}{6} + 2n \Rightarrow t_1 = \frac{43}{360} + \frac{1}{30}n$ or $t_2 = \frac{47}{360} + \frac{1}{30}n$. We must now find the smallest

positive value of t_1 or t_2. $t_1 > 0 \Rightarrow \frac{1}{30}n > -\frac{43}{360} \Rightarrow n > -\frac{43}{12} \approx -3.58$. The last result

indicates that n must be an integer in the set $\{-3, -2, -1, \ldots\}$. If $n = -3$, then

$t_1 = \frac{7}{360}$. Greater values of n yield greater values of t_1. Similarly, for t_2,

$t_2 > 0 \Rightarrow \frac{1}{30}n > -\frac{47}{360} \Rightarrow n > -\frac{47}{12}$. If $n = -3$, then $t_1 = \frac{11}{360}$. Thus, the smallest exact

value of t for which $I(t) = -10$ is $t = \frac{7}{360}$ sec.

$\boxed{81}$ Graph $y = \cos x$ and $y = 0.3$ on the same coordinate plane. See *Figure 81* on the next page. The points of intersection are located at $x \approx 1.27$, 5.02, and $\cos x$ is less than 0.3 between these values. Therefore, $\cos x \ge 0.3$ on $[0, 1.27] \cup [5.02, 2\pi]$.

[0, 2π] by [−2.09, 2.09]

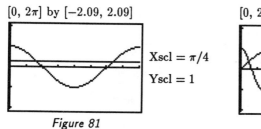

Xscl = π/4
Yscl = 1

Figure 81

[0, 2π] by [−2.09, 2.09]

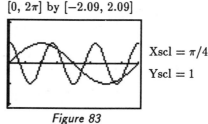

Xscl = π/4
Yscl = 1

Figure 83

⟨83⟩ Graph $y = \cos 3x$ and $y = \sin x$ on the same coordinate plane.

The points of intersection are located at $x \approx 0.39, 1.96, 2.36, 3.53, 5.11, 5.50$.

From the graph, we see that $\cos 3x$ is less than $\sin x$ on

$$(0.39,\ 1.96) \cup (2.36,\ 3.53) \cup (5.11,\ 5.50).$$

⟨85⟩ (a) The largest zero occurs when $x \approx 0.6366$.

(b) As x becomes large, the graph of $f(x) = \cos(1/x)$ approaches the horizontal asymptote $y = 1$.

(c) There appears to be an infinite number of zeros on $[0, c]$ for any $c > 0$.

[0, 3] by [−1.5, 1.5]

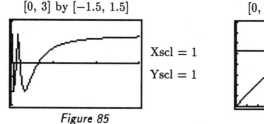

Xscl = 1
Yscl = 1

Figure 85

[0, 12] by [0, 8]

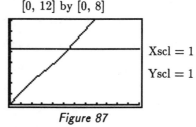

Xscl = 1
Yscl = 1

Figure 87

Note: Exer. 87–90: Graph $Y_1 = M$ and $Y_2 = \theta + e\sin\theta$ and approximate the value of θ such that $Y_1 = Y_2$.

⟨87⟩ Mercury: $Y_1 = 5.241$ and $Y_2 = \theta + 0.206\sin\theta$ intersect when $\theta \approx 5.400$ (radians).

⟨91⟩ Graph $y = \sin 2x$ and $y = 2 - x^2$. From the graph, we see that there are two points of intersection. The x-coordinates of these points are $x \approx -1.48, 1.08$.

[−π, π] by [−2.09, 2.09]

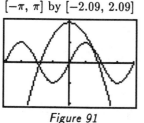

Xscl = π/4
Yscl = 1

Figure 91

[−π, π] by [−2.09, 2.09]

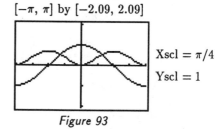

Xscl = π/4
Yscl = 1

Figure 93

⟨93⟩ Graph $y = \ln(1 + \sin^2 x)$ and $y = \cos x$. From the graph, we see that there are two points of intersection. The x-coordinates of these points are $x \approx \pm 1.00$.

$\boxed{97}$ (a) $g = 9.8 \Rightarrow 9.8 = 9.8066(1 - 0.00264\cos 2\theta) \Rightarrow 0.00264\cos 2\theta = 1 - \frac{9.8}{9.8066} \Rightarrow$

$$\cos 2\theta = \frac{0.0066}{(9.8066)(0.00264)} \Rightarrow 2\theta \approx 75.2° \Rightarrow \theta \approx 37.6°$$

(b) At the equator, $g_0 = 9.8066(1 - 0.00264\cos 0°) = 9.8066(0.99736)$.   Since the weight W of a person on the earth's surface is directly proportional to the force of gravity, we have $W = kg$.

At $\theta = 0°$, $W = kg \Rightarrow 150 = kg_0 \Rightarrow k = \frac{150}{g_0} \Rightarrow W = \frac{150}{g_0}g$.

$$W = 150.5 \Rightarrow 150.5 = \frac{150}{g_0}g \Rightarrow 150.5 = \frac{150 \cdot 9.8066(1 - 0.00264\cos 2\theta)}{9.8066(0.99736)} \Rightarrow$$

$$150.5 = \frac{150}{0.99736}(1 - 0.00264\cos 2\theta) \Rightarrow 0.00264\cos 2\theta = 1 - \frac{150.5(0.99736)}{150} \Rightarrow$$

$$\cos 2\theta \approx -0.2593 \Rightarrow 2\theta \approx 105.0° \Rightarrow \theta \approx 52.5°.$$

7.3 Exercises

$\boxed{1}$ *Note:* Use the cofunction formulas with $(\frac{\pi}{2} - u)$ for the argument if you're working with radian measure, $(90° - u)$ if you're working with degree measure.

(a) $\sin 46°37' = \cos(90° - 46°37') = \cos 43°23'$

(b) $\cos 73°12' = \sin(90° - 73°12') = \sin 16°48'$

(c) $\tan\frac{\pi}{6} = \cot(\frac{\pi}{2} - \frac{\pi}{6}) = \cot(\frac{3\pi}{6} - \frac{\pi}{6}) = \cot\frac{2\pi}{6} = \cot\frac{\pi}{3}$

(d) $\sec 17.28° = \csc(90° - 17.28°) = \csc 72.72°$

$\boxed{3}$ (a) $\cos\frac{7\pi}{20} = \sin(\frac{\pi}{2} - \frac{7\pi}{20}) = \sin\frac{3\pi}{20}$ 　　　　(b) $\sin\frac{1}{4} = \cos(\frac{\pi}{2} - \frac{1}{4}) = \cos(\frac{2\pi - 1}{4})$

(c) $\tan 1 = \cot(\frac{\pi}{2} - 1) = \cot(\frac{\pi - 2}{2})$ 　　　　(d) $\csc 0.53 = \sec(\frac{\pi}{2} - 0.53)$

$\boxed{5}$ (a) $\cos\frac{\pi}{4} + \cos\frac{\pi}{6} = \frac{\sqrt{2}}{2} + \frac{\sqrt{3}}{2} = \frac{\sqrt{2} + \sqrt{3}}{2}$

(b) $\cos\frac{5\pi}{12} = \cos(\frac{\pi}{4} + \frac{\pi}{6}) = \cos\frac{\pi}{4}\cos\frac{\pi}{6} - \sin\frac{\pi}{4}\sin\frac{\pi}{6} = \frac{\sqrt{2}}{2} \cdot \frac{\sqrt{3}}{2} - \frac{\sqrt{2}}{2} \cdot \frac{1}{2} = \frac{\sqrt{6} - \sqrt{2}}{4}$

$\boxed{7}$ (a) $\tan 60° + \tan 225° = \sqrt{3} + 1$

(b) $\tan 285° = \tan(60° + 225°) =$

$$\frac{\tan 60° + \tan 225°}{1 - \tan 60°\tan 225°} = \frac{\sqrt{3} + 1}{1 - (\sqrt{3})(1)} \cdot \frac{1 + \sqrt{3}}{1 + \sqrt{3}} = \frac{4 + 2\sqrt{3}}{-2} = -2 - \sqrt{3}$$

$\boxed{9}$ (a) $\sin\frac{3\pi}{4} - \sin\frac{\pi}{6} = \frac{\sqrt{2}}{2} - \frac{1}{2} = \frac{\sqrt{2} - 1}{2}$

(b) $\sin\frac{7\pi}{12} = \sin(\frac{3\pi}{4} - \frac{\pi}{6}) = \sin\frac{3\pi}{4}\cos\frac{\pi}{6} - \cos\frac{3\pi}{4}\sin\frac{\pi}{6} = \frac{\sqrt{2}}{2} \cdot \frac{\sqrt{3}}{2} - (-\frac{\sqrt{2}}{2}) \cdot \frac{1}{2} = \frac{\sqrt{6} + \sqrt{2}}{4}$

$\boxed{11}$ Since the expression is of the form "cos cos plus sin sin", we recognize it as the subtraction formula for the cosine.

$$\cos 48° \cos 23° + \sin 48° \sin 23° = \cos(48° - 23°) = \cos 25°$$

$\boxed{13}$ $\cos 10° \sin 5° - \sin 10° \cos 5° = \sin(5° - 10°) = \sin(-5°)$

$\boxed{15}$ Since we have angle arguments of 2, 3, and -2, we want to change one of them so that we can apply one of the formulas. We recognize that $\cos 2 = \cos(-2)$ and this is probably the simplest change. $\cos 3 \sin(-2) - \cos 2 \sin 3 =$
$$\sin(-2)\cos 3 - \cos(-2)\sin 3 = \sin(-2 - 3) = \sin(-5)$$

$\boxed{17}$ See *Figure 17* for a drawing of angles α and β.

    (a) $\sin(\alpha + \beta) = \sin \alpha \cos \beta + \cos \alpha \sin \beta = \frac{3}{5} \cdot \frac{15}{17} + \frac{4}{5} \cdot \frac{8}{17} = \frac{77}{85}$

    (b) $\cos(\alpha + \beta) = \cos \alpha \cos \beta - \sin \alpha \sin \beta = \frac{4}{5} \cdot \frac{15}{17} - \frac{3}{5} \cdot \frac{8}{17} = \frac{36}{85}$

    (c) Since the sine and cosine of $(\alpha + \beta)$ are positive, $(\alpha + \beta)$ is in QI.

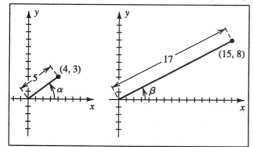

                Figure 17   *Figure 19*

$\boxed{19}$ See *Figure 19* for a drawing of angles α and β.

    (a) $\sin(\alpha + \beta) = \sin \alpha \cos \beta + \cos \alpha \sin \beta = \left(-\frac{4}{5}\right) \cdot \frac{3}{5} + \left(-\frac{3}{5}\right) \cdot \frac{4}{5} = -\frac{24}{25}$

    (b) $\tan(\alpha + \beta) = \dfrac{\tan \alpha + \tan \beta}{1 - \tan \alpha \tan \beta} = \dfrac{\frac{4}{3} + \frac{4}{3}}{1 - \frac{4}{3} \cdot \frac{4}{3}} \cdot \dfrac{9}{9} = \dfrac{12 + 12}{9 - 16} = -\dfrac{24}{7}$

    (c) Since the sine and tangent of $(\alpha + \beta)$ are negative, $(\alpha + \beta)$ is in QIV.

$\boxed{21}$ (a) $\sin(\alpha - \beta) = \sin \alpha \cos \beta - \cos \alpha \sin \beta =$
$$\left(-\frac{\sqrt{21}}{5}\right) \cdot \left(-\frac{3}{5}\right) - \left(-\frac{2}{5}\right) \cdot \left(-\frac{4}{5}\right) = \frac{3\sqrt{21} - 8}{25} \approx 0.23$$

    (b) $\cos(\alpha - \beta) = \cos \alpha \cos \beta + \sin \alpha \sin \beta =$
$$\left(-\frac{2}{5}\right) \cdot \left(-\frac{3}{5}\right) + \left(-\frac{\sqrt{21}}{5}\right) \cdot \left(-\frac{4}{5}\right) = \frac{4\sqrt{21} + 6}{25} \approx 0.97$$

    (c) Since the sine and cosine of $(\alpha - \beta)$ are positive, $(\alpha - \beta)$ is in QI.

$\boxed{23}$ $\sin(\theta + \pi) = \sin \theta \cos \pi + \cos \theta \sin \pi = \sin \theta(-1) + \cos \theta(0) = -\sin \theta$

$\boxed{25}$ $\sin\left(x - \frac{5\pi}{2}\right) = \sin x \cos \frac{5\pi}{2} - \cos x \sin \frac{5\pi}{2} = \sin x(0) - \cos x(1) = -\cos x$

$\boxed{27}$ $\cos(\theta - \pi) = \cos \theta \cos \pi + \sin \theta \sin \pi = \cos \theta(-1) + \sin \theta(0) = -\cos \theta$

$\boxed{29}$ $\cos\left(x + \frac{3\pi}{2}\right) = \cos x \cos \frac{3\pi}{2} - \sin x \sin \frac{3\pi}{2} = \cos x(0) - \sin x(-1) = \sin x$

$\boxed{31}$ $\tan\left(x - \frac{\pi}{2}\right) = \dfrac{\sin\left(x - \frac{\pi}{2}\right)}{\cos\left(x - \frac{\pi}{2}\right)} = \dfrac{\sin x \cos \frac{\pi}{2} - \cos x \sin \frac{\pi}{2}}{\cos x \cos \frac{\pi}{2} + \sin x \sin \frac{\pi}{2}} = \dfrac{-\cos x}{\sin x} = -\cot x$

[33] The tangent of a sum formula won't work here since $\tan \frac{\pi}{2}$ is undefined.

We will use a cofunction identity to verify the identity.

$$\tan\left(\theta + \tfrac{\pi}{2}\right) = \cot\left[\tfrac{\pi}{2} - \left(\theta + \tfrac{\pi}{2}\right)\right] = \cot\left(-\theta\right) = -\cot\theta$$

Alternatively, we could also write $\tan\left(\theta + \tfrac{\pi}{2}\right)$ as $\dfrac{\sin\left(\theta + \tfrac{\pi}{2}\right)}{\cos\left(\theta + \tfrac{\pi}{2}\right)}$ and then simplify.

[37] $\tan\left(u + \tfrac{\pi}{4}\right) = \dfrac{\tan u + \tan\frac{\pi}{4}}{1 - \tan u \tan\frac{\pi}{4}} = \dfrac{\tan u + 1}{1 - \tan u (1)} = \dfrac{1 + \tan u}{1 - \tan u}$

[39] $\cos\left(u + v\right) + \cos\left(u - v\right) = \left(\cos u \cos v - \sin u \sin v\right) + \left(\cos u \cos v + \sin u \sin v\right) =$

$$2\cos u \cos v$$

[41] $\sin\left(u + v\right) \cdot \sin\left(u - v\right) = \left(\sin u \cos v + \cos u \sin v\right) \cdot \left(\sin u \cos v - \cos u \sin v\right)$

$$\{\text{ addition and subtraction formulas for the sine }\}$$

$$= \sin^2 u \cos^2 v - \cos^2 u \sin^2 v$$

$$\{\text{ recognize as the difference of two squares }\}$$

$$= \sin^2 u \left(1 - \sin^2 v\right) - \left(1 - \sin^2 u\right) \sin^2 v$$

$$\{\text{ change to terms only involving sine }\}$$

$$= \sin^2 u - \sin^2 u \sin^2 v - \sin^2 v + \sin^2 u \sin^2 v = \sin^2 u - \sin^2 v$$

[43] $\dfrac{1}{\cot\alpha - \cot\beta} = \dfrac{1}{\dfrac{\cos\alpha}{\sin\alpha} - \dfrac{\cos\beta}{\sin\beta}} = \dfrac{1}{\dfrac{\cos\alpha \sin\beta - \cos\beta \sin\alpha}{\sin\alpha \sin\beta}} = \dfrac{\sin\alpha \sin\beta}{\sin\left(\beta - \alpha\right)}$

[45] $\sin\left(u + v + w\right) = \sin\left[\left(u + v\right) + w\right]$

$$= \sin\left(u + v\right)\cos w + \cos\left(u + v\right)\sin w$$

$$= \left(\sin u \cos v + \cos u \sin v\right)\cos w + \left(\cos u \cos v - \sin u \sin v\right)\sin w$$

$$= \sin u \cos v \cos w + \cos u \sin v \cos w + \cos u \cos v \sin w - \sin u \sin v \sin w$$

[47] The question usually asked here is "why divide by $\sin u \sin v$?" Since we know the

form we want to end up with, we need to "force" the term "$-\sin u \sin v$" to equal

"-1", hence divide all terms by "$\sin u \sin v$."

$$\cot\left(u + v\right) = \dfrac{\cos\left(u + v\right)}{\sin\left(u + v\right)} = \dfrac{\left(\cos u \cos v - \sin u \sin v\right)\left(1/\sin u \sin v\right)}{\left(\sin u \cos v + \cos u \sin v\right)\left(1/\sin u \sin v\right)} = \dfrac{\cot u \cot v - 1}{\cot v + \cot u}$$

[49] $\sin\left(u - v\right) = \sin\left[u + \left(-v\right)\right] = \sin u \cos\left(-v\right) + \cos u \sin\left(-v\right) = \sin u \cos v - \cos u \sin v$

[51] $\dfrac{f(x + h) - f(x)}{h} = \dfrac{\cos\left(x + h\right) - \cos x}{h} = \dfrac{\cos x \cos h - \sin x \sin h - \cos x}{h} =$

$$\dfrac{\cos x \cos h - \cos x}{h} - \dfrac{\sin x \sin h}{h} = \cos x \left(\dfrac{\cos h - 1}{h}\right) - \sin x \left(\dfrac{\sin h}{h}\right)$$

[53] $\sin 4t \cos t = \sin t \cos 4t \Rightarrow \sin 4t \cos t - \sin t \cos 4t = 0 \Rightarrow \sin\left(4t - t\right) = 0 \Rightarrow$

$$\sin 3t = 0 \Rightarrow 3t = \pi n \Rightarrow t = \tfrac{\pi}{3}n. \text{ In } [0, \pi), \; t = 0, \tfrac{\pi}{3}, \tfrac{2\pi}{3}.$$

55 $\cos 5t \cos 2t = -\sin 5t \sin 2t \Rightarrow \cos 5t \cos 2t + \sin 5t \sin 2t = 0 \Rightarrow$

$\cos(5t - 2t) = 0 \Rightarrow \cos 3t = 0 \Rightarrow 3t = \frac{\pi}{2} + \pi n \Rightarrow t = \frac{\pi}{6} + \frac{\pi}{3}n$. In $[0, \pi)$, $t = \frac{\pi}{6}, \frac{\pi}{2}, \frac{5\pi}{6}$.

57 $\tan 2t + \tan t = 1 - \tan 2t \tan t \Rightarrow \dfrac{\tan 2t + \tan t}{1 - \tan 2t \tan t} = 1 \Rightarrow \tan(2t + t) = 1 \Rightarrow$

$\tan 3t = 1 \Rightarrow 3t = \frac{\pi}{4} + \pi n \Rightarrow t = \frac{\pi}{12} + \frac{\pi}{3}n$. In $[0, \pi)$, $t = \frac{\pi}{12}, \frac{5\pi}{12}, \frac{3\pi}{4}$.

However, $\tan 2t$ is undefined if $t = \frac{3\pi}{4}$, so exclude this value of t.

59 (a) $f(x) = \sqrt{3}\cos 2x + \sin 2x$ • $A = \sqrt{(\sqrt{3})^2 + 1^2} = 2$. $\tan C = \dfrac{1}{\sqrt{3}} \Rightarrow C = \frac{\pi}{6}$.

$$f(x) = 2\cos\left(2x - \frac{\pi}{6}\right) = 2\cos\left[2\left(x - \frac{\pi}{12}\right)\right]$$

(b) amplitude $= 2$, period $= \frac{2\pi}{2} = \pi$, phase shift $= \frac{\pi}{12}$

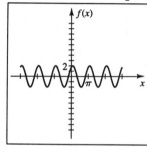

Figure 59

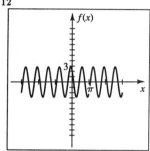

Figure 61

61 (a) $f(x) = 2\cos 3x - 2\sin 3x$ • $A = \sqrt{2^2 + 2^2} = 2\sqrt{2}$. $\tan C = \frac{-2}{2} = -1 \Rightarrow$

$C = -\frac{\pi}{4}$. $f(x) = 2\sqrt{2}\cos\left(3x + \frac{\pi}{4}\right) = 2\sqrt{2}\cos\left[3\left(x + \frac{\pi}{12}\right)\right]$

(b) amplitude $= 2\sqrt{2}$, period $= \frac{2\pi}{3}$, phase shift $= -\frac{\pi}{12}$

63 $y = 50\sin 60\pi t + 40\cos 60\pi t$ • $A = \sqrt{50^2 + 40^2} = 10\sqrt{41}$. $\tan C = \frac{50}{40} \Rightarrow$

$C = \tan^{-1}\frac{5}{4} \approx 0.8961$. $y = 10\sqrt{41}\cos\left(60\pi t - \tan^{-1}\frac{5}{4}\right) \approx 10\sqrt{41}\cos(60\pi t - 0.8961)$.

65 (a) $y = 2\cos t + 3\sin t$; $A = \sqrt{2^2 + 3^2} = \sqrt{13}$; $\tan C = \frac{3}{2} \Rightarrow C \approx 0.98$;

$$y = \sqrt{13}\cos(t - C); \text{ amplitude} = \sqrt{13}, \text{ period} = \frac{2\pi}{1} = 2\pi$$

(b) $y = 0 \Rightarrow \sqrt{13}\cos(t - C) = 0 \Rightarrow t - C = \frac{\pi}{2} + \pi n \Rightarrow$

$t = C + \frac{\pi}{2} + \pi n \approx 2.5536 + \pi n$ for every nonnegative integer n.

67 (a) $p(t) = A\sin\omega t + B\sin(\omega t + \tau)$

$= A\sin\omega t + B(\sin\omega t \cos\tau + \cos\omega t \sin\tau)$

$= A\underline{\sin\omega t} + B\cos\tau \underline{\sin\omega t} + B\sin\tau \underline{\cos\omega t}$

$= (B\sin\tau)\cos\omega t + (A + B\cos\tau)\sin\omega t$

$= a\cos\omega t + b\sin\omega t$ with $a = B\sin\tau$ and $b = A + B\cos\tau$

(b) $C^2 = (B\sin\tau)^2 + (A + B\cos\tau)^2$

$= B^2\sin^2\tau + A^2 + 2AB\cos\tau + B^2\cos^2\tau$

$= A^2 + B^2(\sin^2\tau + \cos^2\tau) + 2AB\cos\tau$

$= A^2 + B^2 + 2AB\cos\tau$

[69] (a) $C^2 = A^2 + B^2 + 2AB \cos \tau \leq A^2 + B^2 + 2AB$, since $\cos \tau \leq 1$ and

$A > 0$, $B > 0$. Thus, $C^2 \leq (A + B)^2$, and hence $C \leq A + B$.

(b) $C = A + B$ if $\cos \tau = 1$, or $\tau = 0$, 2π.

(c) Constructive interference will occur if $C > A$. $C > A \Rightarrow C^2 > A^2 \Rightarrow$

$A^2 + B^2 + 2AB \cos \tau > A^2 \Rightarrow B^2 + 2AB \cos \tau > 0 \Rightarrow B(B + 2A \cos \tau) > 0$.

Since $B > 0$, the product will be positive if $B + 2A \cos \tau > 0$, i.e., $\cos \tau > -\dfrac{B}{2A}$.

[71] Graph $y = 3 \sin 2t + 2 \sin (4t + 1)$. Constructive interference will occur when $y > 3$ or

$y < -3$. From the graph, we see that this occurs on the intervals

$$(-2.97, -2.69), \ (-1.00, -0.37), \ (0.17, 0.46), \ \text{and} \ (2.14, 2.77).$$

$[-\pi, \ \pi]$ by $[-5, \ 5]$

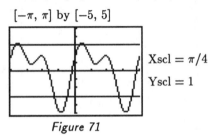

Xscl $= \pi/4$

Yscl $= 1$

Figure 71

7.4 Exercises

[1] From *Figure 1*, $\sin \theta = \frac{4}{5}$ and $\cos \theta = \frac{3}{5}$. Thus, $\sin 2\theta = 2 \sin \theta \cos \theta = 2(\frac{4}{5})(\frac{3}{5}) = \frac{24}{25}$.

$\cos 2\theta = \cos^2\theta - \sin^2\theta = (\frac{3}{5})^2 - (\frac{4}{5})^2 = -\frac{7}{25}$. $\tan 2\theta = \dfrac{\sin 2\theta}{\cos 2\theta} = \dfrac{24/25}{-7/25} = -\dfrac{24}{7}$.

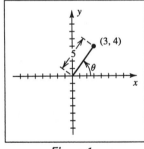

Figure 1

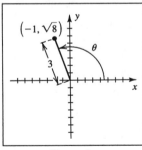

Figure 3

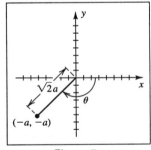

Figure 7

[3] From *Figure 3*, $\sin \theta = \sqrt{8}/3 = \frac{2}{3}\sqrt{2}$ and $\cos \theta = -\frac{1}{3}$.

Thus, $\sin 2\theta = 2 \sin \theta \cos \theta = 2(\frac{2}{3}\sqrt{2})(-\frac{1}{3}) = -\frac{4}{9}\sqrt{2}$.

$\cos 2\theta = \cos^2\theta - \sin^2\theta = (-\frac{1}{3})^2 - (\frac{2}{3}\sqrt{2})^2 = \frac{1}{9} - \frac{8}{9} = -\frac{7}{9}$.

$$\tan 2\theta = \frac{\sin 2\theta}{\cos 2\theta} = \frac{-4\sqrt{2}/9}{-7/9} = \frac{4\sqrt{2}}{7}.$$

$\boxed{5}$ $\sec\theta = \frac{5}{4} \Rightarrow \cos\theta = \frac{4}{5}$. θ acute implies that $\frac{\theta}{2}$ is acute, so all 6 trigonometric functions of $\frac{\theta}{2}$ are positive and we use the " $+$ " sign for the sine and cosine.

$$\sin\frac{\theta}{2} = \sqrt{\frac{1-\cos\theta}{2}} = \sqrt{\frac{1-\frac{4}{5}}{2}} = \sqrt{\frac{\frac{1}{5}}{2}} = \sqrt{\frac{1}{10}\cdot\frac{10}{10}} = \frac{\sqrt{10}}{10}.$$

$$\cos\frac{\theta}{2} = \sqrt{\frac{1+\cos\theta}{2}} = \sqrt{\frac{1+\frac{4}{5}}{2}} = \sqrt{\frac{\frac{9}{5}}{2}} = \sqrt{\frac{9}{10}\cdot\frac{10}{10}} = \frac{3\sqrt{10}}{10}.$$

$$\tan\frac{\theta}{2} = \frac{\sin\frac{\theta}{2}}{\cos\frac{\theta}{2}} = \frac{\sqrt{10}/10}{3\sqrt{10}/10} = \frac{1}{3}.$$

$\boxed{7}$ From *Figure 7* on the preceding page ($a > 0$), $\cos\theta = -\dfrac{a}{\sqrt{2}\,a} = -\dfrac{\sqrt{2}}{2}$ and $\frac{\theta}{2}$ is in QIV.

$$\sin\frac{\theta}{2} = -\sqrt{\frac{1-\cos\theta}{2}} = -\sqrt{\frac{1+\sqrt{2}/2}{2}} = -\sqrt{\frac{2+\sqrt{2}}{4}} = -\tfrac{1}{2}\sqrt{2+\sqrt{2}}.$$

$$\cos\frac{\theta}{2} = \sqrt{\frac{1+\cos\theta}{2}} = \sqrt{\frac{1-\sqrt{2}/2}{2}} = \sqrt{\frac{2-\sqrt{2}}{4}} = \tfrac{1}{2}\sqrt{2-\sqrt{2}}.$$

$$\tan\frac{\theta}{2} = \frac{1-\cos\theta}{\sin\theta} = \frac{1+\sqrt{2}/2}{-\sqrt{2}/2}\cdot\frac{2}{2} = \frac{2+\sqrt{2}}{-\sqrt{2}} = -\sqrt{2}-1.$$

$\boxed{9}$ (a) $\cos 67°30' = \sqrt{\dfrac{1+\cos 135°}{2}} = \sqrt{\dfrac{1-\sqrt{2}/2}{2}} = \sqrt{\dfrac{2-\sqrt{2}}{4}} = \tfrac{1}{2}\sqrt{2-\sqrt{2}}.$

     (b) $\sin 15° = \sqrt{\dfrac{1-\cos 30°}{2}} = \sqrt{\dfrac{1-\sqrt{3}/2}{2}} = \sqrt{\dfrac{2-\sqrt{3}}{4}} = \tfrac{1}{2}\sqrt{2-\sqrt{3}}.$

     (c) $\tan\dfrac{3\pi}{8} = \dfrac{1-\cos\frac{3\pi}{4}}{\sin\frac{3\pi}{4}} = \dfrac{1+\sqrt{2}/2}{\sqrt{2}/2}\cdot\dfrac{2}{2} = \dfrac{2+\sqrt{2}}{\sqrt{2}} = \sqrt{2}+1.$

$\boxed{13}$ We recognize the product of the sine and the cosine as being of the form of the right side of the double-angle formula for the sine, and apply that formula "in reverse."

$$4\sin\tfrac{x}{2}\cos\tfrac{x}{2} = 2\cdot 2\sin\tfrac{x}{2}\cos\tfrac{x}{2} = 2\sin\left(2\cdot\tfrac{x}{2}\right) = 2\sin x$$

$\boxed{15}$ $(\sin t + \cos t)^2 = \sin^2 t + 2\sin t\cos t + \cos^2 t = (\sin^2 t + \cos^2 t) + (2\sin t\cos t) =$

$$1 + \sin 2t$$

$\boxed{17}$ $\sin 3u = \sin(2u+u)$

$$= \sin 2u\cos u + \cos 2u\sin u$$

$$= (2\sin u\cos u)\cos u + (1-2\sin^2 u)\sin u$$

$$= 2\sin u\cos^2 u + \sin u - 2\sin^3 u$$

$$= 2\sin u(1-\sin^2 u) + \sin u - 2\sin^3 u$$

$$= 2\sin u - 2\sin^3 u + \sin u - 2\sin^3 u$$

$$= 3\sin u - 4\sin^3 u$$

$$= \sin u(3 - 4\sin^2 u)$$

$\boxed{19}$ $\cos 4\theta = \cos\left(2\cdot 2\theta\right) = 2\cos^2 2\theta - 1 = 2(2\cos^2\theta - 1)^2 - 1 =$

$$2(4\cos^4\theta - 4\cos^2\theta + 1) - 1 = 8\cos^4\theta - 8\cos^2\theta + 1$$

$\boxed{21}$ $\sin^4 t = (\sin^2 t)^2 \quad = \left(\dfrac{1 - \cos 2t}{2}\right)^2$

$$= \tfrac{1}{4}(1 - 2\cos 2t + \cos^2 2t)$$

$$= \tfrac{1}{4} - \tfrac{1}{2}\cos 2t + \tfrac{1}{4}\left(\dfrac{1 + \cos 4t}{2}\right)$$

$$= \tfrac{1}{4} - \tfrac{1}{2}\cos 2t + \tfrac{1}{8} + \tfrac{1}{8}\cos 4t$$

$$= \tfrac{3}{8} - \tfrac{1}{2}\cos 2t + \tfrac{1}{8}\cos 4t$$

$\boxed{23}$ We do not have a formula for $\sec 2\theta$, so we will write $\sec 2\theta$ in terms of $\cos 2\theta$ in order to apply the double-angle formula for the cosine.

$$\sec 2\theta = \frac{1}{\cos 2\theta} = \frac{1}{2\cos^2\theta - 1} = \frac{1}{2\left(\dfrac{1}{\sec^2\theta}\right) - 1} = \frac{1}{\dfrac{2 - \sec^2\theta}{\sec^2\theta}} = \frac{\sec^2\theta}{2 - \sec^2\theta}$$

$\boxed{25}$ We need to match the arguments of the trigonometric functions involved—that is, either write both of them in terms of $2t$ or in terms of $4t$. Converting $\cos 4t$ to an expression with $2t$ as the angle argument gives us

$$2\sin^2 2t + \cos 4t = 2\sin^2 2t + \cos\left(2\cdot 2t\right) = 2\sin^2 2t + (1 - 2\sin^2 2t) = 1.$$

Alternatively, if we write both arguments in terms of $4t$, we have

$$2\sin^2 2t + \cos 4t = 2\cdot\frac{1 - \cos 4t}{2} + \cos 4t = 1 - \cos 4t + \cos 4t = 1.$$

$\boxed{27}$ $\tan 3u = \tan\left(2u + u\right) = \dfrac{\tan 2u + \tan u}{1 - \tan 2u \tan u}$

$$= \frac{\dfrac{2\tan u}{1 - \tan^2 u} + \tan u}{1 - \dfrac{2\tan u}{1 - \tan^2 u}\cdot\tan u}$$

$$= \frac{\dfrac{2\tan u + \tan u - \tan^3 u}{1 - \tan^2 u}}{\dfrac{1 - \tan^2 u - 2\tan^2 u}{1 - \tan^2 u}}$$

$$= \frac{3\tan u - \tan^3 u}{1 - 3\tan^2 u}$$

$$= \frac{\tan u\left(3 - \tan^2 u\right)}{1 - 3\tan^2 u}$$

$\boxed{29}$ $\cos^4\dfrac{\theta}{2} = \left(\cos^2\dfrac{\theta}{2}\right)^2 = \left(\dfrac{1 + \cos\theta}{2}\right)^2 = \dfrac{1 + 2\cos\theta + \cos^2\theta}{4} = \tfrac{1}{4} + \tfrac{1}{2}\cos\theta + \tfrac{1}{4}\left(\dfrac{1 + \cos 2\theta}{2}\right) =$

$$\tfrac{1}{4} + \tfrac{1}{2}\cos\theta + \tfrac{1}{8} + \tfrac{1}{8}\cos 2\theta = \tfrac{3}{8} + \tfrac{1}{2}\cos\theta + \tfrac{1}{8}\cos 2\theta$$

$\boxed{33}$ $\sin 2t + \sin t = 0 \Rightarrow 2\sin t \cos t + \sin t = 0 \Rightarrow \sin t (2\cos t + 1) = 0 \Rightarrow$

$$\sin t = 0 \text{ or } \cos t = -\tfrac{1}{2} \Rightarrow t = 0,\ \pi \text{ or } \tfrac{2\pi}{3}, \tfrac{4\pi}{3}$$

$\boxed{35}$ $\cos u + \cos 2u = 0 \Rightarrow \cos u + 2\cos^2 u - 1 = 0 \Rightarrow 2\cos^2 u + \cos u - 1 = 0 \Rightarrow$

$$(2\cos u - 1)(\cos u + 1) = 0 \Rightarrow \cos u = \tfrac{1}{2},\ -1 \Rightarrow u = \tfrac{\pi}{3}, \tfrac{5\pi}{3},\ \pi$$

$\boxed{37}$ A first approach uses the concept that if $\tan \alpha = \tan \beta$, then $\alpha = \beta + \pi n$.

$\tan 2x = \tan x \Rightarrow 2x = x + \pi n \Rightarrow x = \pi n \Rightarrow x = 0,\ \pi$.

Another approach is: $\tan 2x = \tan x \Rightarrow \dfrac{\sin 2x}{\cos 2x} = \dfrac{\sin x}{\cos x} \Rightarrow \sin 2x \cos x = \sin x \cos 2x \Rightarrow$

$\sin 2x \cos x - \sin x \cos 2x = 0 \Rightarrow \sin(2x - x) = 0 \Rightarrow \sin x = 0 \Rightarrow x = 0,\ \pi$.

$\boxed{39}$ $\sin \tfrac{1}{2}u + \cos u = 1 \Rightarrow \sin \tfrac{1}{2}u + \cos\left[2 \cdot (\tfrac{1}{2}u)\right] = 1 \Rightarrow \sin \tfrac{1}{2}u + (1 - 2\sin^2 \tfrac{1}{2}u) = 1 \Rightarrow$

$\sin \tfrac{1}{2}u - 2\sin^2 \tfrac{1}{2}u = 0 \Rightarrow \sin \tfrac{1}{2}u\,(1 - 2\sin \tfrac{1}{2}u) = 0 \Rightarrow \sin \tfrac{1}{2}u = 0, \tfrac{1}{2} \Rightarrow$

$$\tfrac{1}{2}u = 0, \tfrac{\pi}{6}, \tfrac{5\pi}{6} \Rightarrow u = 0, \tfrac{\pi}{3}, \tfrac{5\pi}{3}$$

$\boxed{41}$ $\sqrt{a^2 + b^2}\,\sin(u + v) = \sqrt{a^2 + b^2}\,\sin u \cos v + \sqrt{a^2 + b^2}\,\cos u \sin v = a \sin u + b \cos u$

$\{$ equate coefficients of $\sin u$ and $\cos u$ $\} \Rightarrow a = \sqrt{a^2 + b^2}\,\cos v$ and $b = \sqrt{a^2 + b^2}\,\sin v$

$\Rightarrow \cos v = \dfrac{a}{\sqrt{a^2 + b^2}}$ and $\sin v = \dfrac{b}{\sqrt{a^2 + b^2}}$. Since $0 < u < \tfrac{\pi}{2}$, $\sin u > 0$ and $\cos v > 0$.

Now $a > 0$ and $b > 0$ combine with the above to imply that $\cos v > 0$ and $\sin v > 0$.

Thus, $0 < v < \tfrac{\pi}{2}$.

$\boxed{43}$ (a) $\cos 2x + 2\cos x = 0 \Rightarrow 2\cos^2 x + 2\cos x - 1 = 0 \Rightarrow$

$$\cos x = \frac{-2 \pm \sqrt{12}}{4} = \frac{-1 \pm \sqrt{3}}{2} \approx 0.366 \left\{ \cos x \neq \frac{-1 - \sqrt{3}}{2} < -1 \right\}.$$

Thus, $x \approx 1.20$ and 5.09.

(b) $\sin 2x + \sin x = 0 \Rightarrow 2\sin x \cos x + \sin x = 0 \Rightarrow \sin x (2\cos x + 1) = 0 \Rightarrow \sin x = 0$ or

$\cos x = -\tfrac{1}{2} \Rightarrow x = 0,\ \pi,\ 2\pi$ or $\tfrac{2\pi}{3}, \tfrac{4\pi}{3}$. $P(\tfrac{2\pi}{3}, -1.5)$, $Q(\pi, -1)$, $R(\tfrac{4\pi}{3}, -1.5)$

$\boxed{45}$ (a) $\cos 3x - 3\cos x = 0 \Rightarrow 4\cos^3 x - 3\cos x - 3\cos x = 0 \Rightarrow$

$4\cos^3 x - 6\cos x = 0 \Rightarrow 2\cos x\,(2\cos^2 x - 3) = 0 \Rightarrow \cos x = 0,\ \pm\sqrt{3/2} \Rightarrow$

$$x = -\tfrac{3\pi}{2}, -\tfrac{\pi}{2}, \tfrac{\pi}{2}, \tfrac{3\pi}{2} \left\{ \cos x \neq \pm\sqrt{3/2} \text{ since } \sqrt{3/2} > 1 \right\}$$

(b) $\sin 3x - \sin x = 0 \Rightarrow 3\sin x - 4\sin^3 x - \sin x = 0 \Rightarrow 4\sin^3 x - 2\sin x = 0 \Rightarrow$

$2\sin x\,(2\sin^2 x - 1) = 0 \Rightarrow \sin x = 0,\ \pm 1/\sqrt{2} \Rightarrow$

$$x = 0,\ \pm\pi,\ \pm 2\pi,\ \pm\tfrac{\pi}{4},\ \pm\tfrac{3\pi}{4},\ \pm\tfrac{5\pi}{4},\ \pm\tfrac{7\pi}{4}$$

47 (a) Let $y = \overline{BC}$. Form a right triangle with hypotenuse y, side opposite θ, 20, and

side adjacent θ, x. $\sin\theta = \frac{20}{y} \Rightarrow y = \frac{20}{\sin\theta}$. $\cos\theta = \frac{x}{y} \Rightarrow x = y\cos\theta = \frac{20\cos\theta}{\sin\theta}$.

Now $d = (40 - x) + y = 40 - \frac{20\cos\theta}{\sin\theta} + \frac{20}{\sin\theta} = 20\left(\frac{1 - \cos\theta}{\sin\theta}\right) + 40 = 20\tan\frac{\theta}{2} + 40$.

(b) $50 = 20\tan\frac{\theta}{2} + 40 \Rightarrow \tan\frac{\theta}{2} = \frac{1}{2} \Rightarrow \frac{1 - \cos\theta}{\sin\theta} = \frac{1}{2} \Rightarrow 2 - 2\cos\theta = \sin\theta \Rightarrow$

$4 - 8\cos\theta + 4\cos^2\theta = \sin^2\theta = 1 - \cos^2\theta \Rightarrow 5\cos^2\theta - 8\cos\theta + 3 = 0 \Rightarrow$

$(5\cos\theta - 3)(\cos\theta - 1) = 0 \Rightarrow \cos\theta = \frac{3}{5}, 1.$ { $\cos\theta = 1 \Rightarrow \theta = 0$ and 0 is

extraneous }. $\cos\theta = \frac{3}{5} \Rightarrow \sin\theta = \frac{4}{5}$ and $y = \frac{20}{4/5} = 25.$ $\cos\theta = \frac{x}{y}$ and

$\cos\theta = \frac{3}{5} \Rightarrow \frac{x}{25} = \frac{3}{5} \Rightarrow x = 15$, which means that B would be 25 miles from A.

49 (a) From Example 8, the area A of a cross section is

$A = \frac{1}{2}(\text{side})^2(\text{sine of included angle}) = \frac{1}{2}(\frac{1}{2})^2\sin\theta = \frac{1}{8}\sin\theta.$

The volume $V = (\text{length of gutter})(\text{area of cross section}) = 20(\frac{1}{8}\sin\theta) = \frac{5}{2}\sin\theta.$

(b) $V = 2 \Rightarrow \frac{5}{2}\sin\theta = 2 \Rightarrow \sin\theta = \frac{4}{5} \Rightarrow \theta \approx 53.13°.$

51 (a) Let $y = \overline{DB}$ and x denote the distance from D to the midpoint of $\overline{BC}$.

$\sin\frac{\theta}{2} = \frac{b/2}{y} \Rightarrow y = \frac{b}{2} \cdot \frac{1}{\sin(\theta/2)}$ and $\tan\frac{\theta}{2} = \frac{b/2}{x} \Rightarrow x = \frac{b}{2} \cdot \frac{\cos(\theta/2)}{\sin(\theta/2)}.$

$l = (a - x) + y = a - \frac{b}{2} \cdot \frac{\cos(\theta/2)}{\sin(\theta/2)} + \frac{b}{2} \cdot \frac{1}{\sin(\theta/2)} = a + \frac{b}{2} \cdot \frac{1 - \cos(\theta/2)}{\sin(\theta/2)} =$

$$a + \frac{b}{2}\tan\left(\frac{\theta/2}{2}\right) = a + \frac{b}{2}\tan\frac{\theta}{4}.$$

(b) $a = 10$ mm, $b = 6$ mm, and $\theta = 156° \Rightarrow l = 10 + 3\tan 39° \approx 12.43$ mm.

53 The graph of f appears to be that of $y = g(x) = \tan x$.

$$\frac{\sin 2x + \sin x}{\cos 2x + \cos x + 1} = \frac{2\sin x\cos x + \sin x}{(2\cos^2 x - 1) + \cos x + 1} = \frac{\sin x(2\cos x + 1)}{\cos x(2\cos x + 1)} = \frac{\sin x}{\cos x} = \tan x$$

57 Graph $Y_1 = 1/\sin(0.25x + 1)$ and $Y_2 = 1.5 - \cos 2x$ on $[-\pi, \pi]$. There are four points
of intersection. They occur at $x \approx -2.03, -0.72, 0.58, 2.62$.

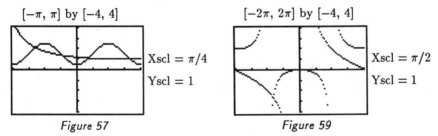

$[-\pi, \pi]$ by $[-4, 4]$ $[-2\pi, 2\pi]$ by $[-4, 4]$

Xscl $= \pi/4$ Xscl $= \pi/2$
Yscl $= 1$ Yscl $= 1$

Figure 57   Figure 59

59 Graph $Y_1 = 2/\tan(.25x)$ and $Y_2 = 1 - 1/\cos(.5x)$ on $[-2\pi, 2\pi]$ in Dot mode. There
is one point of intersection. It occurs at $x \approx -2.59$.

Note: We will reference the product-to-sum formulas as [P1]–[P4] and the sum-to-product formulas as [S1]–[S4] in the order they appear in the text. The formulas $\cos(-kx) = \cos kx$ and $\sin(-kx) = -\sin kx$ will be used without mention.

$\boxed{1}$ $\sin 7t \sin 3t = $ [P4] $\frac{1}{2}[\cos(7t - 3t) - \cos(7t + 3t)] = \frac{1}{2}\cos 4t - \frac{1}{2}\cos 10t$

$\boxed{3}$ $\cos 6u \cos(-4u) = $ [P3] $\frac{1}{2}\{\cos[6u + (-4u)] + \cos[6u - (-4u)]\} = \frac{1}{2}\cos 2u + \frac{1}{2}\cos 10u$

$\boxed{5}$ $2\sin 9\theta \cos 3\theta = $ [P1] $2 \cdot \frac{1}{2}[\sin(9\theta + 3\theta) + \sin(9\theta - 3\theta)] = \sin 12\theta + \sin 6\theta$

$\boxed{7}$ $3\cos x \sin 2x = $ [P2] $3 \cdot \frac{1}{2}[\sin(x + 2x) - \sin(x - 2x)] = \frac{3}{2}\sin 3x - \frac{3}{2}\sin(-x) =$

$$\frac{3}{2}\sin 3x + \frac{3}{2}\sin x$$

$\boxed{9}$ $\sin 6\theta + \sin 2\theta = $ [S1] $2\sin\dfrac{6\theta + 2\theta}{2} \cos\dfrac{6\theta - 2\theta}{2} = 2\sin 4\theta \cos 2\theta$

$\boxed{11}$ $\cos 5x - \cos 3x = $ [S4] $-2\sin\dfrac{5x + 3x}{2} \sin\dfrac{5x - 3x}{2} = -2\sin 4x \sin x$

$\boxed{13}$ $\sin 3t - \sin 7t = $ [S2] $2\cos\dfrac{3t + 7t}{2} \sin\dfrac{3t - 7t}{2} = 2\cos 5t \sin(-2t) = -2\cos 5t \sin 2t$

$\boxed{15}$ $\cos x + \cos 2x = $ [S3] $2\cos\dfrac{x + 2x}{2} \cos\dfrac{x - 2x}{2} = 2\cos\frac{3}{2}x \cos(-\frac{1}{2}x) = 2\cos\frac{3}{2}x \cos\frac{1}{2}x$

$\boxed{17}$ $\dfrac{\sin 4t + \sin 6t}{\cos 4t - \cos 6t} = \dfrac{\text{[S1] } 2\sin 5t \cos(-t)}{\text{[S4] } -2\sin 5t \sin(-t)} = \dfrac{\cos t}{\sin t} = \cot t$

$\boxed{19}$ $\dfrac{\sin u + \sin v}{\cos u + \cos v} = \dfrac{\text{[S1] } 2\sin\frac{1}{2}(u + v) \cos\frac{1}{2}(u - v)}{\text{[S3] } 2\cos\frac{1}{2}(u + v) \cos\frac{1}{2}(u - v)} = \dfrac{\sin\frac{1}{2}(u + v)}{\cos\frac{1}{2}(u + v)} = \tan\frac{1}{2}(u + v)$

$\boxed{21}$ $\dfrac{\sin u - \sin v}{\sin u + \sin v} = \dfrac{\text{[S2] } 2\cos\frac{1}{2}(u + v) \sin\frac{1}{2}(u - v)}{\text{[S1] } 2\sin\frac{1}{2}(u + v) \cos\frac{1}{2}(u - v)} = \cot\frac{1}{2}(u + v) \tan\frac{1}{2}(u - v) = \dfrac{\tan\frac{1}{2}(u - v)}{\tan\frac{1}{2}(u + v)}$

$\boxed{23}$ Since the arguments on the right side are all even multiples of x ($2x$, $4x$, and $6x$), we begin by grouping the terms with the odd multiples of x together, and operate on them with a product-to-sum formula, which will convert these expressions to expressions with even multiples of x.

$$4\cos x \cos 2x \sin 3x = 2\cos 2x (2\sin 3x \cos x)$$

$$= 2\cos 2x ([P1] \sin 4x + \sin 2x)$$

$$= (2\cos 2x \sin 4x) + (2\cos 2x \sin 2x)$$

$$= [[P2] \sin 6x - \sin(-2x)] + ([P2] \sin 4x - \sin 0)$$

$$= \sin 2x + \sin 4x + \sin 6x$$

$\boxed{25}$ $(\sin ax)(\cos bx) = $ [P1] $\frac{1}{2}[\sin(ax+bx) + \sin(ax-bx)] = \frac{1}{2}\sin[(a+b)x] + \frac{1}{2}\sin[(a-b)x]$

$\boxed{27}$ $\sin 5t + \sin 3t = 0 \Rightarrow$ [S1] $2\sin 4t \cos t = 0 \Rightarrow \sin 4t = 0$ or $\cos t = 0 \Rightarrow$

$$4t = \pi n \text{ or } t = \tfrac{\pi}{2} + \pi n \Rightarrow t = \tfrac{\pi}{4}n \text{ \{ which includes } t = \tfrac{\pi}{2} + \pi n \text{ \}}$$

$\boxed{29}$ $\cos x = \cos 3x \Rightarrow \cos x - \cos 3x = 0 \Rightarrow$ [S4] $-2\sin 2x \sin(-x) = 0 \Rightarrow$

$$\sin 2x = 0 \text{ or } \sin x = 0 \Rightarrow 2x = \pi n \text{ or } x = \pi n \Rightarrow x = \tfrac{\pi}{2}n \text{ \{ which includes } x = \pi n \text{ \}}$$

$\boxed{31}$ $\cos 3x + \cos 5x = \cos x \Rightarrow$ [S3] $2\cos 4x \cos(-x) - \cos x = 0 \Rightarrow$

$$\cos x(2\cos 4x - 1) = 0 \Rightarrow \cos x = 0 \text{ or } \cos 4x = \tfrac{1}{2} \Rightarrow$$

$$x = \tfrac{\pi}{2} + \pi n \text{ or } 4x = \tfrac{\pi}{3} + 2\pi n, \tfrac{5\pi}{3} + 2\pi n \Rightarrow x = \tfrac{\pi}{2} + \pi n, \tfrac{\pi}{12} + \tfrac{\pi}{2}n, \tfrac{5\pi}{12} + \tfrac{\pi}{2}n$$

$\boxed{33}$ $\sin 2x - \sin 5x = 0 \Rightarrow$ [S2] $2\cos\tfrac{7}{2}x \sin(-\tfrac{3}{2}x) = 0 \Rightarrow \tfrac{7}{2}x = \tfrac{\pi}{2} + \pi n \text{ or } \tfrac{3}{2}x = \pi n \Rightarrow$

$$x = \tfrac{\pi}{7} + \tfrac{2\pi}{7}n \text{ or } x = \tfrac{2\pi}{3}n$$

$\boxed{35}$ $\cos x + \cos 3x = 0 \Rightarrow$ [S3] $2\cos 2x \cos(-x) = 0 \Rightarrow \cos 2x = 0 \text{ or } \cos x = 0 \Rightarrow$

$$2x = \tfrac{\pi}{2} + \pi n \text{ or } x = \tfrac{\pi}{2} + \pi n \Rightarrow x = \tfrac{\pi}{4} + \tfrac{\pi}{2}n \text{ or } x = \tfrac{\pi}{2} + \pi n \Rightarrow$$

$$x = \tfrac{\pi}{4}, \tfrac{3\pi}{4}, \tfrac{5\pi}{4}, \tfrac{7\pi}{4}, \tfrac{\pi}{2}, \tfrac{3\pi}{2} \text{ for } 0 \le x \le 2\pi$$

$\boxed{37}$ $\sin 3x - \sin x = 0 \Rightarrow$ [S2] $2\cos 2x \sin x = 0 \Rightarrow \cos 2x = 0 \text{ or } \sin x = 0 \Rightarrow$

$$2x = \tfrac{\pi}{2} + \pi n \text{ or } x = \pi n \Rightarrow x = \tfrac{\pi}{4} + \tfrac{\pi}{2}n \text{ or } x = \pi n \Rightarrow$$

$$x = 0, \pm\pi, \pm 2\pi, \pm\tfrac{\pi}{4}, \pm\tfrac{3\pi}{4}, \pm\tfrac{5\pi}{4}, \pm\tfrac{7\pi}{4} \text{ for } -2\pi \le x \le 2\pi.$$

$\boxed{39}$ $f(x) = \sin\left(\tfrac{\pi n}{l}x\right)\cos\left(\tfrac{k\pi n}{l}t\right)$

$$= \text{[P1]} \ \tfrac{1}{2}\left[\sin\left(\tfrac{\pi n}{l}x + \tfrac{k\pi n}{l}t\right) + \sin\left(\tfrac{\pi n}{l}x - \tfrac{k\pi n}{l}t\right)\right]$$

$$= \tfrac{1}{2}\left[\sin\tfrac{\pi n}{l}(x + kt) + \sin\tfrac{\pi n}{l}(x - kt)\right]$$

$$= \tfrac{1}{2}\sin\tfrac{\pi n}{l}(x + kt) + \tfrac{1}{2}\sin\tfrac{\pi n}{l}(x - kt)$$

$\boxed{41}$ (a) Estimating the x-intercepts, we have $x \approx 0, \pm 1.05, \pm 1.57, \pm 2.09, \pm 3.14$.

(b) $\sin 4x + \sin 2x = 2\sin 3x \cos x = 0 \Rightarrow \sin 3x = 0 \text{ or } \cos x = 0$.

$$\sin 3x = 0 \Rightarrow 3x = \pi n \Rightarrow x = \tfrac{\pi}{3}n \Rightarrow x = 0, \pm\tfrac{\pi}{3}, \pm\tfrac{2\pi}{3}, \pm\pi. \ \cos x = 0 \Rightarrow x = \pm\tfrac{\pi}{2}.$$

The x-intercepts are $0, \pm\tfrac{\pi}{3}, \pm\tfrac{\pi}{2}, \pm\tfrac{2\pi}{3}, \pm\pi$.

$[-\pi, \pi]$ by $[-2.09, 2.09]$

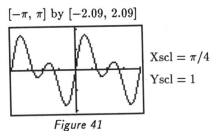

Xscl $= \pi/4$

Yscl $= 1$

Figure 41

43 Graphing on an interval of $[-\pi, \pi]$ gives us a figure that resembles a tangent function. It appears that there is a vertical asymptote at about 0.78. Recognizing that this value is about $\frac{\pi}{4}$, we might make the conjecture that we have "halved" the period of the tangent and that the graph of $f(x) = \dfrac{\sin x + \sin 2x + \sin 3x}{\cos x + \cos 2x + \cos 3x}$ appears to be that of $y = g(x) = \tan 2x$. Verifying this identity, we have

$$\frac{\sin x + \sin 2x + \sin 3x}{\cos x + \cos 2x + \cos 3x} = \frac{\sin 2x + (\sin 3x + \sin x)}{\cos 2x + (\cos 3x + \cos x)} = \frac{\sin 2x + 2\sin 2x \cos x}{\cos 2x + 2\cos 2x \cos x} =$$

$$\frac{\sin 2x(1 + 2\cos x)}{\cos 2x(1 + 2\cos x)} = \frac{\sin 2x}{\cos 2x} = \tan 2x.$$

7.6 Exercises

1 (a) $\sin^{-1}\left(-\frac{\sqrt{2}}{2}\right) = -\frac{\pi}{4}$ since $\sin\left(-\frac{\pi}{4}\right) = -\frac{\sqrt{2}}{2}$ and $-\frac{\pi}{2} \le -\frac{\pi}{4} \le \frac{\pi}{2}$

 (b) $\cos^{-1}\left(-\frac{1}{2}\right) = \frac{2\pi}{3}$ since $\cos\frac{2\pi}{3} = -\frac{1}{2}$ and $0 \le \frac{2\pi}{3} \le \pi$

 (c) $\tan^{-1}(-\sqrt{3}) = -\frac{\pi}{3}$ since $\tan\left(-\frac{\pi}{3}\right) = -\sqrt{3}$ and $-\frac{\pi}{2} < -\frac{\pi}{3} < \frac{\pi}{2}$

3 (a) $\arcsin\frac{\sqrt{3}}{2} = \frac{\pi}{3}$ since $\sin\frac{\pi}{3} = \frac{\sqrt{3}}{2}$ and $-\frac{\pi}{2} \le \frac{\pi}{3} \le \frac{\pi}{2}$

 (b) $\arccos\frac{\sqrt{2}}{2} = \frac{\pi}{4}$ since $\cos\frac{\pi}{4} = \frac{\sqrt{2}}{2}$ and $0 \le \frac{\pi}{4} \le \pi$

 (c) $\arctan\frac{1}{\sqrt{3}} = \frac{\pi}{6}$ since $\tan\frac{\pi}{6} = \frac{1}{\sqrt{3}}$ and $-\frac{\pi}{2} < \frac{\pi}{6} < \frac{\pi}{2}$

5 (a) $\sin^{-1}\frac{\pi}{3}$ is <u>not defined</u> since $\frac{\pi}{3} > 1$, i.e., $\frac{\pi}{3} \notin [-1, 1]$

 (b) $\cos^{-1}\frac{\pi}{2}$ is <u>not defined</u> since $\frac{\pi}{2} > 1$, i.e., $\frac{\pi}{2} \notin [-1, 1]$

 (c) $\tan^{-1}1 = \frac{\pi}{4}$ since $\tan\frac{\pi}{4} = 1$ and $-\frac{\pi}{2} < \frac{\pi}{4} < \frac{\pi}{2}$

Note: Exercises 7–10 refer to the boxed properties of $\sin^{-1}$, $\cos^{-1}$, and $\tan^{-1}$.

7 (a) $\sin[\arcsin(-\frac{3}{10})] = -\frac{3}{10}$ since $-1 \le -\frac{3}{10} \le 1$

 (b) $\cos(\arccos\frac{1}{2}) = \frac{1}{2}$ since $-1 \le \frac{1}{2} \le 1$

 (c) $\tan(\arctan 14) = 14$ since $\tan(\arctan x) = x$ for every x

9 (a) $\sin^{-1}(\sin\frac{\pi}{3}) = \frac{\pi}{3}$ since $-\frac{\pi}{2} \le \frac{\pi}{3} \le \frac{\pi}{2}$ (b) $\cos^{-1}\left[\cos\left(\frac{5\pi}{6}\right)\right] = \frac{5\pi}{6}$ since $0 \le \frac{5\pi}{6} \le \pi$

 (c) $\tan^{-1}\left[\tan\left(-\frac{\pi}{6}\right)\right] = -\frac{\pi}{6}$ since $-\frac{\pi}{2} < -\frac{\pi}{6} < \frac{\pi}{2}$

11 (a) $\arcsin\left(\sin\frac{5\pi}{4}\right) = \arcsin\left(-\frac{\sqrt{2}}{2}\right) = -\frac{\pi}{4}$

 (b) $\arccos\left(\cos\frac{5\pi}{4}\right) = \arccos\left(-\frac{\sqrt{2}}{2}\right) = \frac{3\pi}{4}$ (c) $\arctan\left(\tan\frac{7\pi}{4}\right) = \arctan(-1) = -\frac{\pi}{4}$

$\boxed{13}$ (a) $\sin\left[\cos^{-1}\left(-\tfrac{1}{2}\right)\right] = \sin\tfrac{2\pi}{3} = \dfrac{\sqrt{3}}{2}$ (b) $\cos\left(\tan^{-1}1\right) = \cos\tfrac{\pi}{4} = \dfrac{\sqrt{2}}{2}$

     (c) $\tan\left[\sin^{-1}(-1)\right] = \tan\left(-\tfrac{\pi}{2}\right)$, which is <u>not defined</u>.

$\boxed{15}$ (a) Let $\theta = \sin^{-1}\tfrac{2}{3}$. From *Figure 15(a)*, $\cot\left(\sin^{-1}\tfrac{2}{3}\right) = \cot\theta = \dfrac{x}{y} = \dfrac{\sqrt{5}}{2}$.

     (b) Let $\theta = \tan^{-1}\left(-\tfrac{3}{5}\right)$. From *Figure 15(b)*, $\sec\left[\tan^{-1}\left(-\tfrac{3}{5}\right)\right] = \sec\theta = \dfrac{r}{x} = \dfrac{\sqrt{34}}{5}$.

     (c) Let $\theta = \cos^{-1}\left(-\tfrac{1}{4}\right)$. From *Figure 15(c)*, $\csc\left[\cos^{-1}\left(-\tfrac{1}{4}\right)\right] = \csc\theta = \dfrac{r}{y} = \dfrac{4}{\sqrt{15}}$.

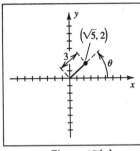
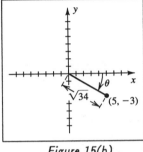
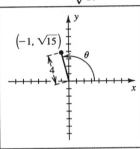

Figure 15(a)   *Figure 15(b)*   *Figure 15(c)*

$\boxed{17}$ (a) $\sin\left(\arcsin\tfrac{1}{2} + \arccos 0\right) = \sin\left(\tfrac{\pi}{6} + \tfrac{\pi}{2}\right) = \sin\tfrac{2\pi}{3} = \dfrac{\sqrt{3}}{2}$.

     (b) Remember, $\arctan\left(-\tfrac{3}{4}\right)$ and $\arcsin\tfrac{4}{5}$ are <u>angles</u>. To abbreviate this solution, we let $\alpha = \arctan\left(-\tfrac{3}{4}\right)$ and $\beta = \arcsin\tfrac{4}{5}$. Using the difference identity for the cosine and figures as in Exercises 15 and 16, we have $\cos\left[\arctan\left(-\tfrac{3}{4}\right) - \arcsin\tfrac{4}{5}\right]$
$$= \cos(\alpha - \beta) = \cos\alpha\,\cos\beta + \sin\alpha\,\sin\beta = \tfrac{4}{5}\cdot\tfrac{3}{5} + \left(-\tfrac{3}{5}\right)\cdot\tfrac{4}{5} = 0.$$

     (c) Let $\alpha = \arctan\tfrac{4}{3}$ and $\beta = \arccos\tfrac{8}{17}$. $\tan\left(\arctan\tfrac{4}{3} + \arccos\tfrac{8}{17}\right) =$
$$\tan(\alpha + \beta) = \frac{\tan\alpha + \tan\beta}{1 - \tan\alpha\,\tan\beta} = \frac{\tfrac{4}{3} + \tfrac{15}{8}}{1 - \tfrac{4}{3}\cdot\tfrac{15}{8}}\cdot\frac{24}{24} = \frac{32 + 45}{24 - 60} = -\frac{77}{36}.$$

$\boxed{19}$ (a) We first recognize this expression as being of the form $\sin(\textit{twice an angle})$, where $\arccos\left(-\tfrac{3}{5}\right)$ is the angle. Let $\alpha = \arccos\left(-\tfrac{3}{5}\right)$. It may help to draw a figure as in Exercise 15 to determine that α is a second quadrant angle and that $\sin\alpha = \tfrac{4}{5}$. Applying the double-angle formula for the sine gives us
$$\sin\left[2\arccos\left(-\tfrac{3}{5}\right)\right] = \sin 2\alpha = 2\sin\alpha\,\cos\alpha = 2\left(\tfrac{4}{5}\right)\left(-\tfrac{3}{5}\right) = -\tfrac{24}{25}.$$

     (b) Let $\alpha = \sin^{-1}\tfrac{15}{17}$. Thus, $\cos\alpha = \tfrac{8}{17}$ and we apply the double-angle formula for the cosine. $\cos\left(2\sin^{-1}\tfrac{15}{17}\right) = \cos 2\alpha = \cos^2\alpha - \sin^2\alpha = \left(\tfrac{8}{17}\right)^2 - \left(\tfrac{15}{17}\right)^2 = -\tfrac{161}{289}$.

     (c) Let $\alpha = \tan^{-1}\tfrac{3}{4}$. Thus, $\tan\alpha = \tfrac{3}{4}$ and we apply the double-angle formula for the tangent. $\tan\left(2\tan^{-1}\tfrac{3}{4}\right) = \tan 2\alpha = \dfrac{2\tan\alpha}{1 - \tan^2\alpha} = \dfrac{2\cdot\tfrac{3}{4}}{1 - \left(\tfrac{3}{4}\right)^2}\cdot\dfrac{16}{16} = \dfrac{24}{16 - 9} = \dfrac{24}{7}$.

21 (a) We first recognize this expression as being of the form $\sin(\text{one-half an angle})$, where $\sin^{-1}(-\frac{7}{25})$ is the angle. Let $\alpha = \sin^{-1}(-\frac{7}{25})$. Hence, α is a fourth quadrant angle and $\cos\alpha = \frac{24}{25}$. $-\frac{\pi}{2} < \alpha < 0 \Rightarrow -\frac{\pi}{4} < \frac{1}{2}\alpha < 0$. We need to know the sign of $\sin\frac{1}{2}\alpha$ in order to correctly apply the half-angle formula for the sine. Since $-\frac{\pi}{4} < \frac{1}{2}\alpha < 0$ {QIV}, $\sin\frac{1}{2}\alpha < 0$.

$$\sin\left[\frac{1}{2}\sin^{-1}\left(-\frac{7}{25}\right)\right] = \sin\frac{1}{2}\alpha = -\sqrt{\frac{1-\cos\alpha}{2}} = -\sqrt{\frac{1-\frac{24}{25}}{2}} = -\sqrt{\frac{1}{50}\cdot\frac{2}{2}} = -\frac{1}{10}\sqrt{2}.$$

(b) Let $\alpha = \tan^{-1}\frac{8}{15}$. $0 < \alpha < \frac{\pi}{2} \Rightarrow 0 < \frac{1}{2}\alpha < \frac{\pi}{4}$ and $\cos\frac{1}{2}\alpha > 0$.

Applying the half-angle formula for the cosine,

$$\cos\left(\frac{1}{2}\tan^{-1}\frac{8}{15}\right) = \cos\frac{1}{2}\alpha = \sqrt{\frac{1+\cos\alpha}{2}} = \sqrt{\frac{1+\frac{15}{17}}{2}} = \sqrt{\frac{16}{17}\cdot\frac{17}{17}} = \frac{4}{17}\sqrt{17}.$$

(c) Let $\alpha = \cos^{-1}\frac{3}{5}$. Applying the half-angle formula for the tangent,

$$\tan\left(\frac{1}{2}\cos^{-1}\frac{3}{5}\right) = \tan\frac{1}{2}\alpha = \frac{1-\cos\alpha}{\sin\alpha} = \frac{1-\frac{3}{5}}{\frac{4}{5}} = \frac{1}{2}.$$

23 We recognize this expression as being of the form $\sin(\text{some angle})$. The angle is *the angle whose tangent is* α. The easiest way to picture this angle is to consider it as being the angle whose ratio of opposite side to adjacent side is x to 1. Let $\alpha = \tan^{-1}x$. From *Figure 23*, $\sin(\tan^{-1}x) = \sin\alpha = \dfrac{x}{\sqrt{x^2+1}}$.

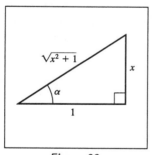

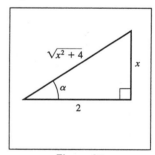

Figure 23   *Figure 25*

25 Let $\alpha = \sin^{-1}\dfrac{x}{\sqrt{x^2+4}}$. From *Figure 25*, $\sec\left(\sin^{-1}\dfrac{x}{\sqrt{x^2+4}}\right) = \sec\alpha = \dfrac{\sqrt{x^2+4}}{2}$.

27 Let $\alpha = \sin^{-1} x$. From *Figure 27*,

$$\sin\left(2\sin^{-1} x\right) = \sin 2\alpha = 2\sin\alpha\cos\alpha = 2 \cdot \frac{x}{1} \cdot \frac{\sqrt{1-x^2}}{1} = 2x\sqrt{1-x^2}.$$

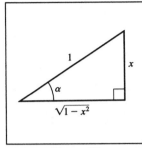

Figure 27

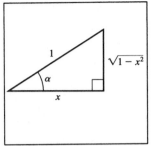

Figure 29

29 Let $\alpha = \arccos x$. $\;0 \le \alpha \le \pi \Rightarrow 0 \le \frac{1}{2}\alpha \le \frac{\pi}{2}$.

Thus $\cos\frac{1}{2}\alpha > 0$ and we use the "+" in the half-angle formula for the cosine.

$$\cos\left(\tfrac{1}{2}\arccos x\right) = \cos\tfrac{1}{2}\alpha = \sqrt{\frac{1+\cos\alpha}{2}} = \sqrt{\frac{1+x}{2}}.$$

31 (a) See text Figure 19. As $x \to -1^+$, $\sin^{-1} x \to$ $-\frac{\pi}{2}$.

     (b) See text Figure 21. As $x \to 1^-$, $\cos^{-1} x \to$ 0 .

     (c) See text Figure 23. As $x \to \infty$, $\tan^{-1} x \to$ $\frac{\pi}{2}$.

33 $y = \sin^{-1} 2x$ •   Horizontally compress $y = \sin^{-1} x$ by a factor of 2.

         Note that the domain changes from $[-1, 1]$ to $[-\frac{1}{2}, \frac{1}{2}]$.

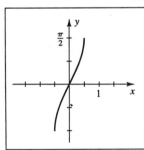

Figure 33

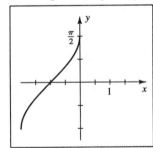

Figure 35

35 $y = \sin^{-1}(x + 1)$ •   Shift $y = \sin^{-1} x$ left 1 unit.

         The domain changes from $[-1, 1]$ to $[-2, 0]$.

37 $y = \cos^{-1}\frac{1}{2}x$ • Horizontally stretch $y = \cos^{-1}x$ by a factor of 2.

The domain changes from $[-1, 1]$ to $[-2, 2]$.

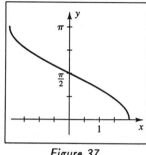

Figure 37

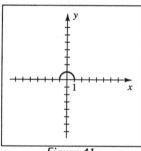

Figure 39

39 $y = 2 + \tan^{-1}x$ • Shift $y = \tan^{-1}x$ up 2 units. The range changes from

$\left(-\frac{\pi}{2}, \frac{\pi}{2}\right)$ to $\left(2 - \frac{\pi}{2}, 2 + \frac{\pi}{2}\right)$, which is approximately $(0.43, 3.57)$.

41 If $\alpha = \arccos x$, then $\cos\alpha = x$, where $0 \le \alpha \le \pi$.

Hence, $y = \sin(\arccos x) = \sin\alpha = \sqrt{1 - \cos^2\alpha} = \sqrt{1 - x^2}$.

Thus, we have the graph of the semicircle $y = \sqrt{1 - x^2}$

on the interval $[-1, 1]$.

Figure 41

43 (a) Since the domain of the arcsine function is $[-1, 1]$, we know that $x - 3$ must be

in that interval. Thus, $-1 \le x - 3 \le 1 \Rightarrow 2 \le x \le 4$.

  (b) The range of the arcsine function is $[-\frac{\pi}{2}, \frac{\pi}{2}]$.

Thus, $-\frac{\pi}{2} \le \sin^{-1}(x-3) \le \frac{\pi}{2} \Rightarrow -\frac{\pi}{4} \le \frac{1}{2}\sin^{-1}(x-3) \le \frac{\pi}{4} \Rightarrow -\frac{\pi}{4} \le y \le \frac{\pi}{4}$.

  (c) $y = \frac{1}{2}\sin^{-1}(x-3) \Rightarrow 2y = \sin^{-1}(x-3) \Rightarrow \sin 2y = x - 3 \Rightarrow x = \sin 2y + 3$

45 (a) $-1 \le \frac{2}{3}x \le 1 \Rightarrow -\frac{3}{2} \le x \le \frac{3}{2}$

  (b) $0 \le \cos^{-1}\frac{2}{3}x \le \pi \Rightarrow 0 \le 4\cos^{-1}\frac{2}{3}x \le 4\pi \Rightarrow 0 \le y \le 4\pi$

  (c) $y = 4\cos^{-1}\frac{2}{3}x \Rightarrow \frac{1}{4}y = \cos^{-1}\frac{2}{3}x \Rightarrow \cos\frac{1}{4}y = \frac{2}{3}x \Rightarrow x = \frac{3}{2}\cos\frac{1}{4}y$

47 $y = -3 - \sin x \Rightarrow y + 3 = -\sin x \Rightarrow -(y+3) = \sin x \Rightarrow x = \sin^{-1}(-y - 3)$

49 $y = 15 - 2\cos x \Rightarrow 2\cos x = 15 - y \Rightarrow \cos x = \frac{1}{2}(15 - y) \Rightarrow x = \cos^{-1}\left[\frac{1}{2}(15 - y)\right]$

51 $\frac{\sin x}{3} = \frac{\sin y}{4} \Rightarrow \sin x = \frac{3}{4}\sin y$. Since $0 < y < \pi$, we know that $\sin y > 0$. Thus, solving

$\sin x = \left(\frac{3}{4}\sin y\right)$ is similar to solving $\sin x = a$, where $0 < a < 1$. Remember that when

solving equations of this form, there are two solutions, a first quadrant angle and a

second quadrant angle. The reference angle for x is $x_R = \sin^{-1}\left(\frac{3}{4}\sin y\right)$, where

$0 < \frac{3}{4}\sin y \le \frac{3}{4} < 1$. If $0 < x < \frac{\pi}{2}$, then $x = x_R$. If $\frac{\pi}{2} < x < \pi$, then $x = \pi - x_R$.

$\boxed{53}$ $\cos^2 x + 2\cos x - 1 = 0 \Rightarrow \cos x = -1 \pm \sqrt{2} \approx 0.4142, \, -2.4142.$

Since $-2.4142 < -1$, $x = \cos^{-1}(-1 + \sqrt{2}) \approx 1.1437$ is one answer.

$$x = 2\pi - \cos^{-1}(-1 + \sqrt{2}) \approx 2\pi - 1.1437 \approx 5.1395 \text{ is the other.}$$

$\boxed{57}$ $15\cos^4 x - 14\cos^2 x + 3 = 0 \Rightarrow (5\cos^2 x - 3)(3\cos^2 x - 1) = 0 \Rightarrow \cos^2 x = \frac{3}{5}, \frac{1}{3} \Rightarrow$

$\cos x = \pm\frac{1}{5}\sqrt{15}, \, \pm\frac{1}{3}\sqrt{3} \Rightarrow x = \cos^{-1}(\pm\frac{1}{5}\sqrt{15}), \, \cos^{-1}(\pm\frac{1}{3}\sqrt{3}).$

$\cos^{-1}\frac{1}{5}\sqrt{15} \approx 0.6847, \, \cos^{-1}(-\frac{1}{5}\sqrt{15}) \approx 2.4569,$

$$\cos^{-1}\tfrac{1}{3}\sqrt{3} \approx 0.9553, \, \cos^{-1}(-\tfrac{1}{3}\sqrt{3}) \approx 2.1863$$

$\boxed{59}$ $6\sin^3\theta + 18\sin^2\theta - 5\sin\theta - 15 = 0 \Rightarrow 6\sin^2\theta(\sin\theta + 3) - 5(\sin\theta + 3) = 0 \Rightarrow$

$$(6\sin^2\theta - 5)(\sin\theta + 3) = 0 \Rightarrow \sin\theta = \pm\tfrac{1}{6}\sqrt{30} \Rightarrow \theta = \sin^{-1}(\pm\tfrac{1}{6}\sqrt{30}) \approx \pm 1.1503$$

$\boxed{61}$ $(\cos x)(15\cos x + 4) = 3 \Rightarrow 15\cos^2 x + 4\cos x - 3 = 0 \Rightarrow (5\cos x + 3)(3\cos x - 1) =$

$0 \Rightarrow \cos x = -\frac{3}{5}, \frac{1}{3} \Rightarrow x = \cos^{-1}(-\frac{3}{5}) \approx 2.2143, \, \cos^{-1}\frac{1}{3} \approx 1.2310.$ These angles are in

the second quadrant and the first quadrant. In $[0, \, 2\pi)$, we must also have a third

quadrant angle and a fourth quadrant angle that satisfy the original equation. These

angles are $2\pi - \cos^{-1}(-\frac{3}{5}) \approx 4.0689$ and $2\pi - \cos^{-1}\frac{1}{3} \approx 5.0522.$

$\boxed{63}$ $3\cos 2x - 7\cos x + 5 = 0 \Rightarrow 3(2\cos^2 x - 1) - 7\cos x + 5 = 0 \Rightarrow$

$6\cos^2 x - 7\cos x + 2 = 0 \Rightarrow (3\cos x - 2)(2\cos x - 1) = 0 \Rightarrow \cos x = \frac{2}{3}, \frac{1}{2} \Rightarrow$

$$x = \cos^{-1}\tfrac{2}{3} \approx 0.8411, \, 2\pi - \cos^{-1}\tfrac{2}{3} \approx 5.4421, \, \tfrac{\pi}{3} \approx 1.0472, \, \tfrac{5\pi}{3} \approx 5.2360.$$

$\boxed{65}$ (a) $S = 4$, $D = 3.5$, $d = 1 \Rightarrow M = \frac{S}{2}\left(1 - \frac{2}{\pi}\tan^{-1}\frac{d}{D}\right) = \frac{4}{2}\left(1 - \frac{2}{\pi}\tan^{-1}\frac{1}{3.5}\right) \approx 1.65$ m

(b) $d = 4 \Rightarrow M = \frac{S}{2}\left(1 - \frac{2}{\pi}\tan^{-1}\frac{d}{D}\right) = \frac{4}{2}\left(1 - \frac{2}{\pi}\tan^{-1}\frac{4}{3.5}\right) \approx 0.92$ m

(c) $d = 10 \Rightarrow M = \frac{S}{2}\left(1 - \frac{2}{\pi}\tan^{-1}\frac{d}{D}\right) = \frac{4}{2}\left(1 - \frac{2}{\pi}\tan^{-1}\frac{10}{3.5}\right) \approx 0.43$ m

$\boxed{67}$ Form a right triangle with the center line of the fairway and half the width of the

fairway as the legs of the triangle.

$$\text{opp} = \tfrac{1}{2}(30) = 15 \text{ and hyp} = 280 \Rightarrow \sin\theta = \tfrac{15}{280} \Rightarrow \theta = \sin^{-1}\tfrac{15}{280} \approx 3.07°$$

$\boxed{69}$ (a) Let β denote the angle by the sailboat with opposite side d and hypotenuse k.

Now $\sin\beta = \frac{d}{k} \Rightarrow \beta = \sin^{-1}\frac{d}{k}.$ Using alternate interior angles,

$$\text{we see that } \alpha + \beta = \theta. \text{ Thus, } \alpha = \theta - \beta = \theta - \sin^{-1}\tfrac{d}{k}.$$

(b) $d = 50$, $k = 210$, and $\theta = 53.4° \Rightarrow \alpha = 53.4° - \sin^{-1}\frac{50}{210} \approx 39.63°$, or $40°$.

Note: The following is a general outline that can be used for verifying trigonometric identities involving inverse trigonometric functions.

(1) Define angles and their ranges—make sure the range of values for one side of the equation is equal to the range of values for the other side.

(2) Choose a trigonometric function T that is one-to-one on the range of values listed in part (1).

(3) Show that $T(\text{LS}) = T(\text{RS})$. Note that $T(\text{LS}) = T(\text{RS}) \not\Rightarrow \text{LS} = \text{RS}$.

(4) Conclude that since T is one-to-one on the range of values, $\text{LS} = \text{RS}$.

$\boxed{71}$ Let $\alpha = \sin^{-1}x$ and $\beta = \tan^{-1}\dfrac{x}{\sqrt{1-x^2}}$ with $-\frac{\pi}{2} < \alpha < \frac{\pi}{2}$ and $-\frac{\pi}{2} < \beta < \frac{\pi}{2}$.

Thus, $\sin\alpha = x$ and $\sin\beta = x$. Since the sine function is one-to-one on $\left(-\frac{\pi}{2}, \frac{\pi}{2}\right)$,

$$\text{we have } \alpha = \beta \text{—that is, } \sin^{-1}x = \tan^{-1}\frac{x}{\sqrt{1-x^2}}.$$

$\boxed{73}$ Let $\alpha = \arcsin(-x)$ and $\beta = \arcsin x$ with $-\frac{\pi}{2} \le \alpha \le \frac{\pi}{2}$ and $-\frac{\pi}{2} \le \beta \le \frac{\pi}{2}$.

Thus, $\sin\alpha = -x$ and $\sin\beta = x$. Consequently, $\sin\alpha = -\sin\beta = \sin(-\beta)$.

Since the sine function is one-to-one on $\left[-\frac{\pi}{2}, \frac{\pi}{2}\right]$,

$$\text{we have } \alpha = -\beta \text{—that is, } \arcsin(-x) = -\arcsin x.$$

$\boxed{75}$ Let $\alpha = \arctan x$ and $\beta = \arctan(1/x)$.

Since $x > 0$, we have $0 < \alpha < \frac{\pi}{2}$ and $0 < \beta < \frac{\pi}{2}$, and hence $0 < \alpha + \beta < \pi$.

Thus, $\tan(\alpha + \beta) = \dfrac{\tan\alpha + \tan\beta}{1 - \tan\alpha\tan\beta} = \dfrac{x + (1/x)}{1 - x\cdot(1/x)} = \dfrac{x + (1/x)}{0}$.

Since the denominator is 0, $\tan(\alpha + \beta)$ is undefined and hence $\alpha + \beta = \frac{\pi}{2}$ since $\frac{\pi}{2}$ is the only value between 0 and π for which the tangent is undefined.

$\boxed{77}$ The domain of $\sin^{-1}(x - 1)$ is $[0, 2]$ and the domain of $\cos^{-1}\frac{1}{2}x$ is $[-2, 2]$.

The domain of f is the intersection of $[0, 2]$ and $[-2, 2]$, i.e., $[0, 2]$.

From the graph, we see that the function is increasing and its range is $\left[-\frac{\pi}{2}, \pi\right]$.

$[-3, 6]$ by $[-2, 4]$ $[-3, 3]$ by $[-2, 2]$

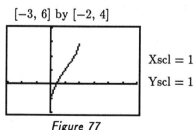   Xscl = 1   Yscl = 1   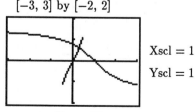   Xscl = 1   Yscl = 1

Figure 77   *Figure 79*

$\boxed{79}$ Graph $y = \sin^{-1}2x$ and $y = \tan^{-1}(1 - x)$.

From the graph, we see that there is one solution at $x \approx 0.29$.

[81] Make the assignments $Y_1 = \sin^{-1}(\sin x/1.52)$, $Y_2 = x - Y_1$, $Y_3 = x + Y_1$, and $Y_4 = .5((\sin Y_2)^2/(\sin Y_3)^2 + (\tan Y_2)^2/(\tan Y_3)^2)$. Now turn *off* Y_1, Y_2, and Y_3—leaving only Y_4 *on* to graph. From the graph, we see that when $f(\theta) = 0.2$, $\theta \approx 1.25$, or approximately $72°$.

$[0,\ \pi/2]$ by $[0,\ 1.05]$

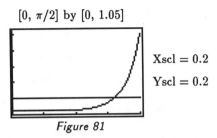

Xscl = 0.2

Yscl = 0.2

Figure 81

[83] Actual distance between x-ticks is equal to $x_A = \frac{3\ \text{units}}{3\ \text{ticks}} = 1$ unit between ticks.

Actual distance between y-ticks is equal to $y_A = \frac{2\ \text{units}}{2\ \text{ticks}} = 1$ unit between ticks.

The ratio is $m_A = \frac{y_A}{x_A} = \frac{1}{1} = 1$. The graph will make an angle of $\theta = \tan^{-1} 1 = 45°$.

$[0,\ 3]$ by $[0,\ 2]$ $[0,\ 3]$ by $[0,\ 4]$

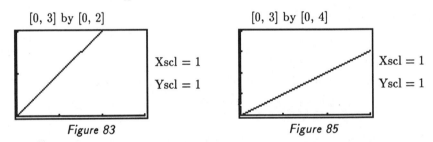

Xscl = 1   Xscl = 1

Yscl = 1   Yscl = 1

Figure 83   *Figure 85*

[85] $x_A = \frac{3\ \text{units}}{3\ \text{ticks}} = 1$, $y_A = \frac{2\ \text{units}}{4\ \text{ticks}} = \frac{1}{2} \Rightarrow m_A = \frac{1/2}{1} = \frac{1}{2} \Rightarrow \theta = \tan^{-1}\frac{1}{2} \approx 26.6°$.

Chapter 7 Review Exercises

[3] $\dfrac{(\sec^2\theta - 1)\cot\theta}{\tan\theta\,\sin\theta + \cos\theta} = \dfrac{(\tan^2\theta)\cot\theta}{\dfrac{\sin\theta}{\cos\theta} \cdot \sin\theta + \cos\theta}$ { Pythagorean and tangent identities }

$= \dfrac{\tan\theta\,(\tan\theta\,\cot\theta)}{\dfrac{\sin^2\theta}{\cos\theta} + \cos\theta}$ { combine terms }

$= \dfrac{\tan\theta}{\dfrac{\sin^2\theta + \cos^2\theta}{\cos\theta}}$ { reciprocal identity, common denominator }

$= \dfrac{\sin\theta/\cos\theta}{1/\cos\theta}$ { Pythagorean and tangent identities }

$= \sin\theta$ { simplify }

$\boxed{5}$ $\dfrac{1}{1+\sin t} = \dfrac{1}{1+\sin t}\cdot\dfrac{1-\sin t}{1-\sin t}$ $\left\{\begin{array}{c}\text{multiply the numerator and the denominator}\\ \text{by conjugate of the denominator}\end{array}\right\}$

$\qquad\qquad = \dfrac{1-\sin t}{1-\sin^2 t}$ $\qquad$ { the difference of two squares }

$\qquad\qquad = \dfrac{1-\sin t}{\cos^2 t}$ $\qquad$ { Pythagorean identity }

$\qquad\qquad = \dfrac{1-\sin t}{\cos t}\cdot\dfrac{1}{\cos t}$ $\qquad$ { break up since we want $\sec t$ on the right side }

$\qquad\qquad = \left(\dfrac{1}{\cos t}-\dfrac{\sin t}{\cos t}\right)\cdot\sec t$ $\quad$ { split up fraction }

$\qquad\qquad = (\sec t - \tan t)\sec t$ $\qquad$ { reciprocal and tangent identities }

$\boxed{6}$ $\dfrac{\sin(\alpha-\beta)}{\cos(\alpha+\beta)} = \dfrac{\sin\alpha\cos\beta - \cos\alpha\sin\beta}{\cos\alpha\cos\beta - \sin\alpha\sin\beta}$ $\left\{\begin{array}{l}\underline{\text{subtraction formula for the sine}}\\ \underline{\text{addition formula for the cosine}}\end{array}\right\}$

The first term in the denominator is $\cos\alpha\cos\beta$, but looking ahead, we see that the first term in the denominator of the expression we want to obtain is 1. Hence, we will divide both the numerator and the denominator by $\cos\alpha\cos\beta$.

$$= \dfrac{(\sin\alpha\cos\beta - \cos\alpha\sin\beta)\,/\,\cos\alpha\cos\beta}{(\cos\alpha\cos\beta - \sin\alpha\sin\beta)\,/\,\cos\alpha\cos\beta}$$

$$= \dfrac{\dfrac{\sin\alpha\cos\beta}{\cos\alpha\cos\beta} - \dfrac{\cos\alpha\sin\beta}{\cos\alpha\cos\beta}}{\dfrac{\cos\alpha\cos\beta}{\cos\alpha\cos\beta} - \dfrac{\sin\alpha\sin\beta}{\cos\alpha\cos\beta}} = \dfrac{\dfrac{\sin\alpha}{\cos\alpha} - \dfrac{\sin\beta}{\cos\beta}}{1 - \dfrac{\sin\alpha}{\cos\alpha}\cdot\dfrac{\sin\beta}{\cos\beta}} = \dfrac{\tan\alpha - \tan\beta}{1 - \tan\alpha\tan\beta}$$

$\boxed{7}$ $\tan 2u = \dfrac{2\tan u}{1-\tan^2 u}$ $\qquad$ { apply the double-angle formula for the tangent }

$\qquad\qquad = \dfrac{2\cdot\dfrac{1}{\cot u}}{1 - \dfrac{1}{\cot^2 u}}$ $\qquad$ { put in terms of cot since it appears on the right side }

$\qquad\qquad = \dfrac{\dfrac{2}{\cot u}}{\dfrac{\cot^2 u - 1}{\cot^2 u}}$ $\qquad$ { combine into one fraction }

$\qquad\qquad = \dfrac{2\cot u}{\cot^2 u - 1}$ $\qquad$ { simplify complex fraction }

$\qquad\qquad = \dfrac{2\cot u}{(\csc^2 u - 1) - 1}$ $\quad$ { Pythagorean identity }

$\qquad\qquad = \dfrac{2\cot u}{\csc^2 u - 2}$ $\qquad$ { simplify }

$\boxed{10}$ $\text{LS} = \dfrac{\sin u + \sin v}{\csc u + \csc v} = \dfrac{\sin u + \sin v}{\dfrac{1}{\sin u} + \dfrac{1}{\sin v}} = \dfrac{\sin u + \sin v}{\dfrac{\sin v + \sin u}{\sin u \, \sin v}} = \sin u \, \sin v$

At this point, there is no apparent "next step." Thus, we will stop working with the left side and try to simplify the right side to the same expression, $\sin u \sin v$.

$\text{RS} = \dfrac{1 - \sin u \, \sin v}{-1 + \csc u \, \csc v} = \dfrac{1 - \sin u \, \sin v}{-1 + \dfrac{1}{\sin u \, \sin v}} = \dfrac{1 - \sin u \, \sin v}{\dfrac{1 - \sin u \, \sin v}{\sin u \, \sin v}} = \sin u \, \sin v$

Since the LS and RS equal the same expression and the steps are reversible,

the identity is verified.

$\boxed{12}$ $\dfrac{\cos \gamma}{1 - \tan \gamma} + \dfrac{\sin \gamma}{1 - \cot \gamma} = \dfrac{\cos \gamma}{1 - \dfrac{\sin \gamma}{\cos \gamma}} + \dfrac{\sin \gamma}{1 - \dfrac{\cos \gamma}{\sin \gamma}} = \dfrac{\cos \gamma}{\dfrac{\cos \gamma - \sin \gamma}{\cos \gamma}} + \dfrac{\sin \gamma}{\dfrac{\sin \gamma - \cos \gamma}{\sin \gamma}} =$

$\dfrac{\cos^2 \gamma}{\cos \gamma - \sin \gamma} + \dfrac{\sin^2 \gamma}{\sin \gamma - \cos \gamma} = \dfrac{\cos^2 \gamma}{\cos \gamma - \sin \gamma} - \dfrac{\sin^2 \gamma}{-(\sin \gamma - \cos \gamma)} =$

$\dfrac{\cos^2 \gamma}{\cos \gamma - \sin \gamma} - \dfrac{\sin^2 \gamma}{\cos \gamma - \sin \gamma} = \dfrac{\cos^2 \gamma - \sin^2 \gamma}{\cos \gamma - \sin \gamma} = \dfrac{(\cos \gamma + \sin \gamma)(\cos \gamma - \sin \gamma)}{\cos \gamma - \sin \gamma} =$

$\cos \gamma + \sin \gamma$

$\boxed{13}$ $\dfrac{\cos(-t)}{\sec(-t) + \tan(-t)} = \dfrac{\cos t}{\sec t + (-\tan t)} = \dfrac{\cos t}{\sec t - \tan t} = \dfrac{\cos t}{\dfrac{1}{\cos t} - \dfrac{\sin t}{\cos t}} = \dfrac{\cos t}{\dfrac{1 - \sin t}{\cos t}} =$

$\dfrac{\cos^2 t}{1 - \sin t} = \dfrac{1 - \sin^2 t}{1 - \sin t} = \dfrac{(1 - \sin t)(1 + \sin t)}{1 - \sin t} = 1 + \sin t$

$\boxed{15}$ In the following solution, we could multiply both the numerator and the denominator by *either* the conjugate of the numerator *or* the conjugate of the denominator. Since the numerator on the right side looks like the numerator on the left side, we'll change the denominator.

$\sqrt{\dfrac{1 - \cos t}{1 + \cos t}} = \sqrt{\dfrac{(1 - \cos t)}{(1 + \cos t)} \cdot \dfrac{(1 - \cos t)}{(1 - \cos t)}} = \sqrt{\dfrac{(1 - \cos t)^2}{1 - \cos^2 t}} = \sqrt{\dfrac{(1 - \cos t)^2}{\sin^2 t}} =$

$\dfrac{\sqrt{(1 - \cos t)^2}}{\sqrt{\sin^2 t}} = \dfrac{|1 - \cos t|}{|\sin t|} = \dfrac{1 - \cos t}{|\sin t|}$, since $(1 - \cos t) \geq 0$.

$\boxed{19}$ We need to break down the angle argument of 4β into terms with only β as their argument. We can do this by using either the double-angle formula for the sine or the addition formula for the sine.

$\frac{1}{4} \sin 4\beta = \frac{1}{4} \sin(2 \cdot 2\beta) = \frac{1}{4}(2 \sin 2\beta \, \cos 2\beta) = \frac{1}{2}(2 \sin \beta \, \cos \beta)(\cos^2 \beta - \sin^2 \beta) =$

$\sin \beta \, \cos^3 \beta - \cos \beta \, \sin^3 \beta$

$\boxed{20}$ $\tan\frac{1}{2}\theta = \dfrac{1-\cos\theta}{\sin\theta}$ { apply the half-angle formula for the tangent }

$\phantom{\tan\frac{1}{2}\theta} = \dfrac{1}{\sin\theta} - \dfrac{\cos\theta}{\sin\theta}$ { split up the fraction }

$\phantom{\tan\frac{1}{2}\theta} = \csc\theta - \cot\theta$ { reciprocal and cotangent identities }

$\boxed{22}$ Let $\alpha = \arctan x$ and $\beta = \arctan\dfrac{2x}{1-x^2}$.

Because $-1 < x < 1$, $\arctan(-1) < \arctan(x) < \arctan(1)$, and hence $-\dfrac{\pi}{4} < \alpha < \dfrac{\pi}{4}$.

Thus, $\tan\alpha = x$ and $\tan\beta = \dfrac{2x}{1-x^2} = \dfrac{2\tan\alpha}{1-\tan^2\alpha} = \tan 2\alpha$.

Since the tangent function is one-to-one on $(-\frac{\pi}{2}, \frac{\pi}{2})$, we have $\beta = 2\alpha$ or, equivalently,

$$\alpha = \tfrac{1}{2}\beta. \text{ In terms of } x, \text{ we have } \arctan x = \tfrac{1}{2}\arctan\dfrac{2x}{1-x^2}.$$

$\boxed{24}$ $2\cos\alpha + \tan\alpha = \sec\alpha \Rightarrow 2\cos\alpha + \dfrac{\sin\alpha}{\cos\alpha} = \dfrac{1}{\cos\alpha} \Rightarrow$ { multiply by the lcd, $\cos\alpha$ }

$2\cos^2\alpha + \sin\alpha = 1 \Rightarrow 2(1-\sin^2\alpha) + \sin\alpha = 1 \Rightarrow 2 - 2\sin^2\alpha + \sin\alpha = 1 \Rightarrow$
$2\sin^2\alpha - \sin\alpha - 1 = 0 \Rightarrow (2\sin\alpha + 1)(\sin\alpha - 1) = 0 \Rightarrow \sin\alpha = -\frac{1}{2}, 1 \Rightarrow \alpha = \frac{7\pi}{6}, \frac{11\pi}{6}, \frac{\pi}{2}$.

Checking these values in the original equation, we see that $\tan\frac{\pi}{2}$ is undefined so

exclude $\frac{\pi}{2}$ and our solution is $\alpha = \frac{7\pi}{6}, \frac{11\pi}{6}$.

$\boxed{25}$ $\sin\theta = \tan\theta \Rightarrow \sin\theta - \dfrac{\sin\theta}{\cos\theta} = 0 \Rightarrow \sin\theta\left(1 - \dfrac{1}{\cos\theta}\right) = 0 \Rightarrow \sin\theta = 0$ or $1 = \dfrac{1}{\cos\theta} \Rightarrow$

$$\sin\theta = 0 \text{ or } \cos\theta = 1 \Rightarrow \theta = 0, \pi \text{ or } \theta = 0 \Rightarrow \theta = 0, \pi$$

$\boxed{27}$ $2\cos^3 t + \cos^2 t - 2\cos t - 1 = 0 \Rightarrow \cos^2 t\,(2\cos t + 1) - 1(2\cos t + 1) = 0 \Rightarrow$

$$(\cos^2 t - 1)(2\cos t + 1) = 0 \Rightarrow \cos t = \pm 1, -\tfrac{1}{2} \Rightarrow t = 0, \pi, \tfrac{2\pi}{3}, \tfrac{4\pi}{3}$$

$\boxed{29}$ $\sin\beta + 2\cos^2\beta = 1 \Rightarrow \sin\beta + 2(1-\sin^2\beta) = 1 \Rightarrow 2\sin^2\beta - \sin\beta - 1 = 0 \Rightarrow$

$$(2\sin\beta + 1)(\sin\beta - 1) = 0 \Rightarrow \sin\beta = -\tfrac{1}{2}, 1 \Rightarrow \beta = \tfrac{7\pi}{6}, \tfrac{11\pi}{6}, \tfrac{\pi}{2}$$

$\boxed{32}$ $\tan 2x \cos 2x = \sin 2x \Rightarrow \sin 2x = \sin 2x$. This is an identity and is true for all values

of x in $[0, 2\pi)$ except those that make $\tan 2x$ undefined, or, equivalently, those that

make $\cos 2x$ equal to 0. $\cos 2x = 0 \Rightarrow 2x = \frac{\pi}{2} + \pi n \Rightarrow x = \frac{\pi}{4} + \frac{\pi}{2}n$.

Hence, the solutions are all x in $[0, 2\pi)$ except $\frac{\pi}{4}, \frac{3\pi}{4}, \frac{5\pi}{4}, \frac{7\pi}{4}$.

$\boxed{33}$ $2\cos 3x \cos 2x = 1 - 2\sin 3x \sin 2x \Rightarrow 2\cos 3x \cos 2x + 2\sin 3x \sin 2x = 1 \Rightarrow$

$$2(\cos 3x \cos 2x + \sin 3x \sin 2x) = 1 \Rightarrow \cos(3x - 2x) = \tfrac{1}{2} \Rightarrow \cos x = \tfrac{1}{2} \Rightarrow x = \tfrac{\pi}{3}, \tfrac{5\pi}{3}$$

$\boxed{35}$ $\cos\pi x + \sin\pi x = 0 \Rightarrow \sin\pi x = -\cos\pi x \Rightarrow \dfrac{\sin\pi x}{\cos\pi x} = \dfrac{-\cos\pi x}{\cos\pi x}$ { divide by $\cos\pi x$ } $\Rightarrow$

$$\tan\pi x = -1 \Rightarrow \pi x = \tfrac{3\pi}{4} + \pi n \Rightarrow x = \tfrac{3}{4} + n \Rightarrow x = \tfrac{3}{4}, \tfrac{7}{4}, \tfrac{11}{4}, \tfrac{15}{4}, \tfrac{19}{4}, \tfrac{23}{4}$$

$\boxed{37}$ $2\cos^2\frac{1}{2}\theta - 3\cos\theta = 0 \Rightarrow 2\left(\dfrac{1+\cos\theta}{2}\right) - 3\cos\theta = 0 \Rightarrow$

$$(1+\cos\theta) - 3\cos\theta = 0 \Rightarrow 1 - 2\cos\theta = 0 \Rightarrow \cos\theta = \tfrac{1}{2} \Rightarrow \theta = \tfrac{\pi}{3}, \tfrac{5\pi}{3}$$

$\boxed{39}$ $\sin 5x = \sin 3x \Rightarrow \sin 5x - \sin 3x = 0 \Rightarrow$ [S2] $2\cos\dfrac{5x+3x}{2}\sin\dfrac{5x-3x}{2} = 0 \Rightarrow$

$\cos 4x \sin x = 0 \Rightarrow 4x = \tfrac{\pi}{2} + \pi n$ or $x = \pi n \Rightarrow x = \tfrac{\pi}{8} + \tfrac{\pi}{4}n$ or $x = 0, \pi \Rightarrow$

$$x = 0, \tfrac{\pi}{8}, \tfrac{3\pi}{8}, \tfrac{5\pi}{8}, \tfrac{7\pi}{8}, \pi, \tfrac{9\pi}{8}, \tfrac{11\pi}{8}, \tfrac{13\pi}{8}, \tfrac{15\pi}{8}$$

$\boxed{40}$ $\cos 3x = -\cos 2x \Rightarrow \cos 3x + \cos 2x = 0 \Rightarrow$ [S3] $2\cos\dfrac{3x+2x}{2}\cos\dfrac{3x-2x}{2} = 0 \Rightarrow$

$\cos\tfrac{5}{2}x \cos\tfrac{1}{2}x = 0 \Rightarrow \tfrac{5}{2}x = \tfrac{\pi}{2} + \pi n$ or $\tfrac{1}{2}x = \tfrac{\pi}{2} + \pi n \Rightarrow x = \tfrac{\pi}{5} + \tfrac{2\pi}{5}n$ or $x = \pi + 2\pi n \Rightarrow$

$$x = \tfrac{\pi}{5}, \tfrac{3\pi}{5}, \pi, \tfrac{7\pi}{5}, \tfrac{9\pi}{5}$$

$\boxed{42}$ $\tan 285° = \tan(225° + 60°) =$

$$\frac{\tan 225° + \tan 60°}{1 - \tan 225° \tan 60°} = \frac{1+\sqrt{3}}{1 - 1\cdot\sqrt{3}} = \frac{1+\sqrt{3}}{1-\sqrt{3}}\cdot\frac{1+\sqrt{3}}{1+\sqrt{3}} = \frac{4+2\sqrt{3}}{-2} = -2 - \sqrt{3}$$

$\boxed{44}$ $\csc\dfrac{\pi}{8} = \dfrac{1}{\csc\frac{\pi}{8}} = \dfrac{1}{\sin\left(\frac{1}{2}\cdot\frac{\pi}{4}\right)} = \dfrac{1}{\sqrt{\dfrac{1-\cos\frac{\pi}{4}}{2}}} = \dfrac{1}{\sqrt{\dfrac{1-\sqrt{2}/2}{2}}} = \dfrac{1}{\sqrt{\dfrac{2-\sqrt{2}}{4}}} = \dfrac{2}{\sqrt{2-\sqrt{2}}}$

$\boxed{45}$ $\csc\theta = \tfrac{5}{3}$ and $\cos\phi = \tfrac{8}{17} \Rightarrow \sin\theta = \tfrac{3}{5}, \cos\theta = \tfrac{4}{5}, \tan\theta = \tfrac{3}{4}$ and $\sin\phi = \tfrac{15}{17}, \tan\phi = \tfrac{15}{8}$.

$$\sin(\theta+\phi) = \sin\theta\cos\phi + \cos\theta\sin\phi = \tfrac{3}{5}\cdot\tfrac{8}{17} + \tfrac{4}{5}\cdot\tfrac{15}{17} = \tfrac{84}{85}$$

$\boxed{48}$ $\tan(\theta-\phi) = \dfrac{\tan\theta - \tan\phi}{1+\tan\theta\tan\phi} = \dfrac{\tfrac{3}{4} - \tfrac{15}{8}}{1 + \tfrac{3}{4}\cdot\tfrac{15}{8}}\cdot\dfrac{32}{32} = \dfrac{24-60}{32+45} = -\dfrac{36}{77}$

$\boxed{49}$ $\sin(\phi-\theta) = \sin\phi\cos\theta - \cos\phi\sin\theta = \tfrac{15}{17}\cdot\tfrac{4}{5} - \tfrac{8}{17}\cdot\tfrac{3}{5} = \tfrac{36}{85}$

$\boxed{50}$ First recognize the relationship to the expression in Exercise 49.

$$\sin(\theta-\phi) = \sin\left[-(\phi-\theta)\right] = -\sin(\phi-\theta) = -\tfrac{36}{85}$$

$\boxed{52}$ $\cos 2\phi = \cos^2\phi - \sin^2\phi = \left(\tfrac{8}{17}\right)^2 - \left(\tfrac{15}{17}\right)^2 = -\tfrac{161}{289}$

$\boxed{54}$ $\sin\tfrac{1}{2}\theta = \sqrt{\dfrac{1-\cos\theta}{2}} = \sqrt{\dfrac{1-\tfrac{4}{5}}{2}} = \sqrt{\dfrac{\tfrac{1}{5}}{2}} = \sqrt{\tfrac{1}{10}\cdot\tfrac{10}{10}} = \tfrac{1}{10}\sqrt{10}$

$\boxed{55}$ $\tan\tfrac{1}{2}\theta = \dfrac{1-\cos\theta}{\sin\theta} = \dfrac{1-\tfrac{4}{5}}{\tfrac{3}{5}} = \dfrac{\tfrac{1}{5}}{\tfrac{3}{5}} = \tfrac{1}{3}$

$\boxed{57}$ (a) $\sin 7t \sin 4t =$ [P4] $\tfrac{1}{2}[\cos(7t-4t) - \cos(7t+4t)] = \tfrac{1}{2}\cos 3t - \tfrac{1}{2}\cos 11t$

    (b) $\cos\tfrac{1}{4}u \cos\left(-\tfrac{1}{6}u\right) =$ [P3] $\tfrac{1}{2}\left\{\cos\left[\tfrac{1}{4}u + \left(-\tfrac{1}{6}u\right)\right] + \cos\left[\tfrac{1}{4}u - \left(-\tfrac{1}{6}u\right)\right]\right\} =$

$$\tfrac{1}{2}\left(\cos\tfrac{2}{24}u + \cos\tfrac{10}{24}u\right) = \tfrac{1}{2}\cos\tfrac{1}{12}u + \tfrac{1}{2}\cos\tfrac{5}{12}u$$

$\boxed{58}$ (b) $\cos 3\theta - \cos 8\theta = $ [S4] $-2\sin\dfrac{3\theta + 8\theta}{2}\sin\dfrac{3\theta - 8\theta}{2} = -2\sin\frac{11}{2}\theta \sin\left(-\frac{5}{2}\theta\right) =$

$$2\sin\tfrac{11}{2}\theta \sin\tfrac{5}{2}\theta$$

(c) $\sin\frac{1}{4}t - \sin\frac{1}{5}t = $ [S2] $2\cos\dfrac{\frac{1}{4}t + \frac{1}{5}t}{2}\sin\dfrac{\frac{1}{4}t - \frac{1}{5}t}{2} = 2\cos\dfrac{\frac{5}{20}t + \frac{4}{20}t}{2}\sin\dfrac{\frac{5}{20}t - \frac{4}{20}t}{2} =$

$$2\cos\tfrac{9}{40}t \sin\tfrac{1}{40}t$$

$\boxed{62}$ $\arccos\left(\tan\frac{3\pi}{4}\right) = \arccos\left(-1\right) = \pi$

$\boxed{63}$ $\arcsin\left(\sin\frac{5\pi}{4}\right) = \arcsin\left(-\dfrac{\sqrt{2}}{2}\right) = -\dfrac{\pi}{4}$ $\boxed{64}$ $\cos^{-1}\left(\cos\frac{5\pi}{4}\right) = \cos^{-1}\left(-\dfrac{\sqrt{2}}{2}\right) = \dfrac{3\pi}{4}$

$\boxed{69}$ Let $\alpha = \sin^{-1}\frac{15}{17}$ and $\beta = \sin^{-1}\frac{8}{17}$.

$\cos\left(\sin^{-1}\frac{15}{17} - \sin^{-1}\frac{8}{17}\right) = \cos\left(\alpha - \beta\right) = \cos\alpha \cos\beta + \sin\alpha \sin\beta = \frac{8}{17}\cdot\frac{15}{17} + \frac{15}{17}\cdot\frac{8}{17} = \frac{240}{289}$.

$\boxed{70}$ Let $\alpha = \sin^{-1}\frac{4}{5}$. $\cos\left(2\sin^{-1}\frac{4}{5}\right) = \cos\left(2\alpha\right) = \cos^2\alpha - \sin^2\alpha = \left(\frac{3}{5}\right)^2 - \left(\frac{4}{5}\right)^2 = -\frac{7}{25}$.

$\boxed{73}$ $y = 1 - \sin^{-1}x = -\sin^{-1}x + 1$ •

Reflect $y = \sin^{-1}x$ through the x-axis and shift it up 1 unit.

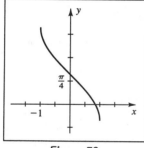

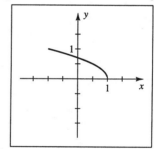

Figure 73   Figure 74

$\boxed{74}$ If $\alpha = \cos^{-1}x$, then $\cos\alpha = x$, where $0 \le \alpha \le \pi$.

Hence, $y = \sin\left(\frac{1}{2}\cos^{-1}x\right) = \sin\frac{1}{2}\alpha = \sqrt{\dfrac{1 - \cos\alpha}{2}} = \sqrt{\dfrac{1 - x}{2}}$.

Thus, we have the graph of the half-parabola $y = \sqrt{\frac{1}{2}(1 - x)}$ on the interval $[-1, 1]$.

$\boxed{75}$ $\cos\left(\alpha + \beta + \gamma\right)$

$$= \cos\left[(\alpha + \beta) + \gamma\right]$$
$$= \cos\left(\alpha + \beta\right)\cos\gamma - \sin\left(\alpha + \beta\right)\sin\gamma$$
$$= \left(\cos\alpha \cos\beta - \sin\alpha \sin\beta\right)\cos\gamma - \left(\sin\alpha \cos\beta + \cos\alpha \sin\beta\right)\sin\gamma$$
$$= \cos\alpha \cos\beta \cos\gamma - \sin\alpha \sin\beta \cos\gamma - \sin\alpha \cos\beta \sin\gamma - \cos\alpha \sin\beta \sin\gamma$$

$\boxed{76}$ (a) $t = -\frac{\pi}{2b} \Rightarrow F = A\left[\cos\left(-\frac{\pi}{2}\right) - a\cos\left(-\frac{3\pi}{2}\right)\right] = A(0 - a \cdot 0) = 0$

$$t = \frac{\pi}{2b} \Rightarrow F = A\left(\cos\frac{\pi}{2} - a\cos\frac{3\pi}{2}\right) = A(0 - a \cdot 0) = 0$$

(b) $a = \frac{1}{3} \Rightarrow \sin 3bt = \sin bt \Rightarrow \sin 3bt - \sin bt = 0 \Rightarrow$

[S2] $2\cos\dfrac{3bt + bt}{2}\sin\dfrac{3bt - bt}{2} = 0 \Rightarrow \cos 2bt \sin bt = 0 \Rightarrow$

$\cos 2bt = 0$ or $\sin bt = 0 \Rightarrow 2bt = \frac{\pi}{2} + \pi n$ or $bt = \pi n \Rightarrow$

$$t = \frac{\pi}{4b} + \frac{\pi}{2b}n \text{ or } t = \frac{\pi}{b}n. \text{ Since } -\frac{\pi}{2b} < t < \frac{\pi}{2b}, \ t = \pm\frac{\pi}{4b}, \ 0.$$

(c) Using the values from part (b), $t = 0 \Rightarrow F = A\left(\cos 0 - \frac{1}{3}\cos 0\right) = A\left(1 - \frac{1}{3}\right) = \frac{2}{3}A.$

$$t = \pm\frac{\pi}{4b} \Rightarrow F = A\left[\cos\left(\pm\frac{\pi}{4}\right) - \frac{1}{3}\cos\left(\pm\frac{3\pi}{4}\right)\right] = A\left(\frac{\sqrt{2}}{2} + \frac{\sqrt{2}}{6}\right) = \frac{4\sqrt{2}}{6}A = \frac{2}{3}\sqrt{2}\,A.$$

The second value is $\sqrt{2}$ times the first, hence $\frac{2}{3}\sqrt{2}\,A$ is the maximum force.

$\boxed{78}$ (a) Bisect θ to form two right triangles. $\tan\frac{1}{2}\theta = \dfrac{\frac{1}{2}x}{d} \Rightarrow x = 2d\tan\frac{1}{2}\theta.$

(b) Using part (a) with $x = 0.5$ ft and $\theta = 0.0005$ radian,

$$\text{we have } d = \frac{x}{2\tan\frac{1}{2}\theta} \approx 1000 \text{ ft, so } d \le 1000 \text{ ft.}$$

$\boxed{79}$ (a) Bisect θ to form two right triangles.

$$\cos\frac{1}{2}\theta = \frac{r}{d+r} \Rightarrow d + r = \frac{r}{\cos\frac{1}{2}\theta} \Rightarrow d = r\sec\frac{1}{2}\theta - r = r\left(\sec\frac{1}{2}\theta - 1\right).$$

(b) $d = 300$ and $r = 4000 \Rightarrow \cos\frac{1}{2}\theta = \dfrac{r}{d+r} = \dfrac{4000}{4300} \Rightarrow \frac{1}{2}\theta \approx 21.5° \Rightarrow \theta \approx 43°.$

$\boxed{80}$ $N = h/w$ and $\tan\theta = N \Rightarrow \tan\theta = \frac{h}{w}.$

(a) $\tan\theta = \frac{h}{w} = \frac{400}{80} = 5 \Rightarrow \theta = \tan^{-1} 5 \approx 78.7°$

(b) $\tan\theta = \frac{h}{w} = \frac{55}{30} = \frac{11}{6} \Rightarrow \theta = \tan^{-1}\frac{11}{6} \approx 61.4°$

Chapter 7 Discussion Exercises

$\boxed{1}$ $\dfrac{\tan x}{1-\cot x}+\dfrac{\cot x}{1-\tan x}=\dfrac{\frac{\sin x}{\cos x}}{1-\frac{\cos x}{\sin x}}+\dfrac{\frac{\cos x}{\sin x}}{1-\frac{\sin x}{\cos x}}=$

$$\dfrac{\sin^2 x}{\cos x\,(\sin x-\cos x)}+\dfrac{\cos^2 x}{\sin x\,(\cos x-\sin x)}=\dfrac{\sin^2 x}{\cos x\,(\sin x-\cos x)}-\dfrac{\cos^2 x}{\sin x\,(\sin x-\cos x)}=$$

$$\dfrac{\sin^3 x-\cos^3 x}{\cos x\,\sin x\,(\sin x-\cos x)}=\dfrac{(\sin x-\cos x)(\sin^2 x+\sin x\,\cos x+\cos^2 x)}{\cos x\,\sin x\,(\sin x-\cos x)}=$$

$$\dfrac{1+\sin x\,\cos x}{\cos x\,\sin x}=\dfrac{1}{\cos x\,\sin x}+1=1+\sec x\,\csc x$$

$\boxed{3}$ *Note:* Graphing on a TI-82/83 doesn't really help to solve this problem.

$3\cos 45x+4\sin 45x=5 \Rightarrow \{\,$ by Example 6 in Section 7.3 $\}$

$5\cos\left(45x-\tan^{-1}\tfrac{4}{3}\right)=5 \Rightarrow \cos\left(45x-\tan^{-1}\tfrac{4}{3}\right)=1 \Rightarrow 45x-\tan^{-1}\tfrac{4}{3}=2\pi n \Rightarrow$

$45x=2\pi n+\tan^{-1}\tfrac{4}{3} \Rightarrow x=\dfrac{2\pi n+\tan^{-1}\frac{4}{3}}{45}.$ $n=0,\ 1,\ \ldots,\ 44$ will yield x values in

$[0,\ 2\pi)$.   *Note:* After using Example 6, you might notice that this is a function with period $2\pi/45$, and it will obtain 45 maximums on an interval of length 2π. The largest value of x is approximately 6.164 $\{$ when $n=44\,\}$.

$\boxed{5}$ Let $\alpha=\tan^{-1}\left(\tfrac{1}{239}\right)$ and $\theta=\tan^{-1}\left(\tfrac{1}{5}\right)$. $\tfrac{\pi}{4}=4\theta-\alpha \Rightarrow \tfrac{\pi}{4}+\alpha=4\theta$. Both sides are acute angles, and we will show that the tangent of each side is equal to the same value, hence proving the identity.

$$\text{LS}=\tan\left(\tfrac{\pi}{4}+\alpha\right)=\dfrac{\tan\frac{\pi}{4}+\tan\alpha}{1-\tan\frac{\pi}{4}\tan\alpha}=\dfrac{1+\frac{1}{239}}{1-1\cdot\frac{1}{239}}=\dfrac{\frac{240}{239}}{\frac{238}{239}}=\dfrac{240}{238}=\dfrac{120}{119}.$$

$$\text{RS}=\tan 4\theta=$$

$$\dfrac{2\tan 2\theta}{1-\tan^2(2\theta)}=\dfrac{2\cdot\frac{2\tan\theta}{1-\tan^2\theta}}{1-\left(\frac{2\tan\theta}{1-\tan^2\theta}\right)^2}=\dfrac{2\cdot\frac{\frac{2}{5}}{1-\frac{1}{25}}}{1-\left(\frac{\frac{2}{5}}{1-\frac{1}{25}}\right)^2}=\dfrac{\frac{\frac{4}{5}}{\frac{24}{25}}}{1-\frac{\frac{4}{25}}{\frac{24\cdot24}{25\cdot25}}}=\dfrac{\frac{5}{6}}{\frac{119}{144}}=\dfrac{120}{119}.$$

Similarly, for the second relationship, we could write $\tfrac{\pi}{4}=\alpha+\beta+\gamma$ and show that $\tan\left(\tfrac{\pi}{4}-\alpha\right)=\tan(\beta+\gamma)=\tfrac{1}{3}$.

Chapter 8: Applications of Trigonometry

1 $\beta = 180° - \alpha - \gamma = 180° - 41° - 77° = 62°.$

$\dfrac{b}{\sin \beta} = \dfrac{a}{\sin \alpha} \Rightarrow b = \dfrac{a \sin \beta}{\sin \alpha} = \dfrac{10.5 \sin 62°}{\sin 41°} \approx 14.1.$

$\dfrac{c}{\sin \gamma} = \dfrac{a}{\sin \alpha} \Rightarrow c = \dfrac{a \sin \gamma}{\sin \alpha} = \dfrac{10.5 \sin 77°}{\sin 41°} \approx 15.6.$

3 $\gamma = 180° - \alpha - \beta = 180° - 27°40' - 52°10' = 100°10'.$

$\dfrac{b}{\sin \beta} = \dfrac{a}{\sin \alpha} \Rightarrow b = \dfrac{a \sin \beta}{\sin \alpha} = \dfrac{32.4 \sin 52°10'}{\sin 27°40'} \approx 55.1.$

$\dfrac{c}{\sin \gamma} = \dfrac{a}{\sin \alpha} \Rightarrow c = \dfrac{a \sin \gamma}{\sin \alpha} = \dfrac{32.4 \sin 100°10'}{\sin 27°40'} \approx 68.7.$

7 $\dfrac{\sin \beta}{b} = \dfrac{\sin \gamma}{c} \Rightarrow \beta = \sin^{-1}\left(\dfrac{b \sin \gamma}{c}\right) = \sin^{-1}\left(\dfrac{12 \sin 81°}{11}\right) \approx \sin^{-1}(1.0775).$ Since 1.0775 is

not in the domain of the inverse sine function, which is $[-1, 1]$, *no triangle exists.*

9 $\dfrac{\sin \alpha}{a} = \dfrac{\sin \gamma}{c} \Rightarrow \alpha = \sin^{-1}\left(\dfrac{a \sin \gamma}{c}\right) = \sin^{-1}\left(\dfrac{140 \sin 53°20'}{115}\right) \approx \sin^{-1}(0.9765) \approx$

$77°30'$ or $102°30'$ {rounded to the nearest 10 minutes}. When using the inverse sine
function to solve for an angle, remember that there are always *two* values if $\theta \ne 90°$.
The first angle is the one you obtain using your calculator, and the second is the
reference angle for the first angle in the second quadrant. After obtaining these
values, we need to check the sum of the angles. If this sum is less than $180°$, a
triangle is formed. If this sum is greater than or equal to $180°$, no triangle is formed.
For this exercise, there are two triangles possible since in either case $\alpha + \gamma < 180°$.

$\beta = (180° - \gamma) - \alpha \approx (180° - 53°20') - (77°30' \text{ or } 102°30') = 49°10' \text{ or } 24°10'.$

$\dfrac{b}{\sin \beta} = \dfrac{c}{\sin \gamma} \Rightarrow b = \dfrac{c \sin \beta}{\sin \gamma} \approx \dfrac{115 \sin (49°10' \text{ or } 24°10')}{\sin 53°20'} \approx 108 \text{ or } 58.7.$

11 $\dfrac{\sin \alpha}{a} = \dfrac{\sin \gamma}{c} \Rightarrow \alpha = \sin^{-1}\left(\dfrac{a \sin \gamma}{c}\right) = \sin^{-1}\left(\dfrac{131.08 \sin 47.74°}{97.84}\right) \approx \sin^{-1}(0.9915) \approx$

$82.54°$ or $97.46°$. There are two triangles possible since in either case $\alpha + \gamma < 180°$.

$\beta = (180° - \gamma) - \alpha \approx (180° - 47.74°) - (82.54° \text{ or } 97.46°) = 49.72° \text{ or } 34.80°.$

$\dfrac{b}{\sin \beta} = \dfrac{c}{\sin \gamma} \Rightarrow b = \dfrac{c \sin \beta}{\sin \gamma} \approx \dfrac{97.84 \sin (49.72° \text{ or } 34.80°)}{\sin 47.74°} \approx 100.85 \text{ or } 75.45.$

13 $\dfrac{\sin \beta}{b} = \dfrac{\sin \alpha}{a} \Rightarrow \beta = \sin^{-1}\left(\dfrac{b \sin \alpha}{a}\right) = \sin^{-1}\left(\dfrac{18.9 \sin 65°10'}{21.3}\right) \approx \sin^{-1}(0.8053) \approx$

$53°40'$ or $126°20'$ {rounded to the nearest 10 minutes}. Reject $126°20'$ because then
$\alpha + \beta \ge 180°$. $\gamma = 180° - \alpha - \beta \approx 180° - 65°10' - 53°40' = 61°10'.$

$\dfrac{c}{\sin \gamma} = \dfrac{a}{\sin \alpha} \Rightarrow c = \dfrac{a \sin \gamma}{\sin \alpha} \approx \dfrac{21.3 \sin 61°10'}{\sin 65°10'} \approx 20.6.$

15 $\dfrac{\sin \gamma}{c} = \dfrac{\sin \beta}{b} \Rightarrow \gamma = \sin^{-1}\left(\dfrac{c \sin \beta}{b}\right) = \sin^{-1}\left(\dfrac{0.178 \sin 121.624°}{0.283}\right) \approx \sin^{-1}(0.5356) \approx$

32.383° or 147.617°. Reject 147.617° because then $\beta + \gamma \geq 180°$.

$\alpha = 180° - \beta - \gamma \approx 180° - 121.624° - 32.383° = 25.993°$.

$$\dfrac{a}{\sin \alpha} = \dfrac{b}{\sin \beta} \Rightarrow a = \dfrac{b \sin \alpha}{\sin \beta} \approx \dfrac{0.283 \sin 25.993°}{\sin 121.624°} \approx 0.146.$$

17 $\angle ABC = 180° - 54°10' - 63°20' = 62°30'$. $\dfrac{\overline{AB}}{\sin 54°10'} = \dfrac{240}{\sin 62°30'} \Rightarrow \overline{AB} \approx 219.36$ yd

19 (a) $\angle ABP = 180° - 65° = 115°$. $\angle APB = 180° - 21° - 115° = 44°$.

$$\dfrac{\overline{AP}}{\sin 115°} = \dfrac{1.2}{\sin 44°} \Rightarrow \overline{AP} \approx 1.57, \text{ or } 1.6 \text{ mi.}$$

(b) $\sin 21° = \dfrac{\text{height of } P}{\overline{AP}} \Rightarrow$ height of $P = \dfrac{1.2 \sin 115° \sin 21°}{\sin 44°}$ { from part (a) } $\approx 0.56,$

or 0.6 mi.

21 Let C denote the base of the balloon and P its projection on the ground.

$\angle ACB = 180° - 24°10' - 47°40' = 108°10'$. $\dfrac{\overline{AC}}{\sin 47°40'} = \dfrac{8.4}{\sin 108°10'} \Rightarrow \overline{AC} \approx 6.5$ mi.

$$\sin 24°10' = \dfrac{\overline{PC}}{\overline{AC}} \Rightarrow \overline{PC} = \dfrac{8.4 \sin 47°40' \sin 24°10'}{\sin 108°10'} \approx 2.7 \text{ mi.}$$

23 $\angle APQ = 57° - 22° = 35°$. $\angle AQP = 180° - (63° - 22°) = 139°$.

$$\angle PAQ = 180° - 139° - 35° = 6°. \quad \dfrac{\overline{AP}}{\sin 139°} = \dfrac{100}{\sin 6°} \Rightarrow \overline{AP} = \dfrac{100 \sin 139°}{\sin 6°} \approx 628 \text{ m.}$$

25 $\angle FAB = 90° - 27°10' = 62°50'$.

$\angle FBA = 90° - 52°40' = 37°20'$.

$\angle AFB = 180° - 62°50' - 37°20' = 79°50'$.

$\dfrac{\overline{AF}}{\sin 37°20'} = \dfrac{6}{\sin 79°50'} \Rightarrow \overline{AF} \approx 3.70$ mi.

$\dfrac{\overline{BF}}{\sin 62°50'} = \dfrac{6}{\sin 79°50'} \Rightarrow \overline{BF} \approx 5.42$ mi.

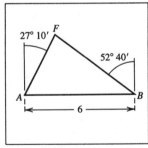

Figure 25

27 Let A denote the base of the hill, B the base of the cathedral, and C the top of the

spire. The angle at the base of the hill is $180° - 48° = 132°$. The angle at the top

of the spire is $180° - 132° - 41° = 7°$. $\dfrac{\overline{AC}}{\sin 41°} = \dfrac{200}{\sin 7°} \Rightarrow \overline{AC} = \dfrac{200 \sin 41°}{\sin 7°} \approx 1077$ ft.

$\angle BAC = 48° - 32° = 16°$. $\angle ACB = 90° - 48° = 42°$. $\angle ABC = 180° - 42° - 16° = 122°$.

$$\dfrac{\overline{BC}}{\sin 16°} = \dfrac{\overline{AC}}{\sin 122°} \Rightarrow \overline{BC} = \dfrac{200 \sin 41° \sin 16°}{\sin 7° \sin 122°} \approx 350 \text{ ft.}$$

$\boxed{29}$ (a) In the triangle that forms the base, the third angle is $180° - 103° - 52° = 25°$.

Let l denote the length of the dashed line. $\dfrac{l}{\sin 103°} = \dfrac{12.0}{\sin 25°} \Rightarrow l \approx 27.7$ units.

Now $\tan 34° = \dfrac{h}{l} \Rightarrow h \approx 18.7$ units.

(b) Draw a line from the 103° angle that is perpendicular to l and call it d.

$\sin 52° = \dfrac{d}{12} \Rightarrow d \approx 9.5$ units. The area of the triangular base is $B = \frac{1}{2}ld$.

The volume V is $\frac{1}{3}(\frac{1}{2}ld)h = 288 \sin 52° \sin^2 103° \tan 34° \csc^2 25° \approx 814$ cubic units.

$\boxed{31}$ Draw a line through P perpendicular to the x-axis. Locate points A and B on this line so that $\angle PAQ = \angle PBR = 90°$. $\overline{AP} = 5127.5 - 3452.8 = 1674.7$, $\overline{AQ} = 3145.8 - 1487.7 = 1658.1$, and $\tan \angle APQ = \frac{1658.1}{1674.7} \Rightarrow \angle APQ \approx 44°43'$. Thus, $\angle BPR \approx 180° - 55°50' - 44°43' = 79°27'$.

By the distance formula, $\overline{PQ} \approx \sqrt{(1674.7)^2 + (1658.1)^2} \approx 2356.7$.

Now $\dfrac{\overline{PR}}{\sin 65°22'} = \dfrac{\overline{PQ}}{\sin(180° - 55°50' - 65°22')} \Rightarrow \overline{PR} = \dfrac{2356.7 \sin 65°22'}{\sin 58°48'} \approx 2504.5$.

$\sin \angle BPR = \dfrac{\overline{BR}}{\overline{PR}} \Rightarrow \overline{BR} \approx (2504.5)(\sin 79°27') \approx 2462.2$. $\cos \angle BPR = \dfrac{\overline{BP}}{\overline{PR}} \Rightarrow$

$\overline{BP} \approx (2504.5)(\cos 79°27') \approx 458.6$. Using the coordinates of P, we see that

$$R(x,\, y) \approx (1487.7 + 2462.2,\, 3452.8 - 458.6) = (3949.9,\, 2994.2).$$

8.2 Exercises

Note: These formulas will be used to solve problems that involve the law of cosines.

(1) $a^2 = b^2 + c^2 - 2bc \cos \alpha \Rightarrow a = \sqrt{b^2 + c^2 - 2bc \cos \alpha}$

(Similar formulas are used for b and c.)

(2) $a^2 = b^2 + c^2 - 2bc \cos \alpha \Rightarrow 2bc \cos \alpha = b^2 + c^2 - a^2 \Rightarrow$

$\cos \alpha = \left(\dfrac{b^2 + c^2 - a^2}{2bc}\right) \Rightarrow \alpha = \cos^{-1}\left(\dfrac{b^2 + c^2 - a^2}{2bc}\right)$

(Similar formulas are used for β and γ.)

$\boxed{1}$ $a = \sqrt{b^2 + c^2 - 2bc \cos \alpha} = \sqrt{20^2 + 30^2 - 2(20)(30) \cos 60°} = \sqrt{700} \approx 26$.

$\beta = \cos^{-1}\left(\dfrac{a^2 + c^2 - b^2}{2ac}\right) = \cos^{-1}\left(\dfrac{700 + 30^2 - 20^2}{2(\sqrt{700})(30)}\right) \approx \cos^{-1}(0.7559) \approx 41°$.

$\gamma = 180° - \alpha - \beta \approx 180° - 60° - 41° = 79°$.

$\boxed{3}$ $b = \sqrt{a^2 + c^2 - 2ac \cos\beta} = \sqrt{23{,}400 + 4500\sqrt{3}} \approx 177$, or 180.

$\alpha = \cos^{-1}\left(\dfrac{b^2 + c^2 - a^2}{2bc}\right) \approx \cos^{-1}(0.9054) \approx 25°10'$, or 25°.

Note: We do not have to worry about an "ambiguous" case of the law of cosines as we did with the law of sines. This is true because the inverse cosine function gives us a unique angle from 0° to 180° for all values in its domain.

$$\gamma = 180° - \alpha - \beta \approx 180° - 25°10' - 150° = 4°50', \text{ or } 5°.$$

$\boxed{7}$ $\alpha = \cos^{-1}\left(\dfrac{b^2 + c^2 - a^2}{2bc}\right) = \cos^{-1}(0.875) \approx 29°.$

$\beta = \cos^{-1}\left(\dfrac{a^2 + c^2 - b^2}{2ac}\right) = \cos^{-1}(0.6875) \approx 47°.$

$$\gamma = 180° - \alpha - \beta \approx 180° - 29° - 47° = 104°.$$

$\boxed{9}$ $\alpha = \cos^{-1}\left(\dfrac{b^2 + c^2 - a^2}{2bc}\right) \approx \cos^{-1}(0.9766) \approx 12°30'.$

$\beta = \cos^{-1}\left(\dfrac{a^2 + c^2 - b^2}{2ac}\right) = \cos^{-1}(-0.725) \approx 136°30'.$

$$\gamma = 180° - \alpha - \beta \approx 180° - 12°30' - 136°30' = 31°00'.$$

Note: When deciding whether to use the law of cosines or the law of sines, it may be helpful to remember that the law of cosines must be used when you have:

(1) 3 sides, or

(2) 2 sides and the angle between them.

$\boxed{11}$ We have two sides and the angle between them. Hence, we apply the law of cosines.

$$\text{Third side} = \sqrt{175^2 + 150^2 - 2(175)(150)\cos 73°40'} \approx 196 \text{ feet.}$$

$\boxed{13}$ 20 minutes $= \frac{1}{3}$ hour $\Rightarrow$ the cars have traveled $60(\frac{1}{3}) = 20$ miles and $45(\frac{1}{3}) = 15$ miles, respectively. The distance d apart is $d = \sqrt{20^2 + 15^2 - 2(20)(15)\cos 84°} \approx 24$ miles.

$\boxed{15}$ The first ship travels $(24)(2) = 48$ miles in two hours. The second ship travels $(18)(1\frac{1}{2}) = 27$ miles in $1\frac{1}{2}$ hours. The angle between the paths is $20° + 35° = 55°.$

$$\overline{AB} = \sqrt{27^2 + 48^2 - 2(27)(48)\cos 55°} \approx 39 \text{ miles.}$$

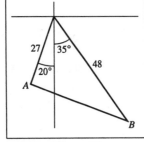

Figure 15

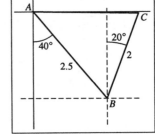

Figure 17

17 $\angle ABC = 40° + 20° = 60°$.

$$\overline{AB} = \left(\frac{1\text{ mile}}{8\text{ min}} \cdot 20\text{ min}\right) = 2.5\text{ miles and } \overline{BC} = \left(\frac{1\text{ mile}}{8\text{ min}} \cdot 16\text{ min}\right) = 2\text{ miles}.$$

$$\overline{AC} = \sqrt{2.5^2 + 2^2 - 2(2)(2.5)\cos 60°} = \sqrt{5.25} \approx 2.3\text{ miles}. \text{ See } Figure\ 17.$$

19 $\gamma = \cos^{-1}\left(\dfrac{2^2 + 3^2 - 4^2}{2 \cdot 2 \cdot 3}\right) = \cos^{-1}(-0.25) \approx 104°29'. \quad \phi \approx 104°29' - 70° = 34°29'.$

The direction that the third side was traversed is approximately

$$N(90° - 34°29')E = N55°31'E.$$

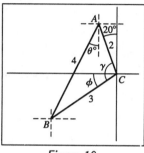

Figure 19

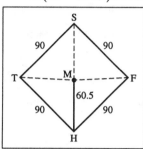

Figure 21

21 Let H denote home plate, M the mound, F first base, S second base, and T third base. $\overline{HS} = \sqrt{90^2 + 90^2} = 90\sqrt{2} \approx 127.3$ ft. $\overline{MS} = 90\sqrt{2} - 60.5 \approx 66.8$ ft.

$\angle MHF = 45°$ so $\overline{MF} = \sqrt{60.5^2 + 90^2 - 2(60.5)(90)\cos 45°} \approx 63.7$ ft.

$$\overline{MT} = \overline{MF}\text{ by the symmetry of the field.}$$

23 $\angle RTP = 21°$ and $\angle RSP = 37°$. $\sin\angle RSP = \dfrac{10{,}000}{\overline{SP}} \Rightarrow \overline{SP} = 10{,}000\csc 37° \approx$

16,616 ft. $\sin\angle RTP = \dfrac{10{,}000}{\overline{TP}} \Rightarrow \overline{TP} = 10{,}000\csc 21° \approx 27{,}904$ ft.

$$\overline{ST} = \sqrt{\overline{SP}^2 + \overline{TP}^2 - 2(\overline{SP})(\overline{TP})\cos 110°} \approx 37{,}039\text{ ft} \approx 7\text{ miles}.$$

25 Let $d = \overline{ES}$. $d^2 = R^2 + R^2 - 2RR\cos\theta \Rightarrow d^2 = 2R^2(1 - \cos\theta) \Rightarrow$

$$d^2 = 4R^2\left(\frac{1 - \cos\theta}{2}\right) \Rightarrow d = 2R\sqrt{\frac{1 - \cos\theta}{2}} \Rightarrow d = 2R\sin\frac{\theta}{2}.$$

$$\text{Since } d = vt,\ t = \frac{d}{v} = \frac{2R}{v}\sin\frac{\theta}{2}.$$

27 (a) $\angle BCP = \frac{1}{2}(\angle BCD) = \frac{1}{2}(72°) = 36°$. $\triangle BPC$ is isosceles so

$\angle BPC = \angle PBC$ and $2\angle BPC = 180° - 36° \Rightarrow \underline{\angle BPC = 72°}$.

$\underline{\angle APB} = 180° - \angle BPC = 180° - 72° = \underline{108°}$.

$$\underline{\angle ABP} = 180° - \angle APB - \angle BAP = 180° - 108° - 36° = \underline{36°}.$$

(b) $\overline{BP} = \sqrt{\overline{BC}^2 + \overline{PC}^2 - 2(\overline{BC})(\overline{PC})\cos 36°} = \sqrt{1^2 + 1^2 - 2(1)(1)\cos 36°} \approx 0.62$.

(c) $\text{Area}_{\text{kite}} = 2(\text{Area of } \triangle BPC) = 2 \cdot \frac{1}{2}(\overline{CB})(\overline{CP})\sin\angle BCP = \sin 36° \approx 0.59$.

$\text{Area}_{\text{dart}} = 2(\text{Area of } \triangle ABP) = 2 \cdot \frac{1}{2}(\overline{AB})(\overline{BP})\sin\angle ABP = \overline{BP}\sin 36° \approx 0.36$.

$$\{\overline{BP}\text{ was found in part (b)}\}$$

Note: Exer. 29–36: $\mathcal{A}$ (the area) is measured in square units.

29 Since α is the angle between sides b and c, we may apply the area of a triangle

formula listed in this section. $\mathcal{A} = \frac{1}{2}bc \sin \alpha = \frac{1}{2}(20)(30) \sin 60° = 300(\sqrt{3}/2) \approx 260.$

31 $\gamma = 180° - \alpha - \beta = 180° - 40.3° - 62.9° = 76.8°.$

$$\frac{a}{\sin \alpha} = \frac{b}{\sin \beta} \Rightarrow a = \frac{b \sin \alpha}{\sin \beta} = \frac{5.63 \sin 40.3°}{\sin 62.9°}. \quad \mathcal{A} = \frac{1}{2}ab \sin \gamma \approx 11.21.$$

33 $\frac{\sin \beta}{b} = \frac{\sin \alpha}{a} \Rightarrow \sin \beta = \frac{b \sin \alpha}{a} = \frac{3.4 \sin 80.1°}{8.0} \approx 0.4187 \Rightarrow \beta \approx 24.8°$ or $155.2°.$

Reject $155.2°$ because then $\alpha + \beta = 235.3° \geq 180°.$ $\gamma \approx 180° - 80.1° - 24.8° = 75.1°.$

$$\mathcal{A} = \frac{1}{2}ab \sin \gamma = \frac{1}{2}(8.0)(3.4) \sin 75.1° \approx 13.1.$$

35 Given the lengths of the 3 sides of a triangle, we compute the semiperimeter and then

use Heron's formula to find the area of the triangle.

$s = \frac{1}{2}(a + b + c) = \frac{1}{2}(25.0 + 80.0 + 60.0) = 82.5.$

$$\mathcal{A} = \sqrt{s(s-a)(s-b)(s-c)} = \sqrt{(82.5)(57.5)(2.5)(22.5)} \approx 516.56, \text{ or } 517.0.$$

37 $s = \frac{1}{2}(a + b + c) = \frac{1}{2}(115 + 140 + 200) = 227.5.$ $\mathcal{A} = \sqrt{s(s-a)(s-b)(s-c)} =$

$$\sqrt{(227.5)(112.5)(87.5)(27.5)} \approx 7847.6 \text{ yd}^2, \text{ or } \mathcal{A}/4840 \approx 1.62 \text{ acres.}$$

39 The area of the parallelogram is twice the area of the triangle formed by the two

sides and the included angle. $\mathcal{A} = 2(\frac{1}{2})(12.0)(16.0) \sin 40° \approx 123.4 \text{ ft}^2.$

8.3 Exercises

1 $|3 - 4i| = \sqrt{3^2 + (-4)^2} = \sqrt{25} = 5$ 5 $|8i| = |0 + 8i| = \sqrt{0^2 + 8^2} = \sqrt{64} = 8$

7 From §2.4, $i^m = i, -1, -i,$ or 1. Since all of these are 1 unit from the origin,

$|i^m| = 1$ for any integer m. As an alternate solution,

$$\left|i^{500}\right| = \left|(i^4)^{125}\right| = \left|(1)^{125}\right| = |1| = |1 + 0i| = \sqrt{1^2 + 0^2} = \sqrt{1} = 1.$$

9 $|0| = |0 + 0i| = \sqrt{0^2 + 0^2} = \sqrt{0} = 0$

11 $4 + 2i$ 13 $3 - 5i$ 15 $-(3 - 6i) = -3 + 6i$

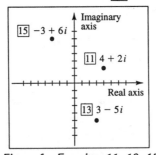

Figure for Exercises 11, 13, 15   Figure for Exercises 17, 19

17 $2i(2 + 3i) = 4i + 6i^2 = 4i - 6 = -6 + 4i$

19 $(1 + i)^2 = 1 + 2(1)(i) + i^2 = 1 + 2i - 1 = 2i$

Note: For each of the following exercises, we need to find r and θ.

If $z = a + bi$, then $r = \sqrt{a^2 + b^2}$. To find θ, we will use the fact that $\tan\theta = \frac{b}{a}$ and our knowledge of what quadrant the terminal side of θ is in.

21 $z = 1 - i \Rightarrow r = \sqrt{1 + (-1)^2} = \sqrt{2}$. $\tan\theta = \frac{-1}{1} = -1$ and θ in QIV $\Rightarrow \theta = \frac{7\pi}{4}$.

Thus, $z = 1 - i = \sqrt{2}\left(\cos\frac{7\pi}{4} + i\sin\frac{7\pi}{4}\right)$, or simply $\sqrt{2}\,\text{cis}\,\frac{7\pi}{4}$.

23 $z = -4\sqrt{3} + 4i \Rightarrow r = \sqrt{(-4\sqrt{3})^2 + 4^2} = \sqrt{64} = 8$.

$\tan\theta = \dfrac{4}{-4\sqrt{3}} = -\dfrac{1}{\sqrt{3}}$ and θ in QII $\Rightarrow \theta = \frac{5\pi}{6}$. $z = 8\,\text{cis}\,\frac{5\pi}{6}$.

27 $z = -4 - 4i \Rightarrow r = \sqrt{(-4)^2 + (-4)^2} = \sqrt{32} = 4\sqrt{2}$.

$\tan\theta = \frac{-4}{-4} = 1$ and θ in QIII $\Rightarrow \theta = \frac{5\pi}{4}$. $z = 4\sqrt{2}\,\text{cis}\,\frac{5\pi}{4}$.

29 $z = -20i \Rightarrow r = 20$. θ on the negative y-axis $\Rightarrow \theta = \frac{3\pi}{2}$. $z = 20\,\text{cis}\,\frac{3\pi}{2}$.

31 $z = 12 \Rightarrow r = 12$. θ on the positive x-axis $\Rightarrow \theta = 0$. $z = 12\,\text{cis}\,0$.

33 $z = -7 \Rightarrow r = 7$. θ on the negative x-axis $\Rightarrow \theta = \pi$. $z = 7\,\text{cis}\,\pi$.

35 $z = 6i \Rightarrow r = 6$. θ on the positive y-axis $\Rightarrow \theta = \frac{\pi}{2}$. $z = 6\,\text{cis}\,\frac{\pi}{2}$.

39 $z = 2 + i \Rightarrow r = \sqrt{2^2 + 1^2} = \sqrt{5}$. $\tan\theta = \frac{1}{2}$ and θ in QI $\Rightarrow \theta = \tan^{-1}\frac{1}{2}$.

$z = \sqrt{5}\,\text{cis}\left(\tan^{-1}\frac{1}{2}\right)$.

41 $z = -3 + i \Rightarrow r = \sqrt{(-3)^2 + 1^2} = \sqrt{10}$.

$\tan\theta = \frac{1}{-3}$ and θ in QII $\Rightarrow \theta = \tan^{-1}\left(-\frac{1}{3}\right) + \pi$. We must add π to $\tan^{-1}\left(-\frac{1}{3}\right)$ because $-\frac{\pi}{2} < \tan^{-1}\left(-\frac{1}{3}\right) < 0$ and we want θ to be in the interval $\left(\frac{\pi}{2}, \pi\right)$.

$z = \sqrt{10}\,\text{cis}\left[\tan^{-1}\left(-\frac{1}{3}\right) + \pi\right]$.

43 $z = -5 - 3i \Rightarrow r = \sqrt{(-5)^2 + (-3)^2} = \sqrt{34}$. $\tan\theta = \frac{-3}{-5} = \frac{3}{5}$ and

θ in QIII $\Rightarrow \theta = \tan^{-1}\frac{3}{5} + \pi$. We must add π to $\tan^{-1}\frac{3}{5}$ because $0 < \tan^{-1}\frac{3}{5} < \frac{\pi}{2}$ and we want θ to be in the interval $\left(\pi, \frac{3\pi}{2}\right)$. $z = \sqrt{34}\,\text{cis}\left(\tan^{-1}\frac{3}{5} + \pi\right)$.

45 $z = 4 - 3i \Rightarrow r = \sqrt{4^2 + (-3)^2} = \sqrt{25} = 5$.

$\tan\theta = \frac{-3}{4}$ and θ in QIV $\Rightarrow \theta = \tan^{-1}\left(-\frac{3}{4}\right) + 2\pi$. We must add 2π to $\tan^{-1}\left(-\frac{3}{4}\right)$ because $-\frac{\pi}{2} < \tan^{-1}\left(-\frac{3}{4}\right) < 0$ and we want θ to be in the interval $\left(\frac{3\pi}{2}, 2\pi\right)$.

$z = 5\,\text{cis}\left[\tan^{-1}\left(-\frac{3}{4}\right) + 2\pi\right]$.

47 $4\left(\cos\frac{\pi}{4} + i\sin\frac{\pi}{4}\right) = 4\left(\frac{\sqrt{2}}{2} + \frac{\sqrt{2}}{2}i\right) = 2\sqrt{2} + 2\sqrt{2}\,i$

51 $5\left(\cos\pi + i\sin\pi\right) = 5\left(-1 + 0i\right) = -5$

$\boxed{53}$ For any given angle θ, where $\theta = \tan^{-1}\frac{y}{x}$, we have $r = \sqrt{x^2 + y^2}$. In this case,

$\theta = \tan^{-1}\frac{3}{5}$, so $r = \sqrt{3^2 + 5^2} = \sqrt{34}$. You may want to draw a figure to represent x,

y, and r, as we did in previous chapters. Also, note that $\cos\theta = \frac{x}{r}$ and $\sin\theta = \frac{y}{r}$.

$$\sqrt{34}\operatorname{cis}\left(\tan^{-1}\tfrac{3}{5}\right) = \sqrt{34}\left[\cos\left(\tan^{-1}\tfrac{3}{5}\right) + i\sin\left(\tan^{-1}\tfrac{3}{5}\right)\right] = \sqrt{34}\left(\frac{5}{\sqrt{34}} + \frac{3}{\sqrt{34}}i\right) = 5 + 3i$$

$\boxed{55}$ $\sqrt{5}\operatorname{cis}\left[\tan^{-1}\left(-\tfrac{1}{2}\right)\right] = \sqrt{5}\left\{\cos\left[\tan^{-1}\left(-\tfrac{1}{2}\right)\right] + i\sin\left[\tan^{-1}\left(-\tfrac{1}{2}\right)\right]\right\} =$

$$\sqrt{5}\left(\frac{2}{\sqrt{5}} - \frac{1}{\sqrt{5}}i\right) = 2 - i$$

Note: For Exercises 57–64, the trigonometric forms for z_1 and z_2 are listed and then used in the theorem in this section. The trigonometric forms can be found as in Exercises 21–46.

$\boxed{57}$ $z_1 = \sqrt{2}\operatorname{cis}\frac{3\pi}{4}$ and $z_2 = \sqrt{2}\operatorname{cis}\frac{\pi}{4}$. $z_1 z_2 = \sqrt{2}\cdot\sqrt{2}\operatorname{cis}\left(\frac{3\pi}{4} + \frac{\pi}{4}\right) = 2\operatorname{cis}\pi = -2 + 0i$.

$$\frac{z_1}{z_2} = \frac{\sqrt{2}}{\sqrt{2}}\operatorname{cis}\left(\frac{3\pi}{4} - \frac{\pi}{4}\right) = 1\operatorname{cis}\frac{\pi}{2} = 0 + i$$

$\boxed{59}$ $z_1 = 4\operatorname{cis}\frac{4\pi}{3}$ and $z_2 = 5\operatorname{cis}\frac{\pi}{2}$. $z_1 z_2 = 4\cdot 5\operatorname{cis}\left(\frac{4\pi}{3} + \frac{\pi}{2}\right) = 20\operatorname{cis}\frac{11\pi}{6} = 10\sqrt{3} - 10i$.

$$\frac{z_1}{z_2} = \frac{4}{5}\operatorname{cis}\left(\frac{4\pi}{3} - \frac{\pi}{2}\right) = \frac{4}{5}\operatorname{cis}\frac{5\pi}{6} = -\frac{2}{5}\sqrt{3} + \frac{2}{5}i.$$

$\boxed{61}$ $z_1 = 10\operatorname{cis}\pi$ and $z_2 = 4\operatorname{cis}\pi$. $z_1 z_2 = 10\cdot 4\operatorname{cis}(\pi + \pi) = 40\operatorname{cis}2\pi = 40 + 0i$.

$$\frac{z_1}{z_2} = \frac{10}{4}\operatorname{cis}(\pi - \pi) = \frac{5}{2}\operatorname{cis}0 = \frac{5}{2} + 0i.$$

$\boxed{63}$ $z_1 = 4\operatorname{cis}0$ and $z_2 = \sqrt{5}\operatorname{cis}\left[\tan^{-1}\left(-\tfrac{1}{2}\right)\right]$. Let $\theta = \tan^{-1}\left(-\tfrac{1}{2}\right)$.

Thus $\cos\theta = \frac{2}{\sqrt{5}}$ and $\sin\theta = -\frac{1}{\sqrt{5}}$.

$$z_1 z_2 = 4\cdot\sqrt{5}\operatorname{cis}(0 + \theta) = 4\sqrt{5}(\cos\theta + i\sin\theta) = 4\sqrt{5}\left(\frac{2}{\sqrt{5}} + \frac{-1}{\sqrt{5}}i\right) = 8 - 4i.$$

$$\frac{z_1}{z_2} = \frac{4}{\sqrt{5}}\operatorname{cis}(0 - \theta) = \frac{4}{\sqrt{5}}\left[\cos(-\theta) + i\sin(-\theta)\right]$$

$$= \frac{4}{\sqrt{5}}(\cos\theta - i\sin\theta) = \frac{4}{\sqrt{5}}\left(\frac{2}{\sqrt{5}} + \frac{1}{\sqrt{5}}i\right) = \frac{8}{5} + \frac{4}{5}i.$$

$\boxed{65}$ Let $z_1 = r_1\operatorname{cis}\theta_1$ and $z_2 = r_2\operatorname{cis}\theta_2$.

$$\frac{z_1}{z_2} = \frac{r_1\operatorname{cis}\theta_1}{r_2\operatorname{cis}\theta_2} = \frac{r_1(\cos\theta_1 + i\sin\theta_1)(\cos\theta_2 - i\sin\theta_2)}{r_2(\cos\theta_2 + i\sin\theta_2)(\cos\theta_2 - i\sin\theta_2)}$$

$\{$ multiplying by the conjugate of the denominator $\}$

$$= \frac{r_1\left[(\cos\theta_1\,\cos\theta_2 + \sin\theta_1\,\sin\theta_2) + i(\sin\theta_1\,\cos\theta_2 - \sin\theta_2\,\cos\theta_1)\right]}{r_2\left[(\cos^2\theta_2 + \sin^2\theta_2) + i(\sin\theta_2\,\cos\theta_2 - \cos\theta_2\,\sin\theta_2)\right]}$$

$$= \frac{r_1\left[\cos(\theta_1 - \theta_2) + i\sin(\theta_1 - \theta_2)\right]}{r_2(1 + 0i)} = \frac{r_1}{r_2}\operatorname{cis}(\theta_1 - \theta_2).$$

$\boxed{67}$ The unknown quantity is V: $I = V/Z \Rightarrow$

$\quad V = IZ = (10 \operatorname{cis} 35°)(3 \operatorname{cis} 20°) = (10 \times 3) \operatorname{cis}(35° + 20°) = 30 \operatorname{cis} 55° \approx 17.21 + 24.57i$.

$\boxed{69}$ The unknown quantity is Z:

$$I = \frac{V}{Z} \Rightarrow Z = \frac{V}{I} = \frac{115 \operatorname{cis} 45°}{8 \operatorname{cis} 5°} = (115 \div 8) \operatorname{cis}(45° - 5°) = 14.375 \operatorname{cis} 40° \approx 11.01 + 9.24i$$

$\boxed{71}$ $Z = 14 - 13i \Rightarrow |Z| = \sqrt{14^2 + (-13)^2} = \sqrt{365} \approx 19.1$ ohms

$\boxed{73}$ $I = \frac{V}{Z} \Rightarrow V = IZ = (4 \operatorname{cis} 90°)[18 \operatorname{cis}(-78°)] = 72 \operatorname{cis} 12° \approx 70.43 + 14.97i$

8.4 Exercises

Note: In this section, it is assumed that the reader can transform complex numbers to
their trigonometric from. If this is not true, see Exercises 8.3.

$\boxed{1}$ $(3 + 3i)^5 = (3\sqrt{2} \operatorname{cis} \frac{\pi}{4})^5 = (3\sqrt{2})^5 \operatorname{cis}(5 \cdot \frac{\pi}{4}) = (3\sqrt{2})^5 \operatorname{cis} \frac{5\pi}{4} =$

$$972\sqrt{2}\left(-\frac{\sqrt{2}}{2} - \frac{\sqrt{2}}{2}i\right) = -972 - 972i$$

$\boxed{3}$ $(1 - i)^{10} = (\sqrt{2} \operatorname{cis} \frac{7\pi}{4})^{10} = (\sqrt{2})^{10} \operatorname{cis}(10 \cdot \frac{7\pi}{4}) = (\sqrt{2})^{10} \operatorname{cis} \frac{35\pi}{2} =$

$$2^5 \operatorname{cis}(16\pi + \frac{3\pi}{2}) = 32 \operatorname{cis} \frac{3\pi}{2} = 32(0 - i) = -32i$$

$\boxed{7}$ $\left(-\frac{\sqrt{2}}{2} + \frac{\sqrt{2}}{2}i\right)^{15} = (1 \operatorname{cis} \frac{3\pi}{4})^{15} = 1^{15} \operatorname{cis} \frac{45\pi}{4} = \operatorname{cis} \frac{5\pi}{4} = -\frac{\sqrt{2}}{2} - \frac{\sqrt{2}}{2}i$

$\boxed{9}$ $\left(-\frac{\sqrt{3}}{2} - \frac{1}{2}i\right)^{20} = (1 \operatorname{cis} \frac{7\pi}{6})^{20} = 1^{20} \operatorname{cis} \frac{70\pi}{3} = \operatorname{cis} \frac{4\pi}{3} = -\frac{1}{2} - \frac{\sqrt{3}}{2}i$

$\boxed{13}$ $1 + \sqrt{3}\,i = 2 \operatorname{cis} 60°$. $w_k = \sqrt{2} \operatorname{cis}\left(\dfrac{60° + 360° k}{2}\right)$ for $k = 0, 1$.

$$w_0 = \sqrt{2} \operatorname{cis} 30° = \sqrt{2}\left(\frac{\sqrt{3}}{2} + \frac{1}{2}i\right) = \frac{\sqrt{6}}{2} + \frac{\sqrt{2}}{2}i.$$

$$w_1 = \sqrt{2} \operatorname{cis} 210° = \sqrt{2}\left(-\frac{\sqrt{3}}{2} - \frac{1}{2}i\right) = -\frac{\sqrt{6}}{2} - \frac{\sqrt{2}}{2}i.$$

$\boxed{15}$ $-1 - \sqrt{3}\,i = 2 \operatorname{cis} 240°$. $w_k = \sqrt[4]{2} \operatorname{cis}\left(\dfrac{240° + 360° k}{4}\right)$ for $k = 0, 1, 2, 3$.

$$w_0 = \sqrt[4]{2} \operatorname{cis} 60° = \sqrt[4]{2}\left(\frac{1}{2} + \frac{\sqrt{3}}{2}i\right) = \frac{\sqrt[4]{2}}{2} + \frac{\sqrt[4]{18}}{2}i.$$

$$\{\text{since } \sqrt[4]{2} \cdot \sqrt{3} = \sqrt[4]{2} \cdot \sqrt[4]{9} = \sqrt[4]{18}\}$$

$$w_1 = \sqrt[4]{2} \operatorname{cis} 150° = \sqrt[4]{2}\left(-\frac{\sqrt{3}}{2} + \frac{1}{2}i\right) = -\frac{\sqrt[4]{18}}{2} + \frac{\sqrt[4]{2}}{2}i.$$

$$w_2 = \sqrt[4]{2} \operatorname{cis} 240° = \sqrt[4]{2}\left(-\frac{1}{2} - \frac{\sqrt{3}}{2}i\right) = -\frac{\sqrt[4]{2}}{2} - \frac{\sqrt[4]{18}}{2}i.$$

$$w_3 = \sqrt[4]{2} \operatorname{cis} 330° = \sqrt[4]{2}\left(\frac{\sqrt{3}}{2} - \frac{1}{2}i\right) = \frac{\sqrt[4]{18}}{2} - \frac{\sqrt[4]{2}}{2}i.$$

$\boxed{17}$ $-27i = 27\operatorname{cis}270°.$ $w_k = \sqrt[3]{27}\operatorname{cis}\left(\dfrac{270° + 360°k}{3}\right)$ for $k = 0,\ 1,\ 2.$

$$w_0 = 3\operatorname{cis}90° = 3(0 + i) = 3i.$$

$$w_1 = 3\operatorname{cis}210° = 3\left(-\frac{\sqrt{3}}{2} - \tfrac{1}{2}i\right) = -\frac{3\sqrt{3}}{2} - \frac{3}{2}i.$$

$$w_2 = 3\operatorname{cis}330° = 3\left(\frac{\sqrt{3}}{2} - \tfrac{1}{2}i\right) = \frac{3\sqrt{3}}{2} - \frac{3}{2}i.$$

$\boxed{19}$ $1 = 1\operatorname{cis}0°.$ $w_k = \sqrt[6]{1}\operatorname{cis}\left(\dfrac{0° + 360°k}{6}\right)$ for $k = 0,\ 1,\ 2,\ 3,\ 4,\ 5.$

$$w_0 = 1\operatorname{cis}0° = 1 + 0i. \qquad\qquad w_1 = 1\operatorname{cis}60° = \tfrac{1}{2} + \frac{\sqrt{3}}{2}i.$$

$$w_2 = 1\operatorname{cis}120° = -\tfrac{1}{2} + \frac{\sqrt{3}}{2}i. \qquad\qquad w_3 = 1\operatorname{cis}180° = -1 + 0i.$$

$$w_4 = 1\operatorname{cis}240° = -\tfrac{1}{2} - \frac{\sqrt{3}}{2}i. \qquad\qquad w_5 = 1\operatorname{cis}300° = \tfrac{1}{2} - \frac{\sqrt{3}}{2}i.$$

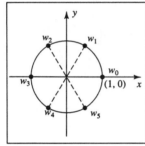

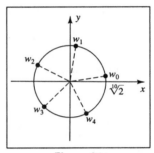

Figure 19   Figure 21

$\boxed{21}$ $1 + i = \sqrt{2}\operatorname{cis}45°.$ $w_k = \sqrt{\sqrt[5]{2}}\operatorname{cis}\left(\dfrac{45° + 360°k}{5}\right)$ for $k = 0,\ 1,\ 2,\ 3,\ 4.$

$$w_k = \sqrt[10]{2}\operatorname{cis}\theta \text{ with } \theta = 9°,\ 81°,\ 153°,\ 225°,\ 297°.$$

$\boxed{23}$ $x^4 - 16 = 0 \Rightarrow x^4 = 16.$ The problem is now to find the 4 fourth roots of 16.

$$16 = 16 + 0i = 16\operatorname{cis}0°.\quad w_k = \sqrt[4]{16}\operatorname{cis}\left(\dfrac{0° + 360°k}{4}\right) \text{ for } k = 0,\ 1,\ 2,\ 3.$$

$$w_0 = 2\operatorname{cis}0° = 2(1 + 0i) = 2. \qquad\qquad w_1 = 2\operatorname{cis}90° = 2(0 + i) = 2i.$$

$$w_2 = 2\operatorname{cis}180° = 2(-1 + 0i) = -2. \qquad\qquad w_3 = 2\operatorname{cis}270° = 2(0 - i) = -2i.$$

$\boxed{25}$ $x^6 + 64 = 0 \Rightarrow x^6 = -64.$ The problem is now to find the 6 sixth roots of $-64.$

$$-64 = -64 + 0i = 64\operatorname{cis}180°.\quad w_k = \sqrt[6]{64}\operatorname{cis}\left(\dfrac{180° + 360°k}{6}\right) \text{ for } k = 0,\ 1,\ \dots,\ 5.$$

$$w_0 = 2\operatorname{cis}30° = 2\left(\frac{\sqrt{3}}{2} + \tfrac{1}{2}i\right) = \sqrt{3} + i.$$

$$w_1 = 2\operatorname{cis}90° = 2(0 + i) = 2i.$$

$$w_2 = 2\operatorname{cis}150° = 2\left(-\frac{\sqrt{3}}{2} + \tfrac{1}{2}i\right) = -\sqrt{3} + i. \qquad\qquad \text{(continued)}$$

$$w_3 = 2\operatorname{cis}210° = 2\left(-\frac{\sqrt{3}}{2} - \frac{1}{2}i\right) = -\sqrt{3} - i.$$

$$w_4 = 2\operatorname{cis}270° = 2(0 - i) = -2i.$$

$$w_5 = 2\operatorname{cis}330° = 2\left(\frac{\sqrt{3}}{2} - \frac{1}{2}i\right) = \sqrt{3} - i.$$

27 $x^3 + 8i = 0 \Rightarrow x^3 = -8i.$ The problem is now to find the 3 cube roots of $-8i$.

$$-8i = 0 - 8i = 8\operatorname{cis}270°.\quad w_k = \sqrt[3]{8}\operatorname{cis}\left(\frac{270° + 360°k}{3}\right) \text{ for } k = 0, 1, 2.$$

$$w_0 = 2\operatorname{cis}90° = 2(0 + i) = 2i.$$

$$w_1 = 2\operatorname{cis}210° = 2\left(-\frac{\sqrt{3}}{2} - \frac{1}{2}i\right) = -\sqrt{3} - i.$$

$$w_2 = 2\operatorname{cis}330° = 2\left(\frac{\sqrt{3}}{2} - \frac{1}{2}i\right) = \sqrt{3} - i.$$

29 $x^5 - 243 = 0 \Rightarrow x^5 = 243.$ The problem is now to find the 5 fifth roots of 243.

$$243 = 243 + 0i = 243\operatorname{cis}0°.\quad w_k = \sqrt[5]{243}\operatorname{cis}\left(\frac{0° + 360°k}{5}\right) \text{ for } k = 0, 1, 2, 3, 4.$$

$$w_k = 3\operatorname{cis}\theta \text{ with } \theta = 0°, 72°, 144°, 216°, 288°.$$

8.5 Exercises

1 $\mathbf{a} + \mathbf{b} = \langle 2, -3\rangle + \langle 1, 4\rangle = \langle 2 + 1, -3 + 4\rangle = \langle 3, 1\rangle.$

$\mathbf{a} - \mathbf{b} = \langle 2, -3\rangle - \langle 1, 4\rangle = \langle 2 - 1, -3 - 4\rangle = \langle 1, -7\rangle.$

$4\mathbf{a} + 5\mathbf{b} = 4\langle 2, -3\rangle + 5\langle 1, 4\rangle = \langle 4(2), 4(-3)\rangle + \langle 5(1), 5(4)\rangle$
$$= \langle 8, -12\rangle + \langle 5, 20\rangle = \langle 8 + 5, -12 + 20\rangle = \langle 13, 8\rangle.$$

From above, $4\mathbf{a} - 5\mathbf{b} = \langle 8, -12\rangle - \langle 5, 20\rangle = \langle 8 - 5, -12 - 20\rangle = \langle 3, -32\rangle.$

3 Simplifying $\mathbf{a}$ and $\mathbf{b}$ first, we have $\mathbf{a} = -\langle 7, -2\rangle = \langle -(7), -(-2)\rangle = \langle -7, 2\rangle$ and
$$\mathbf{b} = 4\langle -2, 1\rangle = \langle 4(-2), 4(1)\rangle = \langle -8, 4\rangle.$$

$\mathbf{a} + \mathbf{b} = \langle -7, 2\rangle + \langle -8, 4\rangle = \langle -7 + (-8), 2 + 4\rangle = \langle -15, 6\rangle.$

$\mathbf{a} - \mathbf{b} = \langle -7, 2\rangle - \langle -8, 4\rangle = \langle -7 - (-8), 2 - 4\rangle = \langle 1, -2\rangle.$

$4\mathbf{a} + 5\mathbf{b} = 4\langle -7, 2\rangle + 5\langle -8, 4\rangle = \langle -28, 8\rangle + \langle -40, 20\rangle = \langle -68, 28\rangle.$

From above, $4\mathbf{a} - 5\mathbf{b} = \langle -28, 8\rangle - \langle -40, 20\rangle = \langle 12, -12\rangle.$

5 $\mathbf{a} + \mathbf{b} = (\mathbf{i} + 2\mathbf{j}) + (3\mathbf{i} - 5\mathbf{j}) = (1 + 3)\mathbf{i} + (2 - 5)\mathbf{j} = 4\mathbf{i} - 3\mathbf{j}.$

$\mathbf{a} - \mathbf{b} = (\mathbf{i} + 2\mathbf{j}) - (3\mathbf{i} - 5\mathbf{j}) = (1 - 3)\mathbf{i} + (2 - (-5))\mathbf{j} = -2\mathbf{i} + 7\mathbf{j}.$

$4\mathbf{a} + 5\mathbf{b} = 4(\mathbf{i} + 2\mathbf{j}) + 5(3\mathbf{i} - 5\mathbf{j}) = (4\mathbf{i} + 8\mathbf{j}) + (15\mathbf{i} - 25\mathbf{j}) = 19\mathbf{i} - 17\mathbf{j}.$

From above, $4\mathbf{a} - 5\mathbf{b} = (4\mathbf{i} + 8\mathbf{j}) - (15\mathbf{i} - 25\mathbf{j}) = -11\mathbf{i} + 33\mathbf{j}.$

$\boxed{7}$ $\mathbf{a} = 3\mathbf{i} + 2\mathbf{j}$ and $\mathbf{b} = -\mathbf{i} + 5\mathbf{j} \Rightarrow \mathbf{a} + \mathbf{b} = 2\mathbf{i} + 7\mathbf{j}$, $2\mathbf{a} = 6\mathbf{i} + 4\mathbf{j}$, and $-3\mathbf{b} = 3\mathbf{i} - 15\mathbf{j}$.

Terminal points of the vectors are $(3,\, 2)$, $(-1,\, 5)$, $(2,\, 7)$, $(6,\, 4)$, and $(3,\, -15)$.

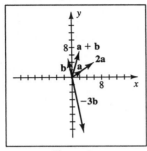

Figure 7

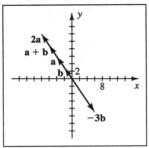

Figure 9

$\boxed{9}$ $\mathbf{a} = \langle -4,\, 6 \rangle$ and $\mathbf{b} = \langle -2,\, 3 \rangle \Rightarrow \mathbf{a} + \mathbf{b} = \langle -6,\, 9 \rangle$, $2\mathbf{a} = \langle -8,\, 12 \rangle$, and $-3\mathbf{b} = \langle 6,\, -9 \rangle$.

Terminal points of the vectors are $(-4,\, 6)$, $(-2,\, 3)$, $(-6,\, 9)$, $(-8,\, 12)$, and $(6,\, -9)$.

$\boxed{11}$ $\mathbf{a} + \mathbf{b} = \langle 2,\, 0 \rangle + \langle -1,\, 0 \rangle = \langle 1,\, 0 \rangle = -\langle -1,\, 0 \rangle = -\mathbf{b}$

$\boxed{13}$ $\mathbf{b} + \mathbf{e} = \langle -1,\, 0 \rangle + \langle 2,\, 2 \rangle = \langle 1,\, 2 \rangle = \mathbf{f}$

$\boxed{15}$ $\mathbf{b} + \mathbf{d} = \langle -1,\, 0 \rangle + \langle 0,\, -1 \rangle = \langle -1,\, -1 \rangle = -\frac{1}{2}\langle 2,\, 2 \rangle = -\frac{1}{2}\mathbf{e}$

$\boxed{17}$ $\mathbf{a} + (\mathbf{b} + \mathbf{c}) = \langle a_1,\, a_2 \rangle + (\langle b_1,\, b_2 \rangle + \langle c_1,\, c_2 \rangle)$

$\qquad\qquad\quad = \langle a_1,\, a_2 \rangle + \langle b_1 + c_1,\, b_2 + c_2 \rangle$

$\qquad\qquad\quad = \langle a_1 + b_1 + c_1,\, a_2 + b_2 + c_2 \rangle$

$\qquad\qquad\quad = \langle a_1 + b_1,\, a_2 + b_2 \rangle + \langle c_1,\, c_2 \rangle$

$\qquad\qquad\quad = (\langle a_1,\, a_2 \rangle + \langle b_1,\, b_2 \rangle) + \langle c_1,\, c_2 \rangle = (\mathbf{a} + \mathbf{b}) + \mathbf{c}$

$\boxed{21}$ $(mn)\mathbf{a} = (mn)\langle a_1,\, a_2 \rangle$

$\qquad\quad = \langle (mn)\,a_1,\, (mn)\,a_2 \rangle$

$\qquad\quad = \langle mna_1,\, mna_2 \rangle$

$\qquad\quad = m\langle na_1,\, na_2 \rangle \qquad$ or $n\langle ma_1,\, ma_2 \rangle$

$\qquad\quad = m(n\langle a_1,\, a_2 \rangle) \qquad$ or $n(m\langle a_1,\, a_2 \rangle)$

$\qquad\quad = m(n\mathbf{a}) \qquad\qquad$ or $n(m\mathbf{a})$

$\boxed{25}$ $-(\mathbf{a} + \mathbf{b}) = -(\langle a_1,\, a_2 \rangle + \langle b_1,\, b_2 \rangle)$

$\qquad\qquad\quad = -(\langle a_1 + b_1,\, a_2 + b_2 \rangle)$

$\qquad\qquad\quad = \langle -(a_1 + b_1),\, -(a_2 + b_2) \rangle$

$\qquad\qquad\quad = \langle -a_1 - b_1,\, -a_2 - b_2 \rangle$

$\qquad\qquad\quad = \langle -a_1,\, -a_2 \rangle + \langle -b_1,\, -b_2 \rangle$

$\qquad\qquad\quad = -\mathbf{a} + (-\mathbf{b}) = -\mathbf{a} - \mathbf{b}$

$\boxed{27}$ $\| 2\mathbf{v} \| = \| 2\langle a,\, b \rangle \| = \| \langle 2a,\, 2b \rangle \| = \sqrt{(2a)^2 + (2b)^2} = \sqrt{4a^2 + 4b^2} =$

$\qquad\qquad\qquad\qquad\qquad\qquad\qquad 2\sqrt{a^2 + b^2} = 2\| \langle a,\, b \rangle \| = 2\| \mathbf{v} \|$

$\boxed{29}$ $\| \mathbf{a} \| = \sqrt{3^2 + (-3)^2} = \sqrt{18} = 3\sqrt{2}$. $\tan\theta = \frac{-3}{3} = -1$ and θ in QIV $\Rightarrow \theta = \frac{7\pi}{4}$.

[31] $\|\mathbf{a}\| = 5$. The terminal side of θ is on the negative x-axis $\Rightarrow \theta = \pi$.

[33] $\|\mathbf{a}\| = \sqrt{41}$. $\tan\theta = \frac{5}{-4}$ and θ in QII $\Rightarrow \theta = \tan^{-1}\left(-\frac{5}{4}\right) + \pi$.

[35] $\|\mathbf{a}\| = 18$. The terminal side of θ is on the negative y-axis $\Rightarrow \theta = \frac{3\pi}{2}$.

Note: Exercises 37–42: Each resultant force is found by completing the parallelogram and then applying the law of cosines.

[37] $\|\mathbf{r}\| = \sqrt{40^2 + 70^2 - 2(40)(70)\cos 135°} = \sqrt{6500 + 2800\sqrt{2}} \approx 102.3$, or 102 lb.

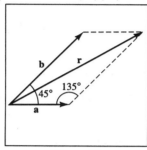

Figure 37

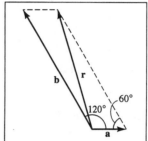

Figure 39

[39] $\|\mathbf{r}\| = \sqrt{2^2 + 8^2 - 2(2)(8)\cos 60°} = \sqrt{68 - 16} = \sqrt{52} \approx 7.2$ kg.

[41] $\|\mathbf{r}\| = \sqrt{90^2 + 60^2 - 2(90)(60)\cos 70°} \approx 89.48$, or 89 kg.

Using the law of cosines, $\alpha = \cos^{-1}\left(\dfrac{90^2 + \|\mathbf{r}\|^2 - 60^2}{2(90)(\|\mathbf{r}\|)}\right) \approx$

$\cos^{-1}(0.7765) \approx 39°$, which is 24° under the negative x-axis.

This angle is 204°, or S66°W.

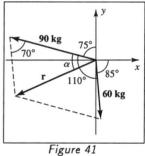

Figure 41

[43] We will use a component approach for this exercise.

(**a**) $= \langle 6\cos 110°, 6\sin 110° \rangle \approx \langle -2.05, 5.64 \rangle$.

(**b**) $= \langle 2\cos 215°, 2\sin 215° \rangle \approx \langle -1.64, -1.15 \rangle$.

$\mathbf{a} + \mathbf{b} \approx \langle -3.69, 4.49 \rangle$ and $\|\mathbf{a} + \mathbf{b}\| \approx 5.8$ lb. $\tan\theta \approx \frac{4.49}{-3.69} \Rightarrow \theta \approx 129°$ since θ is in QII.

[45] Horizontal $= 50\cos 35° \approx 40.96$. Vertical $= 50\sin 35° \approx 28.68$.

[47] Horizontal $= 20\cos 108° \approx -6.18$. Vertical $= 20\sin 108° \approx 19.02$.

[49] (a) $\mathbf{F} = \mathbf{F_1} + \mathbf{F_2} + \mathbf{F_3} = \langle 4, 3 \rangle + \langle -2, -3 \rangle + \langle 5, 2 \rangle = \langle 7, 2 \rangle$.

(b) $\mathbf{F} + \mathbf{G} = \mathbf{0} \Rightarrow \mathbf{G} = -\mathbf{F} = \langle -7, -2 \rangle$.

[51] (a) $\mathbf{F} = \mathbf{F_1} + \mathbf{F_2} = \langle 6\cos 130°, 6\sin 130° \rangle + \langle 4\cos(-120°), 4\sin(-120°) \rangle \approx \langle -5.86, 1.13 \rangle$.

(b) $\mathbf{F} + \mathbf{G} = \mathbf{0} \Rightarrow \mathbf{G} = -\mathbf{F} \approx \langle 5.86, -1.13 \rangle$.

53 The vertical components of the forces must add up to zero for the large ship to move along the line segment AB. The vertical component of the smaller tug is $3200 \sin(-30°) = -1600$. The vertical component of the larger tug is $4000 \sin\theta$.

$$4000 \sin\theta = 1600 \Rightarrow \theta = \sin^{-1}\left(\frac{1600}{4000}\right) = \sin^{-1}(0.4) \approx 23.6°.$$

Note: Exercises 55–60: Measure angles from the positive x-axis.

55 $\mathbf{p} = \langle 200 \cos 40°,\ 200 \sin 40° \rangle \approx \langle 153.21, 128.56 \rangle$. $\mathbf{w} = \langle 40 \cos 0°,\ 40 \sin 0° \rangle = \langle 40,\ 0 \rangle$.

$\mathbf{p} + \mathbf{w} \approx \langle 193.21, 128.56 \rangle$ and $\| \mathbf{p} + \mathbf{w} \| \approx 232.07$, or 232 mi/hr.

$\tan\theta \approx \frac{128.56}{193.21} \Rightarrow \theta \approx 34°$. The true course is then $N(90° - 34°)E$, or $N56°E$.

57 $\mathbf{w} = \langle 50 \cos 90°,\ 50 \sin 90° \rangle = \langle 0,\ 50 \rangle$.

$\mathbf{r} = \langle 400 \cos 200°,\ 400 \sin 200° \rangle \approx \langle -375.88, -136.81 \rangle$, where $\mathbf{r}$ is the desired resultant of $\mathbf{p} + \mathbf{w}$. Since $\mathbf{r} = \mathbf{p} + \mathbf{w}$, $\mathbf{p} = \mathbf{r} - \mathbf{w} \approx \langle -375.88, -186.81 \rangle$. $\| \mathbf{p} \| \approx 419.74$, or 420 mi/hr.

$\tan\theta \approx \frac{-186.81}{-375.88}$ and θ is in QIII $\Rightarrow \theta \approx 206°$ from the positive x-axis,

or 244° using the directional form.

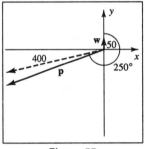

Figure 57

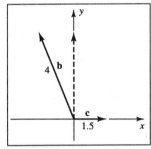

Figure 59

59 Let the vectors $\mathbf{c}$, $\mathbf{b}$, and $\mathbf{r}$ denote the current, the boat, and the resultant, respectively. $\mathbf{c} = \langle 1.5 \cos 0°,\ 1.5 \sin 0° \rangle = \langle 1.5,\ 0 \rangle$. $\mathbf{r} = \langle s \cos 90°,\ s \sin 90° \rangle = \langle 0,\ s \rangle$, where s is the resulting speed. $\mathbf{b} = \langle 4 \cos\theta,\ 4 \sin\theta \rangle$. Also, $\mathbf{b} = \mathbf{r} - \mathbf{c} = \langle -1.5,\ s \rangle$.

$$4 \cos\theta = -1.5 \Rightarrow \theta \approx 112°, \text{ or } N22°W.$$

61 In the figure in the text, suppose $\mathbf{v}_1$ was pointing upwards instead of downwards. If we then write $\mathbf{v}_1$ in terms of its vertical and horizontal components, we have $\mathbf{v}_1 = \| \mathbf{v}_1 \| \cos\theta_1 \mathbf{j} - \| \mathbf{v}_1 \| \sin\theta_1 \mathbf{i}$. We want the negative of this vector since $\mathbf{v}_1$ is actually pointing downward, so using $\| \mathbf{v}_1 \| = 8.2$ and $\theta_1 = 30°$, we obtain $\mathbf{v}_1 = \| \mathbf{v}_1 \| \sin\theta_1 \mathbf{i} - \| \mathbf{v}_1 \| \cos\theta_1 \mathbf{j} = 8.2(\frac{1}{2}) \mathbf{i} - 8.2(\sqrt{3}/2) \mathbf{j} = 4.1 \mathbf{i} - 4.1\sqrt{3} \mathbf{j} \approx 4.1 \mathbf{i} - 7.10 \mathbf{j}$. The angle θ_2 can now be computed using the given relationship in the text.

$$\frac{\| \mathbf{v}_1 \|}{\| \mathbf{v}_2 \|} = \frac{\tan\theta_1}{\tan\theta_2} \Rightarrow \tan\theta_2 = \frac{\| \mathbf{v}_2 \|}{\| \mathbf{v}_1 \|} \tan\theta_1 = \frac{3.8}{8.2} \times \frac{1}{\sqrt{3}} \Rightarrow \tan\theta_2 \approx 0.2676 \Rightarrow \theta_2 \approx 14.98°.$$

We now write $\mathbf{v}_2$ in terms of its horizontal and vertical components, and using

$$\| \mathbf{v}_2 \| = 3.8, \text{ it follows that } \mathbf{v}_2 = \| \mathbf{v}_2 \| \sin\theta_2 \mathbf{i} - \| \mathbf{v}_2 \| \cos\theta_2 \mathbf{j} \approx 0.98 \mathbf{i} - 3.67 \mathbf{j}.$$

43 (a) The horizontal component has magnitude 93×10^6 and the vertical component has magnitude 0.432×10^6.

Thus, $\mathbf{v} = (93 \times 10^6)\mathbf{i} + (0.432 \times 10^6)\mathbf{j}$ and $\mathbf{w} = (93 \times 10^6)\mathbf{i} - (0.432 \times 10^6)\mathbf{j}$.

(b) $\cos\theta = \dfrac{\mathbf{v} \cdot \mathbf{w}}{\|\mathbf{v}\|\|\mathbf{w}\|} = \dfrac{(93 \times 10^6)^2 - (0.432 \times 10^6)^2}{\sqrt{(93 \times 10^6)^2 + (0.432 \times 10^6)^2}\,\sqrt{(93 \times 10^6)^2 + (0.432 \times 10^6)^2}} \approx$

$$0.99995684 \Rightarrow \theta \approx 0.53°.$$

45 $\mathbf{R} = 2(\mathbf{N} \cdot \mathbf{L})\mathbf{N} - \mathbf{L} = 2(\langle 0, 1\rangle \cdot \langle -\frac{4}{5}, \frac{3}{5}\rangle)\langle 0, 1\rangle - \langle -\frac{4}{5}, \frac{3}{5}\rangle = 2(\frac{3}{5})\langle 0, 1\rangle - \langle -\frac{4}{5}, \frac{3}{5}\rangle =$

$$\langle 0, \tfrac{6}{5}\rangle - \langle -\tfrac{4}{5}, \tfrac{3}{5}\rangle = \langle \tfrac{4}{5}, \tfrac{3}{5}\rangle$$

47 Let horizontal ground be represented by $\mathbf{b} = \langle 1, 0\rangle$ (it could be any $\langle a, 0\rangle$).

$$\text{comp}_{\mathbf{b}}\,\mathbf{a} = \frac{\mathbf{a} \cdot \mathbf{b}}{\|\mathbf{b}\|} = \frac{\langle 2.6, 4.5\rangle \cdot \langle 1, 0\rangle}{\|\langle 1, 0\rangle\|} = \frac{2.6}{1} = 2.6 \Rightarrow |\text{comp}_{\mathbf{b}}\,\mathbf{a}| = 2.6$$

49 Let the direction of the ground be represented by $\mathbf{b} = \langle \cos\theta, \sin\theta\rangle = \langle \cos 12°, \sin 12°\rangle$.

$$\text{comp}_{\mathbf{b}}\,\mathbf{a} = \frac{\mathbf{a} \cdot \mathbf{b}}{\|\mathbf{b}\|} = \frac{\langle 25.7, -3.9\rangle \cdot \langle \cos 12°, \sin 12°\rangle}{\|\langle \cos 12°, \sin 12°\rangle\|} \approx \frac{24.33}{1} = 24.33 \Rightarrow |\text{comp}_{\mathbf{b}}\,\mathbf{a}| = 24.33$$

51 From the "Theorem on the Dot Product," we know that $\mathbf{F} \cdot \mathbf{v} = \|\mathbf{F}\|\|\mathbf{v}\|\cos\theta$, so

$$P = \tfrac{1}{550}(\mathbf{F} \cdot \mathbf{v}) = \tfrac{1}{550}\|\mathbf{F}\|\|\mathbf{v}\|\cos\theta = \tfrac{1}{550}(2200)(8)\cos 30° = 16\sqrt{3} \approx 27.7 \text{ horsepower.}$$

Chapter 8 Review Exercises

1 We are given 2 sides of a triangle and the angle between them—so we use the law of cosines to find the third side.

$$a = \sqrt{b^2 + c^2 - 2bc\cos\alpha} = \sqrt{6^2 + 7^2 - 2(6)(7)\cos 60°} = \sqrt{43}.$$

$$\beta = \cos^{-1}\left(\frac{a^2 + c^2 - b^2}{2ac}\right) = \cos^{-1}\left(\frac{43 + 49 - 36}{2\sqrt{43}\,(7)}\right) = \cos^{-1}\left(\frac{4}{\sqrt{43}}\right).$$

$$\gamma = \cos^{-1}\left(\frac{a^2 + b^2 - c^2}{2ab}\right) = \cos^{-1}\left(\frac{43 + 36 - 49}{2\sqrt{43}\,(6)}\right) = \cos^{-1}\left(\frac{5}{2\sqrt{43}}\right).$$

2 $\dfrac{\sin\alpha}{a} = \dfrac{\sin\gamma}{c} \Rightarrow \alpha = \sin^{-1}\left(\dfrac{a\sin\gamma}{c}\right) = \sin^{-1}\left(\dfrac{2\sqrt{3}\cdot\frac{1}{2}}{2}\right) = \sin^{-1}\left(\dfrac{\sqrt{3}}{2}\right) = 60°$ or $120°$.

There are two triangles possible since in either case $\alpha + \gamma < 180°$.

$\beta = (180° - \gamma) - \alpha = (180° - 30°) - (60° \text{ or } 120°) = 90°$ or $30°$.

$$\frac{b}{\sin\beta} = \frac{c}{\sin\gamma} \Rightarrow b = \frac{c\sin\beta}{\sin\gamma} = \frac{2\sin(90° \text{ or } 30°)}{\sin 30°} = 4 \text{ or } 2.$$

$\boxed{3}$ $\gamma = 180° - \alpha - \beta = 180° - 60° - 45° = 75°.$

$\dfrac{a}{\sin \alpha} = \dfrac{b}{\sin \beta} \Rightarrow a = \dfrac{b \sin \alpha}{\sin \beta} = \dfrac{100 \sin 60°}{\sin 45°} = \dfrac{100 \cdot (\sqrt{3}/2)}{\sqrt{2}/2} \cdot \dfrac{\sqrt{2}}{\sqrt{2}} = 50\sqrt{6}.$

$\dfrac{c}{\sin \gamma} = \dfrac{b}{\sin \beta} \Rightarrow c = \dfrac{b \sin \gamma}{\sin \beta} = \dfrac{100 \sin (45° + 30°)}{\sqrt{2}/2} =$

$100\sqrt{2} \left(\sin 45° \cos 30° + \cos 45° \sin 30° \right) = 100\sqrt{2} \left(\dfrac{\sqrt{2}}{2} \cdot \dfrac{\sqrt{3}}{2} + \dfrac{\sqrt{2}}{2} \cdot \dfrac{1}{2} \right) =$

$\dfrac{100}{4} \sqrt{2} (\sqrt{6} + \sqrt{2}) = 25(2\sqrt{3} + 2) = 50(1 + \sqrt{3}).$

$\boxed{4}$ $\alpha = \cos^{-1} \left(\dfrac{b^2 + c^2 - a^2}{2bc} \right) = \cos^{-1} \left(\dfrac{9 + 16 - 4}{2(3)(4)} \right) = \cos^{-1} \left(\dfrac{7}{8} \right).$

$\beta = \cos^{-1} \left(\dfrac{a^2 + c^2 - b^2}{2ac} \right) = \cos^{-1} \left(\dfrac{4 + 16 - 9}{2(2)(4)} \right) = \cos^{-1} \left(\dfrac{11}{16} \right).$

$\gamma = \cos^{-1} \left(\dfrac{a^2 + b^2 - c^2}{2ab} \right) = \cos^{-1} \left(\dfrac{4 + 9 - 16}{2(2)(3)} \right) = \cos^{-1} \left(-\dfrac{1}{4} \right).$

$\boxed{6}$ $\dfrac{\sin \gamma}{c} = \dfrac{\sin \alpha}{a} \Rightarrow \gamma = \sin^{-1} \left(\dfrac{c \sin \alpha}{a} \right) = \sin^{-1} \left(\dfrac{125 \sin 23°30'}{152} \right) \approx \sin^{-1} (0.3279) \approx$

$19°10'$ or $160°50'$ {rounded to the nearest 10 minutes}. Reject $160°50'$ because then

$\alpha + \gamma \geq 180°$. $\beta = 180° - \alpha - \gamma \approx 180° - 23°30' - 19°10' = 137°20'.$

$\dfrac{b}{\sin \beta} = \dfrac{a}{\sin \alpha} \Rightarrow b = \dfrac{a \sin \beta}{\sin \alpha} = \dfrac{152 \sin 137°20'}{\sin 23°30'} \approx 258.3$, or $258.$

$\boxed{9}$ Since we are given two sides and the included angle of a triangle,

we use the formula for the area of a triangle from Section 8.2.

$\mathcal{A} = \frac{1}{2} bc \sin \alpha = \frac{1}{2} (20)(30) \sin 75° \approx 289.8$, or 290 square units.

$\boxed{10}$ Given the three sides of a triangle, we apply Heron's formula to find the area.

$s = \frac{1}{2} (a + b + c) = \frac{1}{2} (4 + 7 + 10) = 10.5.$

$\mathcal{A} = \sqrt{s(s - a)(s - b)(s - c)} = \sqrt{(10.5)(6.5)(3.5)(0.5)} \approx 10.9$ square units.

$\boxed{11}$ $z = -10 + 10i \Rightarrow r = \sqrt{(-10)^2 + 10^2} = \sqrt{200} = 10\sqrt{2}.$

$\tan \theta = \dfrac{10}{-10} = -1$ and θ in QII $\Rightarrow \theta = \dfrac{3\pi}{4}$. $z = 10\sqrt{2} \operatorname{cis} \dfrac{3\pi}{4}.$

$\boxed{13}$ $z = -17 \Rightarrow r = 17.$ θ on the negative x-axis $\Rightarrow \theta = \pi.$ $z = 17 \operatorname{cis} \pi.$

$\boxed{16}$ $z = 4 + 5i \Rightarrow r = \sqrt{4^2 + 5^2} = \sqrt{41}.$ $\tan \theta = \dfrac{5}{4}$ and θ in QI $\Rightarrow \theta = \tan^{-1} \dfrac{5}{4}.$

$z = \sqrt{41} \operatorname{cis} \left(\tan^{-1} \dfrac{5}{4} \right).$

$\boxed{17}$ $20 \left(\cos \dfrac{11\pi}{6} + i \sin \dfrac{11\pi}{6} \right) = 20 \left(\dfrac{\sqrt{3}}{2} - \dfrac{1}{2} i \right) = 10\sqrt{3} - 10i$

$\boxed{19}$ $z_1 = -3\sqrt{3} - 3i = 6\operatorname{cis}\frac{7\pi}{6}$ and $z_2 = 2\sqrt{3} + 2i = 4\operatorname{cis}\frac{\pi}{6}$.

$$z_1 z_2 = 6 \cdot 4 \operatorname{cis}\left(\tfrac{7\pi}{6} + \tfrac{\pi}{6}\right) = 24\operatorname{cis}\tfrac{4\pi}{3} = 24\left(-\tfrac{1}{2} - \tfrac{\sqrt{3}}{2}i\right) = -12 - 12\sqrt{3}\,i.$$

$$\frac{z_1}{z_2} = \frac{6}{4}\operatorname{cis}\left(\tfrac{7\pi}{6} - \tfrac{\pi}{6}\right) = \tfrac{3}{2}\operatorname{cis}\pi = \tfrac{3}{2}(-1 + 0i) = -\tfrac{3}{2}.$$

$\boxed{21}$ $(-\sqrt{3} + i)^9 = (2\operatorname{cis}\tfrac{5\pi}{6})^9 = 2^9\operatorname{cis}\left(9 \cdot \tfrac{5\pi}{6}\right) = 2^9\operatorname{cis}\tfrac{15\pi}{2} = 512\operatorname{cis}\tfrac{3\pi}{2} = 512(0 - i) = -512i$

$\boxed{23}$ $(3 - 3i)^5 = (3\sqrt{2}\operatorname{cis}\tfrac{7\pi}{4})^5 = (3\sqrt{2})^5\operatorname{cis}\tfrac{35\pi}{4} = 972\sqrt{2}\operatorname{cis}\tfrac{3\pi}{4} = 972\sqrt{2}\left(-\tfrac{\sqrt{2}}{2} + \tfrac{\sqrt{2}}{2}i\right) =$
$$-972 + 972i$$

$\boxed{25}$ $-27 + 0i = 27\operatorname{cis}180°$. $\quad w_k = \sqrt[3]{27}\operatorname{cis}\left(\dfrac{180° + 360°k}{3}\right)$ for $k = 0,\ 1,\ 2$.

$$w_0 = 3\operatorname{cis}60° = 3\left(\tfrac{1}{2} + \tfrac{\sqrt{3}}{2}i\right) = \tfrac{3}{2} + \tfrac{3\sqrt{3}}{2}i.$$

$$w_1 = 3\operatorname{cis}180° = 3(-1 + 0i) = -3.$$

$$w_2 = 3\operatorname{cis}300° = 3\left(\tfrac{1}{2} - \tfrac{\sqrt{3}}{2}i\right) = \tfrac{3}{2} - \tfrac{3\sqrt{3}}{2}i.$$

$\boxed{27}$ $x^5 - 32 = 0 \Rightarrow x^5 = 32$. The problem is now to find the 5 fifth roots of 32.

$$32 = 32 + 0i = 32\operatorname{cis}0°. \quad w_k = \sqrt[5]{32}\operatorname{cis}\left(\dfrac{0° + 360°k}{5}\right) \text{ for } k = 0,\ 1,\ 2,\ 3,\ 4.$$

$$w_k = 2\operatorname{cis}\theta \text{ with } \theta = 0°,\ 72°,\ 144°,\ 216°,\ 288°.$$

$\boxed{29}$ (a) $4\mathbf{a} + \mathbf{b} = 4(2\mathbf{i} + 5\mathbf{j}) + (4\mathbf{i} - \mathbf{j}) = 8\mathbf{i} + 20\mathbf{j} + 4\mathbf{i} - \mathbf{j} = 12\mathbf{i} + 19\mathbf{j}$.

(b) $2\mathbf{a} - 3\mathbf{b} = 2(2\mathbf{i} + 5\mathbf{j}) - 3(4\mathbf{i} - \mathbf{j}) = 4\mathbf{i} + 10\mathbf{j} - 12\mathbf{i} + 3\mathbf{j} = -8\mathbf{i} + 13\mathbf{j}$.

(c) $\|\mathbf{a} - \mathbf{b}\| = \|(2\mathbf{i} + 5\mathbf{j}) - (4\mathbf{i} - \mathbf{j})\| = \|-2\mathbf{i} + 6\mathbf{j}\| = \sqrt{(-2)^2 + 6^2} = \sqrt{40} = 2\sqrt{10} \approx$
$$6.32.$$

(d) $\|\mathbf{a}\| - \|\mathbf{b}\| = \|2\mathbf{i} + 5\mathbf{j}\| - \|4\mathbf{i} - \mathbf{j}\| = \sqrt{2^2 + 5^2} - \sqrt{4^2 + (-1)^2} = \sqrt{29} - \sqrt{17} \approx 1.26$.

$\boxed{30}$ $\|\mathbf{r} - \mathbf{a}\| = c \Rightarrow \|\langle x,\ y\rangle - \langle a_1,\ a_2\rangle\| = c \Rightarrow \|\langle x - a_1,\ y - a_2\rangle\| = c \Rightarrow$

$$\sqrt{(x - a_1)^2 + (y - a_2)^2} = c \Rightarrow (x - a_1)^2 + (y - a_2)^2 = c^2.$$

This is a circle with center $(a_1,\ a_2)$ and radius c.

$\boxed{33}$ S60°E is equivalent to 330° and N74°E is equivalent to 16°.

$$\langle 72\cos 330°,\ 72\sin 330°\rangle + \langle 46\cos 16°,\ 46\sin 16°\rangle = \mathbf{r} \approx \langle 106.57,\ -23.32\rangle.$$

$$\|\mathbf{r}\| \approx 109 \text{ kg.} \quad \tan\theta \approx \tfrac{-23.32}{106.57} \Rightarrow \theta \approx -12°, \text{ or equivalently, S78°E.}$$

$\boxed{34}$ $\mathbf{p} = \langle 400 \cos 10°, \, 400 \sin 10° \rangle \approx \langle 393.92, \, 69.46 \rangle$.

$\mathbf{r} = \langle 390 \cos 0°, \, 390 \sin 0° \rangle = \langle 390, \, 0 \rangle$.

$\mathbf{w} = \mathbf{r} - \mathbf{p} \approx \langle -3.92, \, -69.46 \rangle$ and $\| \mathbf{w} \| \approx 69.57$, or 70 mi/hr.

$\tan \theta \approx \frac{-69.46}{-3.92} \Rightarrow \theta \approx 267°$, or in the direction of 183°.

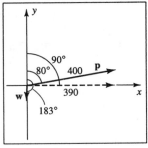

Figure 34

$\boxed{36}$ (a) $(2\mathbf{a} - 3\mathbf{b}) \cdot \mathbf{a} = \left[2(6\mathbf{i} - 2\mathbf{j}) - 3(\mathbf{i} + 3\mathbf{j}) \right] \cdot (6\mathbf{i} - 2\mathbf{j})$

$\qquad\qquad\qquad\quad = (9\mathbf{i} - 13\mathbf{j}) \cdot (6\mathbf{i} - 2\mathbf{j}) = 54 + 26 = 80.$

(b) $\mathbf{c} = \mathbf{a} + \mathbf{b} = (6\mathbf{i} - 2\mathbf{j}) + (\mathbf{i} + 3\mathbf{j}) = 7\mathbf{i} + \mathbf{j}$. The angle between $\mathbf{a}$ and $\mathbf{c}$ is

$$\theta = \cos^{-1} \left(\frac{\mathbf{a} \cdot \mathbf{c}}{\| \mathbf{a} \| \| \mathbf{c} \|} \right) = \cos^{-1} \left(\frac{(6\mathbf{i} - 2\mathbf{j}) \cdot (7\mathbf{i} + \mathbf{j})}{\| 6\mathbf{i} - 2\mathbf{j} \| \| 7\mathbf{i} + \mathbf{j} \|} \right) = \cos^{-1} \left(\frac{40}{\sqrt{40} \, \sqrt{50}} \right) \approx 26°34'.$$

(c) $\text{comp}_{\mathbf{a}}(\mathbf{a} + \mathbf{b}) = \text{comp}_{\mathbf{a}} \mathbf{c} = \frac{\mathbf{c} \cdot \mathbf{a}}{\| \mathbf{a} \|} = \frac{40}{\sqrt{40}} = \sqrt{40} = 2\sqrt{10} \approx 6.32.$

$\boxed{38}$ $\dfrac{\sin \gamma}{150} = \dfrac{\sin 27.4°}{200} \Rightarrow$

$\gamma = \sin^{-1} \left(\dfrac{150 \sin 27.4°}{200} \right) \approx \sin^{-1}(0.3451) \approx 20.2°.$

$\beta = 180° - \alpha - \beta \approx 180° - 27.4° - 20.2° = 132.4°.$

The angle between the hill and the horizontal is then

$180° - 132.4° = 47.6°.$

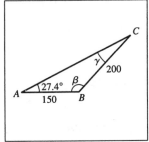

Figure 38

$\boxed{39}$ Let a be the Earth–Venus distance, b be the Earth–sun distance, and c be the Venus–sun distance. Then, by the law of cosines (with a, b, and c in millions), $a^2 = b^2 + c^2 - 2bc \cos \alpha = 93^2 + 67^2 - 2(93)(67) \cos 34° \approx 2807 \Rightarrow$ $a \approx 53$—that is, 53,000,000 miles.

$\boxed{40}$ Let P denote the point at the base of the shorter building, S the point at the top of the shorter building, T the point at the top of the skyscraper, Q the point 50 feet up the side of the skyscraper, and h the height of the skyscraper.

(a) $\angle SPT = 90° - 62° = 28°$. $\angle PST = 90° + 59° = 149°$.

Thus, $\angle STP = 180° - 28° - 149° = 3°$. $\dfrac{\overline{ST}}{\sin 28°} = \dfrac{50}{\sin 3°} \Rightarrow \overline{ST} \approx 448.52$, or 449 ft.

(b) $h = \overline{QT} + 50 = \overline{ST} \sin 59° + 50 \approx 434.45$, or 434 ft.

42 Let E denote the middle point. $\angle CDA = \angle BDC - \angle BDA = 125° - 100° = 25°$.

In $\triangle CAD$, $\angle CAD = 180° - \angle ACD - \angle CDA = 180° - 115° - 25° = 40°$.

$\dfrac{\overline{AD}}{\sin 115°} = \dfrac{120}{\sin 40°} \Rightarrow \overline{AD} \approx 169.20$. $\angle DCB = \angle ACD - \angle ACB = 115° - 92° = 23°$.

In $\triangle DBC$, $\angle DBC = 180° - \angle BDC - \angle DCB = 180° - 125° - 23° = 32°$.

$\dfrac{\overline{BD}}{\sin 23°} = \dfrac{120}{\sin 32°} \Rightarrow \overline{BD} \approx 88.48$.

In $\triangle ADB$, $\overline{AB}^2 = \overline{AD}^2 + \overline{BD}^2 - 2(\overline{AD})(\overline{BD})\cos \angle BDA \Rightarrow$

$$\overline{AB} \approx \sqrt{(169.20)^2 + (88.48)^2 - 2(169.20)(88.48)\cos 100°} \approx 204.1, \text{ or } 204 \text{ ft.}$$

43 If d denotes the distance each girl walks before losing contact with each other,

then $d = 5t$, where t is in hours. Using the law of cosines,

$$10^2 = d^2 + d^2 - 2(d)(d)\cos 105° \Rightarrow 100 = 2d^2(1 - \cos 105°) \Rightarrow d \approx 6.30 \Rightarrow$$

$$t = d/5 \approx 1.26 \text{ hours, or } 1 \text{ hour and } 16 \text{ minutes.}$$

44 (a) Draw a vertical line l through C and label its x-intercept D. Since we have

alternate interior angles, $\angle ACD = \theta_1$. $\angle DCP = 180° - \theta_2$.

$$\text{Thus } \angle ACP = \angle ACD + \angle DCP = \theta_1 + (180° - \theta_2) = 180° - (\theta_2 - \theta_1).$$

(b) Let $k = d(A, P)$. $k^2 = 17^2 + 17^2 - 2(17)(17)\cos\left[180° - (\theta_2 - \theta_1)\right]$.

Since $\cos(180° - \alpha) = \cos 180° \cos\alpha + \sin 180° \sin\alpha = -\cos\alpha$, we have

$k^2 = 578 + 578\cos(\theta_2 - \theta_1) = 578\left[1 + \cos(\theta_2 - \theta_1)\right]$. Using the distance

formula with the points $A(0, 26)$ and $P(x, y)$, we also have $k^2 = x^2 + (y - 26)^2$.

Hence, $578\left[1 + \cos(\theta_2 - \theta_1)\right] = x^2 + (y - 26)^2 \Rightarrow 1 + \cos(\theta_2 - \theta_1) = \dfrac{x^2 + (y - 26)^2}{578}$.

(c) If $x = 25$, $y = 4$, and $\theta_1 = 135°$, then $1 + \cos(\theta_2 - 135°) = \dfrac{25^2 + (-22)^2}{578} = \dfrac{1109}{578} \Rightarrow$

$$\cos(\theta_2 - 135°) = \tfrac{531}{578} \Rightarrow \theta_2 - 135° \approx 23.3° \Rightarrow \theta_2 \approx 158.3°, \text{ or } 158°.$$

45 (a) Let d denote the length of the rescue tunnel. Using the law of cosines,

$d^2 = 45^2 + 50^2 - 2(45)(50)\cos 78° \Rightarrow d \approx 59.91$ ft. Now using the law of sines,

$$\frac{\sin\theta}{45} = \frac{\sin 78°}{d} \Rightarrow \theta = \sin^{-1}\left(\frac{45\sin 78°}{d}\right) \approx 47.28°, \text{ or } 47°.$$

(b) If x denotes the number of hours needed, then

$$d \text{ ft} = (3 \text{ ft/hr})(x \text{ hr}) \Rightarrow x = \tfrac{1}{3}d = \tfrac{1}{3}(59.91) \approx 20 \text{ hr.}$$

$\boxed{46}$ (a) $\angle CBA = 180° - 136° = 44°$ and

$$d = \overline{AC} = \sqrt{22.9^2 + 17.2^2 - 2(22.9)(17.2)\cos 44°} \approx 15.9. \text{ Let } \alpha = \angle BAC.$$

Using the law of sines, $\dfrac{\sin \alpha}{22.9} = \dfrac{\sin 44°}{d} \Rightarrow \alpha = \sin^{-1}\left(\dfrac{22.9\sin 44°}{d}\right) \approx 87.4°.$

Let $\beta = \angle CAD$. Using the law of cosines, $5.7^2 = d^2 + 16^2 - 2(d)(16)\cos\beta \Rightarrow$

$$\beta = \cos^{-1}\left(\frac{d^2 + 16^2 - 5.7^2}{2(d)(16)}\right) \approx 20.6°. \quad \phi \approx 180° - 87.4° - 20.6° = 72°.$$

(b) The area of $ABCD$ is the sum of the areas of $\triangle CBA$ and $\triangle ADC$.

Area $= \frac{1}{2}$(base $\overline{BC}$)(height to A) $+ \frac{1}{2}$(base $\overline{AC}$)(height to D)

$= \frac{1}{2}(\overline{BC})(\overline{BA})\sin \angle CBA + \frac{1}{2}(\overline{AC})(\overline{AD})\sin \angle CAD$

$= \frac{1}{2}(22.9)(17.2)\sin 44° + \frac{1}{2}(15.9)(16)\sin 20.6° \approx 136.8 + 44.8 = 181.6 \text{ ft}^2.$

(c) Let h denote the perpendicular distance from $\overline{BA}$ to C.

$$\sin 44° = \frac{h}{22.9} \Rightarrow h \approx 15.9. \text{ The wing span } \overline{CC'} \text{ is } 2h + 5.8 \approx 37.6 \text{ ft.}$$

Chapter 8 Discussion Exercises

$\boxed{1}$ (a) $\dfrac{a}{\sin \alpha} = \dfrac{c}{\sin \gamma} \Rightarrow \dfrac{a}{c} = \dfrac{\sin \alpha}{\sin \gamma}$ and $\dfrac{b}{\sin \beta} = \dfrac{c}{\sin \gamma} \Rightarrow \dfrac{b}{c} = \dfrac{\sin \beta}{\sin \gamma}.$

Adding the equations yields $\dfrac{a}{c} + \dfrac{b}{c} = \dfrac{\sin \alpha}{\sin \gamma} + \dfrac{\sin \beta}{\sin \gamma} \Rightarrow \dfrac{a+b}{c} = \dfrac{\sin \alpha + \sin \beta}{\sin \gamma}.$

(b) $\dfrac{a+b}{c} = \dfrac{\sin \alpha + \sin \beta}{\sin \gamma} \Rightarrow \dfrac{a+b}{c} \overset{[\text{S}1]}{=} \dfrac{2\sin\frac{1}{2}(\alpha+\beta)\cos\frac{1}{2}(\alpha-\beta)}{2\sin\frac{1}{2}\gamma\cos\frac{1}{2}\gamma}.$

Now $\gamma = 180° - (\alpha + \beta) \Rightarrow \frac{1}{2}\gamma = \left[90° - \frac{1}{2}(\alpha+\beta)\right]$ and

$$\sin\tfrac{1}{2}(\alpha+\beta) = \cos\left[90° - \tfrac{1}{2}(\alpha+\beta)\right] = \cos\tfrac{1}{2}\gamma. \text{ Thus, } \frac{a+b}{c} = \frac{\cos\frac{1}{2}(\alpha-\beta)}{\sin\frac{1}{2}\gamma}.$$

Note: This is an interesting result and gives an answer to the question "How can I check these triangle problems?" Some of my students have written programs for their graphing calculators to utilize this check.

$\boxed{3}$ Example 4 in Section 8.4 illustrates the case $a = 1$.

Algebraic: $\sqrt[3]{a}, \ \sqrt[3]{a}\ \text{cis}\ \frac{2\pi}{3}, \ \sqrt[3]{a}\ \text{cis}\ \frac{4\pi}{3}$

Geometric: All roots lie on a circle of radius $\sqrt[3]{a}$, they are all 120° apart,

one is on the real axis, one is on $\theta = \frac{2\pi}{3}$, and one is on $\theta = \frac{4\pi}{3}$

$\boxed{5}$ (a) $\mathbf{c} = \mathbf{b} + \mathbf{a} = (\|\mathbf{b}\|\cos\alpha\,\mathbf{i} + \|\mathbf{b}\|\sin\alpha\,\mathbf{j}) + (\|\mathbf{a}\|\cos(-\beta)\,\mathbf{i} + \|\mathbf{a}\|\sin(-\beta)\,\mathbf{j})$

$\qquad = \|\mathbf{b}\|\cos\alpha\,\mathbf{i} + \|\mathbf{b}\|\sin\alpha\,\mathbf{j} + \|\mathbf{a}\|\cos\beta\,\mathbf{i} - \|\mathbf{a}\|\sin\beta\,\mathbf{j}$

$\qquad = (\|\mathbf{b}\|\cos\alpha + \|\mathbf{a}\|\cos\beta)\mathbf{i} + (\|\mathbf{b}\|\sin\alpha - \|\mathbf{a}\|\sin\beta)\mathbf{j}$

(b) $\|\mathbf{c}\|^2 = (\|\mathbf{b}\|\cos\alpha + \|\mathbf{a}\|\cos\beta)^2 + (\|\mathbf{b}\|\sin\alpha - \|\mathbf{a}\|\sin\beta)^2$

$\qquad = \|\mathbf{b}\|^2\cos^2\alpha + 2\|\mathbf{a}\|\|\mathbf{b}\|\cos\alpha\cos\beta + \|\mathbf{a}\|^2\cos^2\beta +$

$\qquad\qquad\qquad\qquad \|\mathbf{b}\|^2\sin^2\alpha - 2\|\mathbf{a}\|\|\mathbf{b}\|\sin\alpha\sin\beta + \|\mathbf{a}\|^2\sin^2\beta$

$\qquad = (\|\mathbf{b}\|^2\cos^2\alpha + \|\mathbf{b}\|^2\sin^2\alpha) + (\|\mathbf{a}\|^2\cos^2\beta + \|\mathbf{a}\|^2\sin^2\beta) +$

$\qquad\qquad\qquad\qquad 2\|\mathbf{a}\|\|\mathbf{b}\|\cos\alpha\cos\beta - 2\|\mathbf{a}\|\|\mathbf{b}\|\sin\alpha\sin\beta$

$\qquad = \|\mathbf{b}\|^2 + \|\mathbf{a}\|^2 + 2\|\mathbf{a}\|\|\mathbf{b}\|(\cos\alpha\cos\beta - \sin\alpha\sin\beta)$

$\qquad = \|\mathbf{a}\|^2 + \|\mathbf{b}\|^2 + 2\|\mathbf{a}\|\|\mathbf{b}\|\cos(\alpha + \beta)$

$\qquad = \|\mathbf{a}\|^2 + \|\mathbf{b}\|^2 + 2\|\mathbf{a}\|\|\mathbf{b}\|\cos(\pi - \gamma) \; \{\, \alpha + \beta + \gamma = \pi \,\}$

$\qquad = \|\mathbf{a}\|^2 + \|\mathbf{b}\|^2 - 2\|\mathbf{a}\|\|\mathbf{b}\|\cos\gamma \; \{\, \cos(\pi - \gamma) = -\cos\gamma \,\}$

(c) From part (a), we let $(\|\mathbf{b}\|\sin\alpha - \|\mathbf{a}\|\sin\beta) = 0$.

$\qquad\qquad\qquad$ Thus, $\|\mathbf{b}\|\sin\alpha = \|\mathbf{a}\|\sin\beta$, and $\dfrac{\sin\alpha}{\|\mathbf{a}\|} = \dfrac{\sin\beta}{\|\mathbf{b}\|}$.

Chapter 9: Systems of Equations and Inequalities

Note: The notation E_1 and E_2 refers to the first equation and the second equation.

$\boxed{1}$ Substituting y in E_2 into E_1 yields $2x - 1 = x^2 - 4 \Rightarrow x^2 - 2x - 3 = 0 \Rightarrow$

$(x - 3)(x + 1) = 0 \Rightarrow x = 3, -1$. Substituting $x = 3$ into E_2 gives us $y = 5$,

so $(3, 5)$ is a solution of the system. Similarly, $(-1, -3)$ is a solution.

$\boxed{3}$ Solving E_2 for x, $x = 1 - 2y$, and substituting into E_1 yields $y^2 = 1 - (1 - 2y) \Rightarrow$

$y^2 - 2y = 0 \Rightarrow y(y - 2) = 0 \Rightarrow y = 0, 2; \ x = 1, -3$. $\qquad \star \ (1, 0), (-3, 2)$

$\boxed{7}$ Solving E_1 for x, $x = -2y - 1$, and substituting into E_2 yields

$2(-2y - 1) - 3y = 12 \Rightarrow -7y = 14 \Rightarrow y = -2; \ x = 3.$ $\qquad \star \ (3, -2)$

$\boxed{9}$ Solving E_1 for x, $x = \frac{1}{2} + \frac{3}{2}y$, and substituting into E_2 yields

$-6(\frac{1}{2} + \frac{3}{2}y) + 9y = 4 \Rightarrow -3 = 4$, a contradiction. There are *no solutions.*

$\boxed{11}$ Solving E_1 for x, $x = 5 - 3y$, and substituting into E_2 yields $(5 - 3y)^2 + y^2 = 25 \Rightarrow$

$10y^2 - 30y = 0 \Rightarrow 10y(y - 3) = 0 \Rightarrow y = 0, 3; \ x = 5, -4.$ $\qquad \star \ (-4, 3), (5, 0)$

$\boxed{15}$ Solving E_2 for y, $y = 3x + 2$, and substituting into E_1 yields

$$x^2 + (3x + 2)^2 = 9 \Rightarrow 10x^2 + 12x - 5 = 0 \Rightarrow x = \frac{-6 \pm \sqrt{86}}{10} = -\frac{3}{5} \pm \frac{1}{10}\sqrt{86}.$$

$$y = 3\left(\frac{-6 \pm \sqrt{86}}{10}\right) + 2 = \frac{-18 \pm 3\sqrt{86}}{10} + \frac{20}{10} = \frac{2 \pm 3\sqrt{86}}{10} = \frac{1}{5} \pm \frac{3}{10}\sqrt{86}.$$

$$\star \ \left(-\tfrac{3}{5} + \tfrac{1}{10}\sqrt{86}, \tfrac{1}{5} + \tfrac{3}{10}\sqrt{86}\right), \left(-\tfrac{3}{5} - \tfrac{1}{10}\sqrt{86}, \tfrac{1}{5} - \tfrac{3}{10}\sqrt{86}\right)$$

$\boxed{19}$ Solving E_2 for x and substituting into E_1 yields $(1 - y - 1)^2 + (y + 2)^2 = 10 \Rightarrow$

$y^2 + (y^2 + 4y + 4) = 10 \Rightarrow 2y^2 + 4y - 6 = 0 \Rightarrow 2(y^2 + 2y - 3) = 0 \Rightarrow$

$2(y + 3)(y - 1) = 0 \Rightarrow y = -3, 1; \ x = 4, 0.$ $\qquad \star \ (0, 1), (4, -3)$

$\boxed{21}$ Substituting y in E_1 into E_2 yields $20/x^2 = 9 - x^2$ {multiply by x^2} $\Rightarrow$

$x^4 - 9x^2 + 20 = 0 \Rightarrow (x^2 - 4)(x^2 - 5) = 0 \Rightarrow x^2 = 4, 5 \Rightarrow x = \pm 2, \pm \sqrt{5}; \ y = 5, 4.$

$$\star \ (\pm 2, 5), (\pm \sqrt{5}, 4)$$

$\boxed{23}$ Solving E_1 for y^2, $y^2 = 4x^2 + 4$, and substituting into E_2 yields

$9(4x^2 + 4) + 16x^2 = 140 \Rightarrow 52x^2 = 104 \Rightarrow x^2 = 2 \Rightarrow x = \pm \sqrt{2}; \ y = \pm 2\sqrt{3}.$

There are four solutions. $\qquad \star \ (\sqrt{2}, \pm 2\sqrt{3}), (-\sqrt{2}, \pm 2\sqrt{3})$

$\boxed{25}$ Solving E_1 for x^2 and substituting into E_2 yields $(y^2 + 4) + y^2 = 12 \Rightarrow 2y^2 = 8 \Rightarrow$

$y = \pm 2; \ x = \pm 2\sqrt{2}.$ $\qquad \star \ (2\sqrt{2}, \pm 2), (-2\sqrt{2}, \pm 2)$

$\boxed{27}$ Solving E_2 for y, $y = 2x + z - 9$, and substituting into E_1 and E_3 yields

$$\begin{cases} x + 2(2x + z - 9) - z & = -1 \\ x + 3(2x + z - 9) + 3z & = 6 \end{cases} \Rightarrow \begin{cases} 5x + z & = 17 \quad (E_4) \\ 7x + 6z & = 33 \quad (E_5) \end{cases}$$

Solving E_4 for z, $z = 17 - 5x$, and substituting into E_5 yields $7x + 6(17 - 5x) = 33 \Rightarrow$
$-23x = -69 \Rightarrow x = 3$. Now $z = 17 - 5x = 17 - 5(3) = 2$ and
$y = 2x + z - 9 = 2(3) + 2 - 9 = -1$.   ★ $(3, -1, 2)$

$\boxed{29}$ Solving E_3 for y and substituting into E_2 yields $\begin{cases} x^2 + z^2 & = 5 \quad (E_1) \\ 2x - z & = 0 \quad (E_4) \end{cases}$

Now solve E_4 for z and substitute into E_1 yielding $x^2 + (2x)^2 = 5 \Rightarrow 5x^2 = 5 \Rightarrow$
$x = \pm 1$; $z = 2x = \pm 2$; $y = 1 - z = -1, 3$.   ★ $(1, -1, 2), (-1, 3, -2)$

$\boxed{31}$ The graph of $x^2 + 4y^2 = 20$, or, equivalently, $\dfrac{x^2}{20} + \dfrac{y^2}{5} = 1$, is that of an ellipse with x-

intercepts at $(\pm \sqrt{20}, 0)$ and y-intercepts at $(0, \pm \sqrt{5})$. The graph of $x + 2y = 6$, or,

equivalently, $y = -\frac{1}{2}x + 3$, is that of a line with y-intercept 3 and slope $-\frac{1}{2}$.

Substituting $x = 6 - 2y$ into $x^2 + 4y^2 = 20$ yields $(6 - 2y)^2 + 4y^2 = 20 \Rightarrow$
$8y^2 - 24y + 16 = 0 \Rightarrow 8(y^2 - 3y + 2) = 0 \Rightarrow 8(y - 1)(y - 2) = 0 \Rightarrow y = 1, 2$; $x = 4, 2$.
The two points of intersection are $(2, 2)$ and $(4, 1)$.

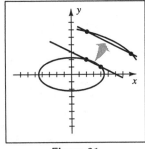

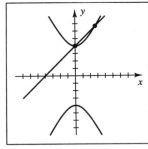

Figure 31   *Figure 33*

$\boxed{33}$ The graph of $y^2 - 4x^2 = 16$, or, equivalently, $\dfrac{y^2}{16} - \dfrac{x^2}{4} = 1$, is that of a hyperbola with

y-intercepts at $(0, \pm 4)$. The graph of $y - x = 4$, or, equivalently, $y = x + 4$, is that

of a line with y-intercept 4 and slope 1. Substituting $y = x + 4$ into $y^2 - 4x^2 = 16$

yields $(x + 4)^2 - 4x^2 = 16 \Rightarrow 3x^2 - 8x = 0 \Rightarrow x(3x - 8) = 0 \Rightarrow x = 0, \frac{8}{3}$; $y = 4, \frac{20}{3}$. The

two points of intersection are $(0, 4)$ and $(\frac{8}{3}, \frac{20}{3})$.

$\boxed{35}$ Using $P = 40 = 2l + 2w$ and $A = 96 = lw$, we have $\begin{cases} 2l + 2w & = 40 \quad (E_1) \\ lw & = 96 \quad (E_2) \end{cases}$

Solving E_1 for l, $l = 20 - w$, and substituting into E_2 yields $(20 - w)w = 96 \Rightarrow$
$20w - w^2 = 96 \Rightarrow w^2 - 20w + 96 = 0 \Rightarrow (w - 8)(w - 12) = 0 \Rightarrow w = 8, 12$; $l = 12, 8$.

In either case, the rectangle is 12 in. × 8 in.

39 $A = 200 = (2\pi r)h \Rightarrow \pi rh = 100.$ $V = 200 = \pi r^2 h \Rightarrow h = 200/(\pi r^2).$

Substituting h into $\pi rh = 100$ yields $200/r = 100 \Rightarrow r = 2$ in.; $h = 50/\pi \approx 15.9$ in.

41 (a) Let $S = 40{,}000$ and $R = 60{,}000$ for the data for 1993 and 1994.

Then let $S = 60{,}000$ and $R = 72{,}000$ for the data for 1994 and 1995.

$$\begin{cases} 60{,}000 = (40{,}000a)/(40{,}000 + b) \\ 72{,}000 = (60{,}000a)/(60{,}000 + b) \end{cases} \Rightarrow \begin{cases} 120{,}000 + 3b = 2a & (\text{E}_1) \\ 360{,}000 + 6b = 5a & (\text{E}_2) \end{cases}$$

Solving E_1 for a and substituting into E_2 yields

$$360{,}000 + 6b = 5(60{,}000 + \tfrac{3}{2}b) \Rightarrow 60{,}000 = \tfrac{3}{2}b \Rightarrow b = 40{,}000; \; a = 120{,}000.$$

(b) Now let $S = 72{,}000$ and thus $R = \dfrac{(120{,}000)(72{,}000)}{72{,}000 + 40{,}000} = \dfrac{540{,}000}{7} \approx 77{,}143.$

43 Let R_1 and R_2 equal 0. The system is then $\begin{cases} 0 = 0.01x(50 - x - y) \\ 0 = 0.02y(100 - y - 0.5x) \end{cases}$

The first equation is zero if $x = 0$ or if $50 - x - y = 0$. The second equation is zero if $y = 0$ or if $100 - y - 0.5x = 0$. Hence, there are 4 possible solutions.

(1) A solution is $x = 0$ and $y = 0$, or $(0, 0)$.

(2) A second solution is $x = 0$ and $100 - y - 0.5x = 0$ $\{y = 100\}$, or $(0, 100)$.

(3) A third solution is $y = 0$ and $50 - x - y = 0$ $\{x = 50\}$, or $(50, 0)$.

(4) A fourth solution occurs if $50 - x - y = 0$ and $100 - y - 0.5x = 0$.

Solve the first equation for y $\{y = 50 - x\}$ and substitute into the second equation: $100 - (50 - x) - 0.5x = 0 \Rightarrow 50 = -\tfrac{1}{2}x \Rightarrow x = -100; \; y = 150.$

This solution is meaningless for this problem since x and y are nonnegative.

45 Since we have an *open* top instead of a closed top, we use $3xy$ instead of $4xy$ for the

surface area formula. $\begin{cases} x^2 y = 2 & \textit{Volume} \\ 2x^2 + 3xy = 8 & \textit{Surface Area} \end{cases}$

Solving E_1 for y and substituting into E_2 yields $2x^2 + 3x(2/x^2) = 8 \Rightarrow$

$2x^2 + (6/x) = 8$ $\{\text{multiply by } x\} \Rightarrow 2x^3 - 8x + 6 = 0 \Rightarrow 2(x^3 - 4x + 3) = 0.$

We know that $x = 1$ is a solution of $x^3 - 4x + 3 = 0$ since the sum of the coefficients

is zero. Continuing, $2(x - 1)(x^2 + x - 3) = 0 \Rightarrow \{x > 0\} \; x = 1, \dfrac{-1 + \sqrt{13}}{2}.$

There are two solutions: $1 \text{ ft} \times 1 \text{ ft} \times 2 \text{ ft}$ or

$$\frac{\sqrt{13} - 1}{2} \text{ ft} \times \frac{\sqrt{13} - 1}{2} \text{ ft} \times \frac{8}{(\sqrt{13} - 1)^2} \text{ ft} \approx 1.30 \text{ ft} \times 1.30 \text{ ft} \times 1.18 \text{ ft}.$$

[47] We eliminate n from the equations to determine all intersection points.

(a) $x^2 + y^2 = n^2$ and $y = n - 1 \Rightarrow x^2 + y^2 = (y+1)^2 \Rightarrow x^2 = 2y + 1 \Rightarrow y = \frac{1}{2}x^2 - \frac{1}{2}$.

The points are on the parabola $y = \frac{1}{2}x^2 - \frac{1}{2}$.

(b) $x^2 + y^2 = (y+2)^2 \Rightarrow x^2 = 4y + 4 \Rightarrow y = \frac{1}{4}x^2 - 1$

[49] (a) The slope of the line from $(-4, -3)$ to the origin is $\frac{3}{4}$ so the slope of the tangent line (which is perpendicular to the line to the origin) is $-\frac{4}{3}$. An equation of the line through $(-4, -3)$ with slope $-\frac{4}{3}$ is $y + 3 = -\frac{4}{3}(x + 4)$, or, equivalently, $4x + 3y = -25$. Letting $y = -50$, we find that $x = 31.25$.

(b) The slope of the line from an arbitrary point (x, y) on the circle to the origin is $\frac{y}{x}$, so the slope of the tangent line is $-\frac{x}{y}$. The line through $(0, -50)$ is $y + 50 = \left(-\frac{x}{y}\right)(x - 0)$, or, equivalently, $y^2 + 50y = -x^2$. The equation of the circle is $x^2 + y^2 = 25$. Substituting $x^2 = 25 - y^2$ into $y^2 + 50y = -x^2$ gives us $50y = -25$, or $y = -\frac{1}{2}$. Substituting $y = -\frac{1}{2}$ into $x^2 = 25 - y^2$ gives us $x^2 = \frac{99}{4}$ and hence $x = \pm\frac{3}{2}\sqrt{11}$. The positive solution corresponds to releasing the hammer from a clockwise spin, whereas the negative solution corresponds to releasing the hammer from a counterclockwise spin as depicted in the figures. Hence, the hammer should be released at $\left(-\frac{3}{2}\sqrt{11}, -\frac{1}{2}\right) \approx (-4.975, -0.5)$.

[51] **Graphically:** $x^2 + y^2 = 4 \Rightarrow y = \pm\sqrt{4 - x^2}$ and $x + y = 1 \Rightarrow y = 1 - x$. Graph $Y_1 = \sqrt{4 - x^2}$, $Y_2 = -Y_1$, and $Y_3 = 1 - x$. There are two points of intersection at approximately $(-0.82, 1.82)$ and $(1.82, -0.82)$.

Algebraically: $y = 1 - x$ and $x^2 + y^2 = 4 \Rightarrow x^2 + (1 - x)^2 = 4 \Rightarrow 2x^2 - 2x - 3 = 0$. Using the quadratic formula, $x = \dfrac{2 \pm \sqrt{4 - 4(2)(-3)}}{4} = \dfrac{1 \pm \sqrt{7}}{2}$. $y = 1 - x \Rightarrow$ $y = 1 - \dfrac{1 \pm \sqrt{7}}{2} = \dfrac{1 \mp \sqrt{7}}{2}$. The points of intersection are $\left(\dfrac{1}{2} \pm \dfrac{\sqrt{7}}{2}, \dfrac{1}{2} \mp \dfrac{\sqrt{7}}{2}\right)$.

The graphical solution approximates the algebraic solution.

$$[-6, 6] \text{ by } [-4, 4]$$

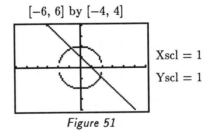

Xscl = 1

Yscl = 1

Figure 51

53 After zooming-in near the region of interest in the second quadrant, we see that the cubic intersects the circle twice. Due to the symmetry, we know there are two more points of intersection in the fourth quadrant. Thus, the six points of intersection are approximately $(\mp 0.56, \pm 1.92)$, $(\mp 0.63, \pm 1.90)$, and $(\pm 1.14, \pm 1.65)$.

$[-6, 6]$ by $[-4, 4]$

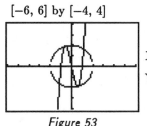

Xscl = 1

Yscl = 1

$[-3, 3]$ by $[-2, 2]$

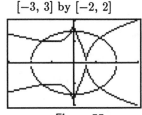

Xscl = 1

Yscl = 1

Figure 53   *Figure 55*

55 $|x + \ln|x|| - y^2 = 0 \Rightarrow y^2 = |x + \ln|x|| \Rightarrow y = \pm\sqrt{|x + \ln|x||}$ and

$\dfrac{x^2}{4} + \dfrac{y^2}{2.25} = 1 \Rightarrow \dfrac{y^2}{2.25} = 1 - \dfrac{x^2}{4} \Rightarrow y^2 = 2.25(1 - x^2/4) \Rightarrow y = \pm 1.5\sqrt{1 - x^2/4}.$

The graph is symmetric with respect to x-axis. There are 8 points of intersection. Their coordinates are approximately

$$(-1.44, \pm 1.04), (-0.12, \pm 1.50), (0.10, \pm 1.50), \text{ and } (1.22, \pm 1.19).$$

57 When $x = 1$, $f(1) = ae^{-b} = 0.80487 \Rightarrow a = 0.80487e^b$. When $x = 2$, $f(2) = ae^{-2b} = 0.53930 \Rightarrow a = 0.53930e^{2b}$. Let $a = y$ and $b = x$ and then graph $Y_1 = 0.80487e^x$ and $Y_2 = 0.53930e^{2x}$. The graphs intersect at approximately $(0.4004, 1.2012) = (b, a)$. Thus, $a \approx 1.2012$, $b \approx 0.4004$, and $f(x) = 1.2012e^{-0.4004x}$. The function f is also accurate at $x = 3, 4$.

$[0, 3]$ by $[0, 2]$

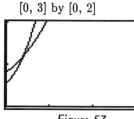

Xscl = 1

Yscl = 1

$[0, 4]$ by $[0, 4]$

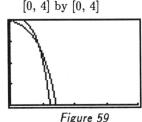

Xscl = 1

Yscl = 1

Figure 57   *Figure 59*

59 When $x = 2$, $f(2) = 4a + e^{2b} = 17.2597 \Rightarrow a = \dfrac{17.2597 - e^{2b}}{4}$. When $x = 3$,

$f(3) = 9a + e^{3b} = 40.1058 \Rightarrow a = \dfrac{40.1058 - e^{3b}}{9}$. Graph $Y_1 = \dfrac{17.2597 - e^{2x}}{4}$ and

$Y_2 = \dfrac{40.1058 - e^{3x}}{9}$. The graphs intersect at approximately $(0.9002, 2.8019)$. Thus,

$a \approx 2.8019$, $b \approx 0.9002$, and $f(x) = 2.8019x^2 + e^{0.9002x}$. The function f is also accurate at $x = 4$.

Note: The notation E_1 and E_2 refers to equation 1 and equation 2. $3\,E_2$ symbolizes "3 times equation 2". After the value of one variable is found, the value(s) of the other variable(s) will be stated and can be found by substituting the known value(s) back into the original equation(s).

1 The easiest choice for multipliers is to take -2 times the second equation and eliminate x. If we wanted to eliminate y, we could choose 2 times the first equation and 3 times the second equation. $-2\,E_2 + E_1 \Rightarrow 7y = -14 \Rightarrow y = -2$; $x = 4$

$\qquad\qquad\qquad\qquad\qquad\qquad\qquad\qquad\qquad\qquad\qquad$ ★ $(4, -2)$

3 $3\,E_1 - 2\,E_2 \Rightarrow 29y = 0 \Rightarrow y = 0$; $x = 8$ $\qquad\qquad\qquad$ ★ $(8, 0)$

7 $3\,E_1 - 5\,E_2 \Rightarrow -53y = -28 \Rightarrow y = \frac{28}{53}$. Instead of substituting into one of the equations to find the value of the other variable, it is usually easier to pick different multipliers and re-solve the system for the other variable.

$\quad 7\,E_1 + 6\,E_2 \Rightarrow 53x = 76 \Rightarrow x = \frac{76}{53}$. $\qquad\qquad\qquad$ ★ $\left(\frac{76}{53}, \frac{28}{53}\right)$

9 We will first eliminate all fractions by multiplying both sides of each equation by its

$\quad$ lcd. $\begin{cases} 6\,E_1 \\ 3\,E_2 \end{cases} \Rightarrow \begin{cases} 2c + 3d = 30 & (E_3) \\ 3c - 2d = -3 & (E_4) \end{cases}$ $\quad 3\,E_3 - 2\,E_4 \Rightarrow 13d = 96 \Rightarrow d = \frac{96}{13}$.

$\quad 2\,E_3 + 3\,E_4 \Rightarrow 13c = 51 \Rightarrow c = \frac{51}{13}$ $\qquad\qquad\qquad$ ★ $\left(\frac{51}{13}, \frac{96}{13}\right)$

11 The least common multiple of 2 and 3 is 6, so we will multiply the first equation by $\sqrt{3}$ and the second equation by $\sqrt{2}$ so that the coefficients of y are $\sqrt{6}$ and $-\sqrt{6}$. $\sqrt{3}\,E_1 + \sqrt{2}\,E_2 \Rightarrow 7x = 8 \Rightarrow x = \frac{8}{7}$. We will now re-solve the system to obtain a value for y. $2\sqrt{2}\,E_1 - \sqrt{3}\,E_2 \Rightarrow -7y = 3\sqrt{6} \Rightarrow y = -\frac{3}{7}\sqrt{6}$.

$\qquad\qquad\qquad\qquad\qquad\qquad\qquad\qquad\qquad\qquad$ ★ $\left(\frac{8}{7}, -\frac{3}{7}\sqrt{6}\right)$

13 $3\,E_1 + E_2 \Rightarrow 0 = 27$; this statement is never true so there are *no solutions.*

15 $2\,E_1 + E_2 \Rightarrow 0 = 0$; this statement is always true so the solution is

$\qquad\qquad\qquad$ all ordered pairs (m, n) such that $3m - 4n = 2$.

19 Using the hint, we let $u = 1/x$ and $v = 1/y$, and the system is $\begin{cases} 2u + 3v = -2 & (E_3) \\ 4u - 5v = 1 & (E_4) \end{cases}$

$\quad -2\,E_3 + E_4 \Rightarrow -11v = 5 \Rightarrow v = -\frac{5}{11}$. $\quad 5\,E_3 + 3\,E_4 \Rightarrow 22u = -7 \Rightarrow u = -\frac{7}{22}$.

$\quad$ Resubstituting, we have $x = 1/u = -\frac{22}{7}$ and $y = 1/v = -\frac{11}{5}$. $\qquad$ ★ $\left(-\frac{22}{7}, -\frac{11}{5}\right)$

21 $\begin{cases} a\,e^{3x} + b\,e^{-3x} = 0 & (E_1) \\ a\,(3e^{3x}) + b\,(-3e^{-3x}) = e^{3x} & (E_2) \end{cases}$

$\quad$ Following the hint in the text, think of this system as a system of two equations in the variables a and b. In E_1, we have "e^{3x}" times a, and in E_2, we have "$3e^{3x}$" times a. Hence, we will multiply E_1 by -3 and add this equation to E_2 to eliminate a and then solve the system for b. $\qquad\qquad\qquad\qquad\qquad\qquad$ (continued)

$-3\,E_1 + E_2 \Rightarrow -3b\,e^{-3x} - 3b\,e^{-3x} = e^{3x} \Rightarrow -6b\,e^{-3x} = e^{3x} \Rightarrow b = -\frac{1}{6}e^{6x}.$

Substituting back into E_1 yields $a\,e^{3x} + (-\frac{1}{6}e^{6x})e^{-3x} = 0 \Rightarrow a\,e^{3x} = \frac{1}{6}e^{3x} \Rightarrow a = \frac{1}{6}.$

$\boxed{23}$ $\begin{cases} a\cos x + b\sin x = 0 & (E_1) \\ -a\sin x + b\cos x = \tan x & (E_2) \end{cases}$

$\sin x$ (E_1) and $\cos x$ (E_2) yield $\begin{cases} a\sin x\cos x + b\sin^2 x = 0 & (E_3) \\ -a\sin x\cos x + b\cos^2 x = \sin x & (E_4) \end{cases}$

$E_3 + E_4 \Rightarrow b\sin^2 x + b\cos^2 x = \sin x \Rightarrow b(\sin^2 x + \cos^2 x) = \sin x \Rightarrow b(1) = \sin x \Rightarrow$

$b = \sin x.$ Substituting back into E_1 yields $a\cos x + \sin^2 x = 0 \Rightarrow$

$$a = -\frac{\sin^2 x}{\cos x} = -\frac{1 - \cos^2 x}{\cos x} = -\frac{1}{\cos x} + \frac{\cos^2 x}{\cos x} = -\sec x + \cos x = \cos x - \sec x.$$

$\boxed{25}$ Let x denote the number of \$1.50 tickets and y the number of \$2.25 tickets.

$\begin{cases} x + y = 450 & \textit{quantity} \\ 1.50x + 2.25y = 777.75 & \textit{value} \end{cases}$

$100\,E_2 - 150\,E_1 \Rightarrow 75y = 10{,}275 \Rightarrow y = 137; \; x = 313$

$\boxed{27}$ The volume for a cylinder is $\pi r^2 h$ and the volume for a cone is $\frac{1}{3}\pi r^2 h$.

The radius of the cylinder is $\frac{1}{2}$ cm. $\begin{cases} x + y = 8 & \textit{length} \\ \pi(\frac{1}{2})^2 x + \frac{1}{3}\pi(\frac{1}{2})^2 y = 5 & \textit{volume} \end{cases}$

Solving E_1 for y and substituting into E_2 yields $\frac{\pi}{4}x + \frac{\pi}{12}(8 - x) = 5 \Rightarrow$

$$\frac{\pi}{6}x = \frac{15 - 2\pi}{3} \Rightarrow x = \frac{30 - 4\pi}{\pi} = \frac{30}{\pi} - 4 \approx 5.55 \text{ cm.}$$

$$y = 8 - \left(\frac{30 - 4\pi}{\pi}\right) = \frac{12\pi - 30}{\pi} = 12 - \frac{30}{\pi} \approx 2.45 \text{ cm.}$$

$\boxed{29}$ The perimeter is composed of 2 sides of the rectangular portion of the table, $2l$, and 2 edges of the semicircular regions, $2 \cdot \frac{1}{2}(2\pi r) = 2\pi r$. Since the radius is $\frac{1}{2}w$, the system

can be represented by $\begin{cases} 2l + 2\pi(\frac{1}{2}w) = 40 & \textit{perimeter} \\ lw = 2\left[\pi(\frac{1}{2}w)^2\right] & \textit{area} \end{cases}$

Solving E_1 for l, $l = \frac{40 - \pi w}{2}$, and substituting into E_2 yields $\left(\frac{40 - \pi w}{2}\right)w = \frac{\pi w^2}{2} \Rightarrow$

$(40 - \pi w)w = \pi w^2 \Rightarrow 40w = 2\pi w^2 \Rightarrow 2\pi w^2 - 40w = 0 \Rightarrow 2w(\pi w - 20) = 0 \Rightarrow$

$w = 0, \frac{20}{\pi}.$ We discard $w = 0$ and use $w = 20/\pi \approx 6.37$ ft.

$$\text{Thus, } l = \frac{40 - \pi(20/\pi)}{2} = \frac{20}{2} = 10 \text{ ft.}$$

$\boxed{31}$ Let x denote the number of adults and y the number of kittens. Thus,

$(\frac{1}{2}x)$ is the number of adult females. $\begin{cases} x + y = 6000 & \textit{total} \\ y = 3(\frac{1}{2}x) & \textit{3 kittens per adult female} \end{cases}$

Substituting y from E_2 into E_1 yields $x + \frac{3}{2}x = 6000 \Rightarrow x = 2400; \; y = 3600.$

[33] Let x denote the number of grams of the 35% alloy and y the number of grams of the

60% alloy.
$$\begin{cases} x + y = 100 & \text{quantity} \\ 0.35x + 0.60y = (0.50)(100) & \text{quality} \end{cases}$$

$$100\,E_2 - 35\,E_1 \Rightarrow 25y = 1500 \Rightarrow y = 60;\ x = 40$$

[35] Let x denote the speed of the plane and y the speed of the wind. Use $d = rt$.

$$\begin{cases} 1200 = (x + y)(2) & \text{with the wind} \\ 1200 = (x - y)(2\frac{1}{2}) & \text{against the wind} \end{cases} \Rightarrow \begin{cases} 600 = x + y \\ 480 = x - y \end{cases}$$

$$E_1 + E_2 \Rightarrow 2x = 1080 \Rightarrow x = 540 \text{ mi/hr};\ y = 60 \text{ mi/hr}$$

[37] $v(2) = 16 \Rightarrow 16 = v_0 + 2a;\ (E_1)$

$v(5) = 25 \Rightarrow 25 = v_0 + 5a;\ (E_2)$ $E_2 - E_1 \Rightarrow 9 = 3a \Rightarrow a = 3;\ v_0 = 10$

[39] Let x denote the number of sofas produced and y the number of recliners produced.

$$\begin{cases} 8x + 6y = 340 & \text{labor hours} \\ 60x + 35y = 2250 & \text{cost of materials} \end{cases}$$

$$6\,E_2 - 35\,E_1 \Rightarrow 80x = 1600 \Rightarrow x = 20;\ y = 30$$

[41] (a) The expression $6x + 5y$ represents the total bill for the plumber's business. This should equal the plumber's income, which is the plumber's number of hours times his/her hourly wage—that is, $(6 + 4)(x) = 10x$. The expression $4x + 6y$ represents the total bill for the electrician's business. This should equal the electrician's income, that is, $(5 + 6)(y) = 11y$.

$$\begin{cases} 6x + 5y = 10x \\ 4x + 6y = 11y \end{cases} \Rightarrow \begin{cases} 5y = 4x \\ 4x = 5y \end{cases} \Rightarrow y = \tfrac{4}{5}x, \text{ or, equivalently, } y = 0.80x$$

(b) The electrician should charge 80% of what the plumber charges—80% of $20 per hour is $16 per hour.

[43] Let $t = 0$ correspond to the year 1891. The average daily maximum can then be approximated by the linear equation $y_1 = 0.011t + 15.1$ and the average daily minimum by the linear equation $y_2 = 0.019t + 5.8$. We must determine t when y_1 and y_2 differ by 9. $y_1 - y_2 = 9 \Rightarrow (0.011t + 15.1) - (0.019t + 5.8) = 9 \Rightarrow$ $-0.008t + 9.3 = 9 \Rightarrow -0.008t = -0.3 \Rightarrow t = 37.5$. $1891 + 37.5 = 1928.5$, or during the year 1928. Also, $t = 37.5 \Rightarrow y_1 = 0.011(37.5) + 15.1 = 15.5125 \approx 15.5\,°C$.

45 (a) Since consumers will buy 1,000,000 T-shirts if the selling price is $10 and they will buy 900,000 if the price is $11, we can model this information with a line through the two points (10, 1,000,000) and (11, 900,000). Using the point-slope formula of a line, we have $Q - 1,000,000 = \frac{900,000 - 1,000,000}{11 - 10}(p - 10) \Rightarrow$

$Q - 1,000,000 = -100,000(p - 10) \Rightarrow Q = -100,000p + 2,000,000.$

(b) As in part (a), use the two points (15, 2,000,000) and (16, 2,150,000).
$K - 2,000,000 = \frac{150,000}{1}(p - 15) \Rightarrow K = 150,000p - 250,000.$

(c) $Q = K \Rightarrow -100,000p + 2,000,000 = 150,000p - 250,000 \Rightarrow$

$$2,250,000 = 250,000p \Rightarrow p = \$9.00$$

9.3 Exercises

Note: Most systems are solved using the back substitution method. The solution for Exercise 7 uses the reduced echelon method. To avoid fractions, some solutions include linear combinations of rows—that is, $m\mathrm{R}_i + n\mathrm{R}_j$.

1
$$\begin{bmatrix} 1 & -2 & -3 & -1 \\ 2 & 1 & 1 & 6 \\ 1 & 3 & -2 & 13 \end{bmatrix} \begin{array}{l} \mathrm{R}_2 - 2\,\mathrm{R}_1 \to \mathrm{R}_2 \\ \mathrm{R}_3 - \mathrm{R}_1 \to \mathrm{R}_3 \end{array}$$

$$\begin{bmatrix} 1 & -2 & -3 & -1 \\ 0 & 5 & 7 & 8 \\ 0 & 5 & 1 & 14 \end{bmatrix} \mathrm{R}_3 - \mathrm{R}_2 \to \mathrm{R}_3$$

$$\begin{bmatrix} 1 & -2 & -3 & -1 \\ 0 & 5 & 7 & 8 \\ 0 & 0 & -6 & 6 \end{bmatrix} -\tfrac{1}{6}\mathrm{R}_3 \to \mathrm{R}_3$$

R_3: $z = -1$

R_2: $5y + 7z = 8 \Rightarrow y = 3$ ★ $(2, 3, -1)$

R_1: $x - 2y - 3z = -1 \Rightarrow x = 2$

5
$$\begin{bmatrix} 2 & 6 & -4 & 1 \\ 1 & 3 & -2 & 4 \\ 2 & 1 & -3 & -7 \end{bmatrix} \mathrm{R}_1 \leftrightarrow \mathrm{R}_2$$

$$\begin{bmatrix} 1 & 3 & -2 & 4 \\ 2 & 6 & -4 & 1 \\ 2 & 1 & -3 & -7 \end{bmatrix} \begin{array}{l} \mathrm{R}_2 - 2\,\mathrm{R}_1 \to \mathrm{R}_2 \\ \mathrm{R}_3 - 2\,\mathrm{R}_1 \to \mathrm{R}_3 \end{array}$$

$$\begin{bmatrix} 1 & 3 & -2 & 4 \\ 0 & 0 & 0 & -7 \\ 0 & -5 & 1 & -15 \end{bmatrix}$$

The second row, $0x + 0y + 0z = -7$, has *no solution.*

7 $\begin{bmatrix} 2 & -3 & 2 & -3 \\ -3 & 2 & 1 & 1 \\ 4 & 1 & -3 & 4 \end{bmatrix}$ $R_1 \leftrightarrow R_2$ and then $R_2 \leftrightarrow R_3$

$\begin{bmatrix} -3 & 2 & 1 & 1 \\ 4 & 1 & -3 & 4 \\ 2 & -3 & 2 & -3 \end{bmatrix}$ $R_1 + R_2 \to R_1$

$\begin{bmatrix} 1 & 3 & -2 & 5 \\ 4 & 1 & -3 & 4 \\ 2 & -3 & 2 & -3 \end{bmatrix}$ $\begin{array}{l} R_2 - 4R_1 \to R_2 \\ R_3 - 2R_1 \to R_3 \end{array}$

$\begin{bmatrix} 1 & 3 & -2 & 5 \\ 0 & -11 & 5 & -16 \\ 0 & -9 & 6 & -13 \end{bmatrix}$ $4R_2 - 5R_3 \to R_2$ ($*$ see note)

$\begin{bmatrix} 1 & 3 & -2 & 5 \\ 0 & 1 & -10 & 1 \\ 0 & -9 & 6 & -13 \end{bmatrix}$ $R_3 + 9R_2 \to R_3$

$\begin{bmatrix} 1 & 3 & -2 & 5 \\ 0 & 1 & -10 & 1 \\ 0 & 0 & -84 & -4 \end{bmatrix}$ $-\frac{1}{84}R_3 \to R_3$

R_3: $z = \frac{1}{21}$

R_2: $y - 10z = 1 \Rightarrow y = \frac{31}{21}$ $\star$ $\left(\frac{2}{3}, \frac{31}{21}, \frac{1}{21}\right)$

R_1: $x + 3y - 2z = 5 \Rightarrow x = \frac{14}{21} = \frac{2}{3}$

($*$) We could have just used $-\frac{1}{11}R_2 \to R_2$, but we would then have to work with many fractions and increase our chance of making a mistake. We will solve this system again, but this time we will obtain the reduced echelon form. The first 4 matrices are exactly the same. Starting with the fourth matrix we have:

$\begin{bmatrix} 1 & 3 & -2 & 5 \\ 0 & 1 & -10 & 1 \\ 0 & -9 & 6 & -13 \end{bmatrix}$ $\begin{array}{l} R_1 - 3R_2 \to R_1 \\ \\ R_3 + 9R_2 \to R_3 \end{array}$

$\begin{bmatrix} 1 & 0 & 28 & 2 \\ 0 & 1 & -10 & 1 \\ 0 & 0 & -84 & -4 \end{bmatrix}$ $-\frac{1}{84}R_3 \to R_3$

$\begin{bmatrix} 1 & 0 & 28 & 2 \\ 0 & 1 & -10 & 1 \\ 0 & 0 & 1 & \frac{1}{21} \end{bmatrix}$ $\begin{array}{l} R_1 - 28R_3 \to R_1 \\ \\ R_2 + 10R_3 \to R_2 \end{array}$

$\begin{bmatrix} 1 & 0 & 0 & \frac{14}{21} \\ 0 & 1 & 0 & \frac{31}{21} \\ 0 & 0 & 1 & \frac{1}{21} \end{bmatrix}$

R_1: $x = \frac{14}{21} = \frac{2}{3}$; R_2: $y = \frac{31}{21}$; R_3: $z = \frac{1}{21}$

Note: Exer. 9–16: There are other forms for the answers; c is any real number.

$\boxed{9}$ $\begin{bmatrix} 1 & 3 & 1 & 0 \\ 1 & 1 & -1 & 0 \\ 1 & -2 & -4 & 0 \end{bmatrix}$ $\begin{matrix} \\ R_2 - R_1 \rightarrow R_2 \\ R_3 - R_1 \rightarrow R_3 \end{matrix}$

$\begin{bmatrix} 1 & 3 & 1 & 0 \\ 0 & -2 & -2 & 0 \\ 0 & -5 & -5 & 0 \end{bmatrix}$ $\begin{matrix} -\frac{1}{2}R_2 \rightarrow R_2 \\ \\ -\frac{1}{5}R_3 \rightarrow R_3 \end{matrix}$

$\begin{bmatrix} 1 & 3 & 1 & 0 \\ 0 & 1 & 1 & 0 \\ 0 & 1 & 1 & 0 \end{bmatrix}$ $\begin{matrix} R_1 - 3R_2 \rightarrow R_1 \\ \\ R_3 - R_2 \rightarrow R_3 \end{matrix}$

$\begin{bmatrix} 1 & 0 & -2 & 0 \\ 0 & 1 & 1 & 0 \\ 0 & 0 & 0 & 0 \end{bmatrix}$

R_1: $x - 2z = 0 \Rightarrow x = 2z$

R_2: $y + z = 0 \Rightarrow y = -z$ $\qquad\qquad\qquad$ ★ $(2c, -c, c)$

$\boxed{11}$ $\begin{bmatrix} 2 & 1 & 1 & 0 \\ 1 & -2 & -2 & 0 \\ 1 & 1 & 1 & 0 \end{bmatrix}$ $R_1 \leftrightarrow R_2$

$\begin{bmatrix} 1 & -2 & -2 & 0 \\ 2 & 1 & 1 & 0 \\ 1 & 1 & 1 & 0 \end{bmatrix}$ $\begin{matrix} \\ R_2 - 2R_1 \rightarrow R_2 \\ R_3 - R_1 \rightarrow R_3 \end{matrix}$

$\begin{bmatrix} 1 & -2 & -2 & 0 \\ 0 & 5 & 5 & 0 \\ 0 & 3 & 3 & 0 \end{bmatrix}$ $\begin{matrix} \frac{1}{5}R_2 \rightarrow R_2 \\ \\ \frac{1}{3}R_3 \rightarrow R_3 \end{matrix}$

$\begin{bmatrix} 1 & -2 & -2 & 0 \\ 0 & 1 & 1 & 0 \\ 0 & 1 & 1 & 0 \end{bmatrix}$ $\begin{matrix} R_1 + 2R_2 \rightarrow R_1 \\ \\ R_3 - R_2 \rightarrow R_3 \end{matrix}$

$\begin{bmatrix} 1 & 0 & 0 & 0 \\ 0 & 1 & 1 & 0 \\ 0 & 0 & 0 & 0 \end{bmatrix}$

R_1: $x = 0$

R_2: $y + z = 0 \Rightarrow y = -z$ $\qquad\qquad\qquad$ ★ $(0, -c, c)$

$\boxed{15}$ $\begin{bmatrix} 4 & -2 & 1 & 5 \\ 3 & 1 & -4 & 0 \end{bmatrix}$ $R_1 - R_2 \rightarrow R_1$

$\begin{bmatrix} 1 & -3 & 5 & 5 \\ 3 & 1 & -4 & 0 \end{bmatrix}$ $\begin{matrix} \\ R_2 - 3R_1 \rightarrow R_2 \end{matrix}$

$\begin{bmatrix} 1 & -3 & 5 & 5 \\ 0 & 10 & -19 & -15 \end{bmatrix}$

R_2: $10y - 19z = -15 \Rightarrow y = \frac{19}{10}z - \frac{3}{2}$

R_1: $x - 3y + 5z = 5 \Rightarrow x = 3(\frac{19}{10}z - \frac{3}{2}) - 5z + 5 = \frac{7}{10}z + \frac{1}{2}$ $\qquad$ ★ $(\frac{7}{10}c + \frac{1}{2}, \frac{19}{10}c - \frac{3}{2}, c)$

$\boxed{19}$ $\begin{bmatrix} 4 & -3 & 1 \\ 2 & 1 & -7 \\ -1 & 1 & -1 \end{bmatrix}$ $R_1 + 3\,R_3 \rightarrow R_1$

$\begin{bmatrix} 1 & 0 & -2 \\ 2 & 1 & -7 \\ -1 & 1 & -1 \end{bmatrix}$ $R_2 - 2\,R_1 \rightarrow R_2$
$R_3 + R_1 \rightarrow R_3$

$\begin{bmatrix} 1 & 0 & -2 \\ 0 & 1 & -3 \\ 0 & 1 & -3 \end{bmatrix}$ $R_3 - R_2 \rightarrow R_3$

$\begin{bmatrix} 1 & 0 & -2 \\ 0 & 1 & -3 \\ 0 & 0 & 0 \end{bmatrix}$ R_2: $y = -3$; R_1: $x = -2$ $\bigstar$ $(-2, -3)$

$\boxed{21}$ $\begin{bmatrix} 2 & 3 & 5 \\ 1 & -3 & 4 \\ 1 & 1 & -2 \end{bmatrix}$ $R_1 \leftrightarrow R_2$

$\begin{bmatrix} 1 & -3 & 4 \\ 2 & 3 & 5 \\ 1 & 1 & -2 \end{bmatrix}$ $R_2 - 2\,R_1 \rightarrow R_2$
$R_3 - R_1 \rightarrow R_3$

$\begin{bmatrix} 1 & -3 & 4 \\ 0 & 9 & -3 \\ 0 & 4 & -6 \end{bmatrix}$ $R_2 - 2\,R_3 \rightarrow R_2$

$\begin{bmatrix} 1 & -3 & 4 \\ 0 & 1 & 9 \\ 0 & 4 & -6 \end{bmatrix}$ $R_1 + 3\,R_2 \rightarrow R_1$
$R_3 - 4\,R_2 \rightarrow R_3$

$\begin{bmatrix} 1 & 0 & 31 \\ 0 & 1 & 9 \\ 0 & 0 & -42 \end{bmatrix}$

There is a contradiction in row 3, therefore there is *no solution.*

$\boxed{23}$ Let x, y, and z denote the number of liters of the 10% acid, 30% acid, and 50% acid.

$\begin{cases} x + y + z = 50 \\ 0.10x + 0.30y + 0.50z = (0.32)(50) \\ z = 2y \end{cases}$ *quantity* (E_1)
quality (E_2)
constraint (E_3)

Substitute $z = 2y$ into E_1 and $100\,E_2$ to obtain

$\begin{cases} x + 3y \quad\ = 50 & (E_4) \\ 10x + 130y = 1600 & (E_5) \end{cases}$

$E_5 - 10\,E_4 \Rightarrow 100y = 1100 \Rightarrow y = 11;\ x = 17;\ z = 22$

25 Let x, y, and z denote the number of hours needed for A, B, and C, respectively, to produce 1000 items. In one hour, A, B, and C produce $\frac{1000}{x}$, $\frac{1000}{y}$, and $\frac{1000}{z}$ items, respectively. In 6 hours, A and B produce $\frac{6000}{x}$ and $\frac{6000}{y}$ items. From the table, this sum must equal 4500. The system of equations is then:

$$\begin{cases} \frac{6000}{x} + \frac{6000}{y} \qquad\qquad = 4500 \\ \frac{8000}{x} \qquad\quad + \frac{8000}{z} = 3600 \\ \qquad\quad \frac{7000}{y} + \frac{7000}{z} = 4900 \end{cases}$$

To simplify, let $a = 1/x$, $b = 1/y$, $c = 1/z$ and divide each equation by its greatest common factor $\{1500, 400, \text{ and } 700\}$.

$$\begin{cases} 4a + 4b \qquad\quad = 3 & (E_1) \\ 20a + \qquad\; + 20c = 9 & (E_2) \\ \qquad 10b + 10c = 7 & (E_3) \end{cases}$$

$E_2 - 5\,E_1 \Rightarrow 20c - 20b = -6 \quad (E_4)$

$E_4 + 2\,E_3 \Rightarrow 40c = 8 \Rightarrow c = \frac{1}{5}; \; b = \frac{1}{2}; \; a = \frac{1}{4}$. Resubstituting, $x = 4$, $y = 2$, and $z = 5$.

27 Let x, y, and z denote the amounts of G_1, G_2, and G_3, respectively.

$$\begin{cases} x + y + z = 600 & quantity & (E_1) \\ 0.30x + 0.20y + 0.15z = (0.25)(600) & quality & (E_2) \\ z = 100 + y & constraint & (E_3) \end{cases}$$

Substitute $z = 100 + y$ into E_1 and $100\,E_2$ to obtain

$$\begin{cases} x + 2y = 500 & (E_4) \\ 30x + 35y = 13{,}500 & (E_5) \end{cases}$$

$$E_5 - 30\,E_4 \Rightarrow -25y = -1500 \Rightarrow y = 60; \; z = 160; \; x = 380.$$

29 (a) $\begin{aligned} &I_1 - I_2 + I_3 = 0 & I_1 = I_2 - I_3 & \quad (E_1) \\ &3\,I_1 + 3\,I_2 = 6 \quad \Rightarrow & I_1 + I_2 = 2 & \quad (E_2) \\ &3\,I_2 + 3\,I_3 = 12 & I_2 + I_3 = 4 & \quad (E_3) \end{aligned}$

Substitute I_1 in E_1 into E_2 to obtain $2\,I_2 - I_3 = 2 \; (E_4)$.

$$E_4 + E_3 \Rightarrow 3\,I_2 = 6 \Rightarrow I_2 = 2; \; I_3 = 2; \; I_1 = 0.$$

(b) $\begin{aligned} &I_1 = I_2 - I_3 & (E_1) \\ &4\,I_1 + I_2 = 6 & (E_2) \\ &I_2 + 4\,I_3 = 12 & (E_3) \end{aligned}$

Substitute I_1 in E_1 into E_2 to obtain $5\,I_2 - 4\,I_3 = 6 \; (E_4)$.

$$E_4 + E_3 \Rightarrow 6\,I_2 = 18 \Rightarrow I_2 = 3; \; I_3 = \frac{9}{4}; \; I_1 = \frac{3}{4}.$$

31 Let x, y, and z denote the amount of Colombian, Brazilian, and Kenyan coffee used, respectively.

$$\begin{cases} x + y + z = 1 & \textit{quantity} & (E_1) \\ 10x + 6y + 8z = (8.50)(1) & \textit{quality} & (E_2) \\ x = 3y & \textit{constraint} & (E_3) \end{cases}$$

Substitute $x = 3y$ into E_1 and E_2 to obtain

$$\begin{cases} 4y + z = 1 & (E_4) \\ 36y + 8z = 8.5 & (E_5) \end{cases}$$ $E_5 - 8E_4 \Rightarrow 4y = \tfrac{1}{2} \Rightarrow y = \tfrac{1}{8};\ z = \tfrac{1}{2};\ x = \tfrac{3}{8}.$

33 (a) A: $x_1 + x_4 = 50 + 25 = 75$, B: $x_1 + x_2 = 100 + 50 = 150$,

C: $x_2 + x_3 = 150 + 75 = 225$, D: $x_3 + x_4 = 100 + 50 = 150$

(b) From C, $x_3 = 100 \Rightarrow x_2 = 225 - 100 = 125$.

From D, $x_3 = 100 \Rightarrow x_4 = 150 - 100 = 50$.

From A, $x_4 = 50 \Rightarrow x_1 = 75 - 50 = 25$.

(c) From D in part (a), $x_3 = 150 - x_4 \Rightarrow x_3 \le 150$ since $x_4 \ge 0$. From C,

$x_3 = 225 - x_2 = 225 - (150 - x_1)$ { from B } $= 75 + x_1 \Rightarrow x_3 \ge 75$ since $x_1 \ge 0$.

35 $t = 2070 - 1990 = 80$. $rt = (0.025)(80) = 2$ and $A = 800$ for E_1,

$(0.015)(80) = 1.2$ and $A = 560$ for E_2, and $(0.01)(0) = 0$ and $A = 340$ for E_3.

Substituting into $A = a + ct + ke^{rt}$ and summarizing as a system, we have:

$$\begin{cases} a + 80c + e^2 k = 800 & (E_1) \\ a + 80c + e^{1.2} k = 560 & (E_2) \\ a + k = 340 & (E_3) \end{cases}$$

We want to find t when A has doubled—that is, $A = 2(340)$. First we find c and k.

$E_1 - E_2 \Rightarrow e^2 k - e^{1.2} k = 240 \Rightarrow k = \dfrac{240}{e^2 - e^{1.2}} \approx 58.98.$

Substituting into E_3 gives $a = 340 - k \approx 281.02$.

Substituting into E_1 gives $c = \dfrac{800 - a - e^2 k}{80} \approx \dfrac{800 - 281.02 - (58.98)e^2}{80} \approx 1.04.$

Thus, $A = 281.02 + 1.04t + 58.98e^{rt}$. If $A = 680$ and $r = 0.01$, then

$680 = 281.02 + 1.04t + 58.98e^{0.01t} \Rightarrow 1.04t + 58.98e^{0.01t} - 398.98 = 0$. Graphing

$y = 1.04t + 58.98e^{0.01t} - 398.98$, we see there is an x-intercept at $x \approx 144.08$.

$1990 + 144.08 = 2134.08$, or during the year 2134.

[0, 1050] by [−350, 350]

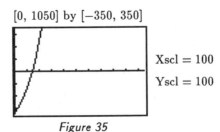

Xscl $= 100$

Yscl $= 100$

Figure 35

37 The parabola has an equation of the form $y = ax^2 + bx + c$. Substituting the x and y values of $P(2, 5)$, $Q(-2, -3)$, and $R(1, 6)$ into this equation yields:

$$\begin{cases} 4a + 2b + c = 5 & P \quad (E_1) \\ 4a - 2b + c = -3 & Q \quad (E_2) \\ a + b + c = 6 & R \quad (E_3) \end{cases}$$

Solving E_3 for c $\{c = 6 - a - b\}$ and substituting into E_1 and E_2 yields:

$$\begin{cases} 3a + b = -1 & (E_4) \\ 3a - 3b = -9 & (E_5) \end{cases}$$

$E_4 - E_5 \Rightarrow 4b = 8 \Rightarrow b = 2$; $a = -1$; $c = 5$. The equation is $y = -x^2 + 2x + 5$.

39 The parabola has an equation of the form $x = ay^2 + by + c$. Substituting the x and y values of $P(-1, 1)$, $Q(11, -2)$, and $R(5, -1)$ into this equation yields:

$$\begin{cases} a + b + c = -1 & P \quad (E_1) \\ 4a - 2b + c = 11 & Q \quad (E_2) \\ a - b + c = 5 & R \quad (E_3) \end{cases}$$

Solving E_3 for c $\{c = 5 - a + b\}$ and substituting into E_1 and E_2 yields:

$$\begin{cases} 2b = -6 & (E_4) \\ 3a - b = 6 & (E_5) \end{cases}$$

$E_4 \Rightarrow b = -3$; $a = 1$; $c = 1$. The equation is $x = y^2 - 3y + 1$.

41 The circle has an equation of the form $x^2 + y^2 + ax + by + c = 0$.

Substituting the x and y values of P, Q, and R into this equation yields:

$$\begin{cases} 2a + b + c = -5 & P \quad (E_1) \\ -a - 4b + c = -17 & Q \quad (E_2) \\ 3a + c = -9 & R \quad (E_3) \end{cases}$$

Solving E_3 for c $\{c = -9 - 3a\}$ and substituting into E_1 and E_2 yields:

$$\begin{cases} -a + b = 4 & (E_4) \\ -4a - 4b = -8 & (E_5) \end{cases} \Rightarrow \begin{cases} -a + b = 4 & (E_6) \\ a + b = 2 & (E_7) \end{cases}$$

$E_6 + E_7 \Rightarrow 2b = 6 \Rightarrow b = 3$; $a = -1$; $c = -6$. The equation is $x^2 + y^2 - x + 3y - 6 = 0$.

43 For $y = f(x) = ax^3 + bx^2 + cx + d$ and the points $(-1, 2)$, $(0.5, 2)$, $(1, 3)$, and $(2, 4.5)$, we obtain the system:

$$\begin{cases} -a + b - c + d = 2 & (-1, 2) \\ 0.125a + 0.25b + 0.5c + d = 2 & (0.5, 2) \\ a + b + c + d = 3 & (1, 3) \\ 8a + 4b + 2c + d = 4.5 & (2, 4.5) \end{cases}$$

Solving the system yields $a = -\frac{4}{9}$, $b = \frac{11}{9}$, $c = \frac{17}{18}$, and $d = \frac{23}{18}$.

Note: The general outline for the solutions in this section is as follows:

1st line)   The expression is shown on the left side of the equation and its decomposition is on the right side.

2nd line)   The equation in the first line is multiplied by its least common denominator and left in factored form.

3rd line and beyond)   Values are substituted into the equation in the second line and the coefficients are found by solving the resulting equations. It will be stated when the method of equating coefficients is used.

3 $\dfrac{x+34}{(x-6)(x+2)} = \dfrac{A}{x-6} + \dfrac{B}{x+2}$

$x + 34 = A(x+2) + B(x-6)$

$x = -2: \ 32 = -8B \Rightarrow B = -4$

$x = 6: \ 40 = 8A \Rightarrow A = 5$

$\bigstar \ \dfrac{5}{x-6} - \dfrac{4}{x+2}$

7 $\dfrac{4x^2 - 5x - 15}{x(x-5)(x+1)} = \dfrac{A}{x} + \dfrac{B}{x-5} + \dfrac{C}{x+1}$

$4x^2 - 5x - 15 = A(x-5)(x+1) + Bx(x+1) + Cx(x-5)$

$x = -1: \ -6 = 6C \Rightarrow C = -1$

$x = 0: \ -15 = -5A \Rightarrow A = 3$

$x = 5: \ 60 = 30B \Rightarrow B = 2$

$\bigstar \ \dfrac{3}{x} + \dfrac{2}{x-5} - \dfrac{1}{x+1}$

11 $\dfrac{19x^2 + 50x - 25}{x^2(3x-5)} = \dfrac{A}{x} + \dfrac{B}{x^2} + \dfrac{C}{3x-5}$

$19x^2 + 50x - 25 = Ax(3x-5) + B(3x-5) + Cx^2$

$x = \frac{5}{3}: \ \frac{1000}{9} = \frac{25}{9}C \Rightarrow C = 40$

$x = 0: \ -25 = -5B \Rightarrow B = 5$

$x = 1: \ 44 = -2A - 2B + C \Rightarrow A = -7$

$\bigstar \ -\dfrac{7}{x} + \dfrac{5}{x^2} + \dfrac{40}{3x-5}$

15 $\dfrac{3x^3 + 11x^2 + 16x + 5}{x(x+1)^3} = \dfrac{A}{x} + \dfrac{B}{x+1} + \dfrac{C}{(x+1)^2} + \dfrac{D}{(x+1)^3}$

$3x^3 + 11x^2 + 16x + 5 = A(x+1)^3 + Bx(x+1)^2 + Cx(x+1) + Dx$

$x = -1: \ -3 = -D \Rightarrow D = 3$

$x = 0: \ 5 = A$

$x = 1: \ 35 = 8A + 4B + 2C + D \quad (E_1)$

$x = -2: \ -7 = -A - 2B + 2C - 2D \quad (E_2)$

Substituting the values for A and D into E_1 and E_2 yields

(continued)

$$\begin{cases}4B+2C &= -8 \\ -2B+2C &= 4\end{cases} \Rightarrow \begin{cases}2B+C &= -4 \quad (E_3) \\ -B+C &= 2 \quad (E_4)\end{cases}$$

$E_3 + 2E_4 \Rightarrow 3C = 0 \Rightarrow C = 0; \ B = -2$

$\bigstar \ \dfrac{5}{x} - \dfrac{2}{x+1} + \dfrac{3}{(x+1)^3}$

$\boxed{19} \ \dfrac{9x^2 - 3x + 8}{x(x^2+2)} = \dfrac{A}{x} + \dfrac{Bx+C}{x^2+2}$

$9x^2 - 3x + 8 = A(x^2+2) + (Bx+C)x$

$x = 0: \ 8 = 2A \Rightarrow A = 4$

$x = 1: \ 14 = 3A + B + C \quad (E_1)$

$x = -1: \ 20 = 3A + B - C \quad (E_2)$

$\bigstar \ \dfrac{4}{x} + \dfrac{5x-3}{x^2+2}$

$E_1 - E_2 \Rightarrow -6 = 2C \Rightarrow C = -3; \ B = 5$

$\boxed{21} \ \dfrac{4x^3 - x^2 + 4x + 2}{(x^2+1)^2} = \dfrac{Ax+B}{x^2+1} + \dfrac{Cx+D}{(x^2+1)^2}$

$$4x^3 - x^2 + 4x + 2 = (Ax+B)(x^2+1) + Cx + D$$
$$= Ax^3 + Bx^2 + (A+C)x + (B+D)$$

Equating coefficients, i.e., the coefficient of x^3 on the left side must equal the coefficient of x^3 on the right side, we have the following:

$x^3 \qquad : A = 4$

$x^2 \qquad : B = -1$

$x \qquad : A + C = 4 \Rightarrow C = 0$

$\bigstar \ \dfrac{4x-1}{x^2+1} + \dfrac{3}{(x^2+1)^2}$

constant $: B + D = 2 \Rightarrow D = 3$

$\boxed{23}$ The degree of the numerator is not lower than the degree of the denominator.

Thus, we must first use long division and then decompose the remaining expression.

Hence, by first dividing and then factoring, we have the following:

$2x + \dfrac{4x^2 - 3x + 1}{(x^2+1)(x-1)} = 2x + \dfrac{Ax+B}{x^2+1} + \dfrac{C}{x-1}$

$4x^2 - 3x + 1 = (Ax+B)(x-1) + C(x^2+1)$

$x = 1: \ 2 = 2C \Rightarrow C = 1$

$x = 0: \ 1 = -B + C \Rightarrow B = 0$

$\bigstar \ 2x + \dfrac{1}{x-1} + \dfrac{3x}{x^2+1}$

$x = -1: \ 8 = 2A - 2B + 2C \Rightarrow A = 3$

$\boxed{25}$ By first dividing and then factoring, we have the following:

$3 + \dfrac{12x - 16}{x(x-4)} = 3 + \dfrac{A}{x} + \dfrac{B}{x-4}$

$12x - 16 = A(x-4) + Bx$

$x = 0: \ -16 = -4A \Rightarrow A = 4$

$\bigstar \ 3 + \dfrac{4}{x} + \dfrac{8}{x-4}$

$x = 4: \ 32 = 4B \Rightarrow B = 8$

$\boxed{27}$ By first dividing and then factoring, we have the following:

$$2x + 3 + \frac{x+5}{(2x+1)(x-1)} = 2x + 3 + \frac{A}{2x+1} + \frac{B}{x-1}$$

$x + 5 = A(x-1) + B(2x+1)$

$x = 1: 6 = 3B \Rightarrow B = 2$ $\bigstar\ 2x + 3 + \dfrac{2}{x-1} - \dfrac{3}{2x+1}$

$x = -\frac{1}{2}: \frac{9}{2} = -\frac{3}{2}A \Rightarrow A = -3$

9.5 Exercises

$\boxed{1}$ $3x - 2y < 6 \Leftrightarrow y > \frac{3}{2}x - 3$. Sketch the graph of $y = \frac{3}{2}x - 3$ with dashes. The point
(0, 0) is clearly on one side of the line, so we will substitute $x = 0$ and $y = 0$ into the
inequality $y > \frac{3}{2}x - 3$. Checking, we have $0 > -3$, a true statement. Hence, we shade
all points that are on the same side of the line as (0, 0).

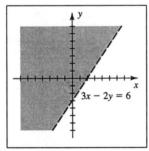

Figure 1

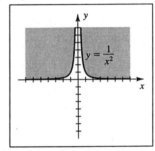

Figure 9

$\boxed{9}$ $yx^2 \geq 1 \Leftrightarrow y \geq 1/x^2 \ \{x \neq 0\}$. Sketch the graph of $y = 1/x^2$ with a solid curve. The
point (1, 0) is clearly not on the graph. Substituting $x = 1$ and $y = 0$ into $y \geq 1/x^2$
yields $0 \geq 1$, a false statement. Hence, we do not shade the region containing (1, 0),
but we do shade the regions in the first and second quadrants that are above the
graph.

Note: The will use the notation V @ (a, b), (c, d), ... to denote the intersection point(s)
of the solution region of the graph. These can be found by solving the system of
equalities that correspond to the given system of inequalities.

11 $\begin{cases} 3x + y < 3 \\ 4 - y < 2x \end{cases} \Leftrightarrow \begin{cases} y < -3x + 3 \\ y > -2x + 4 \end{cases}$ V @ $(-1, 6)$

Solving the system $y = -3x + 3$ and $y = -2x + 4$ gives us the solution $(-1, 6)$. Testing the point $(0, 0)$ in both inequalities, we see that we need to shade under $y = -3x + 3$ and above $y = -2x + 4$, as shown in *Figure 11*.

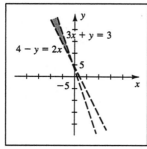

Figure 11

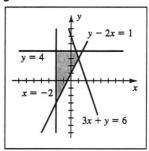

Figure 15

15 See *Figure 15*.

$\begin{cases} 3x + y \le 6 \\ y - 2x \ge 1 \\ x \ge -2 \\ y \le 4 \end{cases} \Leftrightarrow \begin{cases} y \le -3x + 6 \\ y \ge 2x + 1 \\ x \ge -2 \\ y \le 4 \end{cases}$ V @ $(-2, -3)$, $(-2, 4)$, $(\frac{2}{3}, 4)$, $(1, 3)$

21 $|x + 2| \le 1 \Leftrightarrow -1 \le x + 2 \le 1 \Leftrightarrow -3 \le x \le -1$.

This region is bounded by the vertical lines $x = -3$ and $x = -1$, including the lines.

$|y - 3| < 5 \Leftrightarrow -5 < y - 3 < 5 \Leftrightarrow -2 < y < 8$.

This region is bounded by the horizontal lines $y = -2$ and $y = 8$, excluding the lines.

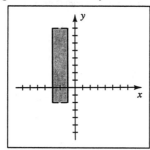

Figure 21

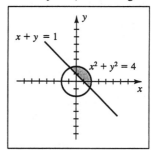

Figure 23

23 $\begin{cases} x^2 + y^2 \le 4 \\ x + y \ge 1 \end{cases} \Leftrightarrow \begin{cases} x^2 + y^2 \le 2^2 \\ y \ge -x + 1 \end{cases}$ V @ $(\frac{1}{2} \mp \frac{1}{2}\sqrt{7}, \frac{1}{2} \pm \frac{1}{2}\sqrt{7})$

$\boxed{27}$ The shaded region lies between $x = 0$ and $x = 3$, including $x = 0$ and excluding $x = 3$. This region is described by $0 \le x < 3$.

     The dashed line goes through $(0, 4)$ and $(4, 0)$. Hence, its slope is -1, its y-intercept is 4, and an equation for it is $y = -x + 4$. Substituting $x = 0$ and $y = 0$ into $y = -x + 4$ gives us $0 = 4$. We replace $=$ by $<$ { the line is dashed } to make this statement a true statement. Hence, we have the inequality $y < -x + 4$.

     The solid line goes through $(0, -4)$ and $(4, 0)$. Hence, its slope is 1, its y-intercept is -4, and an equation for it is $y = x - 4$. Substituting $x = 0$ and $y = 0$ into $y = x - 4$ gives us $0 = -4$. We replace $=$ by $\ge$ { the line is solid } to make this statement a true statement. Hence, we have the inequality $y \ge x - 4$. Thus, the graph may be described by the system $\begin{cases} 0 \le x < 3 \\ y < -x + 4 \\ y \ge x - 4 \end{cases}$

$\boxed{31}$ The center of the circle is $(2, 2)$ and it passes through $(0, 0)$. The radius is the distance from $(0, 0)$ to $(2, 2)$, which is $\sqrt{8}$. An equation of the circle is $(x - 2)^2 + (y - 2)^2 = 8$. The center is clearly inside the circle, and if we substitute $x = 2$ and $y = 2$ in the equation of the circle, we get $0 = 8$. Since we want to make this a true statement and include the boundary, we use $(x - 2)^2 + (y - 2)^2 \le 8$ to describe the shaded circular region and its boundary.

     For the dashed line $y = x$, test the point $(1, 0)$. This test gives us $0 = 1$, which we can make true by replacing $=$ by $<$. Hence $y < x$ describes the shaded region under the line $y = x$. The other inequality may be found as in Exercise 27. Thus, the graph may be described by the system $\begin{cases} y < x \\ y \le -x + 4 \\ (x - 2)^2 + (y - 2)^2 \le 8 \end{cases}$

$\boxed{33}$ For the dashed line through $(-4, 0)$ and $(4, 1)$, we can use the point-slope form to find an equation of the line. $y - 0 = \frac{1 - 0}{4 - (-4)}(x - (-4)) \Leftrightarrow y = \frac{1}{8}x + \frac{1}{2}$. Checking the point $(0, 1)$ { which is in the shaded region } gives us $1 = \frac{1}{2}$, so we change $=$ to $>$ to obtain $y > \frac{1}{8}x + \frac{1}{2}$.

     Similarly, for the solid line with x-intercept -4 and y-intercept 4, the shaded region can be described by $y \le x + 4$.

     Lastly, the solid line passing through $(4, 1)$ has slope $-\frac{3}{4}$, y-intercept 4, and its shaded region can be described by $y \le -\frac{3}{4}x + 4$. Thus, the system is $\begin{cases} y > \frac{1}{8}x + \frac{1}{2} \\ y \le x + 4 \\ y \le -\frac{3}{4}x + 4 \end{cases}$

35 Let x and y denote the number of sets of brand A and brand B, respectively. "Necessary to stock at least twice as many sets of brand A as of brand B" may be symbolized as $x \geq 2y$. "Necessary to have on hand at least 10 sets of brand B" may be symbolized as $y \geq 10$. Consequently, $x \geq 20$ from the first inequality. "Room for not more than 100 sets in the store" may be symbolized as $x + y \leq 100$. We now sketch the system of inequalities and shade the region that they have in common. The graph is the region bounded by the triangle with vertices $(20, 10)$, $(90, 10)$, and $\left(\frac{200}{3}, \frac{100}{3}\right)$.

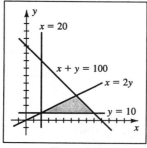

Figure 35

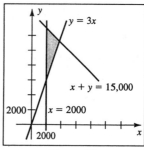

Figure 37

37 If x and y denote the amount placed in the high-risk and low-risk investment, respectively, then a system is $x \geq 2000$, $y \geq 3x$, $x + y \leq 15,000$. The graph is the region bounded by the triangle with vertices $(2000, 6000)$, $(2000, 13,000)$, and $(3750, 11,250)$.

39 A system is $x + y \leq 9$, $y \geq x$, $x \geq 1$. To justify the condition $y \geq x$, start with

$$\frac{\text{cylinder volume}}{\text{total volume}} > 0.75 \Rightarrow \frac{\pi r^2 y}{\pi r^2 y + \frac{1}{3}\pi r^2 x} \geq \frac{3}{4} \Rightarrow 4\pi r^2 y \geq 3\pi r^2 y + \pi r^2 x \Rightarrow$$

$\pi r^2 y \geq \pi r^2 x \Rightarrow y \geq x$. The graph is the region bounded by the triangle with vertices $(1, 1)$, $(1, 8)$, and $\left(\frac{9}{2}, \frac{9}{2}\right)$.

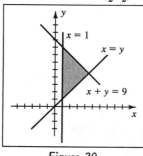

Figure 39

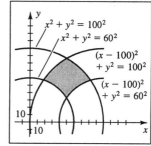

Figure 41

41 If the plant is located at (x, y), then a system is $(60)^2 \leq x^2 + y^2 \leq (100)^2$, $(60)^2 \leq (x - 100)^2 + y^2 \leq (100)^2$, $y \geq 0$. The graph is the region in the first quadrant that lies between the two concentric circles with center $(0, 0)$ and radii 60 and 100, and also between the two concentric circles with center $(100, 0)$ and radii 60 and 100. See *Figure 41* on the previous page.

Equating the different circle equations, we obtain the vertices of the solution region $(50, 50\sqrt{3})$, $(50, 10\sqrt{11})$, $(18, 6\sqrt{91})$, and $(82, 6\sqrt{91})$. For example, to determine where the circle $x^2 + y^2 = 100^2$ intersects the circle $(x - 100)^2 + y^2 = 60^2$, we subtract the first equation from the second to obtain $[(x - 100)^2 + y^2] - (x^2 + y^2) = 60^2 - 100^2 \Rightarrow -200x + 10,000 = -6400 \Rightarrow 200x = 16,400 \Rightarrow x = 82.$ Substituting $x = 82$ into $x^2 + y^2 = 100^2$ gives us $y^2 = 100^2 - 82^2 = 3276 \Rightarrow y = \pm\sqrt{3276} = \pm 6\sqrt{91} \approx \pm 57.2.$

$\boxed{43}$ $64y^3 - x^3 \le e^{1 - 2x} \Rightarrow y \le \frac{1}{4}(e^{1 - 2x} + x^3)^{1/3}$ Graph $y = \frac{1}{4}(e^{1 - 2x} + x^3)^{1/3}$ $\{Y_1\}.$

The solution includes the graph and the region below the graph.

Shading was obtained by using the command Shade(Ymin, Y_1, Xres, Xmin, Xmax).

$[-3.5, 4]$ by $[-1, 4]$ $[-1.5, 1.5]$ by $[-1, 1]$

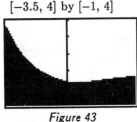
Xscl $= 1$
Yscl $= 1$

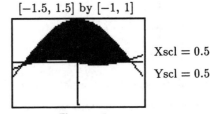
Xscl $= 0.5$
Yscl $= 0.5$

Figure 43   Figure 45

$\boxed{45}$ $5^{1 - y} \ge x^4 + x^2 + 1 \Rightarrow 1 - y \ge \log_5(x^4 + x^2 + 1) \Rightarrow y \le 1 - \log_5(x^4 + x^2 + 1).$

$x + 3y \ge x^{5/3} \Rightarrow y \ge \frac{1}{3}(x^{5/3} - x).$ Graph $y = 1 - \log_5(x^4 + x^2 + 1)$ $\{Y_1\}$ and $y = \frac{1}{3}(x^{5/3} - x)$ $\{Y_2\}.$ The graphs intersect at approximately $(1.21, 0.05)$ and $(-1.32, -0.09)$. The solution is located between the points of intersection. It includes the graphs and the region below the first graph and above the second graph.

Shading was obtained by using the command Shade(Y_2, Y_1, Xres, -1.32, 1.21).

$\boxed{47}$ $x^4 - 2x < 3y \Rightarrow y > \frac{1}{3}(x^4 - 2x).$ $x + 2y < x^3 - 5 \Rightarrow y < \frac{1}{2}(x^3 - x - 5).$

Graph $y = \frac{1}{3}(x^4 - 2x)$ and $y = \frac{1}{2}(x^3 - x - 5).$ The graphs do not intersect.

The solution must be above the first graph and below the second graph.

Since the regions do not intersect, there is no solution.

$[-4.5, 4.5]$ by $[-3, 3]$ $[33, 80]$ by $[0, 50]$

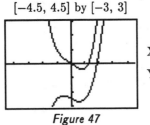
Xscl $= 1$
Yscl $= 1$

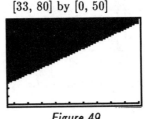
Xscl $= 5$
Yscl $= 5$

Figure 47   Figure 49

[49] (a) $29T - 39P < 450 \Rightarrow 29(37) - 39(21.2) < 450 \Rightarrow 246.2 < 450$ (true).

Yes, forests can grow.

(b) $29T - 39P < 450 \Rightarrow -39P < -29T + 450 \Rightarrow P > \frac{29}{39}T - \frac{150}{13}$.

Graph $Y_1 = \frac{29}{39}x - \frac{150}{13}$. See *Figure 49* on the preceding page.

(c) Forests can grow whenever the point (T, P) lies above the line $P = \frac{29}{39}T - \frac{150}{13}$.

9.6 Exercises

[1] We substitute the x and y values of each vertex into $C = 3x + 2y + 5$ and summarize the results in the following table. We see that there is a maximum of 27 at $(6, 2)$, and a minimum of 9 at $(0, 2)$.

(x, y)	(0, 2)	(0, 4)	(3, 5)	(6, 2)	(5, 0)	(2, 0)
C	9 ∎	13	24	27 ∎	20	11

[3] The region R is sketched in *Figure 3*. The vertices are found by obtaining the intersection points of the system of equations corresponding to the given inequalities.

(x, y)	(0, 0)	(0, 3)	(4, 6)	(6, 3)	(5, 0)
C	0	3	18	21 ∎	15

$C = 3x + y$;

maximum of 21 at $(6, 3)$

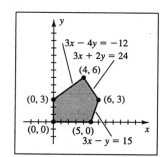

Figure 3

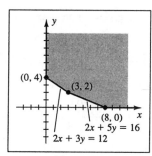

Figure 5

[5]

(x, y)	(8, 0)	(3, 2)	(0, 4)
C	24	21 ∎	24

$C = 3x + 6y$;

minimum of 21 at $(3, 2)$

7

(x, y)	(0, 0)	(0, 4)	(2, 5)	(6, 3)	(8, 0)
C	0	16	24 ■	24 ■	16

As in Example 2, $C = 2x + 4y$ has the maximum value 24 for *any* point on the line segment from (2, 5) to (6, 3). To show that the last statement is true, we will substitute $6 - \frac{1}{2}x$ for y in C. $C = 2x + 4y = 2x + 4(6 - \frac{1}{2}x) = 2x + 24 - 2x = 24.$

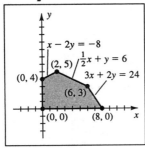

Figure 7

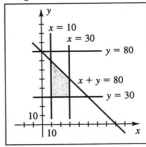

Figure 9

Note: **For the linear programming application exercises, the solution outline is as follows:**

(1) Define the variables used in the problem.

(2) Define the function to be maximized or minimized.

(3) List the system of inequalities that determine the solution region R.

(4) A table of intersection points and values of the function in (2) at those points is given. The maximum or minimum value is denoted by a ■.

(5) A summarizing statement is given.

(6) A figure is shown for the system of inequalities.

9 Let x and y denote the number of oversized and standard rackets, respectively.

Profit function: $P = 15x + 8y$

(x, y)	(10, 30)	(30, 30)	(30, 50)	(10, 70)
P	390	690	850 ■	710

$$\begin{cases} 30 \le y \le 80 \\ 10 \le x \le 30 \\ x + y \le 80 \end{cases}$$

The maximum profit of $850 per day occurs when 30 oversized rackets and 50 standard rackets are manufactured.

[11] Let x and y denote the number of pounds of S and T, respectively.

Cost function: $C = 3x + 4y$

(x, y)	$(0, 4.5)$	$(3.5, 1)$	$(5, 0)$
C	18	14.5 ■	15

$$\begin{cases} 2x + 2y \geq 9 & amount\ of\ \text{I} \\ 4x + 6y \geq 20 & amount\ of\ \text{G} \\ x \geq 0 \\ y \geq 0 \end{cases}$$

The minimum cost of \$14.50 occurs when 3.5 pounds of S and 1 pound of T are used.

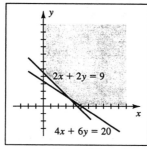

Figure 11

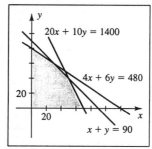

Figure 15

[15] Let x and y denote the number of acres planted with alfalfa and corn, respectively.

Profit function: $P = 110x - 4x - 20x + 150y - 6y - 10y = \underline{86x + 134y}$

$$\begin{cases} 4x + 6y \leq 480 & seed\ cost \\ 20x + 10y \leq 1400 & labor\ cost \\ x + y \leq 90 & area \\ x, y \geq 0 \end{cases}$$

(x, y)	$(70, 0)$	$(50, 40)$	$(30, 60)$	$(0, 80)$	$(0, 0)$
P	6020	9660	10,620	10,720 ■	0

The maximum profit of \$10,720 occurs when

0 acres of alfalfa are planted and 80 acres of corn are planted.

$\boxed{17}$ Let x, y, and z denote the number of ounces of X, Y, and Z, respectively.

Cost function: $C = 0.25x + 0.35y + 0.50z$
$$= 0.25x + 0.35y + 0.50(20 - x - y) = \underline{10 - 0.25x - 0.15y}$$

$$\left\{\begin{array}{l} 0.20x + 0.20y + 0.10z \geq 0.14(20) \quad \text{amount of A} \\ 0.10x + 0.40y + 0.20z \geq 0.16(20) \quad \text{amount of B} \\ 0.25x + 0.15y + 0.25z \geq 0.20(20) \quad \text{amount of C} \end{array}\right. \Rightarrow \left\{\begin{array}{r} x + y \geq 8 \\ x - 2y \leq 8 \\ y \leq 10 \\ x + y \leq 20 \\ 0 \leq x,\, y \leq 20 \end{array}\right.$$

The new restrictions are found by substituting $z = 20 - x - y$ into the 3 inequalities, simplifying, and adding the last 2 inequalities.

(x, y)	(8, 0)	(16, 4)	(10, 10)	(0, 10)	(0, 8)
C	8.00	5.40 ■	6.00	8.50	8.80 ■■

The minimum cost of $5.40 requires 16 oz of X, 4 oz of Y, and 0 oz of Z.

The maximum cost of $8.80 requires 0 oz of X, 8 oz of Y, and 12 oz of Z.

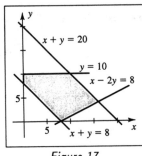

Figure 17

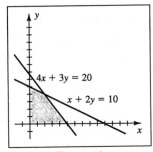

Figure 19

$\boxed{19}$ Let x and y denote the number of vans and buses purchased, respectively.

$$\left\{\begin{array}{r} 10{,}000x + 20{,}000y \leq 100{,}000 \quad \text{purchase} \\ 100x + 75y \leq 500 \quad \text{maintenance} \\ x \geq 0 \\ y \geq 0 \end{array}\right. \Rightarrow \left\{\begin{array}{r} x + 2y \leq 10 \\ 4x + 3y \leq 20 \\ x \geq 0 \\ y \geq 0 \end{array}\right.$$

(x, y)	(5, 0)	(2, 4)	(0, 5)	(0, 0)
P	75	130 ■	125	0

Passenger capacity function: $P = 15x + 25y$

The maximum passenger capacity of 130 would occur if the community purchases 2 vans and 4 buses.

[21] Let x and y denote the number of trout and bass, respectively.

Pound function: $P = 3x + 4y$

$$\begin{cases} x + y \le 5000 & number\ of\ fish \\ 0.50x + 0.75y \le 3000 & cost \\ x \ge 0 \\ y \ge 0 \end{cases} \Rightarrow \begin{cases} x + y \le 5000 \\ 2x + 3y \le 12{,}000 \\ x \ge 0 \\ y \ge 0 \end{cases}$$

(x, y)	(5000, 0)	(3000, 2000)	(0, 4000)	(0, 0)
P	15,000	17,000 ■	16,000	0

The total number of pounds of fish will be a maximum of 17,000 if 3000 trout and 2000 bass are purchased.

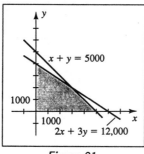

Figure 21

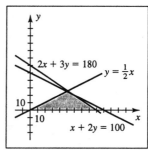

Figure 23

[23] Let x and y denote the number of basic and deluxe units constructed, respectively.

$$\begin{cases} 300x + 600y \le 30{,}000 & cost \\ x \ge 2y & ratio \\ 80x + 120y \le 7200 & area \\ x,\ y \ge 0 \end{cases} \Rightarrow \begin{cases} x + 2y \le 100 \\ y \le \tfrac{1}{2}x \\ 2x + 3y \le 180 \\ x,\ y \ge 0 \end{cases}$$

(x, y)	(90, 0)	(60, 20)	(50, 25)	(0, 0)
R	3600	3900 ■	3875	0

Revenue function: $R = 40x + 75y$

The maximum monthly revenue of $3900 occurs if 60 basic units and 20 deluxe units are constructed.

9.7 Exercises

Note: For Exercises 1–8, $(A + B)$ and $(A - B)$ are possible only if A and B have the same number of rows and columns. { #7 & #8 are not possible } $(2A)$ and $(-3B)$ are always possible by simply multiplying the elements of A and B by 2 and -3, respectively.

$\boxed{1}$ $A + B = \begin{bmatrix} 5 & -2 \\ 1 & 3 \end{bmatrix} + \begin{bmatrix} 4 & 1 \\ -3 & 2 \end{bmatrix} = \begin{bmatrix} 5+4 & -2+1 \\ 1+(-3) & 3+2 \end{bmatrix} = \begin{bmatrix} 9 & -1 \\ -2 & 5 \end{bmatrix}$

$A - B = \begin{bmatrix} 5 & -2 \\ 1 & 3 \end{bmatrix} - \begin{bmatrix} 4 & 1 \\ -3 & 2 \end{bmatrix} = \begin{bmatrix} 5-4 & -2-1 \\ 1-(-3) & 3-2 \end{bmatrix} = \begin{bmatrix} 1 & -3 \\ 4 & 1 \end{bmatrix}$

$2A = 2 \begin{bmatrix} 5 & -2 \\ 1 & 3 \end{bmatrix} = \begin{bmatrix} 2(5) & 2(-2) \\ 2(1) & 2(3) \end{bmatrix} = \begin{bmatrix} 10 & -4 \\ 2 & 6 \end{bmatrix}$

$-3B = -3 \begin{bmatrix} 4 & 1 \\ -3 & 2 \end{bmatrix} = \begin{bmatrix} -3(4) & -3(1) \\ -3(-3) & -3(2) \end{bmatrix} = \begin{bmatrix} -12 & -3 \\ 9 & -6 \end{bmatrix}$

$\boxed{7}$ $A + B$ and $A - B$ are not possible since A and B are different sizes.

$2A = 2 \begin{bmatrix} 3 & -2 & 2 \\ 0 & 1 & -4 \\ -3 & 2 & -1 \end{bmatrix} = \begin{bmatrix} 6 & -4 & 4 \\ 0 & 2 & -8 \\ -6 & 4 & -2 \end{bmatrix}$, $-3B = -3 \begin{bmatrix} 4 & 0 \\ 2 & -1 \\ -1 & 3 \end{bmatrix} = \begin{bmatrix} -12 & 0 \\ -6 & 3 \\ 3 & -9 \end{bmatrix}$

$\boxed{9}$ To find c_{21} in Exercise 15, use the second row of A and the first column of B.

$$c_{21} = (-5)(2) + (2)(0) + (2)(-4) = -10 + 0 - 8 = -18.$$

Note: For Exercises 11–22, AB is possible only if the number of columns of A equal the number of rows of B. { BA is not possible in #21, AB is not possible in #22 } Multiplying an $\boxed{m} \times n$ matrix with an $n \times \boxed{k}$ matrix will result in an $m \times k$ matrix.

$\boxed{11}$ $AB = \begin{bmatrix} 2 & 6 \\ 3 & -4 \end{bmatrix} \begin{bmatrix} 5 & -2 \\ 1 & 7 \end{bmatrix}$

$= \begin{bmatrix} 2(5)+6(1) & 2(-2)+6(7) \\ 3(5)+(-4)(1) & 3(-2)+(-4)(7) \end{bmatrix} = \begin{bmatrix} 10+6 & -4+42 \\ 15-4 & -6-28 \end{bmatrix} = \begin{bmatrix} 16 & 38 \\ 11 & -34 \end{bmatrix}$

$BA = \begin{bmatrix} 5 & -2 \\ 1 & 7 \end{bmatrix} \begin{bmatrix} 2 & 6 \\ 3 & -4 \end{bmatrix}$

$= \begin{bmatrix} 5(2)+(-2)(3) & 5(6)+(-2)(-4) \\ 1(2)+7(3) & 1(6)+7(-4) \end{bmatrix} = \begin{bmatrix} 4 & 38 \\ 23 & -22 \end{bmatrix}$

$\boxed{15}$ $AB = \begin{bmatrix} 4 & -3 & 1 \\ -5 & 2 & 2 \end{bmatrix} \begin{bmatrix} 2 & 1 \\ 0 & 1 \\ -4 & 7 \end{bmatrix}$

$= \begin{bmatrix} 4(2)+(-3)(0)+1(-4) & 4(1)+(-3)(1)+1(7) \\ -5(2)+2(0)+2(-4) & -5(1)+2(1)+2(7) \end{bmatrix} = \begin{bmatrix} 4 & 8 \\ -18 & 11 \end{bmatrix}$

$BA = \begin{bmatrix} 2 & 1 \\ 0 & 1 \\ -4 & 7 \end{bmatrix} \begin{bmatrix} 4 & -3 & 1 \\ -5 & 2 & 2 \end{bmatrix}$

$= \begin{bmatrix} 2(4)+1(-5) & 2(-3)+1(2) & 2(1)+1(2) \\ 0(4)+1(-5) & 0(-3)+1(2) & 0(1)+1(2) \\ -4(4)+7(-5) & -4(-3)+7(2) & -4(1)+7(2) \end{bmatrix} = \begin{bmatrix} 3 & -4 & 4 \\ -5 & 2 & 2 \\ -51 & 26 & 10 \end{bmatrix}$

19 $AB = \begin{bmatrix} -3 & 7 & 2 \end{bmatrix} \begin{bmatrix} 1 \\ 4 \\ -5 \end{bmatrix} = \begin{bmatrix} -3(1) + 7(4) + 2(-5) \end{bmatrix} = \begin{bmatrix} 15 \end{bmatrix}$

$BA = \begin{bmatrix} 1 \\ 4 \\ -5 \end{bmatrix} \begin{bmatrix} -3 & 7 & 2 \end{bmatrix} = \begin{bmatrix} 1(-3) & 1(7) & 1(2) \\ 4(-3) & 4(7) & 4(2) \\ -5(-3) & -5(7) & -5(2) \end{bmatrix} = \begin{bmatrix} -3 & 7 & 2 \\ -12 & 28 & 8 \\ 15 & -35 & -10 \end{bmatrix}$

21 $AB = \begin{bmatrix} 2 & 0 & 1 \\ -1 & 2 & 0 \end{bmatrix} \begin{bmatrix} 1 & -1 & 2 \\ 3 & 1 & 0 \\ 0 & 2 & 1 \end{bmatrix}$

$= \begin{bmatrix} 2(1) + 0(3) + 1(0) & 2(-1) + 0(1) + 1(2) & 2(2) + 0(0) + 1(1) \\ -1(1) + 2(3) + 0(0) & -1(-1) + 2(1) + 0(2) & -1(2) + 2(0) + 0(1) \end{bmatrix}$

$= \begin{bmatrix} 2 & 0 & 5 \\ 5 & 3 & -2 \end{bmatrix}$. BA is a 3×3 matrix times a 2×3 matrix. Since the number

of columns of B, 3, is not equal to the number of rows of A, 2, BA is not possible.

25 $AB = \begin{bmatrix} 2 & 1 & 0 & -3 \\ -7 & 0 & -2 & 4 \end{bmatrix} \begin{bmatrix} 4 & -2 & 0 \\ 1 & 1 & -2 \\ 0 & 0 & 5 \\ -3 & -1 & 0 \end{bmatrix} = \begin{bmatrix} c_{11} & c_{12} & c_{13} \\ c_{21} & c_{22} & c_{23} \end{bmatrix}$, where

$c_{11} = 2(4) + 1(1) + 0(0) + (-3)(-3)$ $\qquad c_{12} = 2(-2) + 1(1) + 0(0) + (-3)(-1)$

$c_{13} = 2(0) + 1(-2) + 0(5) + (-3)(0)$ $\qquad c_{21} = -7(4) + 0(1) + (-2)(0) + 4(-3)$

$c_{22} = -7(-2) + 0(1) + (-2)(0) + 4(-1)$ $\qquad c_{23} = -7(0) + 0(-2) + (-2)(5) + 4(0)$

Hence, $AB = \begin{bmatrix} 18 & 0 & -2 \\ -40 & 10 & -10 \end{bmatrix}$.

27 $(A+B)(A-B) = \begin{bmatrix} 3 & 1 \\ 3 & -2 \end{bmatrix} \begin{bmatrix} -1 & 3 \\ -3 & -4 \end{bmatrix} = \begin{bmatrix} -6 & 5 \\ 3 & 17 \end{bmatrix}$;

$A^2 - B^2 = \begin{bmatrix} 1 & -4 \\ 0 & 9 \end{bmatrix} - \begin{bmatrix} 1 & -3 \\ 9 & -2 \end{bmatrix} = \begin{bmatrix} 0 & -1 \\ -9 & 11 \end{bmatrix}$; $(A+B)(A-B) \neq A^2 - B^2$

29 $A(B+C) = \begin{bmatrix} 1 & 2 \\ 0 & -3 \end{bmatrix} \begin{bmatrix} 5 & 0 \\ 1 & 1 \end{bmatrix} = \begin{bmatrix} 7 & 2 \\ -3 & -3 \end{bmatrix}$;

$AB + AC = \begin{bmatrix} 8 & 1 \\ -9 & -3 \end{bmatrix} + \begin{bmatrix} -1 & 1 \\ 6 & 0 \end{bmatrix} = \begin{bmatrix} 7 & 2 \\ -3 & -3 \end{bmatrix}$

31 $m(A+B) = m \begin{bmatrix} a+p & b+q \\ c+r & d+s \end{bmatrix} = \begin{bmatrix} m(a+p) & m(b+q) \\ m(c+r) & m(d+s) \end{bmatrix}$

$= \begin{bmatrix} ma+mp & mb+mq \\ mc+mr & md+ms \end{bmatrix} = \begin{bmatrix} ma & mb \\ mc & md \end{bmatrix} + \begin{bmatrix} mp & mq \\ mr & ms \end{bmatrix}$

$= m \begin{bmatrix} a & b \\ c & d \end{bmatrix} + m \begin{bmatrix} p & q \\ r & s \end{bmatrix} = mA + mB$

33 $A(B+C) = \begin{bmatrix} a & b \\ c & d \end{bmatrix}\begin{bmatrix} p+w & q+x \\ r+y & s+z \end{bmatrix}$

$= \begin{bmatrix} a(p+w)+b(r+y) & a(q+x)+b(s+z) \\ c(p+w)+d(r+y) & c(q+x)+d(s+z) \end{bmatrix}$

$= \begin{bmatrix} ap+aw+br+by & aq+ax+bs+bz \\ cp+cw+dr+dy & cq+cx+ds+dz \end{bmatrix}$

$= \begin{bmatrix} ap+br & aq+bs \\ cp+dr & cq+ds \end{bmatrix} + \begin{bmatrix} aw+by & ax+bz \\ cw+dy & cx+dz \end{bmatrix}$

$= \begin{bmatrix} a & b \\ c & d \end{bmatrix}\begin{bmatrix} p & q \\ r & s \end{bmatrix} + \begin{bmatrix} a & b \\ c & d \end{bmatrix}\begin{bmatrix} w & x \\ y & z \end{bmatrix} = AB + AC$

Note: For Exercises 35–38, $A = \begin{bmatrix} 3 & -3 & 7 \\ 2 & 6 & -2 \\ 4 & 2 & 5 \end{bmatrix}$ and $B = \begin{bmatrix} -9 & 5 & -8 \\ 3 & -7 & 1 \\ -1 & 2 & 6 \end{bmatrix}$.

35 $A^2 + B^2 = \begin{bmatrix} 31 & -13 & 62 \\ 10 & 26 & -8 \\ 36 & 10 & 49 \end{bmatrix} + \begin{bmatrix} 104 & -96 & 29 \\ -49 & 66 & -25 \\ 9 & -7 & 46 \end{bmatrix} = \begin{bmatrix} 135 & -109 & 91 \\ -39 & 92 & -33 \\ 45 & 3 & 95 \end{bmatrix}$

37 $A^2 - 5B = \begin{bmatrix} 31 & -13 & 62 \\ 10 & 26 & -8 \\ 36 & 10 & 49 \end{bmatrix} - \begin{bmatrix} -45 & 25 & -40 \\ 15 & -35 & 5 \\ -5 & 10 & 30 \end{bmatrix} = \begin{bmatrix} 76 & -38 & 102 \\ -5 & 61 & -13 \\ 41 & 0 & 19 \end{bmatrix}$

39 (a) For the inventory matrix, we have 5 colors and 3 kinds of disks, and for each kind of disk, we have 1 price. We choose a 5×3 matrix A and a 3×1 matrix B.

$$\text{inventory matrix } A = \begin{bmatrix} 400 & 550 & 500 \\ 400 & 450 & 500 \\ 300 & 500 & 600 \\ 250 & 200 & 300 \\ 100 & 100 & 200 \end{bmatrix}, \text{ price matrix } B = \begin{bmatrix} \$0.22 \\ \$0.25 \\ \$0.28 \end{bmatrix}$$

(b) $C = AB = \begin{bmatrix} 400 & 550 & 500 \\ 400 & 450 & 500 \\ 300 & 500 & 600 \\ 250 & 200 & 300 \\ 100 & 100 & 200 \end{bmatrix}\begin{bmatrix} \$0.22 \\ \$0.25 \\ \$0.28 \end{bmatrix} = \begin{bmatrix} \$365.50 \\ \$340.50 \\ \$359.00 \\ \$189.00 \\ \$103.00 \end{bmatrix}$

(c) The $103.00 represents the amount the store would receive if all the yellow disks were sold.

Note: Exer. 1–10: Let A denote the given matrix.

$\boxed{1}$ $\begin{bmatrix} 2 & -4 & | & 1 & 0 \\ 1 & 3 & | & 0 & 1 \end{bmatrix} R_1 - R_2 \to R_1 \Rightarrow \begin{bmatrix} 1 & -7 & | & 1 & -1 \\ 1 & 3 & | & 0 & 1 \end{bmatrix} R_2 - R_1 \to R_2$

$\begin{bmatrix} 1 & -7 & | & 1 & -1 \\ 0 & 10 & | & -1 & 2 \end{bmatrix} \frac{1}{10} R_2 \to R_2 \Rightarrow \begin{bmatrix} 1 & -7 & | & 1 & -1 \\ 0 & 1 & | & -\frac{1}{10} & \frac{2}{10} \end{bmatrix} R_1 + 7 R_2 \to R_1$

$\begin{bmatrix} 1 & 0 & | & \frac{3}{10} & \frac{4}{10} \\ 0 & 1 & | & -\frac{1}{10} & \frac{2}{10} \end{bmatrix} \Rightarrow A^{-1} = \frac{1}{10} \begin{bmatrix} 3 & 4 \\ -1 & 2 \end{bmatrix}$

$\boxed{3}$ $\begin{bmatrix} 2 & 4 & | & 1 & 0 \\ 4 & 8 & | & 0 & 1 \end{bmatrix} \frac{1}{2} R_1 \to R_1 \Rightarrow \begin{bmatrix} 1 & 2 & | & \frac{1}{2} & 0 \\ 4 & 8 & | & 0 & 1 \end{bmatrix} R_2 - 4 R_1 \to R_2$

$\begin{bmatrix} 1 & 2 & | & \frac{1}{2} & 0 \\ 0 & 0 & | & -2 & 1 \end{bmatrix}$

Since the identity matrix cannot be obtained on the left, *no inverse exists.*

$\boxed{5}$ $\begin{bmatrix} 3 & -1 & 0 & | & 1 & 0 & 0 \\ 2 & 2 & 0 & | & 0 & 1 & 0 \\ 0 & 0 & 4 & | & 0 & 0 & 1 \end{bmatrix} R_1 - R_2 \to R_1$

$\begin{bmatrix} 1 & -3 & 0 & | & 1 & -1 & 0 \\ 2 & 2 & 0 & | & 0 & 1 & 0 \\ 0 & 0 & 4 & | & 0 & 0 & 1 \end{bmatrix} R_2 - 2 R_1 \to R_2$

$\begin{bmatrix} 1 & -3 & 0 & | & 1 & -1 & 0 \\ 0 & 8 & 0 & | & -2 & 3 & 0 \\ 0 & 0 & 4 & | & 0 & 0 & 1 \end{bmatrix} \begin{matrix} (1/8) R_2 \to R_2 \\ (1/4) R_3 \to R_3 \end{matrix}$

$\begin{bmatrix} 1 & -3 & 0 & | & 1 & -1 & 0 \\ 0 & 1 & 0 & | & -\frac{2}{8} & \frac{3}{8} & 0 \\ 0 & 0 & 1 & | & 0 & 0 & \frac{1}{4} \end{bmatrix} R_1 + 3 R_2 \to R_1$

$\begin{bmatrix} 1 & 0 & 0 & | & \frac{2}{8} & \frac{1}{8} & 0 \\ 0 & 1 & 0 & | & -\frac{2}{8} & \frac{3}{8} & 0 \\ 0 & 0 & 1 & | & 0 & 0 & \frac{2}{8} \end{bmatrix} \qquad A^{-1} = \frac{1}{8} \begin{bmatrix} 2 & 1 & 0 \\ -2 & 3 & 0 \\ 0 & 0 & 2 \end{bmatrix}$

7 $\begin{bmatrix} -2 & 2 & 3 & | & 1 & 0 & 0 \\ 1 & -1 & 0 & | & 0 & 1 & 0 \\ 0 & 1 & 4 & | & 0 & 0 & 1 \end{bmatrix}$ $R_1 + 2R_2 \leftrightarrow R_2$

$\begin{bmatrix} 1 & -1 & 0 & | & 0 & 1 & 0 \\ 0 & 0 & 3 & | & 1 & 2 & 0 \\ 0 & 1 & 4 & | & 0 & 0 & 1 \end{bmatrix}$ $R_1 + R_3 \rightarrow R_1$

$\begin{bmatrix} 1 & 0 & 4 & | & 0 & 1 & 1 \\ 0 & 0 & 3 & | & 1 & 2 & 0 \\ 0 & 1 & 4 & | & 0 & 0 & 1 \end{bmatrix}$ $\frac{1}{3}R_2 \leftrightarrow R_3$

$\begin{bmatrix} 1 & 0 & 4 & | & 0 & 1 & 1 \\ 0 & 1 & 4 & | & 0 & 0 & 1 \\ 0 & 0 & 1 & | & \frac{1}{3} & \frac{2}{3} & 0 \end{bmatrix}$ $\begin{matrix} R_1 - 4R_3 \rightarrow R_1 \\ R_2 - 4R_3 \rightarrow R_2 \end{matrix}$

$\begin{bmatrix} 1 & 0 & 0 & | & -\frac{4}{3} & -\frac{5}{3} & 1 \\ 0 & 1 & 0 & | & -\frac{4}{3} & -\frac{8}{3} & 1 \\ 0 & 0 & 1 & | & \frac{1}{3} & \frac{2}{3} & 0 \end{bmatrix}$ $A^{-1} = \frac{1}{3}\begin{bmatrix} -4 & -5 & 3 \\ -4 & -8 & 3 \\ 1 & 2 & 0 \end{bmatrix}$

11 $\begin{bmatrix} a & 0 & | & 1 & 0 \\ 0 & b & | & 0 & 1 \end{bmatrix}$ $\begin{matrix} (1/a)R_1 \rightarrow R_1 \\ (1/b)R_2 \rightarrow R_2 \end{matrix}$ $\Rightarrow \begin{bmatrix} 1 & 0 & | & 1/a & 0 \\ 0 & 1 & | & 0 & 1/b \end{bmatrix}$

The inverse is the matrix with main diagonal elements $(1/a)$ and $(1/b)$.

The required conditions are that a and b are nonzero to avoid division by zero.

15 (a) $X = A^{-1}B = \frac{1}{10}\begin{bmatrix} 3 & 4 \\ -1 & 2 \end{bmatrix}\begin{bmatrix} 3 \\ 1 \end{bmatrix} = \frac{1}{10}\begin{bmatrix} 13 \\ -1 \end{bmatrix};$ $(\frac{13}{10}, -\frac{1}{10})$

(b) $X = A^{-1}B = \frac{1}{10}\begin{bmatrix} 3 & 4 \\ -1 & 2 \end{bmatrix}\begin{bmatrix} -2 \\ 5 \end{bmatrix} = \frac{1}{10}\begin{bmatrix} 14 \\ 12 \end{bmatrix};$ $(\frac{7}{5}, \frac{6}{5})$

17 (a) $X = A^{-1}B = \frac{1}{3}\begin{bmatrix} -4 & -5 & 3 \\ -4 & -8 & 3 \\ 1 & 2 & 0 \end{bmatrix}\begin{bmatrix} 1 \\ 3 \\ -2 \end{bmatrix} = \frac{1}{3}\begin{bmatrix} -25 \\ -34 \\ 7 \end{bmatrix};$ $(-\frac{25}{3}, -\frac{34}{3}, \frac{7}{3})$

(b) $X = A^{-1}B = \frac{1}{3}\begin{bmatrix} -4 & -5 & 3 \\ -4 & -8 & 3 \\ 1 & 2 & 0 \end{bmatrix}\begin{bmatrix} -1 \\ 0 \\ 4 \end{bmatrix} = \frac{1}{3}\begin{bmatrix} 16 \\ 16 \\ -1 \end{bmatrix};$ $(\frac{16}{3}, \frac{16}{3}, -\frac{1}{3})$

19 The inverse should be found using some type of computational device. If you are using a TI-82/83, enter the 9 values into the matrix [A]. Change Float to 5 via MODE. Now find $[A]^{-1}$ { be sure to use the $\boxed{x^{-1}}$ key }. Use the right arrow key to see the rightmost elements in [A].

$A = \begin{bmatrix} 2 & -5 & 8 \\ 3 & 7 & -1 \\ 0 & 2 & 1 \end{bmatrix} \Rightarrow A^{-1} \approx \begin{bmatrix} 0.11111 & 0.25926 & -0.62963 \\ -0.03704 & 0.02469 & 0.32099 \\ 0.07407 & -0.04938 & 0.35802 \end{bmatrix}$

23 (a) $AX = B \Leftrightarrow \begin{bmatrix} 4.0 & 7.1 \\ 2.2 & -4.9 \end{bmatrix} \begin{bmatrix} x \\ y \end{bmatrix} = \begin{bmatrix} 6.2 \\ 2.9 \end{bmatrix}$

(b) On your calculator, Find $[A]^{-1}$ and STOre this matrix into matrix [B] to use in part (c).

$$A^{-1} \approx \begin{bmatrix} 0.1391 & 0.2016 \\ 0.0625 & -0.1136 \end{bmatrix}$$

(c) Following the instructions in part (b), enter the 3 values into [C] {a 3×1 matrix}, and then evaluate [B]*[C].

$$X = A^{-1}B \approx \begin{bmatrix} 0.1391 & 0.2016 \\ 0.0625 & -0.1136 \end{bmatrix} \begin{bmatrix} 6.2 \\ 2.9 \end{bmatrix} \approx \begin{bmatrix} 1.4472 \\ 0.0579 \end{bmatrix}.$$

25 (a) $AX = B \Leftrightarrow \begin{bmatrix} 3.1 & 6.7 & -8.7 \\ 4.1 & -5.1 & 0.2 \\ 0.6 & 1.1 & -7.4 \end{bmatrix} \begin{bmatrix} x \\ y \\ z \end{bmatrix} = \begin{bmatrix} 1.5 \\ 2.1 \\ 3.9 \end{bmatrix}$

(b) $A^{-1} \approx \begin{bmatrix} 0.1474 & 0.1572 & -0.1691 \\ 0.1197 & -0.0696 & -0.1426 \\ 0.0297 & 0.0024 & -0.1700 \end{bmatrix}$

(c) $X = A^{-1}B \approx \begin{bmatrix} 0.1474 & 0.1572 & -0.1691 \\ 0.1197 & -0.0696 & -0.1426 \\ 0.0297 & 0.0024 & -0.1700 \end{bmatrix} \begin{bmatrix} 1.5 \\ 2.1 \\ 3.9 \end{bmatrix} \approx \begin{bmatrix} -0.1081 \\ -0.5227 \\ -0.6135 \end{bmatrix}$

27 (a) $f(2) = 4a + 2b + c = 19$; $f(8) = 64a + 8b + c = 59$; $f(11) = 121a + 11b + c = 26$

Solve the 3×3 linear system using the inverse method.

$$\begin{bmatrix} 4 & 2 & 1 \\ 64 & 8 & 1 \\ 121 & 11 & 1 \end{bmatrix} \begin{bmatrix} a \\ b \\ c \end{bmatrix} = \begin{bmatrix} 19 \\ 59 \\ 26 \end{bmatrix} \Rightarrow \begin{bmatrix} a \\ b \\ c \end{bmatrix} \approx \begin{bmatrix} -1.9630 \\ 26.2963 \\ -25.7407 \end{bmatrix}$$

Thus, let $f(x) = -1.9630x^2 + 26.2963x - 25.7407$.

(b)

[1, 12] by [−15, 70]

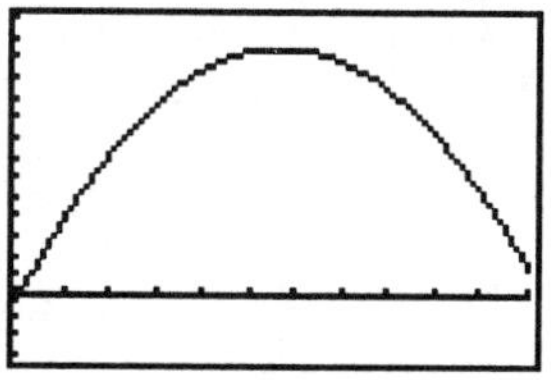

Xscl = 1
Yscl = 5

Figure 27

(c) For June, $f(6) \approx 61\,°F$ and for October, $f(10) \approx 41\,°F$.

Note: The minor M_{ij} and the cofactor A_{ij} are equal if $(i + j)$ is even and of opposite sign if $(i + j)$ is odd.

$\boxed{1}$ The minor M_{11} is obtained by deleting the first row and first column from

$$A = \begin{bmatrix} 7 & -1 \\ 5 & 0 \end{bmatrix}. \quad \text{Thus, } M_{11} = 0 = A_{11}. \quad \text{Similarly, } M_{12} = 5 \text{ and } A_{12} = -5;$$

$M_{21} = -1$ and $A_{21} = 1$; and $M_{22} = 7 = A_{22}$.

$\boxed{3}$ Let $A = \begin{bmatrix} 2 & 4 & -1 \\ 0 & 3 & 2 \\ -5 & 7 & 0 \end{bmatrix}$.

Be sure you understand the note preceding the solution for Exercise 1.

$$M_{11} = \begin{vmatrix} 3 & 2 \\ 7 & 0 \end{vmatrix} = (3)(0) - (7)(2) = 0 - 14 = -14 = A_{11};$$

$$M_{12} = \begin{vmatrix} 0 & 2 \\ -5 & 0 \end{vmatrix} = (0)(0) - (-5)(2) = 0 - (-10) = 10; \; A_{12} = -10;$$

$$M_{13} = \begin{vmatrix} 0 & 3 \\ -5 & 7 \end{vmatrix} = 15 = A_{13}; \qquad M_{21} = \begin{vmatrix} 4 & -1 \\ 7 & 0 \end{vmatrix} = 7; \; A_{21} = -7;$$

$$M_{22} = \begin{vmatrix} 2 & -1 \\ -5 & 0 \end{vmatrix} = -5 = A_{22}; \qquad M_{23} = \begin{vmatrix} 2 & 4 \\ -5 & 7 \end{vmatrix} = 34; \; A_{23} = -34;$$

$$M_{31} = \begin{vmatrix} 4 & -1 \\ 3 & 2 \end{vmatrix} = 11 = A_{31}; \qquad M_{32} = \begin{vmatrix} 2 & -1 \\ 0 & 2 \end{vmatrix} = 4; \; A_{32} = -4;$$

$$M_{33} = \begin{vmatrix} 2 & 4 \\ 0 & 3 \end{vmatrix} = 6 = A_{33}.$$

Note: Exercises 5–20: Let A denote the given matrix.

$\boxed{7}$ Expanding $|A|$ by the first column and using the cofactor values from Exercise 3,

$$\text{we obtain } |A| = a_{11}A_{11} + a_{21}A_{21} + a_{31}A_{31} = 2(-14) + 0(A_{21}) - 5(11) = -83.$$

$\boxed{9}$ By the definition of the determinant of a 2×2 matrix A (on page 680),

$$|A| = \begin{vmatrix} -5 & 4 \\ -3 & 2 \end{vmatrix} = (-5)(2) - (4)(-3) = -10 + 12 = 2.$$

$\boxed{13}$ Expand by the first row.

$$|A| = a_{11}A_{11} + a_{12}A_{12} + a_{13}A_{13}$$

$$= (3)(-1)^{1+1}\begin{vmatrix} 2 & 5 \\ 3 & -1 \end{vmatrix} + (1)(-1)^{1+2}\begin{vmatrix} 4 & 5 \\ -6 & -1 \end{vmatrix} + (-2)(-1)^{1+3}\begin{vmatrix} 4 & 2 \\ -6 & 3 \end{vmatrix}$$

$$= (3)(1)(-17) + (1)(-1)(26) + (-2)(1)(24)$$

$$= -51 - 26 - 48 = -125$$

15 Expand by the third row.

$$|A| = a_{31}A_{31} + a_{32}A_{32} + a_{33}A_{33} = 2(30) + 0(A_{32}) + 6(-2) = 48$$

17 Expand $|A|$ by the third row. $\quad |A| = 6A_{32} = -6M_{32} = -6 \begin{vmatrix} 3 & 2 & 0 \\ 4 & -3 & 5 \\ 1 & -4 & 2 \end{vmatrix}$.

Expand M_{32} by the first row.

$$M_{32} = 3(14) + 2(-3) + 0(-13) = 36 \Rightarrow |A| = -6(36) = -216.$$

21 LS $= ad - bc$; $\quad\quad\quad\quad\quad\quad$ RS $= -(bc - ad) = ad - bc$

23 LS $= adk - bck$; $\quad\quad\quad\quad\quad$ RS $= k(ad - bc) = adk - bck$

25 LS $= ad - bc$; $\quad\quad\quad\quad\quad\quad$ RS $= abk + ad - abk - bc = ad - bc$

27 LS $= ad - bc + af - ce$; $\quad\quad\quad$ RS $= ad + af - bc - ce$

29 Consider the matrix in Exercise 20. If we expanded along the first column, we would obtain a times its cofactor. Expanding along the first column again, we obtain ab times another cofactor. This exercise is similar since all elements in A *above* {rather than below} the main diagonal are zero. We can evaluate the determinant using n expansions by the first row, and obtain $|A| = a_{11}a_{22}\cdots a_{nn}$.

31 (a) $A - xI = \begin{bmatrix} 1 & 2 \\ 3 & 2 \end{bmatrix} - x \begin{bmatrix} 1 & 0 \\ 0 & 1 \end{bmatrix} = \begin{bmatrix} 1-x & 2 \\ 3 & 2-x \end{bmatrix}$.

$$f(x) = |A - xI| = \begin{vmatrix} 1-x & 2 \\ 3 & 2-x \end{vmatrix}$$
$$= (1-x)(2-x) - (3)(2) = (2 - 3x + x^2) - 6 = x^2 - 3x - 4.$$

(b) $x^2 - 3x - 4 = 0 \Rightarrow (x-4)(x+1) = 0 \Rightarrow x = -1, 4$

37 (a) $f(x) = \begin{vmatrix} 0-x & 2 & -2 \\ -1 & 3-x & 1 \\ -3 & 3 & 1-x \end{vmatrix} \quad$ {Expand by the first row.}

$$= (-x)[(3-x)(1-x) - 3] - 2[(x-1) + 3] - 2[-3 + 3(3-x)]$$
$$= (-x)[(3 - 4x + x^2) - 3] - 2(x+2) - 2(-3x + 6)$$
$$= (-x^3 + 4x^2) - 2x - 4 + 6x - 12$$
$$= -x^3 + 4x^2 + 4x - 16$$

(b) By trying possible rational roots, we determine that 2 is a zero of f.

Thus, $-x^3 + 4x^2 + 4x - 16 = 0 \Rightarrow (x-2)(-x^2 + 2x + 8) = 0 \Rightarrow$
$$(x+2)(x-2)(-x+4) = 0 \Rightarrow x = -2, 2, 4.$$

39 Expand the determinant by the first row.

$$\begin{vmatrix} i & j & k \\ 2 & -1 & 6 \\ -3 & 5 & 1 \end{vmatrix} = i \begin{vmatrix} -1 & 6 \\ 5 & 1 \end{vmatrix} - j \begin{vmatrix} 2 & 6 \\ -3 & 1 \end{vmatrix} + k \begin{vmatrix} 2 & -1 \\ -3 & 5 \end{vmatrix} = -31i - 20j + 7k$$

47 (a) $f(x) = |A - xI| = \begin{vmatrix} 1-x & 0 & 1 \\ 0 & 2-x & 1 \\ 1 & 1 & -2-x \end{vmatrix} = -x^3 + x^2 + 6x - 7$

(b) The characteristic values of A are equal to the zeros of f. From the graph,

we see that the zeros are approximately -2.51, 1.22, and 2.29.

$[-10, 11]$ by $[-12, 2]$

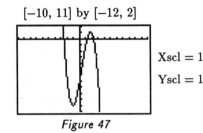

Xscl $= 1$

Yscl $= 1$

Figure 47

9.10 Exercises

1 R_2 and R_3 are interchanged. The determinant value is negated.

3 R_3 is replaced by $(R_3 - R_1)$. There is no change in the determinant value.

5 The number 2 can be factored out of R_1, yielding $2 \begin{vmatrix} 1 & 2 & 1 \\ 1 & 2 & 4 \\ 2 & 6 & 4 \end{vmatrix}$.

Next, the number 2 can be factored out of R_3, yielding $4 \begin{vmatrix} 1 & 2 & 1 \\ 1 & 2 & 4 \\ 1 & 3 & 2 \end{vmatrix}$.

7 R_1 and R_3 are identical. The determinant is 0.

9 The number -1 can be factored out of R_2, yielding the determinant on the right side.

11 Every number in C_2 is 0. The determinant is zero.

13 C_3 is replaced by $(2C_1 + C_3)$. There is no change in the determinant value.

Note: The notation $\{R_i (C_i)\}$ means expand the determinant by the ith row (column).

15 There are many possibilities for introducing zeros. In this case, obtaining a zero in the third row, second column would lead to an easy evaluation using the second column.

$\begin{vmatrix} 3 & 1 & 0 \\ -2 & 0 & 1 \\ 1 & 3 & -1 \end{vmatrix} R_3 - 3R_1 \rightarrow R_3 = \begin{vmatrix} 3 & 1 & 0 \\ -2 & 0 & 1 \\ -8 & 0 & -1 \end{vmatrix} \{C_2\} = (-1) \begin{vmatrix} -2 & 1 \\ -8 & -1 \end{vmatrix} =$

$(-1)(2 + 8) = -10$

19 $\begin{vmatrix} 2 & 2 & -3 \\ 3 & 6 & 9 \\ -2 & 5 & 4 \end{vmatrix}$ $\{3$ is a common factor of $R_2\} = (3)\begin{vmatrix} 2 & 2 & -3 \\ 1 & 2 & 3 \\ -2 & 5 & 4 \end{vmatrix}$

Since there is a "1" in the second row, first column, we will obtain zeros in the other two locations of the first column.

$(3)\begin{vmatrix} 2 & 2 & -3 \\ 1 & 2 & 3 \\ -2 & 5 & 4 \end{vmatrix}\begin{matrix} R_1 - 2R_2 \to R_1 \\ \\ R_3 + 2R_2 \to R_3 \end{matrix} = (3)\begin{vmatrix} 0 & -2 & -9 \\ 1 & 2 & 3 \\ 0 & 9 & 10 \end{vmatrix}\{C_1\}$

$= (3)(-1)\begin{vmatrix} -2 & -9 \\ 9 & 10 \end{vmatrix} = (-3)(-20 + 81) = -183$

21 $\begin{vmatrix} 3 & 1 & -2 & 2 \\ 2 & 0 & 1 & 4 \\ 0 & 1 & 3 & 5 \\ -1 & 2 & 0 & -3 \end{vmatrix}\begin{matrix} \\ \\ R_3 - R_1 \to R_3 \\ R_4 - 2R_1 \to R_4 \end{matrix} = \begin{vmatrix} 3 & 1 & -2 & 2 \\ 2 & 0 & 1 & 4 \\ -3 & 0 & 5 & 3 \\ -7 & 0 & 4 & -7 \end{vmatrix}\{C_2\}$

$= (-1)\begin{vmatrix} 2 & 1 & 4 \\ -3 & 5 & 3 \\ -7 & 4 & -7 \end{vmatrix}\begin{matrix} \\ R_2 - 5R_1 \to R_2 \\ R_3 - 4R_1 \to R_3 \end{matrix} = (-1)\begin{vmatrix} 2 & 1 & 4 \\ -13 & 0 & -17 \\ -15 & 0 & -23 \end{vmatrix}\{C_2\}$

$= (-1)(-1)\begin{vmatrix} -13 & -17 \\ -15 & -23 \end{vmatrix} = (1)(299 - 255) = 44$

23 $\begin{vmatrix} 2 & -2 & 0 & 0 & -3 \\ 3 & 0 & 3 & 2 & -1 \\ 0 & 1 & -2 & 0 & 2 \\ -1 & 2 & 0 & 3 & 0 \\ 0 & 4 & 1 & 0 & 0 \end{vmatrix}\begin{matrix} \\ \\ C_2 - 4C_3 \to C_2 \\ \\ \end{matrix} = \begin{vmatrix} 2 & -2 & 0 & 0 & -3 \\ 3 & -12 & 3 & 2 & -1 \\ 0 & 9 & -2 & 0 & 2 \\ -1 & 2 & 0 & 3 & 0 \\ 0 & 0 & 1 & 0 & 0 \end{vmatrix}\{R_5\}$

$= (1)\begin{vmatrix} 2 & -2 & 0 & -3 \\ 3 & -12 & 2 & -1 \\ 0 & 9 & 0 & 2 \\ -1 & 2 & 3 & 0 \end{vmatrix}\begin{matrix} R_1 + 2R_4 \to R_1 \\ R_2 + 3R_4 \to R_2 \\ \\ \end{matrix} = \begin{vmatrix} 0 & 2 & 6 & -3 \\ 0 & -6 & 11 & -1 \\ 0 & 9 & 0 & 2 \\ -1 & 2 & 3 & 0 \end{vmatrix}\{C_1\}$

$= (1)\begin{vmatrix} 2 & 6 & -3 \\ -6 & 11 & -1 \\ 9 & 0 & 2 \end{vmatrix}\begin{matrix} R_1 - 3R_2 \to R_1 \\ \\ R_3 + 2R_2 \to R_3 \end{matrix} = \begin{vmatrix} 20 & -27 & 0 \\ -6 & 11 & -1 \\ -3 & 22 & 0 \end{vmatrix}\{C_3\}$

$= (1)\begin{vmatrix} 20 & -27 \\ -3 & 22 \end{vmatrix} = (1)(440 - 81) = 359$

25
$$\begin{vmatrix} 1 & 1 & 1 \\ a & b & c \\ a^2 & b^2 & c^2 \end{vmatrix} \quad \begin{matrix} C_1 - C_2 \rightarrow C_1 \\ \\ C_3 - C_2 \rightarrow C_3 \end{matrix}$$

$$= \begin{vmatrix} 0 & 1 & 0 \\ a-b & b & c-b \\ a^2-b^2 & b^2 & c^2-b^2 \end{vmatrix} \quad \begin{matrix} a-b \text{ is a common factor of } C_1 \\ \\ c-b \text{ is a common factor of } C_3 \end{matrix}$$

$$= (a-b)(c-b) \begin{vmatrix} 0 & 1 & 0 \\ 1 & b & 1 \\ a+b & b^2 & c+b \end{vmatrix} \quad \{R_1\}$$

$$= (a-b)(c-b)(-1) \begin{vmatrix} 1 & 1 \\ a+b & c+b \end{vmatrix}$$

$$= (a-b)(b-c)(c+b-a-b) = \underline{(a-b)(b-c)(c-a)}$$

27
$$\begin{vmatrix} a_{11} & a_{12} & a_{13} & a_{14} \\ 0 & a_{22} & a_{23} & a_{24} \\ 0 & 0 & a_{33} & a_{34} \\ 0 & 0 & 0 & a_{44} \end{vmatrix} \{C_1\} = (a_{11}) \begin{vmatrix} a_{22} & a_{23} & a_{24} \\ 0 & a_{33} & a_{34} \\ 0 & 0 & a_{44} \end{vmatrix} \{C_1\}$$

$$= (a_{11})(a_{22}) \begin{vmatrix} a_{33} & a_{34} \\ 0 & a_{44} \end{vmatrix} = (a_{11}a_{22})(a_{33}a_{44} - 0) = a_{11}a_{22}a_{33}a_{44}$$

29 $|AB| = \begin{vmatrix} a_{11}b_{11} + a_{12}b_{21} & a_{11}b_{12} + a_{12}b_{22} \\ a_{21}b_{11} + a_{22}b_{21} & a_{21}b_{12} + a_{22}b_{22} \end{vmatrix}$

$$= (a_{11}b_{11} + a_{12}b_{21})(a_{21}b_{12} + a_{22}b_{22}) - (a_{11}b_{12} + a_{12}b_{22})(a_{21}b_{11} + a_{22}b_{21})$$

$$= \quad a_{11}b_{11}a_{21}b_{12} + a_{11}b_{11}a_{22}b_{22} + a_{12}b_{21}a_{21}b_{12} + a_{12}b_{21}a_{22}b_{22}$$

$$- a_{11}b_{12}a_{21}b_{11} - a_{11}b_{12}a_{22}b_{21} - a_{12}b_{22}a_{21}b_{11} - a_{12}b_{22}a_{22}b_{21}$$

$$= a_{11}a_{22}b_{11}b_{22} - a_{11}a_{22}b_{21}b_{12} - a_{21}a_{12}b_{11}b_{22} + a_{21}a_{12}b_{21}b_{12}$$

$$= (a_{11}a_{22} - a_{21}a_{12})(b_{11}b_{22} - b_{21}b_{12}) = |A||B|$$

31 Expanding by the first row yields $Ax + By + C = 0$ { an equation of a line } where A, B, and C are constants. To show that the line contains (x_1, y_1) and (x_2, y_2), we must show that these points are solutions of the equation. Substituting x_1 for x and y_1 for y, we obtain two identical rows and the determinant is zero. Hence, (x_1, y_1) is a solution of the equation and a similar argument can be made for (x_2, y_2).

33 For the system $\begin{cases} 2x + 3y = 2 \\ x - 2y = 8 \end{cases}$, $|D| = \begin{vmatrix} 2 & 3 \\ 1 & -2 \end{vmatrix} = -4 - 3 = -7.$

Since $|D| = -7 \neq 0$, we may solve the system using Cramer's rule.

$|D_x| = \begin{vmatrix} 2 & 3 \\ 8 & -2 \end{vmatrix} = -4 - 24 = -28.$ $|D_y| = \begin{vmatrix} 2 & 2 \\ 1 & 8 \end{vmatrix} = 16 - 2 = 14.$

$x = \dfrac{|D_x|}{|D|} = \dfrac{-28}{-7} = 4.$ $y = \dfrac{|D_y|}{|D|} = \dfrac{14}{-7} = -2.$ $\bigstar (4, -2)$

37 $|D| = \begin{vmatrix} 2 & -3 \\ -6 & 9 \end{vmatrix} = 18 - 18 = 0$, so Cramer's rule cannot be used.

$\boxed{41}$ $|D| = \begin{vmatrix} 5 & 2 & -1 \\ 1 & -2 & 2 \\ 0 & 3 & 1 \end{vmatrix}$ $\{R_3\} = -3(11) + 1(-12) = -45$

$|D_x| = \begin{vmatrix} -7 & 2 & -1 \\ 0 & -2 & 2 \\ 17 & 3 & 1 \end{vmatrix}$ $\{C_1\} = -7(-8) + 17(2) = 90$

$|D_y| = \begin{vmatrix} 5 & -7 & -1 \\ 1 & 0 & 2 \\ 0 & 17 & 1 \end{vmatrix}$ $\{R_2\} = -1(10) - 2(85) = -180$

$|D_z| = \begin{vmatrix} 5 & 2 & -7 \\ 1 & -2 & 0 \\ 0 & 3 & 17 \end{vmatrix}$ $\{C_1\} = 5(-34) - 1(55) = -225$

$x = \dfrac{|D_x|}{|D|} = \dfrac{90}{-45} = -2$; $y = \dfrac{|D_y|}{|D|} = \dfrac{-180}{-45} = 4$; $z = \dfrac{|D_z|}{|D|} = \dfrac{-225}{-45} = 5$.

$\bigstar$ $(-2, 4, 5)$

Chapter 9 Review Exercises

$\boxed{3}$ Solve E_2 for y, $y = -2x - 1$, and substitute into E_1 to yield $x^2 + 2x - 3 = 0 \Rightarrow$

$(x + 3)(x - 1) = 0 \Rightarrow x = -3, 1$ and $y = 5, -3$. $\bigstar$ $(-3, 5), (1, -3)$

$\boxed{5}$ $4\,E_2 + E_1 \Rightarrow 13x^2 = 156 \Rightarrow x = \pm 2\sqrt{3}$ and $y = \pm\sqrt{2}$.

There are four solutions: $(2\sqrt{3}, \pm\sqrt{2}), (-2\sqrt{3}, \pm\sqrt{2})$.

$\boxed{6}$ From E_3, $x^2 = xz \Rightarrow x^2 - xz = 0 \Rightarrow x(x - z) = 0 \Rightarrow x = 0$ or $x = z$.

If $x = 0$, E_1 is $0 = y^2 + 3z$ and E_2 is $z = 1 - y^2$. Substituting z into E_1 yields

$2y^2 = 3$ or $y = \pm\frac{1}{2}\sqrt{6}$ and z is $-\frac{1}{2}$ for both values of y.

If $x = z$, E_1 is $0 = y^2 + z$ and E_2 is $y^2 = 1$.

Thus, $y = \pm 1$ and in either case, $x = z = -1$.

There are four solutions: $(-1, \pm 1, -1), (0, \pm\frac{1}{2}\sqrt{6}, -\frac{1}{2})$.

$\boxed{8}$ Treat 3^{y+1} as $3 \cdot 3^y$ and 2^{x+1} as $2 \cdot 2^x$.

Thus, E_1 is $1(2^x) + 3(3^y) = 10$ and E_2 is $2(2^x) - 1(3^y) = 5$.

Now $E_1 + 3\,E_2 \Rightarrow 7 \cdot 2^x = 25 \Rightarrow 2^x = \frac{25}{7} \Rightarrow x = \log_2 \frac{25}{7} = \dfrac{\log\frac{25}{7}}{\log 2} \approx 1.84$.

Resolving the original system for y, we have

$-2\,E_1 + E_2 \Rightarrow -7 \cdot 3^y = -15 \Rightarrow 3^y = \frac{15}{7} \Rightarrow y = \log_3 \frac{15}{7} = \dfrac{\log\frac{15}{7}}{\log 3} \approx 0.69$.

$\boxed{9}$ Solve E_3 for z, $z = 4x + 5y + 2$, and substitute into E_1 and E_2 to yield

$\begin{cases} 3x + y - 2(4x + 5y + 2) = -1 \\ 2x - 3y + (4x + 5y + 2) = 4 \end{cases} \Rightarrow \begin{cases} -5x - 9y = 3 & (E_4) \\ 6x + 2y = 2 & (E_5) \end{cases}$

$6\,E_4 + 5\,E_5 \Rightarrow -44y = 28 \Rightarrow y = -\frac{7}{11}$;

$2\,E_4 + 9\,E_5 \Rightarrow 44x = 24 \Rightarrow x = \frac{6}{11}$; $z = 1$ $\bigstar$ $\left(\frac{6}{11}, -\frac{7}{11}, 1\right)$

11 Solve E_2 for x, $x = y + z$, and substitute into E_1 and E_3 to yield

$$\begin{cases} 4(y+z) - 3y - z = 0 \\ 3(y+z) - y + 3z = 0 \end{cases} \Rightarrow \begin{cases} y + 3z = 0 & (E_4) \\ 2y + 6z = 0 & (E_5) \end{cases}$$

Now E_5 is $2\,E_4$, hence $y = -3z$ and $x = y + z = -3z + z = -2z$.

The general solution is $(-2c,\ -3c,\ c)$ for any real number c.

13 $E_1 - E_2 \Rightarrow x - 5z = -1 \Rightarrow x = 5z - 1$;

Substitute this value into E_1 to obtain $y = \dfrac{-19z + 5}{2}$; $\left(5c - 1,\ \dfrac{-19c + 5}{2},\ c \right)$ is

the general solution, where c is any real number.

15 Let $a = 1/x$, $b = 1/y$, and $c = 1/z$ to obtain the system

$$\begin{cases} 4a + b + 2c = 4 & (E_1) \\ 2a + 3b - c = 1 & (E_2) \\ a + b + c = 4 & (E_3) \end{cases}$$

Solving E_2 for c and substituting into E_1 and E_3 yields

$$\begin{cases} 4a + b + 2(2a + 3b - 1) = 4 \\ a + b + (2a + 3b - 1) = 4 \end{cases} \Rightarrow \begin{cases} 8a + 7b = 6 & (E_4) \\ 3a + 4b = 5 & (E_5) \end{cases}$$

$3\,E_4 - 8\,E_5 \Rightarrow -11b = -22 \Rightarrow b = 2$

$4\,E_4 - 7\,E_5 \Rightarrow 11a = -11 \Rightarrow a = -1,\ c = 3;\ (x,\ y,\ z) = (-1,\ \tfrac{1}{2},\ \tfrac{1}{3})$

18 By first dividing and then factoring, we have the following:

$$\frac{2x^2 + 7x + 9}{x^2 + 2x + 1} = 2 + \frac{3x + 7}{(x+1)^2} = 2 + \frac{A}{x+1} + \frac{B}{(x+1)^2}$$

$3x + 7 = A(x + 1) + B$

$x = -1:\ 4 = B \Rightarrow B = 4$ $\bigstar\ 2 + \dfrac{3}{x+1} + \dfrac{4}{(x+1)^2}$

$x = 0:\ 7 = A + B \Rightarrow A = 3$

19 $\dfrac{x^2 + 14x - 13}{x^3 + 5x^2 + 4x + 20} = \dfrac{A}{x+5} + \dfrac{Bx + C}{x^2 + 4}$

$x^2 + 14x - 13 = A(x^2 + 4) + (Bx + C)(x + 5)$

$x = -5:\ -58 = 29A \Rightarrow A = -2$

$x = 0:\ -13 = 4A + 5C \Rightarrow C = -1$ $\bigstar\ -\dfrac{2}{x+5} + \dfrac{3x - 1}{x^2 + 4}$

$x = 1:\ 2 = 5A + 6B + 6C \Rightarrow B = 3$

[21] $\begin{cases} x^2 + y^2 < 16 \\ y - x^2 > 0 \end{cases} \Leftrightarrow \begin{cases} x^2 + y^2 < 4^2 \\ y > x^2 \end{cases}$

$V @ (\pm\sqrt{-\frac{1}{2}+\frac{1}{2}\sqrt{65}}, \; -\frac{1}{2}+\frac{1}{2}\sqrt{65}) \approx (\pm 1.88, \; 3.53)$

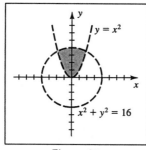

Figure 21

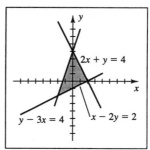

Figure 23

[23] $\begin{cases} x - 2y \le 2 \\ y - 3x \le 4 \\ 2x + y \le 4 \end{cases} \Leftrightarrow \begin{cases} y \ge \frac{1}{2}x - 1 \\ y \le 3x + 4 \\ y \le -2x + 4 \end{cases}$ $V @ (-2, -2), (0, 4), (2, 0)$

[27] $\begin{bmatrix} 2 & 0 \\ 1 & 4 \\ -2 & 3 \end{bmatrix} \begin{bmatrix} 0 & 2 & -3 \\ 4 & 5 & 1 \end{bmatrix} =$

$$\begin{bmatrix} 2(0)+0(4) & 2(2)+0(5) & 2(-3)+0(1) \\ 1(0)+4(4) & 1(2)+4(5) & 1(-3)+4(1) \\ -2(0)+3(4) & -2(2)+3(5) & -2(-3)+3(1) \end{bmatrix} = \begin{bmatrix} 0 & 4 & -6 \\ 16 & 22 & 1 \\ 12 & 11 & 9 \end{bmatrix}$$

[29] $2\begin{bmatrix} 0 & -1 & -4 \\ 3 & 2 & 1 \end{bmatrix} - 3\begin{bmatrix} 4 & -2 & 1 \\ 0 & 5 & -1 \end{bmatrix} =$

$$\begin{bmatrix} 0 & -2 & -8 \\ 6 & 4 & 2 \end{bmatrix} + \begin{bmatrix} -12 & 6 & -3 \\ 0 & -15 & 3 \end{bmatrix} = \begin{bmatrix} -12 & 4 & -11 \\ 6 & -11 & 5 \end{bmatrix}$$

[33] $\begin{bmatrix} 1 & 2 \\ 3 & 4 \end{bmatrix}\left\{\begin{bmatrix} 2 & -4 \\ 3 & 7 \end{bmatrix} + \begin{bmatrix} 1 & 5 \\ -2 & -3 \end{bmatrix}\right\} = \begin{bmatrix} 1 & 2 \\ 3 & 4 \end{bmatrix}\begin{bmatrix} 3 & 1 \\ 1 & 4 \end{bmatrix} = \begin{bmatrix} 5 & 9 \\ 13 & 19 \end{bmatrix}$

[34] Don't multiply, just remember that $A A^{-1} = I_3$.

Note: Let A denote each of the matrices in Exercises 35–50.

[35] $\begin{bmatrix} 5 & -4 & | & 1 & 0 \\ -3 & 2 & | & 0 & 1 \end{bmatrix} 2R_1 + 3R_2 \to R_1 \Rightarrow \begin{bmatrix} 1 & -2 & | & 2 & 3 \\ -3 & 2 & | & 0 & 1 \end{bmatrix} R_2 + 3R_1 \to R_2$

$\begin{bmatrix} 1 & -2 & | & 2 & 3 \\ 0 & -4 & | & 6 & 10 \end{bmatrix} -\frac{1}{4}R_2 \to R_2 \Rightarrow \begin{bmatrix} 1 & -2 & | & 2 & 3 \\ 0 & 1 & | & -\frac{3}{2} & -\frac{5}{2} \end{bmatrix} R_1 + 2R_2 \to R_1$

$\begin{bmatrix} 1 & 0 & | & -1 & -2 \\ 0 & 1 & | & -\frac{3}{2} & -\frac{5}{2} \end{bmatrix}$ $A^{-1} = -\frac{1}{2}\begin{bmatrix} 2 & 4 \\ 3 & 5 \end{bmatrix}$

37 $\begin{bmatrix} 1 & 0 & 0 & | & 1 & 0 & 0 \\ 0 & 4 & 7 & | & 0 & 1 & 0 \\ 0 & 1 & 2 & | & 0 & 0 & 1 \end{bmatrix}$ $R_3 \leftrightarrow R_2$

$\begin{bmatrix} 1 & 0 & 0 & | & 1 & 0 & 0 \\ 0 & 1 & 2 & | & 0 & 0 & 1 \\ 0 & 4 & 7 & | & 0 & 1 & 0 \end{bmatrix}$ $R_3 - 4R_2 \to R_3$

$\begin{bmatrix} 1 & 0 & 0 & | & 1 & 0 & 0 \\ 0 & 1 & 2 & | & 0 & 0 & 1 \\ 0 & 0 & -1 & | & 0 & 1 & -4 \end{bmatrix}$ $-R_3 \to R_3$

$\begin{bmatrix} 1 & 0 & 0 & | & 1 & 0 & 0 \\ 0 & 1 & 2 & | & 0 & 0 & 1 \\ 0 & 0 & 1 & | & 0 & -1 & 4 \end{bmatrix}$ $R_2 - 2R_3 \to R_2$

$\begin{bmatrix} 1 & 0 & 0 & | & 1 & 0 & 0 \\ 0 & 1 & 0 & | & 0 & 2 & -7 \\ 0 & 0 & 1 & | & 0 & -1 & 4 \end{bmatrix}$ $\qquad A^{-1} = \begin{bmatrix} 1 & 0 & 0 \\ 0 & 2 & -7 \\ 0 & -1 & 4 \end{bmatrix}$

39 $X = A^{-1}B = -\frac{1}{2}\begin{bmatrix} 2 & 4 \\ 3 & 5 \end{bmatrix}\begin{bmatrix} 30 \\ -16 \end{bmatrix} = -\frac{1}{2}\begin{bmatrix} -4 \\ 10 \end{bmatrix} = \begin{bmatrix} 2 \\ -5 \end{bmatrix}$; $(x, y) = (2, -5)$

41 $A = \begin{bmatrix} -6 \end{bmatrix} \Rightarrow |A| = -6$.

45 $\{R_1\}$ $|A| = 2(-7) + 3(-5) + 5(-11) = -84$

47 From Exercise 29 of §9.9, the determinant of A is the product of the main diagonal elements of A—that is, $|A| = (5)(-3)(-4)(2) = 120$.

48 $\begin{vmatrix} 1 & 2 & 0 & 3 & 1 \\ -2 & -1 & 4 & 1 & 2 \\ 3 & 0 & -1 & 0 & -1 \\ 2 & -3 & 2 & -4 & 2 \\ -1 & 1 & 0 & 1 & 3 \end{vmatrix}$ $\begin{matrix} R_2 + 4R_3 \to R_2 \\ \\ R_4 + 2R_3 \to R_4 \end{matrix}$ $= \begin{vmatrix} 1 & 2 & 0 & 3 & 1 \\ 10 & -1 & 0 & 1 & -2 \\ 3 & 0 & -1 & 0 & -1 \\ 8 & -3 & 0 & -4 & 0 \\ -1 & 1 & 0 & 1 & 3 \end{vmatrix}$ $\{C_3\}$

$= (-1)\begin{vmatrix} 1 & 2 & 3 & 1 \\ 10 & -1 & 1 & -2 \\ 8 & -3 & -4 & 0 \\ -1 & 1 & 1 & 3 \end{vmatrix}$ $\begin{matrix} R_2 + 2R_1 \to R_2 \\ \\ R_4 - 3R_1 \to R_4 \end{matrix}$ $= (-1)\begin{vmatrix} 1 & 2 & 3 & 1 \\ 12 & 3 & 7 & 0 \\ 8 & -3 & -4 & 0 \\ -4 & -5 & -8 & 0 \end{vmatrix}$ $\{C_4\}$

$= (-1)(-1)\begin{vmatrix} 12 & 3 & 7 \\ 8 & -3 & -4 \\ -4 & -5 & -8 \end{vmatrix}$ $\{4 \text{ is a common factor of } C_1 \text{ and } -1 \text{ of } R_3\}$

$= (-4)\begin{vmatrix} 3 & 3 & 7 \\ 2 & -3 & -4 \\ 1 & 5 & 8 \end{vmatrix}$ $\begin{matrix} R_1 - 3R_3 \to R_1 \\ R_2 - 2R_3 \to R_2 \end{matrix}$ $= (-4)\begin{vmatrix} 0 & -12 & -17 \\ 0 & -13 & -20 \\ 1 & 5 & 8 \end{vmatrix}$ $\{C_1\}$

$= (-4)\begin{vmatrix} -12 & -17 \\ -13 & -20 \end{vmatrix} = (-4)(240 - 221) = -76$

49 C_2 and C_4 are identical, so $|A| = 0$.

$\boxed{51}$ $\begin{vmatrix} 2-x & 3 \\ 1 & -4-x \end{vmatrix} = 0 \Rightarrow (2-x)(-4-x) - 3 = 0 \Rightarrow$

$$x^2 + 2x - 11 = 0 \Rightarrow x = \frac{-2 \pm \sqrt{4+44}}{2} = \frac{-2 \pm 4\sqrt{3}}{2} = -1 \pm 2\sqrt{3}$$

$\boxed{54}$ Interchange R_1 with R_2 and then R_2 with R_3 to obtain the determinant on the right.

The effect is to multiply by -1 twice.

$\boxed{55}$ This is an extension of Exercise 29 of §9.9. Expanding by C_1, only a_{11} is not 0.

Expanding by the new C_1 again, only a_{22} is not 0. Repeating this process yields

$$|A| = a_{11}a_{22}a_{33} \cdots a_{nn}, \text{ the product of the main diagonal elements.}$$

$\boxed{56}$ $\begin{vmatrix} 1 & a & b+c \\ 1 & b & a+c \\ 1 & c & a+b \end{vmatrix}$ $C_3 + C_2 \rightarrow C_2$

$= \begin{vmatrix} 1 & a & a+b+c \\ 1 & b & a+b+c \\ 1 & c & a+b+c \end{vmatrix}$ $C_3 - (a+b+c)C_1 \rightarrow C_3$

$= \begin{vmatrix} 1 & a & 0 \\ 1 & b & 0 \\ 1 & c & 0 \end{vmatrix} = 0$, since C_3 consists of all zeros.

$\boxed{59}$ Let x and y denote the length and width, respectively, of the rectangle. A diagonal of the field is 100 ft. We can use the Pythagorean theorem to formulate an equation that relates the sides and a diagonal.

$$\begin{cases} xy = 4000 & area & (E_1) \\ x^2 + y^2 = 100^2 & diagonal & (E_2) \end{cases}$$

Solve E_1 for y, $y = 4000/x$, and substitute into E_2.

$x^2 + \dfrac{4000^2}{x^2} = 100^2 \Rightarrow x^4 - 10{,}000x^2 + 16{,}000{,}000 = 0 \Rightarrow$

$(x^2 - 2000)(x^2 - 8000) = 0 \Rightarrow x = 20\sqrt{5},\, 40\sqrt{5}$ and $y = 40\sqrt{5},\, 20\sqrt{5}$.

The dimensions are $20\sqrt{5}$ ft $\times\, 40\sqrt{5}$ ft.

$\boxed{60}$ Following the hint, we let $y = mx + 3$ and substitute into $x^2 + y^2 = 1$, obtaining

$x^2 + (mx+3)^2 = 1 \Rightarrow x^2 + (m^2x^2 + 6mx + 9) = 1 \Rightarrow (m^2+1)x^2 + (6m)x + (8) = 0 \Rightarrow$

$$x = \frac{-6m \pm \sqrt{(6m)^2 - 4(m^2+1)(8)}}{2(m^2+1)} = \frac{-6m \pm \sqrt{36m^2 - 32m^2 - 32}}{2(m^2+1)}.$$

If there is to be only one solution to the system, i.e., one point of intersection between the circle and the line, then the discriminant must equal 0.

$4m^2 - 32 = 0 \Rightarrow m = \pm 2\sqrt{2}$ and the equations of the lines are $y = \pm 2\sqrt{2}\, x + 3$.

[62] Let r_1 and r_2 denote the inside radius and the outside radius, respectively.

Inside distance = 90% (outside distance) $\Rightarrow 2\pi r_1 = 0.90\,(2\pi r_2) \Rightarrow$

$r_1 = 0.90\,(r_1 + 10)$ { since $r_2 = r_1 + 10$ } $\Rightarrow 0.1r_1 = 9 \Rightarrow r_1 = 90$ ft and $r_2 = 100$ ft

[64] Let x and y denote the number of desks shipped from the western warehouse and the eastern warehouse, respectively.

$$\begin{cases} x + y = 150 & quantity \\ 24x + 35y = 4205 & price \end{cases} \qquad E_2 - 24\,E_1 \Rightarrow 11y = 605 \Rightarrow y = 55;\ x = 95$$

[65] If x and y denote the length and the width, respectively, then a system is $x \le 12$, $y \le 8$, $y \ge \frac{1}{2}x$. The graph is the region bounded by the quadrilateral with vertices $(0, 0)$, $(0, 8)$, $(12, 8)$, and $(12, 6)$.

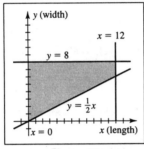

Figure 65

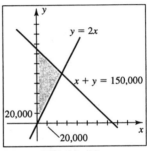

Figure 68

[68] Let x and y denote the amount in the high- and low-risk investments, respectively.

The amount in bonds is $150{,}000 - x - y$.

Profit function: $P = 0.15x + 0.10y + 0.08(150{,}000 - x - y) = \underline{12{,}000 + 0.07x + 0.02y}$

(x, y)	$(0, 0)$	$(0, 150{,}000)$	$(50{,}000, 100{,}000)$
P	12,000	15,000	17,500 ∎

$$\begin{cases} x + y \le 150{,}000 \\ y \ge 2x \\ x,\ y \ge 0 \end{cases}$$

The maximum return of $17,500 occurs when

$50,000 is invested in the high-risk investment,

$100,000 is invested in the low-risk investment, and $0 is invested in bonds.

$\boxed{1}$ (a) For $b = 1.99$, we get $x = 204$ and $y = -100$. For $b = 1.999$, we get $x = 2004$ and $y = -1000$.

(b) Solving $\begin{cases} x + 2y = 4 \\ x + by = 5 \end{cases}$ for y gives us $y = \dfrac{1}{b-2}$ and then solving for x we

obtain $x = \dfrac{4b - 10}{b - 2}$. Both x and y are rational functions of b and as $b \to 2^-$,

$x \to \infty$ and $y \to -\infty$.

(c) If b gets very large, (x, y) gets close to $(4, 0)$.

$\boxed{3}$ If we let A be an $m \times n$ matrix $(m \neq n)$, then B would have to be an $n \times m$ matrix so that AB and BA are both defined. But then $AB = I_m$ and $BA = I_n$, different orders of the identity matrix, which we can't have (they must be the same).

$\boxed{5}$ Synthetically dividing $x^4 + ax^2 + bx + c$ by $x + 1$, $x - 2$, and $x - 3$ yields the remainders $a - b + c + 1$, $4a + 2b + c + 16$, and $9a + 3b + c + 81$, respectively. Setting each of the remainders equal to 0 gives us the following system of equations in matrix form.

$$AX = B \Leftrightarrow \begin{bmatrix} 1 & -1 & 1 \\ 4 & 2 & 1 \\ 9 & 3 & 1 \end{bmatrix} \begin{bmatrix} a \\ b \\ c \end{bmatrix} = \begin{bmatrix} -1 \\ -16 \\ -81 \end{bmatrix} \Rightarrow X = A^{-1}B = \begin{bmatrix} -15 \\ 10 \\ 24 \end{bmatrix}.$$

Hence, $a = -15$, $b = 10$, $c = 24$, and we graph $Y_1 = x^4 - 15x^2 + 10x + 24$.

The roots of Y_1 are -1, 2, 3, and -4.

Chapter 10: Sequences, Series, and Probability

Note: For Exercises 1–16, the answers are listed in the order a_1, a_2, a_3, a_4; and a_8. Simply substitute 1, 2, 3, 4, and 8 for n in the formula for a_n to obtain the results.

$\boxed{3}$ $a_n = \dfrac{3n-2}{n^2+1}$ • $a_1 = \dfrac{3(1)-2}{(1)^2+1} = \dfrac{1}{2}$, $a_2 = \dfrac{3(2)-2}{(2)^2+1} = \dfrac{4}{5}$, $a_3 = \dfrac{3(3)-2}{(3)^2+1} = \dfrac{7}{10}$,

$a_4 = \dfrac{3(4)-2}{(4)^2+1} = \dfrac{10}{17}$; $a_8 = \dfrac{3(8)-2}{(8)^2+1} = \dfrac{22}{65}$ ★ $\dfrac{1}{2}, \dfrac{4}{5}, \dfrac{7}{10}, \dfrac{10}{17}; \dfrac{22}{65}$

$\boxed{7}$ $a_n = 2 + (-0.1)^n$ • $a_1 = 2 + (-0.1)^1 = 2 - 0.1 = 1.9$,

$a_2 = 2 + (-0.1)^2 = 2 + 0.01 = 2.01$,

$a_3 = 2 + (-0.1)^3 = 2 - 0.001 = 1.999$,

$a_4 = 2 + (-0.1)^4 = 2 + 0.0001 = 2.0001$;

$a_8 = 2 + (-0.1)^8 = 2 + 0.00000001 = 2.00000001$

★ 1.9, 2.01, 1.999, 2.0001; 2.00000001

$\boxed{9}$ $a_n = (-1)^{n-1}\left(\dfrac{n+7}{2n}\right)$ •

$a_1 = (-1)^{1-1}\left(\dfrac{1+7}{2(1)}\right) = (-1)^0 \left(\dfrac{8}{2}\right) = (1)(4) = 4$,

$a_2 = (-1)^{2-1}\left(\dfrac{2+7}{2(2)}\right) = (-1)^1 \left(\dfrac{9}{4}\right) = -\dfrac{9}{4}$,

$a_3 = (-1)^{3-1}\left(\dfrac{3+7}{2(3)}\right) = (-1)^2 \left(\dfrac{10}{6}\right) = \dfrac{5}{3}$,

$a_4 = (-1)^{4-1}\left(\dfrac{4+7}{2(4)}\right) = (-1)^3 \left(\dfrac{11}{8}\right) = -\dfrac{11}{8}$;

$a_8 = (-1)^{8-1}\left(\dfrac{8+7}{2(8)}\right) = (-1)^7 \left(\dfrac{15}{16}\right) = -\dfrac{15}{16}$ ★ $4, -\dfrac{9}{4}, \dfrac{5}{3}, -\dfrac{11}{8}; -\dfrac{15}{16}$

$\boxed{13}$ $a_n = \dfrac{2^n}{n^2+2}$ • $a_1 = \dfrac{2^1}{1^2+2} = \dfrac{2}{3}$, $a_2 = \dfrac{2^2}{2^2+2} = \dfrac{4}{6} = \dfrac{2}{3}$, $a_3 = \dfrac{2^3}{3^2+2} = \dfrac{8}{11}$,

$a_4 = \dfrac{2^4}{4^2+2} = \dfrac{16}{18} = \dfrac{8}{9}$; $a_8 = \dfrac{2^8}{8^2+2} = \dfrac{256}{66} = \dfrac{128}{33}$ ★ $\dfrac{2}{3}, \dfrac{2}{3}, \dfrac{8}{11}, \dfrac{8}{9}; \dfrac{128}{33}$

$\boxed{15}$ a_n is the number of decimal places in $(0.1)^n$. •

$(0.1)^1 = 0.1$ { 1 decimal place } $\Rightarrow a_1 = 1$,

$(0.1)^2 = 0.01$ { 2 decimal places } $\Rightarrow a_2 = 2$,

$(0.1)^3 = 0.001$ { 3 decimal places } $\Rightarrow a_3 = 3$,

$(0.1)^4 = 0.0001$ { 4 decimal places } $\Rightarrow a_4 = 4$;

$(0.1)^8 = 0.00000001$ { 8 decimal places } $\Rightarrow a_8 = 8$ ★ 1, 2, 3, 4; 8

17 $a_1 = 2,$ $\qquad a_{k+1} = 3a_k - 5$ •

$a_2 = a_{1+1}\ \{k = 1\} = 3a_1 - 5 = 3(2) - 5 = 1,$ $\quad a_3 = 3a_2 - 5 = 3(1) - 5 = -2,$

$a_4 = 3a_3 - 5 = 3(-2) - 5 = -11,$ $\qquad\qquad a_5 = 3a_4 - 5 = 3(-11) - 5 = -38$

21 $a_1 = 5,$ $\qquad a_{k+1} = ka_k$ •

$a_2 = a_{1+1}\ \{k = 1\} = 1\,a_1 = 1(5) = 5,$ $\qquad a_3 = 2\,a_2 = 2(5) = 10,$

$a_4 = 3\,a_3 = 3(10) = 30,$ $\qquad\qquad a_5 = 4\,a_4 = 4(30) = 120$

23 $a_1 = 2,$ $\qquad a_{k+1} = (a_k)^k$ •

$a_2 = a_{1+1}\ \{k = 1\} = (a_1)^1 = (2)^1 = 2,$ $\qquad a_3 = (a_2)^2 = (2)^2 = 4,$

$a_4 = (a_3)^3 = (4)^3 = 4^3 \text{ or } 64,$ $\qquad a_5 = (a_4)^4 = (4^3)^4 = 4^{12} \text{ or } 16{,}777{,}216$

25 $a_n = 3 + \frac{1}{2}n$ • $S_1 = a_1 = 3 + \frac{1}{2} = \frac{7}{2}.$ $S_2 = S_1 + a_2 = \frac{7}{2} + 4 = \frac{15}{2}.$ Alternatively, to find S_2 we could use $a_1 + a_2$. However, in most cases, it is easier to compute S_{n+1} by using $S_n + a_{n+1}$. $S_3 = S_2 + a_3 = \frac{15}{2} + \frac{9}{2} = 12.$ $S_4 = S_3 + a_4 = 12 + 5 = 17.$

27 $a_n = (-1)^n\, n^{-1/2}$ • $S_1 = a_1 = -1.$

$S_2 = S_1 + a_2 = -1 + 1/\sqrt{2}.$

$S_3 = S_2 + a_3 = -1 + 1/\sqrt{2} - 1/\sqrt{3}.$

$S_4 = S_3 + a_4 = -1 + 1/\sqrt{2} - 1/\sqrt{3} + \frac{1}{2} = -\frac{1}{2} + 1/\sqrt{2} - 1/\sqrt{3}.$

29 Substitute $k = 1, 2, 3, 4,$ and 5 into $2k - 7$ and then add the results.

$$\sum_{k=1}^{5} (2k - 7) = (-5) + (-3) + (-1) + 1 + 3 = -5$$

31 $\displaystyle\sum_{k=1}^{4} (k^2 - 5) = (-4) + (-1) + 4 + 11 = 10$

33 $\displaystyle\sum_{k=0}^{5} k(k - 2) = 0 + (-1) + 0 + 3 + 8 + 15 = 25$

37 $\displaystyle\sum_{k=1}^{5} (-3)^{k-1} = 1 + (-3) + 9 + (-27) + 81 = 61$

39 Using part (1) of the theorem on the sum of a constant, we have

$$\sum_{k=1}^{100} 100 = 100(100) = 10{,}000.$$

41 Using part (2) of the theorem on the sum of a constant, we have

$$\sum_{k=253}^{571} \tfrac{1}{3} = (571 - 253 + 1)(\tfrac{1}{3}) = 319(\tfrac{1}{3}) = \tfrac{319}{3}.$$

43 Note that j, not k, is the summation variable. Hence, we again use part (1) of the theorem on the sum of a constant, obtaining $\displaystyle\sum_{j=1}^{7} \tfrac{1}{2}k^2 = 7(\tfrac{1}{2}k^2) = \tfrac{7}{2}k^2.$

$\boxed{45}$ $\displaystyle\sum_{k=1}^{n} (a_k - b_k) = (a_1 - b_1) + (a_2 - b_2) + \cdots + (a_n - b_n)$

$\qquad\qquad = (a_1 + a_2 + \cdots + a_n) + (-b_1 - b_2 - \cdots - b_n)$

$\qquad\qquad = (a_1 + a_2 + \cdots + a_n) - (b_1 + b_2 + \cdots + b_n)$

$\qquad\qquad = \displaystyle\sum_{k=1}^{n} a_k - \sum_{k=1}^{n} b_k$

$\boxed{51}$ (a) $a_1 = 1$, $a_2 = 1$, $a_3 = a_2 + a_1 = 1 + 1 = 2$,

$\qquad\quad a_4 = a_3 + a_2 = 2 + 1 = 3$,

$\qquad\quad a_5 = a_4 + a_3 = 3 + 2 = 5$,

$\qquad\quad a_6 = a_5 + a_4 = 5 + 3 = 8$,

$\qquad\quad a_7 = a_6 + a_5 = 8 + 5 = 13$,

$\qquad\quad a_8 = a_7 + a_6 = 13 + 8 = 21$,

$\qquad\quad a_9 = a_8 + a_7 = 21 + 13 = 34$,

$\qquad\quad a_{10} = a_9 + a_8 = 34 + 21 = 55$

$\quad$ (b) $r_1 = \frac{1}{1} = 1$, $\qquad\qquad r_2 = \frac{2}{1} = 2$, $\qquad\qquad r_3 = \frac{3}{2} = 1.5$,

$\qquad\quad r_4 = \frac{5}{3} = 1.\overline{6}$, $\qquad\quad r_5 = \frac{8}{5} = 1.6$, $\qquad\qquad r_6 = \frac{13}{8} = 1.625$,

$\qquad\quad r_7 = \frac{21}{13} \approx 1.6153846$, $\quad r_8 = \frac{34}{21} \approx 1.6190476$, $\quad r_9 = \frac{55}{34} \approx 1.6176471$,

$\qquad\qquad\qquad\qquad\qquad\qquad\qquad\qquad\quad$ and $r_{10} = \frac{89}{55} \approx 1.6181818$.

$\boxed{53}$ (a) Since the amount of chlorine decreases by a factor of 0.20 each day {retain 80%}, $a_n = 0.8 a_{n-1}$, where a_0 is the initial amount of chlorine in the pool.

$\quad$ (b) Let $a_0 = 7$ and $a_n = 0.8 a_{n-1}$.

Day (n)	0	1	2	3	4	5
Chlorine (a_n)	7.00	5.60	4.48	3.58	2.87	2.29

$\qquad$ The chlorine level will drop below 3 ppm during the fourth day.

$\boxed{55}$ $N = 5$ and $x_1 = \frac{5}{2} \Rightarrow x_2 = \frac{1}{2}\left(x_1 + \frac{N}{x_1}\right) = \frac{1}{2}\left(2.5 + \frac{5}{2.5}\right) = 2.25 \Rightarrow$

$x_3 = \frac{1}{2}\left(2.25 + \frac{5}{2.25}\right) \approx 2.236111 \Rightarrow x_4 \approx 2.236068 \Rightarrow x_5 \approx 2.236068$.

Thus, $\sqrt{5} \approx 2.236068$.

$\qquad\qquad$ *Note:* TI-users—see the key sequence at the end of the solution of #59.

$\boxed{57}$ $x_1 = 2$ and $x_2 = \frac{1}{3}\sqrt[3]{x_1} + 2 \Rightarrow x_2 \approx 2.419974 \Rightarrow x_3 \approx 2.447523 \Rightarrow x_4 \approx 2.449215 \Rightarrow$

$x_5 \approx 2.449319 \Rightarrow x_6 \approx 2.449325$. The root is approximately 2.4493.

$\qquad\qquad$ *Note:* TI-users—see the key sequence at the end of the solution of #59.

$\boxed{59}$ (a) $f(1) = (1 - 2 + \log 1) = (1 - 2 + 0) = -1 < 0$ and

$\quad\quad f(2) = (2 - 2 + \log 2) = \log 2 \approx 0.30 > 0$.

$\quad\quad\quad\quad$ Thus, f assumes both positive and negative values on $[1, 2]$.

$\quad$ (b) Solving $\underline{x - 2 + \log x = 0}$ for x in terms of x gives us $\underline{x = 2 - \log x}$.

$\quad\quad x_1 = 1.5 \Rightarrow x_2 = 2 - \log x_1 = 2 - \log 1.5 \approx 1.823909 \Rightarrow x_3 \approx 1.738997 \Rightarrow$

$\quad\quad x_4 \approx 1.759701 \Rightarrow x_5 \approx 1.754561 \Rightarrow x_6 \approx 1.755832 \Rightarrow x_7 \approx 1.755517$.

The zero is approximately 1.76. *Note:* If you are using a TI-8x, type 1.5 and press the ENTER key to store 1.5 in the memory location ANS. Now type $2 - \text{LOG ANS}$ and then successively press the ENTER key to obtain the approximations for x_1, x_2, ... (see the display below).

$$1.5\; \boxed{\text{ENTER}}\; \boxed{2}\; \boxed{-}\; \boxed{\text{LOG}}\; \boxed{\text{ANS}}\; \boxed{\text{ENTER}}\; \boxed{\text{ENTER}}\; \ldots$$

Similar key sequences for **Exercises 55 and 57**, respectively, are:

$$2.5\; \boxed{\text{ENTER}}\; \boxed{.5}\; \boxed{(}\; \boxed{\text{ANS}}\; \boxed{+}\; \boxed{5}\; \boxed{\div}\; \boxed{\text{ANS}}\; \boxed{)}\; \boxed{\text{ENTER}}\; \boxed{\text{ENTER}}\; \ldots$$

$$2\; \boxed{\text{ENTER}}\; \boxed{\text{MATH}}\; \boxed{4}\; \boxed{\text{ANS}}\; \boxed{\div}\; \boxed{3}\; \boxed{+}\; \boxed{2}\; \boxed{\text{ENTER}}\; \boxed{\text{ENTER}}\; \ldots$$

Alternatively, we can use the memory location X instead of ANS by using the $\boxed{\text{STO} \triangleright}$ key as shown in *Figure 59*.

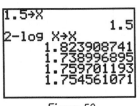

$[1, 100]$ by $[0, 3]$

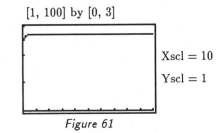

Xscl $= 10$

Yscl $= 1$

Figure 59 $\quad\quad\quad\quad$ *Figure 61*

$\boxed{61}$ Graph $y = \left(1 + \dfrac{1}{x} + \dfrac{1}{2x^2}\right)^x$ on the interval $[1, 100]$.

The graph approaches the horizontal asymptote $y \approx 2.718 \approx e$.

$\quad\quad$ For increasing values of n, the terms of the sequence appear to approximate e.

$\boxed{65}$ *Note:* **TI-82 users:** Under MODE, make sure you are in Seq and Dot mode.

$\quad\quad$ Under Y $=$, assign $1.7\text{U}n - 1 + .5$ to Un $\left\{ \text{U}n - 1 \text{ is } \boxed{\text{2nd}}\; \boxed{7} \right\}$.

$\quad\quad$ Under WINDOW, let UnStart $= .25$, nStart $= 1$, nMin $= 1$, nMax $= 20$,

$\quad\quad$ and the rest of the window parameters as shown in *Figure 65*.

TI-83 users: Under MODE, make sure you are in Seq mode.

Under $Y =$, let nMin $= 1$, u(nMin) $= .25$, and u(n) $= 1.7$u($n-1) + .5$

$\left\{\text{u is }\boxed{\text{2nd}}\ \boxed{7}\ \text{and } n \text{ is }\boxed{\text{X,T,}\theta\text{,}n}\right\}.$

Under WINDOW, let nMin $= 1$, nMax $= 20$, PlotStart $= 1$, PlotStep $= 1$,

and the rest of the window parameters as shown in the figure.

By tracing the graph we see that $a_9 \approx 66.55$ and $a_{10} \approx 113.64$. Thus, $k = 10$.

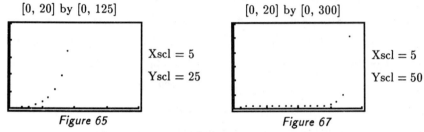

[0, 20] by [0, 125]   [0, 20] by [0, 300]

Xscl $= 5$ Xscl $= 5$
Yscl $= 25$ Yscl $= 50$

Figure 65   Figure 67

67 By tracing the graph we see that $a_{18} \approx 50.39$ and $a_{19} \approx 255.96$. Thus, $k = 19$.

69 (a) Since $c = 0.5$, let the sequence be defined by $a_{k+1} = 0.5a_k(1 - a_k)$. Then,

$a_1 = 0.25$, $a_2 = 0.5a_1(1 - a_1) = 0.5(0.25)(1 - 0.25) = 0.09375$. In a similar

manner, $a_3 \approx 0.04248$, $a_4 \approx 0.02034$, ... , $a_{10} \approx 0.00031$, $a_{11} \approx 0.00015$,

$a_{12} \approx 0.00008$. The insect population is initially $1000a_1 = 1000(0.25) = 250$. It

then becomes approximately 94, 42, 20, and so on, until the population decreases

to zero.

(b) The sequence determined is $a_1 = 0.25$, $a_2 \approx 0.28125$, $a_3 \approx 0.30322$, $a_4 \approx 0.31692$,

... , $a_{18} \approx 0.33333$, $a_{19} \approx 0.33333$, $a_{20} \approx 0.33333$. The insect population is

initially 250. It then becomes approximately 281, 303, 317, and so on, until the

population stabilizes at 333.

(c) The sequence determined is $a_1 = 0.25$, $a_2 \approx 0.51563$, $a_3 \approx 0.68683$, $a_4 \approx 0.59151$,

... , $a_{40} \approx 0.63636$, $a_{41} \approx 0.63636$, $a_{42} \approx 0.63636$. The insect population is

initially 250. It then becomes approximately 516, 687, 592, and so on, until the

population stabilizes at 636.

10.2 Exercises

1 To show that the given sequence, $-6, -2, 2, ..., 4n - 10, ...,$ is arithmetic, we must

show that $a_{k+1} - a_k$ is equal to some constant, which is the common difference.

$a_n = 4n - 10 \Rightarrow a_{k+1} - a_k = [4(k+1) - 10] - [4(k) - 10] = 4k + 4 - 10 - 4k + 10 = 4$

Note: For Exercises 3–10, we will find the nth term first, and then use that term to find a_5 and a_{10}.

5 The common difference can be found by *subtracting* any term from its successor.

$d = 2.7 - 3 = -0.3$; $a_n = a_1 + (n-1)d = 3 + (n-1)(-0.3) = -0.3n + 3.3$;

$$a_5 = -0.3(5) + 3.3 = 1.8;\ a_{10} = -0.3(10) + 3.3 = 0.3$$

7 $d = -3.9 - (-7) = 3.1$; $a_n = -7 + (n-1)(3.1) = 3.1n - 10.1$; $a_5 = 5.4$; $a_{10} = 20.9$

9 An equivalent sequence is $\ln 3$, $\ln 3^2$, $\ln 3^3$, $\ln 3^4$, ..., which is also equivalent to the sequence $\ln 3$, $2\ln 3$, $3\ln 3$, $4\ln 3$, ...; $d = 2\ln 3 - \ln 3 = \ln 3$;

$$a_n = \ln 3 + (n-1)(\ln 3) = n\ln 3 \text{ or } \ln 3^n;\ a_5 = 5\ln 3 \text{ or } \ln 3^5;\ a_{10} = 10\ln 3 \text{ or } \ln 3^{10}$$

11 $a_6 = a_1 + 5d$ and $a_2 = a_1 + d \Rightarrow a_6 - a_2 = (a_1 + 5d) - (a_1 + d) = 4d$.

But $a_6 - a_2 = -11 - 21 = -32$. Hence, $4d = -32$ and $d = -8$.

13 Given a_1 and a_2, we can find the difference d. $d = a_2 - a_1 = 7.5 - 9.1 = -1.6$.

The twelfth term is equal to the first term plus eleven differences.

$$a_{12} = 9.1 + (11)(-1.6) = -8.5.$$

15 $d = a_7 - a_6 = 5.2 - 2.7 = 2.5$; $a_6 = a_1 + 5d \Rightarrow 2.7 = a_1 + 5(2.5) \Rightarrow a_1 = -9.8$

17 As in the solution for Exercise 11, $a_3 = 7$ and $a_{20} = 43 \Rightarrow 17d = 36 \Rightarrow d = \frac{36}{17}$.

$$a_{15} = a_3 + 12d = 7 + 12\left(\frac{36}{17}\right) = \frac{551}{17}.$$

Note: To find the sum in Exercises 19–26, we use the formulas

$$S_n = \frac{n}{2}\left[2a_1 + (n-1)(d)\right] \quad \text{and} \quad S_n = \frac{n}{2}(a_1 + a_n).$$

19 $a_1 = 40$, $d = -3$, $n = 30 \Rightarrow S_{30} = \frac{30}{2}\left[2(40) + (29)(-3)\right] = -105.$

21 $a_1 = -9$, $a_{10} = 15$, $n = 10 \Rightarrow S_{10} = \frac{10}{2}(-9 + 15) = 30.$

23 $\displaystyle\sum_{k=1}^{20}(3k - 5)$ • Using the second formula for S_n with $n = 20$, $a_1 = 3(1) - 5 = -2$,

and $a_{20} = 3(20) - 5 = 55$, we obtain $S_{20} = \frac{20}{2}(-2 + 55) = 530.$

25 $\displaystyle\sum_{k=1}^{18}\left(\tfrac{1}{2}k + 7\right)$ • $a_1 = \frac{15}{2}$, $a_{18} = 16$, $n = 18 \Rightarrow S_{18} = \frac{18}{2}\left(\frac{15}{2} + 16\right) = \frac{423}{2}.$

27 $1 + 3 + 5 + 7$. Since the difference in terms is 2, the general term is

$$a_1 + (n-1)d = 1 + (n-1)2 = 2n - 1. \quad \sum_{n=1}^{4}(2n-1)$$

29 $1 + 3 + 5 + \cdots + 73$. From Exercise 27, the general term is $2n - 1$ with

n starting at 1. $2n - 1 = 73 \Rightarrow n = 37$, the largest value. $\displaystyle\sum_{n=1}^{37}(2n-1)$

31 $\frac{3}{7} + \frac{6}{11} + \frac{9}{15} + \frac{12}{19} + \frac{15}{23} + \frac{18}{27}$. The numerators increase by 3, the denominators increase

by 4. The general terms are $3 + (n-1)3 = 3n$ and $7 + (n-1)4 = 4n + 3$. $\displaystyle\sum_{n=1}^{6} \frac{3n}{4n+3}$

33 $a_1 = -2$, $d = \frac{1}{4}$, $S = 21$, and $S_n = \frac{n}{2}[2a_1 + (n-1)(d)] \Rightarrow$

$21 = \frac{n}{2}\left[2(-2) + (n-1)(\frac{1}{4})\right] \Rightarrow 42 = n(\frac{1}{4}n - \frac{17}{4}) \Rightarrow 168 = n^2 - 17n \Rightarrow$

$n^2 - 17n - 168 = 0 \Rightarrow (n-24)(n+7) = 0 \Rightarrow n = 24, -7$.

Since n can not be negative, $n = 24$.

35 If we insert five arithmetic means between 2 and 10, there will be 6 differences that

span the distance from 2 to 10. Hence, $6d = 10 - 2 \Rightarrow d = \frac{4}{3}$;

The terms are $2, \frac{10}{3}, \frac{14}{3}, 6, \frac{22}{3}, \frac{26}{3}, 10$.

37 (a) The first integer greater than 32 that is divisible by 6 is 36 $\{6 \cdot 6\}$

and the last integer less than 395 that is divisible by 6 is 390 $\{65 \cdot 6\}$.

The number of terms is then $65 - 6 + 1 = 60$.

(b) The sum is $S_{60} = \frac{60}{2}(36 + 390) = 12{,}780$.

39 There are $(24 - 10 + 1) = 15$ layers. Model this problem as an arithmetic sequence

with $a_1 = 10$ and $a_{15} = 24$. $S_{15} = \frac{15}{2}(10 + 24) = 255$.

41 This is similar to inserting 9 arithmetic means between 4 and 24.

$10d = 20 \Rightarrow d = 2$. The circumference of each ring is πD with $D = 4, 6, 8, \ldots, 24$.

$a_1 = 4\pi$ and $a_{11} = 24\pi \Rightarrow S_{11} = \frac{11}{2}(4\pi + 24\pi) = 154\pi$ ft.

43 $n = 5$, $S_5 = 5000$, $d = -100 \Rightarrow$

$5000 = \frac{5}{2}\left[2a_1 + 4(-100)\right] \Rightarrow 2000 = 2a_1 - 400 \Rightarrow a_1 = \$1200$

45 The sequence $16, 48, 80, 112, \ldots$, is an arithmetic sequence with $a_1 = 16$ and

$d = 48 - 16 = 32$. The total distance traveled in n seconds is

$a_1 + a_2 + \cdots + a_n = \frac{n}{2}[2a_1 + (n-1)d] = \frac{n}{2}[2(16) + (n-1)(32)] = \frac{n}{2}(32n) = 16n^2$.

47 If the nth term is $\frac{1}{x_n}$ and $x_{n+1} = \frac{x_n}{1+x_n}$, then the $(n+1)$st term is

$$\frac{1}{x_{n+1}} = \frac{1}{\dfrac{x_n}{1+x_n}} = \frac{1+x_n}{x_n} = \frac{1}{x_n} + \frac{x_n}{x_n} = 1 + \frac{1}{x_n},$$

which is 1 greater than the nth term and therefore the sequence is arithmetic.

49 (a) $T_8 = 1 + 2 + \cdots + 8 = 36$. $A_k = \frac{n - k + 1}{T_n} \Rightarrow A_1 = \frac{8 - 1 + 1}{36} = \frac{8}{36}$.

$A_2 = \frac{7}{36}$, $A_3 = \frac{6}{36}$, $A_4 = \frac{5}{36}$, $A_5 = \frac{4}{36}$, $A_6 = \frac{3}{36}$, $A_7 = \frac{2}{36}$, $A_8 = \frac{1}{36}$.

(b) $d = A_{k+1} - A_k = -\frac{1}{36}$ for $k = 1, 2, \ldots, 7$. $S_8 = \displaystyle\sum_{k=1}^{8} A_k = \frac{8}{36} + \frac{7}{36} + \cdots + \frac{1}{36} = 1$.

(c) $\$1000\left(\frac{8}{36} + \frac{7}{36} + \frac{6}{36} + \frac{5}{36}\right) \approx \$722.22$

10.3 Exercises

1 To show that the given sequence, $5, -\frac{5}{4}, \frac{5}{16}, \ldots, 5(-\frac{1}{4})^{n-1}, \ldots$, is geometric,

we must show that $\frac{a_{k+1}}{a_k}$ is equal to some constant, which is the common ratio.

$$a_n = a_1 r^{n-1} = 5(-\tfrac{1}{4})^{n-1} \Rightarrow \frac{a_{k+1}}{a_k} = \frac{5(-\frac{1}{4})^{(k+1)-1}}{5(-\frac{1}{4})^{k-1}} = -\frac{1}{4}$$

Note: For Exercises 3–14, we will find the nth term first and then use that term to find

a_5 and a_8.

3 The common ratio can be found by *dividing* any term by its successor. $r = \frac{4}{8} = \frac{1}{2}$;

$$a_n = a_1 r^{n-1} = 8(\tfrac{1}{2})^{n-1} = 2^3 (2^{-1})^{n-1} = 2^3 2^{1-n} = 2^{4-n};$$

$$a_5 = 2^{4-5} = 2^{-1} = \tfrac{1}{2}; \qquad\qquad a_8 = 2^{4-8} = 2^{-4} = \tfrac{1}{16}$$

5 $r = \frac{-30}{300} = -0.1;\ a_n = a_1 r^{n-1} = 300(-0.1)^{n-1}$;

$$a_5 = 300(-0.1)^4 = 0.03; \qquad\qquad a_8 = 300(-0.1)^7 = -0.00003$$

11 $r = \frac{-x^2}{1} = -x^2;\ a_n = a_1 r^{n-1} = 1(-x^2)^{n-1} = (-1)^{n-1} x^{2n-2}$;

$$a_5 = (-1)^4 x^{10-2} = x^8; \qquad\qquad a_8 = (-1)^7 x^{16-2} = -x^{14}$$

13 $r = \frac{2^{x+1}}{2} = 2^x;\ a_n = a_1 r^{n-1} = 2(2^x)^{n-1} = 2^{(n-1)x+1};\ a_5 = 2^{4x+1};\ a_8 = 2^{7x+1}$

15 $\frac{a_6}{a_4} = \frac{9}{3} = 3$ and $\frac{a_6}{a_4} = \frac{a_1 r^5}{a_1 r^3} = r^2$. Hence, $r^2 = 3$ and $r = \pm\sqrt{3}$.

17 $r = \frac{6}{4} = \frac{3}{2};\ a_6 = a_1 r^{n-1} = 4(\frac{3}{2})^5 = \frac{243}{8}$

19 $\frac{a_7}{a_4} = \frac{12}{4} = 3$ and $\frac{a_7}{a_4} = \frac{a_1 r^6}{a_1 r^3} = r^3$. Hence, $r^3 = 3$ and $r = \sqrt[3]{3}$. Since the tenth term

can be obtained by multiplying the seventh term by the common ratio three times,

$a_{10} = a_7 r^3 = 12(3) = 36$.

21 Using the formula for S_n with $n = 10$, $r = 3$, and $a_1 = 3^1 = 3$, we get

$$\sum_{k=1}^{10} 3^k = 3 \cdot \frac{1 - 3^{10}}{1 - 3} = 3 \cdot \frac{-59{,}048}{-2} = 88{,}572.$$

23 $\displaystyle\sum_{k=0}^{9} (-\tfrac{1}{2})^{k+1} = \sum_{k=1}^{10} (-\tfrac{1}{2})^k = -\frac{1}{2} \cdot \frac{1 - (-\frac{1}{2})^{10}}{1 - (-\frac{1}{2})} = -\frac{1}{2} \cdot \frac{\frac{1023}{1024}}{\frac{3}{2}} = -\frac{1023}{3072}$

25 $2 + 4 + 8 + 16 + 32 + 64 + 128 = 2^1 + 2^2 + 2^3 + 2^4 + 2^5 + 2^6 + 2^7 = \displaystyle\sum_{n=1}^{7} 2^n$

27 $\frac{1}{4} - \frac{1}{12} + \frac{1}{36} - \frac{1}{108} = \frac{1}{4} - \frac{1}{4} \cdot \frac{1}{3^1} + \frac{1}{4} \cdot \frac{1}{3^2} - \frac{1}{4} \cdot \frac{1}{3^3} = \sum_{n=1}^{4} (-1)^{n+1} \frac{1}{4} \left(\frac{1}{3}\right)^{n-1}$

29 $1 - \frac{1}{2} + \frac{1}{4} - \frac{1}{8} + \cdots$ • $a_1 = 1, r = -\frac{1}{2}, S = \frac{a}{1-r} = \frac{1}{1-(-\frac{1}{2})} = \frac{1}{\frac{3}{2}} = \frac{2}{3}$

33 $\sqrt{2} - 2 + \sqrt{8} - 4 + \cdots$ •

The ratio $r = \frac{-2}{\sqrt{2}} = -\sqrt{2}$, but since $|r| = \sqrt{2} > 1$, the sum does not exist.

35 $256 + 192 + 144 + 108 + \cdots$ • $a_1 = 256, r = \frac{192}{256} = \frac{3}{4}, S = \frac{a}{1-r} = \frac{256}{1-\frac{3}{4}} = 1024$

37 $0.\overline{23}$ • If we write $0.\overline{23}$ as $\frac{23}{100} + \frac{23}{10,000} + \cdots$, we see that the first term is $\frac{23}{100}$ and

the common ratio is $\frac{1}{100}$. $a_1 = 0.23, r = 0.01, S = \frac{0.23}{1-0.01} = \frac{23}{99}$

41 $5.\overline{146}$ • In this problem, we work only with $0.\overline{146}$ and add its rational number

representation to 5. $a_1 = 0.146, r = 0.001$,
$$S = \frac{0.146}{1-0.001} = \frac{146}{999}; \; 5.\overline{146} = 5 + \frac{146}{999} = \frac{5141}{999}$$

43 $1.\overline{6124}$ • $a_1 = 0.6124, r = 0.0001, S = \frac{0.6124}{1-0.0001} = \frac{6124}{9999};$
$$1.\overline{6124} = 1 + \frac{6124}{9999} = \frac{16,123}{9999}$$

45 The geometric mean of 12 and 48 is $\sqrt{12 \cdot 48} = \sqrt{576} = 24$.

47 Inserting 2 geometric means results in a sequence that looks like 4, x, y, 500.

Now $4r = x, 4r^2 = y$, and $4r^3 = 500$. Thus, $4r^3 = 500 \Rightarrow r^3 = \frac{500}{4} = 125 \Rightarrow r = 5$.

The terms are 4, 20, 100, and 500.

49 Let $a_1 = x$ { the original amount of air in the container } and $r = \frac{1}{2}$.

$a_{11} = x\left(\frac{1}{2}\right)^{10} = \frac{1}{1024}x$. This is $\left(\frac{1}{1024} \cdot 100\right)\%$ or $\frac{25}{256}\%$ or approximately 0.1% of x.

51 Let $a_1 = 10,000$ and $r = 1.2$, i.e., 120% every hour.

(a) $N(1) = a_2 = 10,000(1.2)^1, N(2) = a_3 = 10,000(1.2)^2, \ldots,$
$$N(t) = a_{t+1} = 10,000(1.2)^t$$

(b) $N(10) = a_{11} = 10,000(1.2)^{10} \approx 61,917.$

53 See Example 7 in the text.

55 Total amount of local spending $= 2,000,000(0.60) + 2,000,000(0.60)^2 + \cdots$. We have

$a_1 = 2,000,000(0.60) = 1,200,000$ and $r = 0.60$, so $S = \frac{1,200,000}{1-0.60} = \$3,000,000.$

$\boxed{57}$ (a) A half-life of 2 hours means there will be $(\frac{1}{2})(\frac{1}{2}D) = \frac{1}{4}D$ after 4 hours. The amount remaining after n doses $\{\text{not hours}\}$ for a given dose is $a_n = D(\frac{1}{4})^{n-1}$. Since D mg are administered every 4 hours, the amount of the drug in the bloodstream after n doses is $\sum_{k=1}^{n} a_k = \sum_{k=1}^{n} D(\frac{1}{4})^{k-1} = D + \frac{1}{4}D + \cdots + (\frac{1}{4})^{n-1}D$. Since $r = \frac{1}{4} < 1$, S_n may be approximated by $S = \frac{a_1}{1-r}$ for large n.

$$S = \frac{D}{1 - \frac{1}{4}} = \frac{4}{3}D.$$

(b) Since the amount of the drug in the bloodstream is given by $\frac{4}{3}D$, and this amount must be less than or equal to 500 mg, we have $\frac{4}{3}D \le 500$ mg, or, equivalently, $D \le 375$ mg.

$\boxed{59}$ (a) From the figure on the right, we see that

$$(\tfrac{1}{4}a_k)^2 + (\tfrac{3}{4}a_k)^2 = (a_{k+1})^2 \Rightarrow \tfrac{10}{16}a_k^2 = a_{k+1}^2 \Rightarrow a_{k+1} = \tfrac{1}{4}\sqrt{10}\,a_k.$$

(b) From part (a), the common ratio $r = \frac{a_{k+1}}{a_k} = \frac{1}{4}\sqrt{10}$, so

$a_n = a_1 r^{n-1} = a_1(\frac{1}{4}\sqrt{10})^{n-1}$. For the square S_{k+1}, the area $A_{k+1} = a_{k+1}^2$. But from part (a), $a_{k+1}^2 = \frac{10}{16}a_k^2$. Since a_k^2 is the area A_k of square S_k, we have $A_{k+1} = \frac{5}{8}A_k$, and hence $A_n = (\frac{5}{8})^{n-1}A_1$. Similarly, for the perimeter of the square S_k, $P_{k+1} = 4a_{k+1} = 4 \cdot \frac{1}{4}\sqrt{10}\,a_k = \sqrt{10}\,a_k = \sqrt{10}(\frac{1}{4}P_k)$, and hence $P_n = (\frac{1}{4}\sqrt{10})^{n-1}P_1$.

(c) $\sum_{n=1}^{\infty} P_n$ is an infinite geometric series with first term P_1 and $r = \frac{1}{4}\sqrt{10}$.

$$S = \frac{P_1}{1 - \frac{1}{4}\sqrt{10}} \cdot \frac{4}{4} = \frac{4P_1}{4 - \sqrt{10}} = \frac{16a_1}{4 - \sqrt{10}}.$$

$\boxed{61}$ (a) The sequence is 1, 3, 9, 27, 81, ... Thus, $a_k = 3^{k-1}$ for $k = 1, 2, 3, \ldots$.

(b) $a_{15} = 3^{14} = 4{,}782{,}969$

(c) The area of the triangle removed first is $\frac{1}{4}$. During the next step, 3 triangles with an area of $\frac{1}{16}$ are removed. Then, 9 triangles with an area of $\frac{1}{64}$ are removed. At each step the number of triangles increase by a factor of 3, while the area of each triangle decreases by a factor of 4. Thus, $b_k = \frac{3^{k-1}}{4^k} = \frac{1}{4}\left(\frac{3}{4}\right)^{k-1}$.

(d) $b_7 = \frac{1}{4}\left(\frac{3}{4}\right)^6 = \frac{729}{16{,}384} \approx 0.0445 = 4.45\%$.

63 Let $a_k = 100\left(1 + \frac{0.06}{12}\right)^k = 100(1.005)^k$, where k represents the number of compounding

periods for each deposit. For the first deposit, $k = 18 \cdot 12 = 216$. For the last

deposit, $k = 1$. $\quad S_{216} = a_1 + a_2 + \cdots + a_{216}$

$$= 100(1.005)^1 + 100(1.005)^2 + \cdots + 100(1.005)^{216}$$

$$= 100(1.005)\left(\frac{1 - (1.005)^{216}}{1 - (1.005)}\right) \approx \$38,929.00$$

67 (a) $A_1 = \frac{2}{5}\left(1 - \frac{2}{5}\right)^{1-1} = \frac{2}{5}\left(\frac{3}{5}\right)^0 = \frac{2}{5}$.

$A_2 = \frac{2}{5}\left(\frac{3}{5}\right)^1 = \frac{6}{25}$, $A_3 = \frac{2}{5}\left(\frac{3}{5}\right)^2 = \frac{18}{125}$, $A_4 = \frac{2}{5}\left(\frac{3}{5}\right)^3 = \frac{54}{625}$, $A_5 = \frac{2}{5}\left(\frac{3}{5}\right)^4 = \frac{162}{3125}$.

(b) $r = A_{k+1}/A_k = \frac{3}{5}$ for $k = 1, 2, 3, 4$.

$$S_5 = \sum_{k=1}^{5} A_k = A_1 + A_2 + A_3 + A_4 + A_5 = \frac{2}{5} \cdot \frac{1 - \left(\frac{3}{5}\right)^5}{1 - \frac{3}{5}} = 1 - \left(\frac{3}{5}\right)^5 = \frac{2882}{3125} = 0.92224.$$

(c) $\$25,000\left(\frac{2}{5} + \frac{6}{25}\right) = \$25,000\left(\frac{16}{25}\right) = \$16,000$

10.4 Exercises

1 (1) P_1 is true, since $2(1) = 1(1+1) = 2$.

(2) Assume P_k is true:

$2 + 4 + 6 + \cdots + 2k = k(k+1)$. Hence,

$2 + 4 + 6 + \cdots + 2k + 2(k+1) = k(k+1) + 2(k+1)$

$$= (k+1)(k+2)$$

$$= (k+1)(k+1+1).$$

Thus, P_{k+1} is true, and the proof is complete.

5 (1) P_1 is true, since $5(1) - 3 = \frac{1}{2}(1)[5(1) - 1] = 2$.

(2) Assume P_k is true:

$2 + 7 + 12 + \cdots + (5k - 3) = \frac{1}{2}k(5k - 1)$. Hence,

$2 + 7 + 12 + \cdots + (5k - 3) + 5(k+1) - 3 = \frac{1}{2}k(5k - 1) + 5(k+1) - 3$

$$= \frac{5}{2}k^2 + \frac{9}{2}k + 2$$

$$= \frac{1}{2}(5k^2 + 9k + 4)$$

$$= \frac{1}{2}(k+1)(5k+4)$$

$$= \frac{1}{2}(k+1)[5(k+1) - 1].$$

Thus, P_{k+1} is true, and the proof is complete.

7 (1) P_1 is true, since $1 \cdot 2^{1-1} = 1 + (1-1) \cdot 2^1 = 1$.

(2) Assume P_k is true:

$1 + 2 \cdot 2 + 3 \cdot 2^2 + \cdots + k \cdot 2^{k-1} = 1 + (k-1) \cdot 2^k$. Hence,

$$1 + 2 \cdot 2 + 3 \cdot 2^2 + \cdots + k \cdot 2^{k-1} + (k+1) \cdot 2^k = 1 + (k-1) \cdot 2^k + (k+1) \cdot 2^k$$
$$= 1 + k \cdot 2^k - 2^k + k \cdot 2^k + 2^k$$
$$= 1 + k \cdot 2^1 \cdot 2^k$$
$$= 1 + [(k+1) - 1] \cdot 2^{k+1}.$$

Thus, P_{k+1} is true, and the proof is complete.

9 (1) P_1 is true, since $(1)^1 = \dfrac{1(1+1)[2(1)+1]}{6} = 1$.

(2) Assume P_k is true:

$1^2 + 2^2 + 3^2 + \cdots + k^2 = \dfrac{k(k+1)(2k+1)}{6}$. Hence,

$$1^2 + 2^2 + 3^2 + \cdots + k^2 + (k+1)^2 = \frac{k(k+1)(2k+1)}{6} + (k+1)^2$$
$$= (k+1)\left[\frac{k(2k+1)}{6} + \frac{6(k+1)}{6}\right]$$
$$= \frac{(k+1)(2k^2 + 7k + 6)}{6}$$
$$= \frac{(k+1)(k+2)(2k+3)}{6}.$$

Thus, P_{k+1} is true, and the proof is complete.

11 (1) P_1 is true, since $\dfrac{1}{1(1+1)} = \dfrac{1}{1+1} = \dfrac{1}{2}$.

(2) Assume P_k is true:

$\dfrac{1}{1 \cdot 2} + \dfrac{1}{2 \cdot 3} + \dfrac{1}{3 \cdot 4} + \cdots + \dfrac{1}{k(k+1)} = \dfrac{k}{k+1}$. Hence,

$$\frac{1}{1 \cdot 2} + \frac{1}{2 \cdot 3} + \frac{1}{3 \cdot 4} + \cdots + \frac{1}{k(k+1)} + \frac{1}{(k+1)(k+2)} = \frac{k}{k+1} + \frac{1}{(k+1)(k+2)}$$
$$= \frac{k}{k+1} + \frac{1}{(k+1)(k+2)}$$
$$= \frac{k(k+2)+1}{(k+1)(k+2)}$$
$$= \frac{k^2 + 2k + 1}{(k+1)(k+2)}$$
$$= \frac{k+1}{(k+1)+1}.$$

Thus, P_{k+1} is true, and the proof is complete.

13 (1) P_1 is true, since $3^1 = \frac{3}{2}(3^1 - 1) = 3$.

    (2) Assume P_k is true:

$$3 + 3^2 + 3^3 + \cdots + 3^k = \tfrac{3}{2}(3^k - 1). \text{ Hence,}$$

$$
\begin{aligned}
3 + 3^2 + 3^3 + \cdots + 3^k + 3^{k+1} &= \tfrac{3}{2}(3^k - 1) + 3^{k+1} \\
&= \tfrac{3}{2} \cdot 3^k - \tfrac{3}{2} + 3 \cdot 3^k \\
&= \tfrac{9}{2} \cdot 3^k - \tfrac{3}{2} \\
&= \tfrac{3}{2}(3 \cdot 3^k - 1) \\
&= \tfrac{3}{2}(3^{k+1} - 1).
\end{aligned}
$$

Thus, P_{k+1} is true, and the proof is complete.

15 (1) P_1 is true, since $1 < 2^1$.

    (2) Assume P_k is true: $k < 2^k$. Now $k + 1 < k + k = 2(k)$ for $k > 1$.

       From P_k, we see that $2(k) < 2(2^k) = 2^{k+1}$ and conclude that $k + 1 < 2^{k+1}$.

Thus, P_{k+1} is true, and the proof is complete.

17 (1) P_1 is true, since $1 < \frac{1}{8}[2(1) + 1]^2 = \frac{9}{8}$.

    (2) Assume P_k is true: $1 + 2 + 3 + \cdots + k < \frac{1}{8}(2k + 1)^2$. Hence,

$$
\begin{aligned}
1 + 2 + 3 + \cdots + k + (k + 1) &< \tfrac{1}{8}(2k + 1)^2 + (k + 1) \\
&= \tfrac{1}{2}k^2 + \tfrac{3}{2}k + \tfrac{9}{8} \\
&= \tfrac{1}{8}(4k^2 + 12k + 9) \\
&= \tfrac{1}{8}(2k + 3)^2 \\
&= \tfrac{1}{8}[2(k + 1) + 1]^2.
\end{aligned}
$$

Thus, P_{k+1} is true, and the proof is complete.

19 (1) For $n = 1$, $n^3 - n + 3 = 3$ and 3 is a factor of 3.

    (2) Assume 3 is a factor of $k^3 - k + 3$. The $(k + 1)$st term is

$$
\begin{aligned}
(k + 1)^3 - (k + 1) + 3 &= k^3 + 3k^2 + 2k + 3 \\
&= (k^3 - k + 3) + 3k^2 + 3k \\
&= (k^3 - k + 3) + 3(k^2 + k).
\end{aligned}
$$

By the induction hypothesis, 3 is a factor of $k^3 - k + 3$ and 3 is a factor of $3(k^2 + k)$, so 3 is a factor of the $(k + 1)$st term. Thus, P_{k+1} is true, and the proof is complete.

$\boxed{21}$ (1) For $n = 1$, $5^n - 1 = 4$ and 4 is a factor of 4.

   (2) Assume 4 is a factor of $5^k - 1$. The $(k+1)$st term is

   $$5^{k+1} - 1 = 5 \cdot 5^k - 1$$
   $$= 5 \cdot 5^k - 5 + 4$$
   $$= 5(5^k - 1) + 4.$$

   By the induction hypothesis, 4 is a factor of $5^k - 1$ and 4 is a factor of 4, so 4 is a factor of the $(k+1)$st term. Thus, P_{k+1} is true, and the proof is complete.

$\boxed{23}$ (1) If $a > 1$, then $a^1 = a > 1$, so P_1 is true.

   (2) Assume P_k is true: $a^k > 1$.

   Multiply both sides by a to obtain $a^{k+1} > a$, but since $a > 1$, we have $a^{k+1} > 1$.

   Thus, P_{k+1} is true, and the proof is complete.

$\boxed{25}$ (1) For $n = 1$, $a - b$ is a factor of $a^1 - b^1$.

   (2) Assume $a - b$ is a factor of $a^k - b^k$. Following the hint for the $(k+1)$st term, $a^{k+1} - b^{k+1} = a^k \cdot a - b \cdot a^k + b \cdot a^k - b^k \cdot b = a^k(a - b) + (a^k - b^k)b$. Since $(a - b)$ is a factor of $a^k(a - b)$ and since by the induction hypothesis $a - b$ is a factor of $(a^k - b^k)$, it follows that $a - b$ is a factor of the $(k+1)$st term. Thus, P_{k+1} is true, and the proof is complete.

Note: For Exercises 27–32 in this section and Exercises 47–48 in the Chapter Review, there are several ways to find j. Possibilities include: solve the inequality, sketch the graphs of functions representing each side, and trial and error. Trial and error may be the easiest to use.

$\boxed{27}$ For j: $n^2 \geq n + 12 \Rightarrow n^2 - n - 12 \geq 0 \Rightarrow (n - 4)(n + 3) \geq 0 \Rightarrow n \geq 4 \; \{n > 0\}$

   (1) P_4 is true, since $4 + 12 \leq 4^2$.

   (2) Assume P_k is true: $k + 12 \leq k^2$. Hence,

   $$(k + 1) + 12 = (k + 12) + 1 \leq (k^2) + 1 < k^2 + 2k + 1 = (k + 1)^2.$$

   Thus, P_{k+1} is true, and the proof is complete.

$\boxed{29}$ For j: By sketching $y = 5 + \log_2 x$ and $y = x$, we see that the solution for $x > 1$

must be larger than 5. By trial and error, $j = 8$.

(1) P_8 is true, since $5 + \log_2 8 \leq 8$.

(2) Assume P_k is true: $5 + \log_2 k \leq k$. Hence,

$$5 + \log_2 (k + 1) < 5 + \log_2 (k + k)$$
$$= 5 + \log_2 2k$$
$$= 5 + \log_2 2 + \log_2 k$$
$$= (5 + \log_2 k) + 1$$
$$\leq k + 1.$$

Thus, P_{k+1} is true, and the proof is complete.

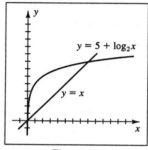

Figure 29

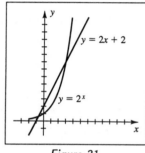

Figure 31

$\boxed{31}$ For j: By sketching $y = 2x + 2$ and $y = 2^x$, we see there is one positive solution.

By trial and error, $j = 3$.

(1) P_3 is true, since $2(3) + 2 \leq 2^3$.

(2) Assume P_k is true: $2k + 2 \leq 2^k$. Hence,

$$2(k + 1) + 2 = (2k + 2) + 2 \leq 2^k + 2^k = 2 \cdot 2^k = 2^{k+1}.$$

Thus, P_{k+1} is true, and the proof is complete.

$\boxed{33}$ Following the hint in the text:

$$\sum_{k=1}^{n} (k^2 + 3k + 5) = \sum_{k=1}^{n} k^2 + 3 \sum_{k=1}^{n} k + \sum_{k=1}^{n} 5$$
$$= \frac{n(n+1)(2n+1)}{6} + 3\left[\frac{n(n+1)}{2}\right] + 5n$$
$$= \frac{n(n+1)(2n+1) + 9n(n+1) + 30n}{6}$$
$$= \frac{2n^3 + 12n^2 + 40n}{6}$$
$$= \frac{n^3 + 6n^2 + 20n}{3}$$

$\boxed{35}$ $\displaystyle\sum_{k=1}^{n} (2k-3)^2 = \sum_{k=1}^{n} (4k^2 - 12k + 9) = 4\sum_{k=1}^{n} k^2 - 12\sum_{k=1}^{n} k + \sum_{k=1}^{n} 9$

$$= 4\left[\frac{n(n+1)(2n+1)}{6}\right] - 12\left[\frac{n(n+1)}{2}\right] + 9n = \frac{4n^3 - 12n^2 + 11n}{3}$$

$\boxed{37}$ (a) $n = 1 \Rightarrow a(1)^3 + b(1)^2 + c(1) = 1^2 \Rightarrow a + b + c = 1$

$n = 2 \Rightarrow a(2)^3 + b(2)^2 + c(2) = 1^2 + 2^2 \Rightarrow 8a + 4b + 2c = 5$

$n = 3 \Rightarrow a(3)^3 + b(3)^2 + c(3) = 1^2 + 2^2 + 3^2 \Rightarrow 27a + 9b + 3c = 14$

$$AX = B \Rightarrow \begin{bmatrix} 1 & 1 & 1 \\ 8 & 4 & 2 \\ 27 & 9 & 3 \end{bmatrix} \begin{bmatrix} a \\ b \\ c \end{bmatrix} = \begin{bmatrix} 1 \\ 5 \\ 14 \end{bmatrix} \Rightarrow X = A^{-1}B = \begin{bmatrix} 1/3 \\ 1/2 \\ 1/6 \end{bmatrix}.$$

(b) $a = \frac{1}{3}$, $b = \frac{1}{2}$, $c = \frac{1}{6} \Rightarrow 1^2 + 2^2 + 3^2 + \cdots + n^2 = \frac{1}{3}n^3 + \frac{1}{2}n^2 + \frac{1}{6}n = \dfrac{n(n+1)(2n+1)}{6}$,

which is the formula found in Exercise 9. This method does not verify the

formula for all n but only for $n = 1, 2, 3$. Mathematical induction should be

used to verify the formula for all n as in Exercise 9.

$\boxed{41}$ (1) For $n = 1$, $\left[r\left(\cos\theta + i\sin\theta\right)\right]^1 = r^1\left[\cos\left(1\theta\right) + i\sin\left(1\theta\right)\right]$.

(2) Assume P_k is true: $\left[r\left(\cos\theta + i\sin\theta\right)\right]^k = r^k\left(\cos k\theta + i\sin k\theta\right)$. Hence,

$$\left[r\left(\cos\theta + i\sin\theta\right)\right]^{k+1} = \left[r\left(\cos\theta + i\sin\theta\right)\right]^k\left[r\left(\cos\theta + i\sin\theta\right)\right]$$

$$= r^k\left[\cos k\theta + i\sin k\theta\right]\left[r\left(\cos\theta + i\sin\theta\right)\right]$$

$$= r^{k+1}\left[(\cos k\theta\cos\theta - \sin k\theta\sin\theta)\right.$$
$$\left. + i\left(\sin k\theta\cos\theta + \cos k\theta\sin\theta\right)\right]$$

{ Use the addition formulas for the sine and cosine. }

$$= r^{k+1}\left[\cos\left(k+1\right)\theta + i\sin\left(k+1\right)\theta\right].$$

Thus, P_{k+1} is true, and the proof is complete.

10.5 Exercises

$\boxed{1}$ $2!6! = (2 \cdot 1) \cdot (6 \cdot 5 \cdot 4 \cdot 3 \cdot 2 \cdot 1) = 2 \cdot 720 = 1440$

$\boxed{3}$ $7!0! = (7 \cdot 6 \cdot 5 \cdot 4 \cdot 3 \cdot 2 \cdot 1) \cdot (1)$ { Remember that $0! = 1$. } $= 5040$

$\boxed{5}$ $\frac{8!}{5!} = \frac{8 \cdot 7 \cdot 6 \cdot 5!}{5!}$ { cancel $5!$ } $= 8 \cdot 7 \cdot 6 = 336$ $\boxed{9}$ $\binom{7}{5} = \frac{7!}{5!\,2!}$ { cancel $5!$ } $= \frac{7 \cdot 6}{2} = 21$

$\boxed{11}$ $\binom{13}{4} = \frac{13!}{4!\,9!}$ { cancel $9!$ } $= \frac{13 \cdot 12 \cdot 11 \cdot 10}{4 \cdot 3 \cdot 2} = 715$

$\boxed{13}$ $\frac{(2n+2)!}{(2n)!} = \frac{(2n+2)(2n+1)(2n)!}{(2n)!}$ { cancel $(2n)!$ } $= (2n+2)(2n+1)$

$\boxed{15}$ We use the binomial theorem formula with $a = 4x$, $b = -y$, and $n = 3$.

$$(4x - y)^3 = \binom{3}{0}(4x)^3(-y)^0 + \binom{3}{1}(4x)^2(-y)^1 + \binom{3}{2}(4x)^1(-y)^2 + \binom{3}{3}(4x)^0(-y)^3$$

$$= (1)(64x^3)(1) + (3)(16x^2)(-y) + (3)(4x)(y^2) + (1)(1)(-y^3)$$

$$= 64x^3 - 48x^2y + 12xy^2 - y^3$$

17 $(a+b)^6 = a^6 + \binom{6}{1}a^5b^1 + \binom{6}{2}a^4b^2 + \binom{6}{3}a^3b^3 + \binom{6}{4}a^2b^4 + \binom{6}{5}a^1b^5 + b^6$

$$= a^6 + 6a^5b + 15a^4b^2 + 20a^3b^3 + 15a^2b^4 + 6ab^5 + b^6$$

21 We use the binomial theorem formula with $a = 3x$, $b = -5y$, and $n = 4$.

$$(3x - 5y)^4 = \binom{4}{0}(3x)^4(-5y)^0 + \binom{4}{1}(3x)^3(-5y)^1 + \binom{4}{2}(3x)^2(-5y)^2 +$$

$$\binom{4}{3}(3x)^1(-5y)^3 + \binom{4}{4}(3x)^0(-5y)^4$$

$$= (1)(81x^4)(1) + (4)(27x^3)(-5y) + (6)(9x^2)(25y^2) +$$

$$(4)(3x)(-125y^3) + (1)(1)(625y^4)$$

$$= 81x^4 - 540x^3y + 1350x^2y^2 - 1500xy^3 + 625y^4$$

25 We use the binomial theorem formula with $a = x^{-2}$, $b = 3x$, and $n = 6$.

$$\left(\frac{1}{x^2} + 3x\right)^6 = (x^{-2} + 3x)^6$$

$$= \binom{6}{0}(x^{-2})^6(3x)^0 + \binom{6}{1}(x^{-2})^5(3x)^1 + \binom{6}{2}(x^{-2})^4(3x)^2 + \binom{6}{3}(x^{-2})^3(3x)^3 +$$

$$\binom{6}{4}(x^{-2})^2(3x)^4 + \binom{6}{5}(x^{-2})^1(3x)^5 + \binom{6}{6}(x^{-2})^0(3x)^6$$

$$= (1)(x^{-12})(1) + (6)(x^{-10})(3x^1) + (15)(x^{-8})(9x^2) + (20)(x^{-6})(27x^3) +$$

$$(15)(x^{-4})(81x^4) + (6)(x^{-2})(243x^5) + (1)(1)(729x^6)$$

$$= x^{-12} + 18x^{-9} + 135x^{-6} + 540x^{-3} + 1215 + 1458x^3 + 729x^6$$

27 We use the binomial theorem formula with $a = x^{1/2}$, $b = -x^{-1/2}$, and $n = 5$.

$$\left(\sqrt{x} - \frac{1}{\sqrt{x}}\right)^5 = (x^{1/2} - x^{-1/2})^5$$

$$= \binom{5}{0}(x^{1/2})^5(-x^{-1/2})^0 + \binom{5}{1}(x^{1/2})^4(-x^{-1/2})^1 + \binom{5}{2}(x^{1/2})^3(-x^{-1/2})^2 +$$

$$\binom{5}{3}(x^{1/2})^2(-x^{-1/2})^3 + \binom{5}{4}(x^{1/2})^1(-x^{-1/2})^4 + \binom{5}{5}(x^{1/2})^0(-x^{-1/2})^5$$

$$= (1)(x^{5/2})(1) + (5)(x^2)(-x^{-1/2}) + (10)(x^{3/2})(x^{-1}) +$$

$$(10)(x^1)(-x^{-3/2}) + (5)(x^{1/2})(x^{-2}) + (1)(1)(-x^{-5/2})$$

$$= x^{5/2} - 5x^{3/2} + 10x^{1/2} - 10x^{-1/2} + 5x^{-3/2} - x^{-5/2}$$

29 For the binomial expression $(3c^{2/5} + c^{4/5})^{25}$, the first three terms are

$$\sum_{k=0}^{2} \binom{25}{k}(3c^{2/5})^{25-k}(c^{4/5})^k$$

$$= \binom{25}{0}(3c^{2/5})^{25}(c^{4/5})^0 + \binom{25}{1}(3c^{2/5})^{24}(c^{4/5})^1 + \binom{25}{2}(3c^{2/5})^{23}(c^{4/5})^2$$

$$= (1)(3^{25}c^{10})(1) + (25)(3^{24}c^{48/5})(c^{4/5}) + (300)(3^{23}c^{46/5})(c^{8/5})$$

$$= 3^{25}c^{10} + 25 \cdot 3^{24}c^{52/5} + 300 \cdot 3^{23}c^{54/5}$$

$\boxed{31}$ For the binomial expression $(4b^{-1} - 3b)^{15}$, the last three terms are

$$\sum_{k=13}^{15} \binom{15}{k}(4b^{-1})^{15-k}(-3b)^k$$

$$= \binom{15}{13}(4b^{-1})^2(-3b)^{13} + \binom{15}{14}(4b^{-1})^1(-3b)^{14} + \binom{15}{15}(4b^{-1})^0(-3b)^{15}$$

$$= (105)(16b^{-2})(-3^{13}b^{13}) + (15)(4b^{-1})(3^{14}b^{14}) + (1)(1)(-3^{15}b^{15})$$

$$= -1680 \cdot 3^{13}b^{11} + 60 \cdot 3^{14}b^{13} - 3^{15}b^{15}$$

Note: For the following exercises, the general formula for the

$$\underline{k\text{th term of } (a+b)^n} \text{ is } \binom{n}{k-1}(a)^{n-(k-1)}(b)^{k-1} = \boxed{\binom{n}{k-1}(a)^{n-k+1}(b)^{k-1}}.$$

There is also a formula for the $(k+1)$st term of an expansion on page 735

following the statement of the binomial theorem.

$\boxed{33}$ $\left(\dfrac{3}{c} + \dfrac{c^2}{4}\right)^7$; sixth term $= \binom{7}{5}\left(\dfrac{3}{c}\right)^2\left(\dfrac{c^2}{4}\right)^5 = 21\left(\dfrac{9}{c^2}\right)\left(\dfrac{c^{10}}{1024}\right) = \dfrac{189}{1024}c^8$

$\boxed{37}$ Since there are 9 terms, the middle term is the fifth term $\left\{\dfrac{9+1}{2} = 5\right\}$.

Using the formula in the *Note* above, we obtain the 5th term of $(x^{1/2} + y^{1/2})^8$,

$$\binom{8}{4}(x^{1/2})^4(y^{1/2})^4 = 70x^2y^2.$$

$\boxed{39}$ $(2y + x^2)^8$; term that contains x^{10} •

Consider only the variable x in the expansion: $(x^2)^{k-1} = x^{10} \Rightarrow 2k - 2 = 10 \Rightarrow k = 6$;

$$6\text{th term} = \binom{8}{5}(2y)^3 (x^2)^5 = 448y^3x^{10}$$

$\boxed{41}$ $(3b^3 - 2a^2)^4$; term that contains b^9 •

Consider only the variable b in the expansion:

$$(b^3)^{4-k+1} = b^9 \Rightarrow 15 - 3k = 9 \Rightarrow k = 2; \text{ 2nd term} = \binom{4}{1}(3b^3)^3(-2a^2)^1 = -216b^9a^2$$

$\boxed{43}$ $\left(3x - \dfrac{1}{4x}\right)^6$; term that does not contain x •

Consider only the variable x in the expansion:

$$x^{6-k+1}(x^{-1})^{k-1} = x^0 \Rightarrow x^{8-2k} = x^0 \Rightarrow k = 4; \text{ 4th term} = \binom{6}{3}(3x)^3\left(-\dfrac{1}{4x}\right)^3 = -\dfrac{135}{16}$$

$\boxed{45}$ The first three terms in the expansion of $(1 + 0.2)^{10}$ are

$$\sum_{k=0}^{2} \binom{10}{k}(1)^{10-k}(0.2)^k = \binom{10}{0}(1)^{10}(0.2)^0 + \binom{10}{1}(1)^9(0.2)^1 + \binom{10}{2}(1)^8(0.2)^2$$

$$= (1)(1)(1) + (10)(1)(0.2) + (45)(1)(0.04) = 1 + 2 + 1.8 = 4.8.$$

The calculator result for $(1.2)^{10}$ is approximately 6.19.

$\boxed{47}$ $\dfrac{(x+h)^4 - x^4}{h} = \dfrac{(x^4 + 4x^3h + 6x^2h^2 + 4xh^3 + h^4) - x^4}{h} = \dfrac{h(4x^3 + 6x^2h + 4xh^2 + h^3)}{h} =$

$$4x^3 + 6x^2h + 4xh^2 + h^3$$

49 $\binom{n}{1} = \dfrac{n!}{(n-1)!\,1!} = n$ and $\binom{n}{n-1} = \dfrac{n!}{[n-(n-1)]!\,(n-1)!} = \dfrac{n!}{1!\,(n-1)!} = n$

10.6 Exercises

1 $P(7,\,3) = \dfrac{7!}{4!} = \dfrac{7\cdot 6\, \vdots\, 5\cdot 4!}{4!} = 7\cdot 6\cdot 5 = 210$

5 $P(5,\,5) = \dfrac{5!}{0!} = \dfrac{5\cdot 4\cdot 3\cdot 2\cdot 1}{1} = 120$

7 $P(6,\,1) = \dfrac{6!}{5!} = \dfrac{6\cdot 5!}{5!} = 6$

9 (a) We can think of the three-digit numbers as filling 3 slots. There are 5 digits to pick from for filling the first slot. There are 4 remaining digits to pick from to fill the second slot. There are 3 remaining digits to pick from to fill the third slot. By the fundamental counting principle, there are a total of $5\cdot 4\cdot 3 = 60$ three-digit numbers.

    (b) The difference between parts (a) and (b) is that we can use any of the 5 digits for all 3 slots. Hence, there are a total of $5\cdot 5\cdot 5 = 125$ three-digit numbers if repetitions are allowed.

11 There are 4 one digit numbers; $4\cdot 3 = 12$ two digit numbers;

$4\cdot 3\cdot 2 = 24$ three digit numbers; $4\cdot 3\cdot 2\cdot 1 = 24$ four digit numbers.

Total is $4 + 12 + 24 + 24 = 64$.

17 (a) By the fundamental counting principle, $26\cdot 9\cdot 10^4 = 2{,}340{,}000$.

    (b) By the fundamental counting principle, $24\cdot 9\cdot 10^4 = 2{,}160{,}000$.

19 (a) $P(10,\,6) = \dfrac{10!}{4!} = 10\cdot 9\cdot 8\cdot 7\cdot 6\cdot 5 = 151{,}200$

    (b) Boy-girl: $6\cdot 4\cdot 5\cdot 3\cdot 4\cdot 2 = 2880$. Girl-boy: $4\cdot 6\cdot 3\cdot 5\cdot 2\cdot 4 = 2880$.

Total $= 2880 + 2880 = 5760$

21 There are 2 choices for each of the 10 questions. 2 times itself 10 times $= 2^{10} = 1024$

27 (a) The number of choices for each letter are: $\underline{2}\,\cdot\,\underline{25}\,\cdot\,\underline{24}\,\cdot\,\underline{23} = 27{,}600$

    (b) The number of choices for each letter are: $\underline{2}\,\cdot\,\underline{26}\,\cdot\,\underline{26}\,\cdot\,\underline{26} = 35{,}152$

33 There are 3! ways to choose the couples and 2 ways for each couple to sit. $3!\cdot 2^3 = 48$

35 $P(10,\,10) = \dfrac{10!}{0!} = 10! = 3{,}628{,}800$

37 (a) There are 9 choices for the first digit, 10 for the second, 10 for the third,

and 1 for the fourth and fifth. $9\cdot 10\cdot 10\cdot 1\cdot 1 = 900$

    (b) If n is even, we need to select the first $\frac{n}{2}$ digits. $9\cdot 10^{(n/2)-1}$

        If n is odd, we need to select the first $\frac{n+1}{2}$ digits. $9\cdot 10^{(n-1)/2}$

39 (a) There is a horizontal asymptote of $y = 1$.

(b) $\dfrac{n!\,e^n}{n^n\,\sqrt{2\pi n}} \approx 1 \Rightarrow n! \approx \dfrac{n^n\,\sqrt{2\pi n}}{e^n}$.

Example: $50! \approx \dfrac{50^{50}\,\sqrt{2\pi(50)}}{e^{50}} \approx 3.0363 \times 10^{64}$.

The actual value is closer to 3.0414×10^{64}.

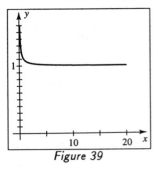

Figure 39

10.7 Exercises

1 $C(7,\,3) = \dfrac{7!}{4!\,3!} = \dfrac{7\cdot 6\cdot 5\cdot 4!}{4!\,3!} = \dfrac{7\cdot 6\cdot 5}{3!} = \dfrac{7\cdot 6\cdot 5}{6} = 7\cdot 5 = 35$

5 $C(n,\,n-1) = \dfrac{n!}{[n-(n-1)]!\,(n-1)!} = \dfrac{n!}{1!\,(n-1)!} = n$

7 $C(7,\,0) = \dfrac{7!}{7!\,0!} = \dfrac{1}{0!} = \dfrac{1}{1} = 1$

9 There are $12!$ total permutations. We want the number of *distinguishable*

permutations of the 12 disks. Using the theorem on distinguishable permutations,

$$\text{we have } \dfrac{(5+3+2+2)!}{5!\,3!\,2!\,2!} = \dfrac{12!}{5!\,3!\,2!\,2!} = 166{,}320.$$

13 There are $C(10,\,5)$ ways to pick the first team.

The second team is determined once the first team is selected. $C(10,\,5) = 252$

15 Two points determine a unique line. $C(8,\,2) = 28$

17 There are $3!$ ways to order the categories. $(5!\cdot 4!\cdot 8!)\cdot 3! = 696{,}729{,}600$

19 Pick the center, $C(3,\,1)$; two guards, $C(10,\,2)$;

two tackles from the 8 remaining linemen, $C(8,\,2)$; two ends, $C(4,\,2)$;

two halfbacks, $C(6,\,2)$; the quarterback, $C(3,\,1)$; and the fullback, $C(4,\,1)$.

$$3\cdot C(10,\,2)\cdot C(8,\,2)\cdot C(4,\,2)\cdot C(6,\,2)\cdot 3\cdot 4 = 4{,}082{,}400$$

21 There are $C(12,\,3) = 220$ ways to pick the men and $C(8,\,2) = 28$ ways to pick the

women. By the fundamental counting principle, the total number of ways to pick the

committee is $220\cdot 28 = 6160$.

23 We need 3 U's out of 8 moves. $C(8,\,3) = 56$

27 Let n denote the number of players. $C(n,\,2) = 45 \Rightarrow \dfrac{n!}{(n-2)!\,2!} = 45 \Rightarrow$

$$n(n-1) = 90 \Rightarrow (n-10)(n+9) = 0 \Rightarrow \{\,n > 0\,\}\ n = 10.$$

29 Each team must win 3 of the first 6 games for the series to be extended to a

7th game. $C(6,\,3) = 20$

31 They may have computed $C(31, 3)$, which is 4495.

33 (a) $S_1 = \binom{1}{1} + \binom{1}{3} + \binom{1}{5} + \cdots = 1 + 0 + 0 + \cdots = 1.$

$S_2 = \binom{2}{1} + \binom{2}{3} + \binom{2}{5} + \cdots = 2 + 0 + 0 + \cdots = 2.$

$S_3 = 3 + 1 + 0 + \cdots = 4.$ $S_4 = 4 + 4 + 0 + \cdots = 8.$

$S_5 = 16,\ S_6 = 32,\ S_7 = 64,\ S_8 = 128,\ S_9 = 256,\ S_{10} = 512.$

(b) It appears that $S_n = 2^{n-1}.$

35 (a) Graph the values of $C(10, 1),\ C(10, 2),\ C(10, 3),\ \dots,\ C(10, 10).$

(b) The maximum value of $C(10, r)$ is 252 and occurs at $r = 5.$

[0, 10] by [0, 300]   [0, 19] by [0, 1×10^5]

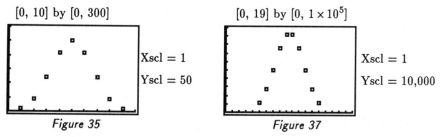

Xscl = 1   Xscl = 1

Yscl = 50   Yscl = 10,000

Figure 35   *Figure 37*

37 (a) Graph the values of $C(19, 1),\ C(19, 2),\ C(19, 3),\ \dots,\ C(19, 19).$

(b) The maximum value of $C(19, r)$ is 92,378 and occurs at $r = 9, 10.$

10.8 Exercises

1 The sample space is the 52-card deck, thus, $n(S) = 52.$ Using $p(E) = \dfrac{n(E)}{n(S)},$

we have (a) $\frac{4}{52} = \frac{1}{13}$ (b) $\frac{4}{52} + \frac{4}{52} = \frac{8}{52} = \frac{2}{13}$ (c) $\frac{4}{52} + \frac{4}{52} + \frac{4}{52} = \frac{12}{52} = \frac{3}{13}.$

3 $S = \{1, 2, 3, 4, 5, 6\}$ and $n(S) = 6.$ (a) $\frac{1}{6}$ (b) $\frac{1}{6}$ (c) $\frac{1}{6} + \frac{1}{6} = \frac{2}{6} = \frac{1}{3}$

5 $n(S) = 5 + 6 + 4 = 15$ (a) $\frac{5}{15} = \frac{1}{3}$ (b) $\frac{6}{15} = \frac{2}{5}$ (c) $\frac{5}{15} + \frac{4}{15} = \frac{9}{15} = \frac{3}{5}$

7 *Note:* The table gives the results for the sum of two dice being tossed.

Sum of two dice	2	3	4	5	6	7	8	9	10	11	12
# of ways to obtain	1	2	3	4	5	6	5	4	3	2	1

(a) $\frac{2}{36} = \frac{1}{18}$ (b) $\frac{5}{36}$ (c) $\frac{2}{36} + \frac{5}{36} = \frac{7}{36}$

9 $n(S) = 6 \times 6 \times 6 = 216.$

There are 6 ways to make a sum of 5 (3 with 1, 1, 3 and 3 with 1, 2, 2). $\frac{6}{216} = \frac{1}{36}$

Note: For Exercises 13–18, there are $C(52, 5) = 2,598,960$ ways to draw 5 cards.

13 There are 13 denominations to pick from and any one of them could be combined

with any one of the remaining 48 cards. $\dfrac{48 \cdot 13}{C(52, 5)} = \dfrac{1}{4165} \approx 0.00024$

15 Pick 4 of the 13 diamonds and 1 of the 13 spades.

$$\frac{C(13,\,4)\cdot C(13,\,1)}{C(52,\,5)}=\frac{143}{39{,}984}\approx 0.00358$$

17 Pick 5 of the 13 cards in one suit. There are 4 suits. $\dfrac{C(13,\,5)\cdot 4}{C(52,\,5)}=\dfrac{33}{16{,}660}\approx 0.00198$

19 Let E_1 be the event that the number is odd, E_2 that the number is prime.

$E_1=\{\,1,\,3,\,5\,\}$ and $E_2=\{\,2,\,3,\,5\,\}$.

$$P(E_1\cup E_2)=P(E_1)+P(E_2)-P(E_1\cap E_2)=\tfrac{3}{6}+\tfrac{3}{6}-\tfrac{2}{6}=\tfrac{4}{6}=\tfrac{2}{3}.$$

21 The probability that the player does *not* get a hit is $1-0.326=0.674$. The probability that the player does not get a hit four times in a row is $(0.674)\cdot(0.674)\cdot(0.674)\cdot(0.674)=(0.674)^4\approx 0.2064$.

23 (a) $P(E_2)=P(2)+P(3)+P(4)=0.10+0.15+0.20=0.45$

(b) $P(E_1\cap E_2)=P(2)=0.10$

(c) $P(E_1\cup E_2)=P(E_1)+P(E_2)-P(E_1\cap E_2)=0.35+0.45-0.10=0.70$

(d) $P(E_2\cup E_3')=P(E_2)+P(E_3')-P(E_2\cap E_3')$. $E_3'=\{\,1,\,2,\,3,\,5\,\}$ and

$$E_2\cap E_3'=\{\,2,\,3\,\}\Rightarrow P(E_2\cup E_3')=0.45+0.75-0.25=0.95.$$

Note: For Exercises 25–26, there are $C(60,\,5)=5{,}461{,}512$ ways to draw 5 chips.

25 (a) We want 5 blue and 0 non-blue. There are 20 blue and 40 non-blue chips in the

box. $\dfrac{C(20,\,5)\cdot C(40,\,0)}{C(60,\,5)}=\dfrac{34}{11{,}977}\approx 0.0028.$

(b) $P(\text{at least 1 green})=1-P(\text{no green})=$

$$1-\frac{C(30,\,0)\cdot C(30,\,5)}{C(60,\,5)}=1-\frac{117}{4484}=\frac{4367}{4484}\approx 0.9739.$$

(c) $P(\text{at most 1 red})=P(0\text{ red})+P(1\text{ red})=$

$$\frac{C(10,\,0)\cdot C(50,\,5)}{C(60,\,5)}+\frac{C(10,\,1)\cdot C(50,\,4)}{C(60,\,5)}=\frac{26{,}320}{32{,}509}\approx 0.8096.$$

27 (a) $\dfrac{C(8,\,8)}{2^8}=\dfrac{1}{256}\approx 0.00391$ (b) $\dfrac{C(8,\,7)}{2^8}=\dfrac{1}{32}=0.03125$

(c) $\dfrac{C(8,\,6)}{2^8}=\dfrac{7}{64}=0.109375$ (d) $\dfrac{C(8,\,6)+C(8,\,7)+C(8,\,8)}{2^8}=\dfrac{37}{256}\approx 0.14453$

29 $P(\text{obtaining at least one ace})=1-P(\text{no aces})=$

$$1-\frac{C(48,\,5)}{C(52,\,5)}=1-\frac{35{,}673}{54{,}145}=\frac{18{,}472}{54{,}145}\approx 0.34116$$

31 (a) We may use ordered pairs to represent the outcomes of the sample space S of the experiment. A representative outcome is (nine of clubs, 3). The number of outcomes in the sample space S is $n(S)=52\cdot 6=312$.

(b) For each integer k, where $2 \leq k \leq 6$, there are 4 ways to obtain an outcome of the form (k, k) since there are 4 suits. Because there are 5 values of k, $n(E_1) = 5 \cdot 4 = 20$. $n(E_1') = n(S) - n(E_1) = 312 - 20 = 292$.

$$P(E_1) = \frac{n(E_1)}{n(S)} = \frac{20}{312} = \frac{5}{78}.$$

(c) No, if E_2 or E_3 occurs, then the other event may occur.

Yes, the occurrence of either E_2 or E_3 has no effect on the other event.

$$P(E_2) = \frac{n(E_2)}{n(S)} = \frac{12 \cdot 6}{312} = \frac{72}{312} = \frac{3}{13}. \qquad P(E_3) = \frac{n(E_3)}{n(S)} = \frac{52 \cdot 3}{312} = \frac{156}{312} = \frac{1}{2}.$$

Since E_2 and E_3 are indep., $P(E_2 \cap E_3) = P(E_2) \cdot P(E_3) = \frac{3}{13} \cdot \frac{1}{2} = \frac{3}{26} = \frac{36}{312}.$

$P(E_2 \cup E_3) = P(E_2) + P(E_3) - P(E_2 \cap E_3) = \frac{72}{312} + \frac{156}{312} - \frac{36}{312} = \frac{192}{312} = \frac{8}{13}.$

(d) Yes, if E_1 or E_2 occurs, then the other event cannot occur. No, the occurrence of either E_1 or E_2 influences the occurrence of the other event.

Remember, (non-empty) *mutually exclusive events* **cannot** *be independent events.*

Since E_1 and E_2 are mutually exclusive, $P(E_1 \cap E_2) = 0$ and

$$P(E_1 \cup E_2) = P(E_1) + P(E_2) = \frac{20}{312} + \frac{72}{312} = \frac{92}{312} = \frac{23}{78}.$$

33 Let k denote the sum.

$$P(k > 5) = 1 - P(k \leq 5) = 1 - \left(\frac{1}{36} + \frac{2}{36} + \frac{3}{36} + \frac{4}{36}\right) = 1 - \frac{10}{36} = \frac{26}{36} = \frac{13}{18}.$$

37 (a) $\dfrac{C(4, 4)}{4!} = \dfrac{1}{24} \approx 0.04167$ (b) $\dfrac{C(4, 2)}{4!} = \dfrac{1}{4} = 0.25$

39 (a) The 3, 4, or 5 on the left die would need to combine with a 4, 3, or 2 on the right die to sum to 7, but the right die only has 1, 5, or 6. The probability is 0.

(b) To obtain 8, we would need a 3 on the left die and a 5 on the right.

$$\tfrac{1}{3} \cdot \tfrac{1}{3} = \tfrac{1}{9} = 0.\overline{1}$$

43 (a) The ball must take 4 "lefts". $\frac{1}{2} \cdot \frac{1}{2} \cdot \frac{1}{2} \cdot \frac{1}{2} = \frac{1}{16} = 0.0625$

(b) We need two "lefts". $\dfrac{C(4, 2)}{2^4} = \dfrac{6}{16} = \dfrac{3}{8} = 0.375$

45 For one ticket, $P(E) = \dfrac{n(E)}{n(S)} = \dfrac{C(6, 6)}{C(54, 6)} = \dfrac{1}{25{,}827{,}165}.$

For two tickets, $P(E) = \dfrac{2 \times 1}{25{,}827{,}165}$, or about 1 chance in 13 million.

47 The probability that the first bulb is not defective is $\frac{195}{200}$ since 195 of the 200 bulbs are not defective. If the first bulb is not replaced and not defective, then there are 199 bulbs left, and 194 of them are not defective. The probability that the second bulb is not defective is then $\frac{194}{199}$. Thus, the probability that both bulbs are not defective is $\frac{195}{200} \times \frac{194}{199} = \frac{37{,}830}{39{,}800} \approx 0.9505$. The event that either light bulb is defective is the complement of the event that neither bulb is defective. The probability that the sample will be rejected is $1 - \frac{37{,}830}{39{,}800} = \frac{1970}{39{,}800} \approx 0.0495$.

49 (a) $P(7 \text{ or } 11) = P(7) + P(11) = \frac{6}{36} + \frac{2}{36} = \frac{8}{36}$

(b) To win with a 4 on the first roll, we must first get a 4, and then get another 4 before a 7. The probability of getting a 4 is $\frac{3}{36}$. The probability of getting another 4 before a 7 is $\frac{3}{3+6}$ since there are 3 ways to get a 4, 6 ways to get a 7, and numbers other than 4 and 7 are immaterial.

$$\text{Thus, } P(\text{winning with 4}) = \frac{3}{36} \cdot \frac{3}{3+6} = \frac{1}{36}.$$

(c) Let $P(k)$ denote the probability of winning a pass line bet with the number k.

We first note that $P(4) = P(10)$, $P(5) = P(9)$, and $P(6) = P(8)$.

$P(\text{winning}) = 2 \cdot P(4) + 2 \cdot P(5) + 2 \cdot P(6) + P(7) + P(11)$

$\qquad = 2 \cdot \frac{3}{36} \cdot \frac{3}{3+6} + 2 \cdot \frac{4}{36} \cdot \frac{4}{4+6} + 2 \cdot \frac{5}{36} \cdot \frac{5}{5+6} + \frac{6}{36} + \frac{2}{36}$

$\qquad = 2 \cdot \frac{1}{36} + 2 \cdot \frac{2}{45} + 2 \cdot \frac{25}{396} + \frac{1}{6} + \frac{1}{18} = \frac{488}{990} = \frac{244}{495} \approx 0.4929$

51 (a) $p = P((S_1 \cap S_2) \cup (S_3 \cap S_4))$

$\qquad = P(S_1 \cap S_2) + P(S_3 \cap S_4) - P((S_1 \cap S_2) \cap (S_3 \cap S_4))$

$\qquad = P(S_1) \cdot P(S_2) + P(S_3) \cdot P(S_4) - P(S_1 \cap S_2) \cdot P(S_3 \cap S_4)$

$\qquad = P(S_1) \cdot P(S_2) + P(S_3) \cdot P(S_4) - P(S_1) \cdot P(S_2) \cdot P(S_3) \cdot P(S_4)$

Let $P(S_k) = x$. Then $p = x \cdot x + x \cdot x - x \cdot x \cdot x \cdot x = -x^4 + 2x^2$.

$$x = 0.9 \Rightarrow p = 0.9639.$$

[−2.25, 2.25] by [−2, 1]

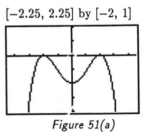

Xscl = 0.5
Yscl = 0.5

Figure 51(a)

[0.96, 1.05] by [−0.03, 0.04]

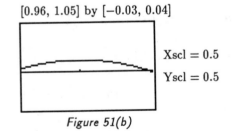

Xscl = 0.5
Yscl = 0.5

Figure 51(b)

(b) $p = 0.99 \Rightarrow -x^4 + 2x^2 = 0.99$. The graph of $y = -x^4 + 2x^2 - 0.99$ is shown in *Figure 51(a)*. The region near $x = 1$ is enlarged in *Figure 51(b)* to show that the graph is above the x-axis for some values of x. The approximate x-intercepts are ± 0.95, ± 1.05. Since $0 \le P(S_k) \le 1$, $P(S_k) = 0.95$.

53 (a) The number of ways that n people can all have a different birthday is $P(365, n)$. The number of outcomes in the sample space is 365^n.

$$\text{Thus, } p = \frac{P(365, n)}{365^n} = \frac{365!}{365^n (365 - n)!}.$$

(b) $n = 32 \Rightarrow p = \dfrac{365!}{365^{32}\, 333!} \Rightarrow \ln p = \ln \dfrac{365!}{365^{32}\, 333!} =$

$\ln 365! - \ln 365^{32} - \ln 333! \approx (365 \ln 365 - 365) - (32 \ln 365) - (333 \ln 333 - 333) \approx$

-1.45. Thus, $p \approx e^{-1.45} \approx 0.24$.

The probability that two or more people have the same birthday is $1 - p \approx 0.76$.

55 From Exercise 49(c), the payoff amount of \$2 has a probability of $\frac{244}{495}$. Hence,

$EV = 2 \cdot \frac{244}{495} = \frac{488}{495} \approx \$0.986$, or about \$0.99. Of course, after paying \$1 to play, you can expect to lose about \$7 for every 495 one-dollar pass line bets.

57 $EV = 1{,}000{,}000 \cdot \frac{1}{20{,}000{,}000} + 100{,}000 \cdot \frac{10}{20{,}000{,}000} + 10{,}000 \cdot \frac{100}{20{,}000{,}000} + 1000 \cdot \frac{1000}{20{,}000{,}000}$

$= \$0.20$ { less than the cost of a first class stamp }

Chapter 10 Review Exercises

3 $a_n = 1 + \left(-\frac{1}{2}\right)^{n-1}$ •

$a_1 = 1 + \left(-\frac{1}{2}\right)^0 = 1 + 1 = 2,$ $\qquad a_2 = 1 + \left(-\frac{1}{2}\right)^1 = 1 - \frac{1}{2} = \frac{1}{2}$

$a_3 = 1 + \left(-\frac{1}{2}\right)^2 = 1 + \frac{1}{4} = \frac{5}{4},$ $\qquad a_4 = 1 + \left(-\frac{1}{2}\right)^3 = 1 - \frac{1}{8} = \frac{7}{8}$

$a_7 = 1 + \left(-\frac{1}{2}\right)^6 = 1 + \frac{1}{64} = \frac{65}{64}$ $\qquad\qquad$ ★ $2, \frac{1}{2}, \frac{5}{4}, \frac{7}{8}; \frac{65}{64}$

8 $a_1 = 1, \; a_{k+1} = (1 + a_k)^{-1}$ •

$a_2 = a_{1+1} \{k = 1\} = (1 + a_1)^{-1} = (1 + 1)^{-1} = (2)^{-1} = \frac{1}{2}$

$a_3 = a_{2+1} = (1 + a_2)^{-1} = \left(1 + \frac{1}{2}\right)^{-1} = \left(\frac{3}{2}\right)^{-1} = \frac{2}{3}$

$a_4 = a_{3+1} = (1 + a_3)^{-1} = \left(1 + \frac{2}{3}\right)^{-1} = \left(\frac{5}{3}\right)^{-1} = \frac{3}{5}$

$a_5 = a_{4+1} = (1 + a_4)^{-1} = \left(1 + \frac{3}{5}\right)^{-1} = \left(\frac{8}{5}\right)^{-1} = \frac{5}{8}$ $\qquad$ ★ $1, \frac{1}{2}, \frac{2}{3}, \frac{3}{5}, \frac{5}{8}$

10 $\displaystyle\sum_{k=2}^{6} \frac{2k - 8}{k - 1} = (-4) + (-1) + 0 + \frac{1}{2} + \frac{4}{5} = -\frac{37}{10}$

14 We first note that $4 + 2 + 1 + \frac{1}{2} + \frac{1}{4} + \frac{1}{8}$ is the sum of six terms of a geometric sequence with first term 4 and ratio $\frac{1}{2}$—that is, $a_n = a_1 r^{n-1} = 4\left(\frac{1}{2}\right)^{n-1}$.

$4 + 2 + 1 + \frac{1}{2} + \frac{1}{4} + \frac{1}{8} = \displaystyle\sum_{n=1}^{6} 4\left(\tfrac{1}{2}\right)^{n-1} = \sum_{n=1}^{6} 2^2 (2^{-1})^{n-1} = \sum_{n=1}^{6} 2^2 2^{1-n} = \sum_{n=1}^{6} 2^{3-n}$

17 $\frac{1}{2} + \frac{2}{5} + \frac{3}{8} + \frac{4}{11}$. The numerators increase by 1, the denominators by 3. $\displaystyle\sum_{n=1}^{4} \frac{n}{3n - 1}$

20 $1 - \frac{1}{2} + \frac{1}{3} - \frac{1}{4} + \frac{1}{5} - \frac{1}{6} + \frac{1}{7}$. The terms have alternating signs, the numerator is 1, and the denominators increase by 1. $\displaystyle\sum_{n=1}^{7} (-1)^{n-1} \frac{1}{n}$

21 $a_0 + a_1 x^4 + a_2 x^8 + \cdots + a_{25} x^{100}$ { the exponents are multiples of 4 } $= \displaystyle\sum_{n=0}^{25} a_n x^{4n}$

24 $1 + x + \dfrac{x^2}{2} + \dfrac{x^3}{3} + \cdots + \dfrac{x^n}{n}.$

The pattern begins with the second term and the general term is listed. $1 + \displaystyle\sum_{k=1}^{n} \dfrac{x^k}{k}$

25 $d = 3 - (4 + \sqrt{3}) = -1 - \sqrt{3}; \; a_{10} = (4 + \sqrt{3}) + (9)(-1 - \sqrt{3}) = -5 - 8\sqrt{3};$

$$S_{10} = \tfrac{10}{2}\Big[(4 + \sqrt{3}) + (-5 - 8\sqrt{3})\Big] = -5 - 35\sqrt{3}.$$

27 $a_5 = a_1 + 4d = 5$ and $a_{13} = a_1 + 12d = 77.$ $a_{13} - a_5 = (a_1 + 12d) - (a_1 + 4d) = 8d$ and

$a_{13} - a_5 = 77 - 5 = 72 \Rightarrow 8d = 72 \Rightarrow d = 9.$ $a_5 = a_1 + 4d \Rightarrow 5 = a_1 + 36 \Rightarrow a_1 = -31.$

$$a_{10} = -31 + 9(9) = 50.$$

28 Four arithmetic means $\Rightarrow 5d = -10 - 20 \Rightarrow d = -6.$

The terms are 20, 14, 8, 2, -4, and -10.

30 We can divide the fourth term by the third term to obtain the common ratio,

$r = \dfrac{-0.3}{3} = -0.1.$ If we multiply the third term by r five times,

we will get the eighth term. $a_8 = a_3 r^5 = 3(-0.1)^5 = -0.00003.$

31 The geometric mean of 4 and 8 is $\sqrt{4 \cdot 8} = \sqrt{32} = 4\sqrt{2}.$

33 $402 = \tfrac{12}{2}(a_1 + 50) \Rightarrow a_1 = 17.$ $a_{12} = a_1 + 11d \Rightarrow d = (50 - 17)/11 = 3.$

34 $a_5 = a_1 r^4 \Rightarrow \tfrac{1}{16} = a_1 \left(\tfrac{3}{2}\right)^4 \Rightarrow a_1 = \tfrac{1}{16} \cdot \tfrac{16}{81} = \tfrac{1}{81}.$

$$S_5 = \frac{1}{81} \cdot \frac{1 - \left(\tfrac{3}{2}\right)^5}{1 - \tfrac{3}{2}} = \frac{1}{81} \cdot \frac{1 - \tfrac{243}{32}}{-\tfrac{1}{2}} = \frac{1}{81} \cdot \frac{-\tfrac{211}{32}}{-\tfrac{1}{2}} = \frac{211}{1296}.$$

37 This sum can be written as a sum of a geometric sequence and an arithmetic

sequence. $\displaystyle\sum_{k=1}^{10}\left(2^k - \tfrac{1}{2}\right) = \sum_{k=1}^{10} 2^k - \sum_{k=1}^{10} \tfrac{1}{2} = 2 \cdot \dfrac{1 - 2^{10}}{1 - 2} - 10(\tfrac{1}{2}) = 2046 - 5 = 2041$

39 $a_1 = 1, \; r = -\tfrac{2}{5} \Rightarrow S = \dfrac{1}{1 - \left(-\tfrac{2}{5}\right)} = \dfrac{1}{\tfrac{7}{5}} = \dfrac{5}{7}.$

41 (1) P_1 is true, since $3(1) - 1 = \dfrac{1[3(1) + 1]}{2} = 2$.

(2) Assume P_k is true: $2 + 5 + 8 + \cdots + (3k - 1) = \dfrac{k(3k+1)}{2}$. Hence,

$$2 + 5 + 8 + \cdots + (3k - 1) + 3(k + 1) - 1 = \dfrac{k(3k+1)}{2} + 3(k + 1) - 1$$

$$= \dfrac{3k^2 + k + 6k + 4}{2}$$

$$= \dfrac{3k^2 + 7k + 4}{2}$$

$$= \dfrac{(k + 1)(3k + 4)}{2}$$

$$= \dfrac{(k + 1)[3(k + 1) + 1]}{2}.$$

Thus, P_{k+1} is true, and the proof is complete.

43 (1) P_1 is true, since $\dfrac{1}{[2(1) - 1][2(1) + 1]} = \dfrac{1}{2(1) + 1} = \dfrac{1}{3}$.

(2) Assume P_k is true:

$$\dfrac{1}{1 \cdot 3} + \dfrac{1}{3 \cdot 5} + \dfrac{1}{5 \cdot 7} + \cdots + \dfrac{1}{(2k - 1)(2k + 1)} = \dfrac{k}{2k + 1}. \text{ Hence,}$$

$$\dfrac{1}{1 \cdot 3} + \dfrac{1}{3 \cdot 5} + \dfrac{1}{5 \cdot 7} + \cdots + \dfrac{1}{(2k - 1)(2k + 1)} + \dfrac{1}{(2k + 1)(2k + 3)}$$

$$= \dfrac{k}{2k + 1} + \dfrac{1}{(2k + 1)(2k + 3)}$$

$$= \dfrac{k(2k + 3) + 1}{(2k + 1)(2k + 3)}$$

$$= \dfrac{2k^2 + 3k + 1}{(2k + 1)(2k + 3)}$$

$$= \dfrac{(2k + 1)(k + 1)}{(2k + 1)(2k + 3)}$$

$$= \dfrac{k + 1}{2(k + 1) + 1}.$$

Thus, P_{k+1} is true, and the proof is complete.

45 (1) For $n = 1$, $n^3 + 2n = 3$ and 3 is a factor of 3.

(2) Assume 3 is a factor of $k^3 + 2k$. The $(k + 1)$st term is

$$(k + 1)^3 + 2(k + 1) = k^3 + 3k^2 + 5k + 3$$

$$= (k^3 + 2k) + (3k^2 + 3k + 3)$$

$$= (k^3 + 2k) + 3(k^2 + k + 1).$$

By the induction hypothesis, 3 is a factor of $k^3 + 2k$ and 3 is a factor of $3(k^2 + k + 1)$, so 3 is a factor of the $(k + 1)$st term. Thus, P_{k+1} is true, and the proof is complete.

46 (1) P_5 is true, since $5^2 + 3 < 2^5$.

(2) Assume P_k is true: $k^2 + 3 < 2^k$. Hence, $(k+1)^2 + 3 = k^2 + 2k + 4 =$

$$(k^2 + 3) + (k+1) < 2^k + (k+1) < 2^k + 2^k = 2 \cdot 2^k = 2^{k+1}.$$

Thus, P_{k+1} is true, and the proof is complete.

48 For j: $10^n \le n^n \Rightarrow \left(\frac{n}{10}\right)^n \ge 1$. This is true if $\frac{n}{10} \ge 1$ or $n \ge 10$. Thus, $j = 10$.

(1) P_{10} is true, since $10^{10} \le 10^{10}$.

(2) Assume P_k is true: $10^k \le k^k$. Hence,

$$10^{k+1} = 10 \cdot 10^k \le 10 \cdot k^k < (k+1) \cdot k^k < (k+1) \cdot (k+1)^k = (k+1)^{k+1}.$$

Thus, P_{k+1} is true, and the proof is complete.

50 We use the binomial theorem formula with $a = 2a$, $b = b^3$, and $n = 4$.

$$(2a + b^3)^4 = \binom{4}{0}(2a)^4(b^3)^0 + \binom{4}{1}(2a)^3(b^3)^1 + \binom{4}{2}(2a)^2(b^3)^2 + \binom{4}{3}(2a)^1(b^3)^3 +$$

$$\binom{4}{4}(2a)^0(b^3)^4$$

$$= (1)(16a^4)(1) + (4)(8a^3)(b^3) + (6)(4a^2)(b^6) + (4)(2a)(b^9) + (1)(1)(b^{12})$$

$$= 16a^4 + 32a^3b^3 + 24a^2b^6 + 8ab^9 + b^{12}$$

53 $(4a^2 - b)^7$; term that contains a^{10} •

Consider only the variable a in the expansion:

$$(a^2)^{7-k+1} = a^{10} \Rightarrow 16 - 2k = 10 \Rightarrow k = 3; \text{ 3rd term} = \binom{7}{2}(4a^2)^5(-b)^2 = 21{,}504a^{10}b^2$$

55 (a) $S_5 = 10 \Rightarrow 10 = \frac{5}{2}(2a_1 + 4d) \Rightarrow 4 = 2a_1 + 4d \Rightarrow 4d = 4 - 2a_1 \Rightarrow$

$$d = 1 - \tfrac{1}{2}a_1. \text{ Since } a_1 \text{ is positive, } 1 - \tfrac{1}{2}a_1 \text{ is less than 1 ft.}$$

(b) $a_1 = \frac{1}{2} \Rightarrow d = 1 - \frac{1}{2}(\frac{1}{2}) = \frac{3}{4}.$

The lengths of the other four pieces are $1\frac{1}{4}$, 2, $2\frac{3}{4}$, and $3\frac{1}{2}$ ft.

57 If $s_1 = 1$, then $s_2 = f$, $s_3 = f^2$, $\ldots$. There are two of each of the s_k's.

Since $0 < f < 1$, and f is the common ratio, we can sum the infinite sequence.

The sum of the s_k's is $2(1 + f + f^2 + \cdots) = 2\left(\frac{1}{1-f}\right) = \frac{2}{1-f}.$

58 $\text{Time}_{\text{total}} = \text{Time}_{\text{down}} + \text{Time}_{\text{up}}$

$$= \left[\frac{\sqrt{10}}{4} + \frac{\sqrt{10 \cdot \frac{3}{4}}}{4} + \frac{\sqrt{10 \cdot (\frac{3}{4})^2}}{4} + \cdots\right] + \left[\frac{\sqrt{10 \cdot \frac{3}{4}}}{4} + \frac{\sqrt{10 \cdot (\frac{3}{4})^2}}{4} + \cdots\right]$$

$$= \frac{\sqrt{10}}{4} + 2\left[\frac{\sqrt{10 \cdot \frac{3}{4}}}{4} + \frac{\sqrt{10 \cdot (\frac{3}{4})^2}}{4} + \cdots\right] = \frac{\sqrt{10}}{4} + 2 \cdot \frac{\frac{\sqrt{30}}{4}/4}{1 - \sqrt{\frac{3}{4}}} = \tfrac{1}{4}\sqrt{10} + 2 \cdot \frac{\sqrt{30}/8}{1 - \frac{1}{2}\sqrt{3}}$$

$$= \tfrac{1}{4}\sqrt{10} + \frac{\sqrt{30}}{2(2 - \sqrt{3})} \cdot \frac{2 + \sqrt{3}}{2 + \sqrt{3}} = \tfrac{1}{4}\sqrt{10} + (\sqrt{30} + \tfrac{3}{2}\sqrt{10}) = \tfrac{7}{4}\sqrt{10} + \sqrt{30} \approx 11.01 \text{ sec.}$$

$\boxed{59}$ (a) $P(52, 13) \approx 3.954 \times 10^{21}$

(b) $P(13, 5) \cdot P(13, 3) \cdot P(13, 3) \cdot P(13, 2) \approx 7.094 \times 10^{13}$

$\boxed{62}$ $\dfrac{(6+5+4+2)!}{6!\,5!\,4!\,2!} = \dfrac{17!}{6!\,5!\,4!\,2!} = 85{,}765{,}680$

$\boxed{64}$ (a) We need 4 of the 26 cards of one color. There are 2 colors.

$$\frac{P(26,\ 4) \cdot 2}{P(52,\ 4)} = \frac{92}{833} \approx 0.1104$$

(b) We need R-B-R-B. $\dfrac{26^2 \cdot 25^2}{P(52,\ 4)}$ or $\dfrac{26}{52} \cdot \dfrac{26}{51} \cdot \dfrac{25}{50} \cdot \dfrac{25}{49} = \dfrac{325}{4998} \approx 0.0650$

$\boxed{67}$ (a) $P(\text{passing}) = P(4, 5, \text{ or } 6 \text{ correct}) =$

$$\frac{C(6,\ 4) + C(6,\ 5) + C(6,\ 6)}{2^6} = \frac{15+6+1}{64} = \frac{22}{64} = \frac{11}{32}$$

(b) $P(\text{failing}) = 1 - P(\text{passing}) = 1 - \dfrac{22}{64} = \dfrac{42}{64}$

$\boxed{68}$ (a) $\dfrac{1}{6} \cdot \dfrac{1}{52} = \dfrac{1}{312} \approx 0.0032$

(b) $\dfrac{1}{6} + \dfrac{1}{52} - \dfrac{1}{312} = \dfrac{52+6-1}{312} = \dfrac{57}{312} = \dfrac{19}{104} \approx 0.1827$

$\boxed{69}$ Let O denote the event that the individual is over 60 and F denote the event that the individual is female.

$$P(O \cup F) = P(O) + P(F) - P(O \cap F) = \frac{1000}{5000} + \frac{2000}{5000} - \frac{0.40(2000)}{5000} = \frac{2200}{5000} = 0.44$$

$\boxed{70}$ There are 2 ways to obtain 10 (6, 4 and 4, 6), 2 ways for 11 (6, 5 and 5, 6), 1 way for

12, 16, 20, and 24 (double 3's, 4's, 5's, and 6's). $\dfrac{2+2+1+1+1+1}{36} = \dfrac{8}{36} = \dfrac{2}{9} = 0.\overline{2}$

[71] The two teams, A and B, can play as few as 4 games or as many as 7 games. Since the two teams are equally matched, the probability that either team wins any particular game is $\frac{1}{2}$.

P(team A wins in 4 games) $= \frac{1}{2} \cdot \frac{1}{2} \cdot \frac{1}{2} \cdot \frac{1}{2} = \left(\frac{1}{2}\right)^4 = \frac{1}{16} = 0.0625$.

P(team A wins in 5 games)

$= P$(team A wins 3 of the first 4 games { losing 1 game } and then wins game 5)

$= \binom{4}{3}\left(\frac{1}{2}\right)^3 \cdot \left(\frac{1}{2}\right)^1 \cdot \frac{1}{2} = \binom{4}{3}\left(\frac{1}{2}\right)^5 = \frac{4}{32} = 0.125$. In a similar fashion,

$$P\text{(team A wins in 6 games)} = \binom{5}{3}\left(\frac{1}{2}\right)^6 = \frac{10}{64} = \frac{5}{32} = 0.15625$$

and $\quad P$(team A wins in 7 games) $= \binom{6}{3}\left(\frac{1}{2}\right)^7 = \frac{20}{128} = \frac{5}{32} = 0.15625$.

Thus, the probability that team A wins the series is

$$0.0625 + 0.125 + 0.15625 + 0.15625 = 0.5,$$

which agrees with our common sense. Since the probabilities for team B winning the series are the same, the expected number of games is

$$4\left(2 \cdot \frac{1}{16}\right) + 5\left(2 \cdot \frac{4}{32}\right) + 6\left(2 \cdot \frac{5}{32}\right) + 7\left(2 \cdot \frac{5}{32}\right) = 5\frac{13}{16} = 5.8125.$$

Chapter 10 Discussion Exercises

[1] Probably the easiest way to describe this sequence is by using a piecewise-defined function. If $1 \le n \le 4$, then $2n$ describes the sequence 2, 4, 6, 8. For $n = 5$, we need to obtain a as the general term—one way to symbolize this is $(n-4)a$. Hence, one

possibility is $a_n = \begin{cases} 2n & \text{if } 1 \le n \le 4 \\ (n-4)a & \text{if } n \ge 5 \end{cases}$

A more sophisticated (and complicated) approach is to add a term to $2n$—call this term k—so that k is 0 for $n = 1, 2, 3, 4$, but equal to $(a - 10)$ for $n = 5$ { the *minus* 10 is needed to compensate for the *plus* 10 from the $2n$ term }. We can do this by starting with the expression $(n-1)(n-2)(n-3)(n-4)$, which is zero for $n = 1, 2, 3, 4$. For $n = 5$, this expression is $4 \cdot 3 \cdot 2 \cdot 1 = 24$, so we will divide it by 24 and multiply it by $(a - 10)$ to obtain $(a - 10)$. Thus, another possibility is

$$a_n = 2n + \frac{(n-1)(n-2)(n-3)(n-4)(a-10)}{24}.$$

3 (a) Following the pattern from Exercises 37–38 in Section 10.4, write

$$1^4 + 2^4 + 3^4 + \cdots + n^4 = an^5 + bn^4 + cn^3 + dn^2 + en.$$ Then, it follows that:

$n = 1 \Rightarrow a(1)^5 + b(1)^4 + c(1)^3 + d(1)^2 + e(1) = 1^4 \Rightarrow a + b + c + d + e = 1,$

$n = 2 \Rightarrow a(2)^5 + b(2)^4 + c(2)^3 + d(2)^2 + e(2) = 1^4 + 2^4 \Rightarrow$

$$32a + 16b + 8c + 4d + 2e = 17,$$

$n = 3 \Rightarrow a(3)^5 + b(3)^4 + c(3)^3 + d(3)^2 + e(3) = 1^4 + 2^4 + 3^4 \Rightarrow$

$$243a + 81b + 27c + 9d + 3e = 98,$$

$n = 4 \Rightarrow a(4)^5 + b(4)^4 + c(4)^3 + d(4)^2 + e(4) = 1^4 + 2^4 + 3^4 + 4^4 \Rightarrow$

$$1024a + 256b + 64c + 16d + 4e = 354,$$

$n = 5 \Rightarrow a(5)^5 + b(5)^4 + c(5)^3 + d(5)^2 + e(5) = 1^4 + 2^4 + 3^4 + 4^4 + 5^4 \Rightarrow$

$$3125a + 625b + 125c + 25d + 5e = 979.$$

$$AX = B \Rightarrow \begin{bmatrix} 1 & 1 & 1 & 1 & 1 \\ 32 & 16 & 8 & 4 & 2 \\ 243 & 81 & 27 & 9 & 3 \\ 1024 & 256 & 64 & 16 & 4 \\ 3125 & 625 & 125 & 25 & 5 \end{bmatrix} \begin{bmatrix} a \\ b \\ c \\ d \\ e \end{bmatrix} = \begin{bmatrix} 1 \\ 17 \\ 98 \\ 354 \\ 979 \end{bmatrix} \Rightarrow X = A^{-1}B = \begin{bmatrix} 1/5 \\ 1/2 \\ 1/3 \\ 0 \\ -1/30 \end{bmatrix}$$

Thus, $1^4 + 2^4 + 3^4 + \cdots + n^4 = \frac{1}{5}n^5 + \frac{1}{2}n^4 + \frac{1}{3}n^3 - \frac{1}{30}n.$

This formula must be verified.

(b) Let P_n be the statement that $1^4 + 2^4 + 3^4 + \cdots + n^4 = \frac{1}{5}n^5 + \frac{1}{2}n^4 + \frac{1}{3}n^3 - \frac{1}{30}n.$

(1) $1^4 = \frac{1}{5} + \frac{1}{2} + \frac{1}{3} - \frac{1}{30} = 1$ and P_1 is true.

(2) Assume that P_k is true. We must show that P_{k+1} is true.

$P_k \Rightarrow 1^4 + 2^4 + 3^4 + \cdots + k^4 = \frac{1}{5}k^5 + \frac{1}{2}k^4 + \frac{1}{3}k^3 - \frac{1}{30}k \Rightarrow$

$1^4 + 2^4 + 3^4 + \cdots + k^4 + (k+1)^4 = \frac{1}{5}k^5 + \frac{1}{2}k^4 + \frac{1}{3}k^3 - \frac{1}{30}k + (k+1)^4 \Rightarrow$

$1^4 + 2^4 + 3^4 + \cdots + k^4 + (k+1)^4 =$

$$\frac{1}{5}k^5 + \frac{1}{2}k^4 + \frac{1}{3}k^3 - \frac{1}{30}k + (k^4 + 4k^3 + 6k^2 + 4k + 1) \Rightarrow$$

$1^4 + 2^4 + 3^4 + \cdots + k^4 + (k+1)^4 = \frac{1}{5}k^5 + \frac{3}{2}k^4 + \frac{13}{3}k^3 + 6k^2 + \frac{119}{30}k + 1.$

$P_{k+1} \Rightarrow 1^4 + 2^4 + 3^4 + \cdots + k^4 + (k+1)^4$

$= \frac{1}{5}(k+1)^5 + \frac{1}{2}(k+1)^4 + \frac{1}{3}(k+1)^3 - \frac{1}{30}(k+1)$

$= \frac{1}{5}(k^5 + 5k^4 + 10k^3 + 10k^2 + 5k + 1) + \frac{1}{2}(k^4 + 4k^3 + 6k^2 + 4k + 1) +$

$$\frac{1}{3}(k^3 + 3k^2 + 3k + 1) - \frac{1}{30}(k+1)$$

$= \frac{1}{5}k^5 + \frac{3}{2}k^4 + \frac{13}{3}k^3 + 6k^2 + \frac{119}{30}k + 1.$ Thus, the formula is true for all n.

⑤ Examine the number of digits in the exponent of the value in scientific notation. The TI-82/3 can compute 69!, but not 70!, since 70! is larger than a 2-digit exponent. The TI-85 can compute 449!, but not 450!, since 450! is larger than a 3-digit exponent.

⑦ There are $8 \times 36 = 288$ contestants daily and $288 \times 30 = 8640$ contestants for the month. The probability that a contestant wins any particular prize for the daily tournament and the monthly tournament is $p_1 = \frac{1}{288}$ and $p_2 = \frac{1}{8640}$, respectively. If the game is to be fair, then the total expected value should equal the entry fee.

$EV_1(\text{daily}) = 250p_1 + 100p_1 + 50p_1 = 400p_1.$

$EV_2(\text{month}) = 4000p_2 + 2000p_2 + 1500p_2 + 1000p_2 + 800p_2 +$
$\qquad\qquad 600p_2 + 500p_2 + 400p_2 + 300p_2 + 200p_2 +$
$\qquad\qquad 100(40p_2) + 75(50p_2) + 50(200p_2) + 25(200p_2) = 34{,}050p_2.$

Thus, $EV_{\text{total}} = EV_1 + EV_2$
$\qquad\qquad = 400p_1 + 34{,}050p_2 = 400 \cdot \frac{1}{288} + 34{,}050 \cdot \frac{1}{8640} = \frac{46{,}050}{8640} = \frac{1535}{288} \approx \$5.33.$

⑨ Since we can have 0 to 5 toppings on a pizza, the number of ways to order one pizza is $\sum_{k=0}^{5} \binom{n}{k}$. Because there are two pizzas, we have $\left[\sum_{k=0}^{5} \binom{n}{k} \right]^2 = 1{,}048{,}576 \Rightarrow$

$\sum_{k=0}^{5} \binom{n}{k} = 1024.$ By trial and error, we find that $n = 11$.

On the TI-82, store 11 in N and use "sum seq(N nCr R,R,0,5,1)" to obtain 1024.

TO THE OWNER OF THIS BOOK:

I hope that you have found the *Student Solutions Manual for Swokowski and Cole's Fundamentals of Algebra and Trigonometry,* Ninth Edition, useful. So that this book can be improved in a future edition, would you take the time to complete this sheet and return it? Thank you.

School and address: _____

Instructor's name: _____

Course name: _____

Your name (optional): _____

E-mail (optional): _____

1. What I like most about this manual is: _____

    _____

    _____

    _____

    _____

2. What I like least about this manual is: _____

    _____

    _____

    _____

    _____

3. Please indicate if you have come across any possible inaccuracies. Please be specific.

    _____

    _____

    _____

    _____

4. On a separate sheet of paper, please write specific suggestions for improving this manual and anything else you'd care to share about your experience in using it.

Optional:

Your name: _____ Date: _____

May Brooks/Cole quote you, either in promotion for *Student Solutions Manual for Swokowski and Cole's Fundamentals of Algebra and Trigonometry*, Ninth Edition, or in future publishing ventures?

Yes: _____ No: _____

Sincerely,

Jeffery A. Cole

FOLD HERE

FOLD HERE